Lecture Notes in Computer Science

Lecture Notes in Artificial Intelligence 15370

Founding Editor

Jörg Siekmann

AF167717

Series Editors

Randy Goebel, *University of Alberta, Edmonton, Canada*
Wolfgang Wahlster, *DFKI, Berlin, Germany*
Zhi-Hua Zhou, *Nanjing University, Nanjing, China*

The series Lecture Notes in Artificial Intelligence (LNAI) was established in 1988 as a topical subseries of LNCS devoted to artificial intelligence.

The series publishes state-of-the-art research results at a high level. As with the LNCS mother series, the mission of the series is to serve the international R & D community by providing an invaluable service, mainly focused on the publication of conference and workshop proceedings and postproceedings.

Mehwish Alam · Marco Rospocher ·
Marieke van Erp · Laura Hollink ·
Genet Asefa Gesese
Editors

Knowledge Engineering and Knowledge Management

24th International Conference, EKAW 2024
Amsterdam, The Netherlands, November 26–28, 2024
Proceedings

 Springer

Editors
Mehwish Alam (iD)
Télécom Paris, Institut Polytechnique de Paris
Palaiseau, France

Marco Rospocher (iD)
Università degli Studi di Verona
Verona, Italy

Marieke van Erp (iD)
KNAW Humanities Cluster
Amsterdam, The Netherlands

Laura Hollink (iD)
Centrum Wiskunde and Informatica
Amsterdam, The Netherlands

Genet Asefa Gesese (iD)
FIZ Karlsruhe
Eggenstein-Leopoldshafen, Germany

ISSN 0302-9743   ISSN 1611-3349 (electronic)
Lecture Notes in Artificial Intelligence
ISBN 978-3-031-77791-2   ISBN 978-3-031-77792-9 (eBook)
https://doi.org/10.1007/978-3-031-77792-9

LNCS Sublibrary: SL7 – Artificial Intelligence

Preface

This volume contains the proceedings of the 24th International Conference on Knowledge Engineering and Knowledge Management (EKAW 2024), held in Amsterdam, the Netherlands during November 26–28, 2024. The conference addressed the pivotal role of knowledge in constructing systems and services for the semantic web, knowledge management, knowledge discovery, information integration, natural language processing, intelligent systems, e-business, e-health, humanities, cultural heritage, and beyond. In addition to the topics above, EKAW 2024 introduced a special theme related to "Knowledge and Language Models", calling for articles focusing on algorithms, tools, methodologies, and applications that leverage the interplay between knowledge and language models.

We invited three types of papers: research papers, in-use papers, and position papers. Overall, we received 138 abstract submissions, of which 115 turned into full paper submissions, which were reviewed by 123 members of the Programme Committee and 19 additional reviewers. The review process was single-blind, i.e., the authors were known to the reviewers, while the reviewers remained anonymous to the authors. Each paper received three to four reviews, and discussions were encouraged by both program chairs for papers that exhibited strongly divergent opinions. In total, 28 papers were accepted for publication in this volume, of which 24 are research papers, two are in-use papers, and two are position papers. In addition to paper presentations, the conference featured a Poster and Demo session, two keynote speeches, four workshops, and four tutorials.

The First Workshop on Knowledge Management for Numerical Modeling, Measurement & Simulation was organized by Benno Kruit (VU Amsterdam), Victoria Deleger (University of Amsterdam), and João Moreira (University of Twente). The eXtraction and eXploitation of long-TAIL Knowledge with LLMs and KGs (X-TAIL) workshop was organized by Arianna Graciotti (University of Bologna), Alba Morales Tirado (Open University), Valentina Presutti (University of Bologna), and Enrico Motta (Open University). The Workshop on Evaluation of Language Models in Knowledge Engineering (ELMKE) was organized by Bohui Zhang (King's College London), Yuan He (University of Oxford), and Reham Alharbi (University of Liverpool). The first Workshop on Structured Knowledge in Newsrooms (K-MiN) was organized by Reshmi G. Pillai (Vrije University Amsterdam) and Laurence Dierickx (University of Bergen/Université Libre de Bruxelles).

The tutorial on "Conversational Knowledge Capture Using the KNOW Ontology" was organized by Tolga Çöplü (Haltia.AI), Arto Bendiken (Haltia.AI), and Andrii Skomorokhov (Haltia.AI). The tutorial on "Enabling Meaning and Understanding in Human-centric AI (MUHAI)" was organized by Ilaria Tiddi (Vrije Universiteit Amsterdam), Rachel Ringe (University of Bremen), Carlo Santagiustina (SciencesPo Medialab Paris), and Remi van Trijp (Sony CSL Paris). The tutorial on "Creating and Accessing Knowledge Graphs for Action Parameterisation" was organized by Michaela Kümpel (University of Bremen), Jan-Philipp Töberg (Bielefeld University), Ilaria Tiddi (Vrije Universiteit Amsterdam), Philipp Cimiano (Bielefeld University), Enrico Motta (Open University), and Michael Beetz (University of Bremen). The tutorial on "Semantic Knowledge Modeling - Ontologies & Vocabularies" was organized by Peter Haase (metaphacts GmbH) and Irina Schmidt (metaphacts GmbH).

We would like to express our gratitude towards the Organizing Committee and the Program Committee. A specific thank you goes to the "emergency reviewers" who provided additional reviews within a short time span in cases where previously provided reviews did not lead to a clear decision to reject or accept a paper. We would also like to thank our keynote speakers (Fabian M. Suchanek and Suzan Verberne) for accepting our invitations without hesitation and bringing their insights into the importance of knowledge engineering in today's world. Finally, our gratitude goes also to the sponsors of the conference, Haltia.AI, Metaphacts, Artificial Intelligence Journal, and Ontotext, and to the local organization team for making the event successful.

October 2024

Mehwish Alam
Marco Rospocher
Marieke van Erp
Laura Hollink
Genet Asefa Gesese

Organization

General Chairs

Laura Hollink   Centrum Wiskunde & Informatica,
The Netherlands

Marieke Van Erp   KNAW Humanities Cluster, The Netherlands

Program Committee Chairs

Mehwish Alam   Télécom Paris, Institut Polytechnique de Paris,
France

Marco Rospocher   Università degli Studi di Verona, Italy

Posters and Demos Chairs

Inna Novalija   Jožef Stefan Institute, Slovenia

Carlos Badenes-Olmedo   Universidad Politécnica de Madrid, Spain

Workshops and Tutorials Chairs

Lise Stork   University of Amsterdam, the Netherlands

Enrico Daga   Open University, UK

Publicity Chairs

Eleni Ilkou   L3S Research Center, Germany

Andrea Schimmenti   University of Bologna, Italy

Proceedings Chair

Genet Asefa Gesese   FIZ Karlsruhe, Germany

Sponsorship Chair

Joe Raad   University of Paris-Saclay, France

Local/PhD Chairs

Delfina Sol Martinez Pandiani   University of Amsterdam, Netherlands
Andrei Nesterov   Centrum Wiskunde & Informatica, Netherlands

Event Support

Martine Anholt Gunzeln   Centrum Wiskunde & Informatica, Netherlands

Program Committee

Nathalie Abadie   Université Gustave Eiffel, France
Albin Ahmeti   Semantic Web Company, TU Wien, Austria
Bradley Allen   University of Amsterdam, Netherlands
Vito Walter Anelli   Politecnico di Bari, Italy
Gianluca Apriceno   FBK and University of Trento, Italy
Nathalie Aussenac-Gilles   IRIT CNRS, France
Flavio Bertini   University of Parma, Italy
Giovanni Maria Biancofiore   Polytechnic University of Bari, Italy
Russa Biswas   Aalborg University, Denmark
Eva Blomqvist   Linköping University, Sweden
Marco Bombieri   University of Verona, Italy
Stefano Borgo   CNR-ISTC, Italy
Grégory Bourguin   LISIC/ULCO, France
Loris Bozzato   Università degli Studi dell'Insubria, Italy
Adrian M. P. Brasoveanu   MODUL Technology GmbH, Austria
Oleksandra Bruns   FIZ Karlsruhe & KIT, Germany
Davide Buscaldi   Université Sorbonne Paris Nord, France
Valentina Anita Carriero   Cefriel, Italy
Davide Ceolin   CWI, Netherlands
Philipp Cimiano   Bielefeld University, Germany
Oscar Corcho   Universidad Politécnica de Madrid, Spain
Francesco Corcoglioniti   Free University of Bozen-Bolzano, Italy
Miguel Couceiro   INESC-ID, IST, University of Lisbon, Portugal
Claudia D'Amato   University of Bari, Italy

Emilio M. Sanfilippo	ISTC-CNR, Italy
Bruno Sartini	Ludwig-Maximilians-Universität München, Germany
Simon Scheider	Utrecht University, Netherlands
Stefan Schlobach	Vrije Universiteit Amsterdam, Netherlands
Christoph Schuetz	Johannes Kepler University Linz, Austria
Luciano Serafini	Fondazione Bruno Kessler, Italy
Cogan Shimizu	Wright State University, USA
Lina F. Soualmia	Université de Rouen Normandie, France
Armando Stellato	University of Rome Tor Vergata, Italy
Vojtěch Svátek	Prague University of Economics and Business, Czech Republic
Valentina Tamma	University of Liverpool, UK
Andrea Tettamanzi	Université Côte d'Azur, France
Ilaria Tiddi	Vrije Universiteit Amsterdam, Netherlands
Tabea Tietz	FIZ Karlsruhe, Germany
Holger Timinger	University of Applied Sciences Landshut, Germany
Cassia Trojahn	UT2J & IRIT, France
Nicolas Troquard	Gran Sasso Science Institute, Italy
Fabio Vitali	University of Bologna, Italy
Paulo Ricardo Viviurka Do Carmo	HTWK Leipzig, Germany
Fouad Zablith	American University of Beirut, Lebanon
Hamada Zahera	Paderborn University, Germany
Ondřej Zamazal	Prague University of Economics and Business, Czech Republic
Haifa Zargayouna	Université Sorbonne Paris Nord, CNRS, France
Amal Zouaq	Polytechnique Montréal, Canada

Additional Reviewers

Mark Adamik
Fatima Zahra Amara
Roberto Barile
Robert David
Alessandro De Bellis
Rémi Felin
Martin Glauer
Nele Köhler
Anelia Kurteva
Delfina Sol Martinez Pandiani

Margherita Martorana
Inna Novalija
Riccardo Pozzi
Célian Ringwald
Andrea Schimmenti
Sarah Binta Alam Shoilee
Nikit Srivastava
Mirjam Stappel
Ziwei Xu

Contents

Examining LGBTQ+-Related Concepts in the Semantic Web: Link Discovery, Concept Drift, Ambiguity, and Multilingual Information Reuse

Shuai Wang[(✉)] and Maria Adamidou

Vrije Universiteit Amsterdam, Amsterdam, The Netherlands
shuai.wang@vu.nl, m.adamidou@student.vu.nl

Abstract. In recent years, there is a notable increase in the use of LGBTQ+ ontologies and structured vocabularies in library systems, digital archives, online databases, heritages, etc. Many were published as linked data, including Homosaurus, QLIT, GSSO, etc. However, little has been reported about the links between the concepts captured by them. To study their interconnection, we retrieve all their published mappings as well as relevant information to form an integrated knowledge graph. We evaluate its usefulness with respect to three aspects. First, taking advantage of its weakly connected components, we study the discovery of missing links between entities. Second, we analyze concept drift and change by providing examples of concept convergence, ambiguity, and scope change. Moreover, we study how multilingual information from other resources can enrich entities using Homosaurus as an example. Finally, we discuss potential challenges and practical implications of our findings in the community.

Keywords: Homosaurus · concept drift · identity management

1 Introduction

Although the LGBTQ+ demographic group has gained more visibility and acceptance, there is still a considerable disparity between the intricate nature of their lived experiences and the current structured vocabulary and other knowledge representation initiatives. Even with the advancements made by several projects with different goals in recent years, many still fall short of capturing the diverse and nuanced realities of the evolving vocabulary used in the community, leaving gaps when aligning concepts between these attempts. This may lead to interoperability issues when projects become outdated, requiring revisions of their compatibility with other projects.

In this paper, we use the term *conceptual models*[1] as the umbrella term for thesauri, structured vocabularies, ontologies, subject headings, and knowledge

---

[1] An alternative umbrella term used in the community is 'knowledge organization systems' (KOS).

M. Alam et al. (Eds.): EKAW 2024, LNAI 15370, pp. 1–17, 2025.
https://doi.org/10.1007/978-3-031-77792-9_1

bases. More specifically, we focus on conceptual models published as linked data in the semantic web. Concepts are represented as entities with (multilingual) *labels*. *Links* between entities are in the form of a triple with subject, predicate, and object. A *mapping* is a set of links between entities from two different conceptual models. By *community*, we specifically refer to the individuals who are engaged in the creation and development of the conceptual models to be studied below. In this paper, we focus on mappings between selected conceptual models and some related links as well as their multilingual labels. Other aspects such as comments, scope notes, and definitions are excluded from this study.

Homosaurus[2] [16] was initiated by librarians in the IHLIA LGBTI Heritage[3] and has become a popular conceptual models in LGBTQ+ libraries and archives. There are mappings between Homosaurus, the QLIT thesaurus [3], the Gender, Sex, and Sexual Orientation (GSSO) ontology [9], the Library of Congress Subject Headings (LCSH) [14], and the Wikidata knowledge base [17]. Many of these terms have evolved or are mistakenly linked to reclaimed terms. For example, 'wolves' is reclaimed as a slang term[4] for 'masculine gay men who are often characterized as having hairy bodies and facial hair'. However, it has a link of `skos:exactmatch` to a term[5] in LCSH, which is about the animal wolf. Moreover, since these entities and links are published in the semantic web, mistakes and accumulated subtle changes can result in errors and complex concept drift which demands careful examination from multiple parties, especially in a multilingual setting. Recently, the need for LGBTQ+ conceptual models that support multiple languages has grown, such as Spanish [13] and Chinese [6]. Considering the suboptimal efficacy of machine translation [8], experts must put considerable effort into manually translating terms and their scope notes. It remains an open question whether the reuse of pre-existing terms in specific languages from relevant resources could facilitate the development of bilingual or multilingual conceptual models. Therefore, in this paper, we construct a knowledge graph based on links between conceptual models and relevant information on identity management. Using it, we study the evolution and drifting of LGBTQ+-related concepts in the semantic web. Finally, we explore how multilingual information from one conceptual model can be used to enrich entities of another.

We retrieve the mappings and related links that contain identifying information, as well as links about concept replacement and redirection. We aim for a preliminary quantitative and qualitative analysis. We demonstrate a detailed examination of LGBTQ+-related concepts and their links in the semantic web and their multilingual information from three aspects: link discovery, concept drift and (multilingual) ambiguity, as well as the reuse of multilingual information. We provide evidence for improving the quality of data by developers and

---

[2] https://homosaurus.org/.
[3] https://ihlia.nl/en/collection/homosaurus/.
[4] https://homosaurus.org/v3/homoit0001508.
[5] http://id.loc.gov/authorities/subjects/sh85147257.

the community. Moreover, we provide our code and all the data[6] using Wikidata and QLIT to reproduce our experimental results and future use. Our main aim is knowledge representation and engineering. Bias, sensitivity, ethics, and political issues are beyond our expertise and are not covered.

2 LGBTQ+ Conceptual Models and Related Work

As far as we are aware, there existst no explicit analysis of concept drift and multilinguality of LGBTQ+ concepts in the semantic web. Here we will summarize the updates of terms in the release notes of conceptual models and present some related research, but not all are exclusively about LGBTQ+ concepts.

Homosaurus is a linked data vocabulary focusing on LGBTQ+ terminology, aimed at enriching general subject term vocabularies, and it undergoes updates every six months. It was intended as a companion to LCSH [16]. Captured concepts are instances of `skos:Concept` and are related to each other using `skos:broader`. Recently, three versions of Homosaurus have been released with updates every half a year. It contains English terms along with their corresponding translations in Dutch, offering a bilingual dimension. Serving as a robust, state-of-the-art conceptual model used in libraries and heritages, Homosaurus significantly improves the findability of LGBTQ+ resources and information. Furthermore, it offers a SPARQL endpoint.[7] At each release, information on updates of "labels" and newly added terms are provided on the website.[8] Despite some statistical analysis on terms in Homosaursus and how they overlap with others such as LCSH [5], to our knowledge, there is no systematic literature on the evolution of terms in Homosaurus and how its links to other conceptual models change.

The QLIT (Queer Literature Indexing Thesaurus) [11] is a recent Swedish thesaurus dedicated to indexing literature with LGBTQI themes. It was mainly used in Queerlit[9] [3], a bibliographic database on Swedish fiction. More than half of the terms in QLIT were translated from the English terms of Homosaurus

---

[6] The Python scripts, SPARQL queries, detailed explanation of experiments, annotated links by Swedish-speaking experts from QLIT, and the analytical results are in the supplementary material on Zenodo with DOI: `10.5281/zenodo.12684869`. Wikidata and QLIT are with the license CC0. Thus, only datasets extracted from them are provided. Due to the strict CC-BY-NC-ND license of GSSO and Homosaurus, the remaining data files are only accessible upon request from IHLIA, the developers of GSSO and QLIT, and the authors. To reproduce the results or extend this work, see the instructions in the GitHub repositor: https://github.com/Multilingual-LGBTQIA-Vocabularies/Examing_LGBTQ_Concepts.

[7] https://data.ihlia.nl/PoolParty/sparql/homosaurus. Note that this endpoint may be delayed compared to the latest release on the Homosaurus website.

[8] See for example https://homosaurus.org/releases/show/3. The latest release is Homosaurus v3.5 released in January 2024 with 255 newly added terms, one term replaced, and 24 changed terms.

[9] https://queerlit.dh.gu.se/.

(v3.3). QLIT has mappings to the two main Swedish library thesauri: Svenska ämnesord (SAO) and Barnämnesord (Barn) [11].

The Gender, Sex, and Sexual Orientation (GSSO)[10] ontology was designed to facilitate communication in gender, sex, and sexual orientation research and assist knowledge discovery in literature [9]. Its second version includes 10,060 entries, an increase from 6,250 in its first version. Its application ranges from clinical studies [10] to archives [15].

The three conceptual models mentioned above have links to LCSH (Library of Congress Subject Headings) [14]. Although LCSH includes some LGBTQ+-related terms, it was reported to have flaws and can be influenced by politics [21], which is beyond our expertise. Wikidata [17] contains identifiers of GSSO, QLIT, and Homosaurus, but to the best of our knowledge, have not been analyzed from this perspective.

These conceptual models serve distinct purposes, were revised at various times, are managed by teams with different expertise, and have not always been developed with full awareness of the changes of each other. Therefore, a perfectly unified representation of concepts and their relations is not possible. Braquet [4] briefly examines the provision of support for LGBTQ+ patrons within library settings, offering insights through various library-based scenarios. Dobreski et al. compared the overlap of the Homosaurus, LCSH, and Library of Congress Demographic Group Terms (LCDGT) [5]. They examined an old version of Homosaurus with 1,754 terms and found 618 terms related to identity. They reported 153 matches in the LCSH (exact matches and closest matches). Similarly, they found 176 matches in LCDGT, including faceted matches. Furthermore, it has been reported that there are outdated terms in LCSH, which leads to problems with the mapping of terms between Homosaurus and LCSH [5,16]. However, to our knowledge, there is no systematic report on the quality and reliability of these links and what kind of consequences would there be following erroneous links or involving ambiguous entities. A comprehensive comparison of all the entities released and their relations in these conceptual models is missing.

We observed redirection when resolving URIs of Homosaurus. Prior examinations of entities indicate frequent redirections among entities in identity graphs, with an estimate of 45% to 83% maintaining the semantics of identity [12]. However, no research has been done to study how many URIs have been redirected among those corresponding to LGBTQ+-related concepts.

Concept drift refers to the phenomenon where the meanings or nuances of terms, concepts, or language evolve over time [18]. In the context of LGBTQ+ vocabularies, concept drift occurs as societal attitudes, understandings, and discussions about gender identity, sexual orientation, and related topics change and progress. An example is a term like "queer" which has changed in meaning over time. "Queer" was used as a slang for homosexuals as well as a term for homophobic abuse, but it has been reclaimed as an umbrella term for a coalition of culturally marginal sexual self-identifications in recent years [7]. The term "homosexual" is now considered somewhat "clinical" [4]. When it comes to

---

[10] https://github.com/Superraptor/GSSO. This paper used release v2.0.10.

the analysis of concept drift at scale, a method for large knowledge bases with instances of classes was proposed by Wang et al., but it does not apply to our data due to the missing of instances [18]. As far as we are aware, there is no systematic report on the concept drift and change in the field.

3   Data Engineering

Details of selected conceptual models and links extracted are in Sect. 3.1. Section 3.2 includes details of multilingual labels extracted for the study of information reuse. Moreover, we capture some outdated URIs and how they are redirected in Sect. 3.3. Finally, we integrate all the links in Sect. 3.4.

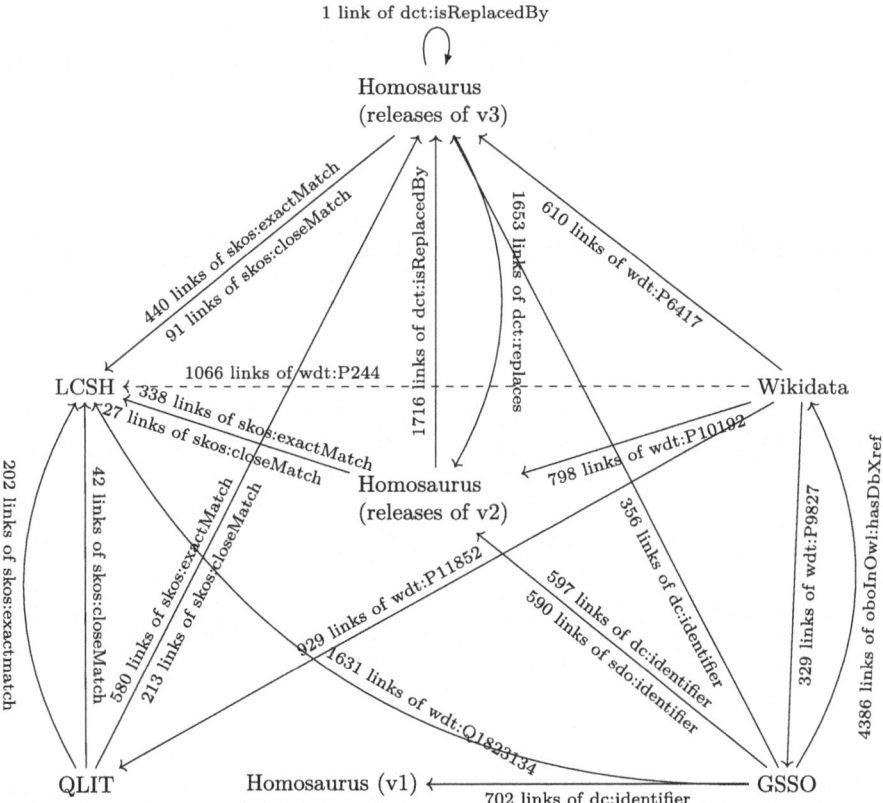

Fig. 1. Conceptual models and their extracted links. The dashed edge indicates that only edges about LCSH entities that appear in the rest of the selected concept models were chosen in this study for further integration and analysis.

3.1 Dataset Selection and Data Preprocessing

Since its version 2.1 in June 2020[11], Homosaurus experienced 8 updates (on average twice per year). In this study, we focus on the last release of version 2 (v2.3) and the latest release (January 2024, v3.5), which captures 3,149 terms. It was noticed that some URIs[12] were no longer maintained in version 2 and were therefore replaced. Although replacement does not necessarily imply equivalence, we include this type of relation in our study to capture the evolution of conceptual models. Replacement was captured in Homosaurus using the relation `dct:isReplacedBy` and its inverse `dct:replaces`. Each entity is accompanied by an identifier (e.g. `homoit0002950`), a preferred label using `skos:prefLabel` and some using `skos:altLabel` as well as a scope note using `rdfs:comment`. Additionally, we noticed that some outdated URIs were redirected to newer ones. However, this information was not explicitly stated in the latest version. In Sect. 3.3, we extract these relations by resolving the URIs and capture this redirection relation for further analysis. Homosaurus has links to LCSH[13]. They are being used together in libraries and heritages. Our examination shows that none of the entities in LCSH is outdated.

Recall that QLIT[14] was developed mostly based on translating terms in Homosaurus. Among its 914 terms, 774 exhibit an exact match (`skos:exactMatch`) or close match `skos:closeMatch` to terms in Homosaurus, with an additional 140 terms not mapped to terms in Homosaurus, some of which are exclusive to QLIT. Only 244 links were found when examined against LCSH. Missing links will be discussed in Sect. 4.1. Moreover, its scope is limited to LGBTQI instead of LGBTQ+, thus the translation misses '+' in its Swedish labels.

Next, we extract links between GSSO and Homosaurus. It was noticed that it uses both `sdo:identifier` and `dc:identifier`. Since its publication in September 2022, the links from GSSO to Homosaurus version 2.2 and 3.1 have not been updated. GSSO has 597 links of `dc:identifier` and 590 `sdo:identifier` to version 2 of Homosaurus. In comparison, there are only 356 `dc:identifier` links to version 3. Moreover, these links are using version 2.2 and version 3.1, which are outdated. This is because GSSO has not been updated since September 2022. Moreover, GSSO has 1,830 links to LCSH. The links of Wikidata and the two versions of Homosaurus are mostly asserted and maintained by experts and members of the Wikidata community. This demands significant human labor.

---

[11] Version 1 is no longer available on the official website. v2.1, v2.2, v3.0, v3.1, and v3.2 are no longer available on their website.

[12] In this paper, we use the prefix `h2` for the namespace http://homosaurus.org/v2/ and `h3` for the namespace https://homosaurus.org/v3/.

[13] The N-Triple file of LCSH was obtained on 9th May, 2024 from https://id.loc.gov/authorities/subjects.html. For fast analysis of its entities, it was converted to its HDT format. Both were included in the supplementary material. We use the prefix `lcsh` for the namespace http://id.loc.gov/authorities/subjects/.

[14] We use `qlit` for the namespace https://queerlit.dh.gu.se/qlit/v1/.

Links from Wikidata[15] to Homosaurus were provided using specified relations: `wdt:P10192` for entities in version 3 and `wdt:P6417` for entities in version 2. There are 610 and 798 links between Wikidata and versions 2 and 3 of Homosaurus, respectively. Similarly, for links from Wikidata to GSSO, a specific relation `wdt:P9827` was used. Only 329 links were found. As far as we know, links from Wikidata to GSSO and Homosaurus are maintained by hand by members of the Wikidata community without any use of automation. In total, 929 links were found for entities in QLIT, corresponding to the Wikidata property `wdt:P11852`. A total of 55,980 links were found between Wikidata and LCSH. Given that we study only LGBTQ+-related concepts, we restrict the entities to only those that appear in the links between conceptual models. Thus, only 1,066 links from Wikidata to LCSH were to be integrated and studied in the next steps. We obtain complete URIs for the entities in GSSO, Homosaurus v2 and v3, LCSH, and QLIT.[16]

3.2   Multilingual Information Extraction

GSSO consists of labels of 77 languages, while entities of Wikidata in the integrated graph (see Sect. 3.4) are associated with 507 languages. For the study of reuse of multilingual information to be presented in Sect. 4.3, we extract also multilingual information in GSSO[17] regarding labels (`rdfs:label`), paradigmatic synonyms (using `oboinowl:hasSynonym`, `oboinowl:hasExactSynonym`, and `oboinowl:hasRelatedSynonym`), short names (`wdt:P1813`, about "short name"), (`wdt:P5191`, about "derived from lexeme"), replaces (`dct:replaces`), `sdo:alternateName`, and `owl:annotatedTarget` in all its languages. Similarly, 595,167 multilingual labels using `rdfs:label` and `skos:altLabel` about the entities in the integrated graph were extracted from Wikidata.

3.3   Redirection

Our analysis showed that Homosaurus switched its protocol from HTTP to HTTPS. Thus, 1,738 URIs were redirected to their HTTPS equivalent in version 2. No redirection was found from version 2 to 3. Another 63[18]

---

[15] We use the prefix `wdt` for the namespace http://www.wikidata.org/prop/direct/ and `wd` for the namespace http://www.wikidata.org/entity/.

[16] Using the Wikidata SPARQL endpoint (https://query.wikidata.org/sparql), we can obtain the corresponding identifiers of Homosaurus, QLIT, and GSSO. To obtain the full URI, we process these identifiers using the "frommatter" as specified on their pages. Take GSSO for example, 002171 is the identifier. Using the formmater http://purl.obolibrary.org/obo/GSSO_$1, we can replace the place-holder and get the full URI: http://purl.obolibrary.org/obo/GSSO_002171. In this paper, we use the prefix `obo` for the namespace http://purl.obolibrary.org/obo/. The preparation of Wikidata and links was done between 5th May and 8th May, 2024.

[17] We use the prefix `oboinowl` for the namespace http://www.geneontology.org/formats/oboInOwl# and `sdo` for the namespace https://schema.org/.

[18] We include the entities that no longer exist in the latest version of Homosaurus but were still referenced in GSSO.

Table 1. Extracted relations from sources and the number of triples

Source	Relation	#Triples	Comments
Homosaurus	`dct:isReplacedBy` and `dct:replaces`	3,370	Mostly links about replacing between version 2 and version 3.
	`skos:exactMatch` and `skos:closeMatch`	896	Links to entities in LCSH extracted from Homosaurus v2 and v3.
	`meta:redirecsTo`	63	Links representing redirection between entities in Homosaurus v3. Redirects for v2 were not included.
GSSO	`wd:Q1823134`	1,827	Links from entities in GSSO to subject headers in LCSH. It is mistaken to use `wd:Q1823134`. It was replaced by `wdt:P244` in the integrated graph.
	`oboInOwl:hasDbXref`	4,643	Links from entities in GSSO to entities in Wikidata
	`dc:identifier` and `sdo:identifier`	2,245	Links from entities in GSSO to entities in Homosaurus (all three versions)
QLIT	`skos:exactMatch` and `skos:closeMatch`	793	There are only links to Homosaurus v3.
	`skos:exactMatch` and `skos:closeMatch`	244	Links from QLIT to LCSH
Wikidata	`wdt:P244`	1,066	Selected links from Wikidata to LCSH
	`wdt:P6417` and `wdt:P10192`	1,408	Links from Wikidata to Homosaurus 2 and 3
	`wdt:P11852`	929	links from Wikidata to QLIT
	`wdt:P9827`	328	links from Wikidata to GSSO
Overall		17,812	The integrated graph involves 19,200 entities.

redirections were found in version 3.[19] None was covered by existing replace relations (`dct:replaces` or `dct:isReplacedBy`). Among them, 61 Homosaurus entities could be from outdated release(s) of Homosaurus and were redirected to entities in the latest release (v3). Only 2 redirections were found between entities in Homosaurus (v3). Moreover, we noticed that some URIs in version 2 cannot be resolved anymore, such as `h2:aromantic`. We use the redirection relation `https://krr.triply.cc/krr/metalink/def/redirectedTo` (`meta:redirectedTo` in short) [20].

3.4   Integrating Extracted Links

Table 1 presents the components of the integrated graph. We noticed a mistake that, for links from GSSO to LCSH, `wd:Q1823134` (representing the LCSH controlled vocabulary, rather than a property about links to their identifiers) was mistakenly used as a property. For consistency with the representation elsewhere, we change it to `wdt:P244`. When examining its links to Wikidata, it was also observed that GSSO has the URLs of webpages rather than the entities in Wikidata.[20] For example, `http://www.wikidata.org/entity/Q190845` should not have been used as https://www.wikidata.org/wiki/Q190845 in its published data. We have corrected it in the integrated dataset. We further double-checked that all the 2,085 LCSH entities in the integrated file are in the latest version of LCSH except one due to a suffix of '.html' from GSSO, which was corrected in the

---

[19] We retrieved the redirection using the *webdriver* of the *selenium* package (https://selenium-python.readthedocs.io/) on 30th April, 2024.

[20] See `https://www.wikidata.org/wiki/Wikidata:Identifiers` for details.

integrated file. The integrated graph involves 19,200 entities with 17,812 links. Its N-Triple file is 2.4MB. Considering only entities with the above-mentioned links are in the integrated graph, no singleton is present.

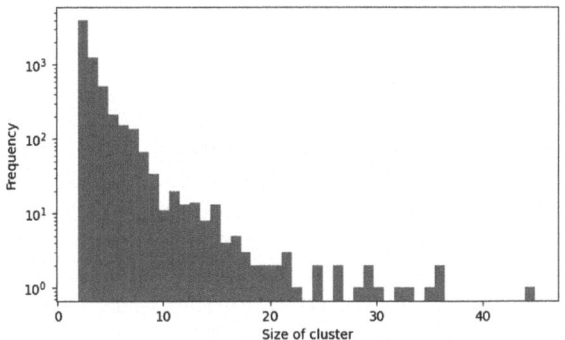

Fig. 2. Frequency histogram of the size of clusters

3.5 Clustering

We compute its *Weakly Connected Component* (WCC) for our integrated directed graph. A WCC of a graph is a subgraph with maximal entities where there is a path between any of its entities regardless of the direction of edges. In this paper, WCCs are clusters of entities that mostly share a similar or related meaning. For our integrated graph, there are 6,406 WCCs. Figure 2 shows that the largest four WCCs consist of 45, 36, 36, and 35 entities respectively. The largest one consists of 12 entities from Wikidata, 6 from GSSO, 5 from QLIT, 7 from Homosaurus v2, 6 from Homosaurus v3, 4 from LCSH, etc. More specifically, it involves related concepts about "human sexuality" (e.g. `wd:Q154136`), "Sexual intercourse" (e.g. `h3:homoit0000662` and `lcsh:sh85120739`), "Sex (Act)" (e.g. `h3:homoit0001267`), "fucking" (e.g. `h2:fucking`), "gender" (`h2:gender`), and "sexuality" (e.g. `wd:Q3043188`). This example shows how the ambiguity of these entities accumulates into bigger clusters. The size of these clusters is within the capability of manual revision, but the number of WCCs remains significant.

4 Evaluation

We assess the integrated data's usefulness by showcasing three scenarios, illustrating how it can ease the maintenance of their conceptual models and their links for developers in the community.

4.1   Scenario 1: Link Discovery

Relying on the WCCs, we propose to discover missing links between two conceptual models. The intuition is that, if entities from different concept models are in the same WCC and they are unique of its conceptual model in the WCC, they are likely to refer to the same or related things. A link could be added after manual examination. In practice, for the case of Homosaurus, we take into account exceptions such as redirection, replacement, and entities no longer maintained. Next, we report our findings for Homosaurus and QLIT. Only 531 links were observed between Homosaurus (v3) and LCSH. In contrast, 2,085 LCSH entities were found in the integrated graph.[21] Using the method mentioned above, 25 links were found. These discovered links have been submitted to the experts in Homosaurus for consideration before the next release. Similarly, QLIT has only 244 links to LCSH. Using the same method, we found 105 potential missing links between them. Further review by Swedish-speaking experts from the QLIT team shows that 78 (72.38%) suggested links should be included: 38 (36.19%) can be included using `skos:exactMatch` and another 38 (36.19%) using `skos:closeMatch`. 28 (26.67%) suggested links are incorrect. Moreover, one suggested link is uncertain for experts. Finally, given that some entities were redirected in Homosaurus v3, we also found an outdated link in the latest Homosaurus (see `qlit:oj77yj15` in Fig. 3).

4.2   Scenario 2: Concept Drift and Change

We addressed in related work that existing measures do not apply to this multilingual case. The change of concepts was captured by different conceptual models at various levels, resulting in a complex co-evolution. Next, we present three examples for concept convergence, ambiguity, and scope change, respectively.

First, we use an example in Fig. 3 to demonstrate how missing identity links, redirection relations, evolving concepts, as well as no longer maintained URIs can result in ambiguity and difficulty in identifying erroneous links. It was observed that `h3:homoit00442` replaced `h2:fetishism` and is the target of a few redirected URIs, including `h3:homoit0000102`, which was linked by many. It could be that `h3:homoit00442` is a merge of the concept of 'BDSM' and 'fetishism'. As a result, two clusters of entities about BDSM and fetish/kinks are in the same connected component. Further examination shows that, 'BDSM' is now an alternative label (`skos:altLabel`) for `h3:homoit00442` in the latest version. It was also noticed that a few URIs (highlighted in red) no longer appear in the latest version of Homosaurus, which leads to missing label information and relations. This example shows how multiple parties can have different views on concepts as conceptual models develop and the consequences of such changes.

In practice, it was observed that concept drift is often coupled with version changes, errors, and ambiguity, leading to a complicated network that requires

---

[21] This little overlap of concepts has been considered evidence by many that Homosaurus can be used as a complementary conceptual model of LCSH.

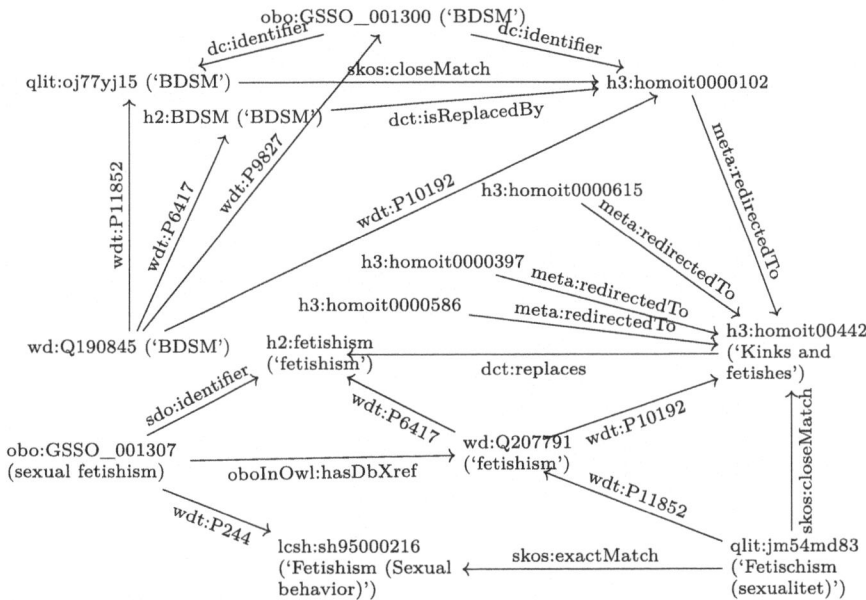

Fig. 3. A subgraph of the seventh largest WCC with 30 entities and 44 edges including concepts related to BDSM and sexual fetish. Labels that can be found are included. Some links and entities were omitted for clear visualisation. Entities in red are no longer in the latest version of Homosaurus. (Color figure online)

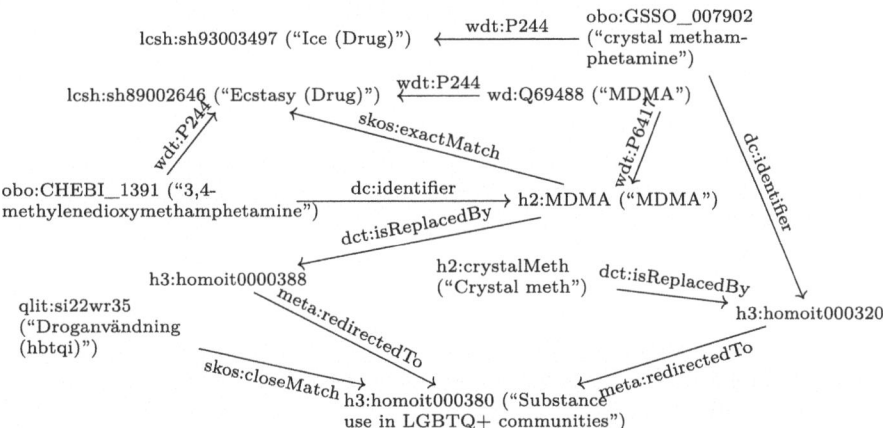

Fig. 4. An example of concept drift and change involving 3,4-methylenedioxymethamphetamine, MDMA, Crystal Meth, Ecstasy, Ice, Substance use in LGBTQ+ communities, etc. Some entities and links are not included for clear visualization. Highlighted in red are two entities no longer in the latest Homosaurus. (Color figure online)

careful manual evaluation. Next, we showcase the complexity by focusing on entities in GSSO and Homosaurus. Figure 4 is an example of concept drift involves the GSSO term "3,4-methylenedioxymethamphetamine" with the URI `obo:CHEBI_1391`, linking to the Homosaurus version 2 term "MDMA" with the URI `h2:MDMA`, which is replaced by the URI `h3:homoit0000388` in Homosaurus version 3. However, the URI no longer exists in the latest version of Homosaurus, and was automatically redirected to the identifier `h3:homoit0000380`, which corresponds to the term "Substance use in LGBTQ+ communities" with a few other labels (`skos:altLabel`) including "Drug use (LGBTQ)", "Alcohol use in LGBTQ+ communities", etc. Therefore, the term "3,4-methylenedioxymetha-mphetamine" from GSSO is the same cluster as the term "Substance use in LGBTQ+ communities" in Homosaurus version 3, which can be inaccurate and misleading. MDMA is an abbreviated form of methylenedioxymethamphetamine and is the main component of the popular party drug ecstasy. Crystal metham-phetamine (a.k.a. Ice) is a different drug. Figure 4 illustrates the complexity and ambiguity by taking into account all related entities and their interconnected links. Moreover, when updating outdated links in conceptual models relying on translated Homosaurus terms, if such information were used, the subsequent con-ceptual model would inherit this problem. In the case above, `h3:homoit0000380` has a `skos:closeMatch` link to the QLIT term "Droganvändning (HBTQI)" ("Drug use (LGBTQ+)" in English) with the identifier `qlit:si22wr35`. Should a connection be established between this Swedish term in QLIT and GSSO through Homosaurus, the established link would be problematic and inaccurate, and following these links would result in confusion and incorrect labels.

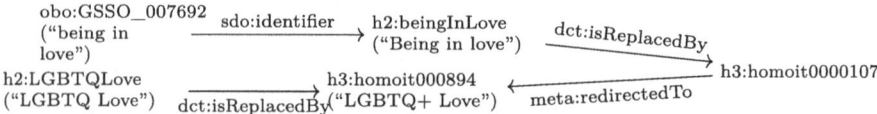

Fig. 5. Convergence of "Being in love" and "LGBTQ love" to "LGBTQ+ love". Fol-lowing the links could lead to a change in scope. Highlighted in red is an entity no longer maintained in Homosaurus v3. Not all entities and links in the WCC were illustrated. (Color figure online)

Figure 5 shows how concepts' scope can change. In GSSO, the entity labelled "being in love" (`obo:GSSO_007692`) has note "The state in which a person is when they are in love.". Its scope is not limited to the LGBTQ+ community. It was linked to `h2:beingInLove` ("Being in love"), which is a broader term than "LGBTQ Love". Moreover, this entity in GSSO links to `h3:homoit0000107`, which was redirected to `h3:homoit0000894` ("LGBTQ+ love"). It was observed that `v2:LGBTQLove` was replaced by `h3:homoit0000894`. This convergence leads to a change in scope.

4.3   Scenario 3: Reusing Multilingual Information

Next, we study the reuse of multilingual labels for entities involved in WCCs where there is a "one-to-one mapping" to entities of Homosaurus v3 (to avoid cases of ambiguity and concept drift). For GSSO, we study how much information can be reused for a total of nine relations about labels, synonyms, and alternative labels (see Sect. 3.2). We compare it against that of Wikidata.

There are 1,779 GSSO entities in the integrated graph. 1,681 are the unique entity of GSSO in the WCC. When further restricting to exactly one entity of Homosaurus v3 in the WCC (with consideration of redirection and replacement), only 48 entities remain. The three languages with the most labels are English (356 labels for 48 entities), Danish (26 labels for 9 entities), and French (24 labels for 9 entities). The corresponding labels per entity are 7.42, 2.89, and 2.67 for English, Danish, and French, respectively.

As for Wikidata, there are 5,769 entities in the integrated graph, among which 4,939 are unique in their WCCs. When further restricting to the correspondence of exactly one entity in Homosaurus v3, only 429 entities remain. The four languages with the most labels are English (1,692 labels for 429 entities), Spanish (951 labels for 333 entities), Chinese (893 labels for 287 entities), and Portuguese (881 labels for 299 entities). The corresponding labels per entity are 3,94, 2.86, 3.11, and 2.95 for English, Spanish, Chinese, and Portuguese, respectively.

It was noticed that the number of entities that has a one-to-one mapping between GSSO and Homosaurus v3 is significantly smaller than that of Wikidata. Despite that the entities take more relations into consideration when retrieving multilingual labels from GSSO, the number of multilingual labels is remarkably larger for Wikidata (with the exception that GSSO can provide more labels in English per entity in this setting). The average number of labels is 2.56 per entity for the top 20 languages with the most labels, with an average of 268.7 entities per language for Wikidata. This indicates that Wikidata can be a better choice when considering reusing its multilingual labels to enrich Homosaurus with manual examination. Nevertheless, these multilingual labels as suggestions cannot be directly used but remain to be assessed after manual revisions by experts for each language. Take `h3:homoit0000295` ("Coming out") for example, it has label "sortie du placard" in GSSO. Moreover, "sortir du placard" and "coming out" are synonyms for relation `oboInOwl:hasRelatedSynonym` and `oboInOwl:hasSynonym`, with three labels using `owl:annotatedTarget`: "sortir du placard", "coming out", and "sortie du placard". The labels retrieved as suggestion from Wikidata are similar: there is a label of "coming out" using `rdfs:label`; some more labels using `skos:altLabel`: "coming-out", "sortie du placard", "sortie de placard", and "sortir du placard". This shows that GSSO and Wikidata have overlaps in the labels they provide. Ultimately, it is the responsibility of the developers to determine the preferred label, the alternative label(s), and recognize incorrect ones.

Similarly, we can extract labels from Wikidata as suggestions for QLIT. There are 914 entities with one prefLabel each but only a total of 480 altLabels. Using the method above, we extracted 775 labels in Wikidata (524 prefLabels and 251 altLabels) for 524 entities. It was noticed that, in many cases, the difference is minor, either in the upper/lower case of the first character or the upper/lower case of 'hbtqi' (the Swedish word for LGBTQI). It remains a question if Wikidata has taken advantage of QLIT for its entries, or these terms are likely to be free from concept change and ambiguity due to the restriction of exactly one WCC.

5   Discussion

In Sect. 4.1, we demonstrated how we can find missing links using the WCCs. A further manual examination found some correspondence of terms between QLIT and Homosaurus. For example, a link between `qlit:tm80vg73` ("Pappor till homosexuella") and `h3:homoit0000427` ("Fathers of queer people") could be included. Similarly, `qlit:iq08ee58` ("Masters (hbtqi)") could be linked to `h3:homoit0000999` ("LGBTQ+ dominants"). These missing links can lead to discrepancies between conceptual models if not fixed before new versions of QLIT are released. Our proposed WCC-based method could reduce the effort of manually finding links between conceptual models. However, our approach is limited to the entities that are linked to at least one other entity. Moreover, assessing these links requires considerable manual effort. All the links in Sect. 4.1 have been submitted to the corresponding communities for manual revision by experts. For this reason, at the time of submission, the quality of newly found links in our approach remains unknown. This is also the case for our use case 3. Given that drift and change in the concept are mixed with ambiguity and errors, it is also difficult to evaluate the output in Sect. 4.2 where a significant amount of work is required for coordination between subcommunities.

In our study, `dct:replaces` and `dct:isReplacedBy` were included in the integrated graph, despite not necessarily implying equivalence relations. The value lies in the study of the dynamics and evolution of entities in Homosaurus and the impact on other linked entities. It was noticed that the concept change in Homosaurus is partially reflected by such links. Concept drift and change can result in potential duplicate terms (see Homosaurus terms in Fig. 3 for example) that could violate the Unique Name Assumption [20]. This requires further manual examination for each concept model. Using the relation `owl:differentFrom` for different entities can ease future automated examination.

Section 4.3 showed that GSSO cannot suggest as many labels as Wikidata. This could be due to the restriction using WCCs. A further experiment with a relaxed condition only considering redirection and replacement shows that there are suggesting labels for 115 entities for Danish and 47 for French.

6   Conclusion and Future Work

In this paper, we studied the properties of LGBTQ+-related concepts and their links in the semantic web. With the links extracted, we constructed an integrated graph and evaluated its use in three scenarios. We demonstrated how to find missing links and outdated links. Experts' review confirmed that 78 links between QLIT and LCSH should have been included. We addressed the issue of concept drift and showed how WCCs can assist the community in easily locating entities with potential issues. Finally, we showcased the reuse of multilingual information for Homosaurus. Our findings indicate that Wikidata offers a substantially greater number of multilingual labels than GSSO. Handling such tasks remains semi-automatic. Manual checking and community input are essential to evaluate our approach before implementation in reality.

In practice, experts can overlook the implications of diverging or converging concepts. This paper demonstrates how we can provide such insight to the specialists. If any results require the intervention of others, discussion in the community would be advantageous. Our code could be reused for automatic detection of outdated links in the future. Given the significant amount of manual work, an interface that supports manual revision of the WCCs could be helpful.

Heterogeneous use of relations between entities, especially that for mappings was observed (see Table 1). It could benefit the community, especially for interoperability, if they consider adapting a common FAIR Implementation Profile, a structured representation of the community's decision on knowledge representation languages, semantic models, metadata, etc. [19].

This paper presents briefly the analysis of concept drift in selected conceptual models. Their labels, comments, and scopenotes could be taken into consideration for further manual analysis. Drifting concepts could be analysed by studying their use in various contexts with natural language processing techniques [18].

Mistakes identified in GSSO should be corrected in upcoming versions (see Sect. 3.4). The 63 redirection links in version 3 presented in Sect. 3.3 remain to be checked to ensure that they maintain the original semantics [12]. If correct, they can be included (e.g. using the replacement relation) in the future release of Homosaurus. Some other conceptual models in the semantic web, such as LCDGT [5] and DBpedia [2], have been reported to contain some LGBTQ+ terms and could be included in some follow-up research together with the mappings from QLIT to some Swedish library thesauri [11]. Other links such as skos:narrower, skos:broader, and skos:relatedMatch can be explored.

Given the nature of the field, there is no perfect conceptual model [5]. As addressed in Sect. 2, there is no systematic report on the quality of links, the drift of concepts, and the potential harm of outdated links. Our data could serve to ease the manual work for this task. A good starting point is a list of historical terms (see h3:homoit0000878) and reclaimed terms (see h3:homoit0001559) in Homosaurus. Additional resources can be used as references (e.g. a list of terms about transgender and diversity in LCSH [1]). While the analysis presented in this paper is limited to selected popular conceptual models, entities of LGBTQ+-related concepts exist widely in the semantic web (e.g. DBpedia, Wikidata), exhibiting a long-tail distribution. A complete examination requires a crawl of the semantic web and some accurate filtering, which are for future work.

Acknowledgments. The authors appreciate the help from Jack van der Wel (IHLIA/Homosaurus), Clair Kronk (GSSO), and the QLIT team: Siska Humlesjö and Olov Kriström. The authors used TeXGPT to rephrase some sentences through Write-full on Overleaf. The LaTeX   code of Fig. 1 was generated with the help of ChatGPT. Neither was used to generate complete sentences or paragraphs.

References

1. Trans & Gender Diverse LCSH (2024). https://translcsh.com/. Accessed 25 May 2024
2. Auer, S., Bizer, C., Kobilarov, G., Lehmann, J., Cyganiak, R., Ives, Z.: DBpedia: a nucleus for a web of open data. In: Aberer, K., et al. (eds.) ASWC/ISWC -2007. LNCS, vol. 4825, pp. 722–735. Springer, Heidelberg (2007). https://doi.org/10.1007/978-3-540-76298-0_52
3. Bergenmar, J., Golub, K., Humelsjö, S.: Queerlit database: making Swedish LGBTQI literature easily accessible. In: DHNB 2022: The 6th Digital Humanities in the Nordic and Baltic Countries Conference 2022, pp. 433–437. CEUR-WS. org (2022)
4. Braquet, D.: Chapter 2 LGBTQ+ Terminology, Scenarios and Strategies, and Relevant Web-based Resources in the 21st Century: A Glimpse, pp. 49–61 (2019). https://doi.org/10.1108/S0065-283020190000045009
5. Dobreski, B., Snow, K., Moulaison-Sandy, H.: On overlap and otherness: a comparison of three vocabularies' approaches to LGBTQ+ identity. Cataloging Classif. Q. **60**(6–7), 490–513 (2022). https://doi.org/10.1080/01639374.2022.2090040
6. Ihrmark, D.O., Golub, K., Tan, X.: Subject indexing of LGBTQ+ fiction in Sweden and china. In: Knowledge Organization for Resilience in Times of Crisis: Challenges and Opportunities, pp. 379–384. Ergon-Verlag (2024)
7. Jagose, A.: Queer theory: An introduction. NYU Press (1996)
8. Kazarian, A.M., Wang, S.: Evaluating Automated Machine Translation of LGBTQ+ Terms: Towards Multilingual Homosaurus (2024). https://doi.org/10.5281/zenodo.10523283
9. Kronk, C.A., Dexheimer, J.W.: Development of the gender, sex, and sexual orientation ontology: evaluation and workflow. J. Am. Med. Inform. Assoc. **27**(7), 1110–1115 (2020)
10. Lynch, K.E., Alba, P.R., Patterson, O.V., Viernes, B., Coronado, G., DuVall, S.L.: The utility of clinical notes for sexual minority health research. Am. J. Prev. Med. **59**(5), 755–763 (2020). https://doi.org/10.1016/j.amepre.2020.05.026, https://www.sciencedirect.com/science/article/pii/S0749379720302774
11. Matsson, A., Kriström, O.: Building and serving the Queerlit thesaurus as linked open data. Digital Humanit. Nord. Baltic Countries Publ. **5**(1), 29–39 (2023)
12. Nasim, I., Wang, S., Raad, J., Bloem, P., van Harmelen, F.: What does it mean when your URIs are redirected? Examining identity and redirection in the LOD cloud. In: Proceedings of the 8th Workshop on Managing the Evolution and Preservation of the Data Web (MEPDaW) (2022)
13. Office of Communications and Marketing: An LGTBQ language thesaurus is translated to Spanish (2024). https://www.gc.cuny.edu/news/lgtbq-language-thesaurus-translated-spanish. Accessed 19 May 2024
14. Peterson, R.: Library of congress subject headings for LGBT studies (2023). https://guides.libraries.emory.edu/main/queerlcsh

15. Tai, J.: Cultural humility as a framework for anti-oppressive archival description. Reinventing the Museum: Relevance, Inclusion, and Global Responsibilities, p. 349 (2023)
16. The Homosaurus editorial Board: Homosaurus vocabulary site (2024). https://homosaurus.org/about. Its documentation was last accessed on 24th May, 2024
17. Vrandečić, D., Krötzsch, M.: Wikidata: a free collaborative knowledgebase. Commun. ACM **57**(10), 78–85 (2014)
18. Wang, S., Schlobach, S., Klein, M.: Concept drift and how to identify it. J. Web Semant. **9**(3), 247–265 (2011). https://doi.org/10.1016/j.websem.2011.05.003, semantic Web Dynamics Semantic Web Challenge, 2010
19. Wang, S., Maineri, A., Singh, N.K., Kuhn, T.: FAIR implementation profiles for social science. In: Garoufallou, E., Sartori, F. (eds.) Metadata and Semantic Research. MTSR 2023. Communications in Computer and Information Science, vol. 2048, pp. 284–290. Springer, Cham (2024). https://doi.org/10.1007/978-3-031-65990-4_26
20. Wang, S., Raad, J., Bloem, P., van Harmelen, F.: Refining large integrated identity graphs using the unique name assumption. In: Pesquita, C., et al. The Semantic Web. ESWC 2023. LNCS, vol. 13870, pp. 55–71. Springer, Cham (2023). https://doi.org/10.1007/978-3-031-33455-9_4
21. Watson, B.M.: there was sex but no sexuality* critical cataloging and the classification of asexuality in LCSH. Cataloging Classif. Q. **58**(6), 547–565 (2020)

Capturing the Viewpoint Dynamics in the News Domain

Enrico Motta[1,2](✉) 📧, Francesco Osborne[1,3] 📧, Martino M. L. Pulici[1] 📧, Angelo Salatino[1] 📧, and Iman Naja[1] 📧

[1] Knowledge Media Institute, The Open University, Milton Keynes, UK
enrico.motta@open.ac.uk
[2] MediaFutures Centre, University of Bergen, Bergen, Norway
[3] Department of Business and Law, University of Milano Bicocca, Milan, Italy

Abstract. Despite the seismic changes brought about by the web and social media, mainstream news sources still play a crucial role in democratic societies. In particular, a healthy democracy requires a balanced and diverse media landscape, able to provide an arena in which the various *topics* and *viewpoints* relevant to the political discourse of the day are presented and discussed. Unfortunately, there is currently little effective computational support available to the various classes of users, who are interested in monitoring the topic and viewpoint dynamics in the news—e.g., for regulatory or research purposes. As a result, current analyses by researchers and practitioners tend to be small scale and, by and large, rely on manual investigations of topic and viewpoint coverage. To address this issue, we have developed a hybrid human-machine approach, which uses a Large Language Model (LLM) first to help analysts to identify the range of viewpoints relevant to the debate around a given topic, and then to classify the claims expressed in the news corpus of interest with respect to the identified viewpoints. We tested a variety of LLMs on a benchmark corpus of news items drawn from British media sources. Our results indicate that GPT4o outperforms the other alternatives and can already provide effective support for this classification task, even when run in a zero-shot learning modality.

Keywords: News Analytics · News Classification · Large Language Models · Viewpoints

1 Introduction

Despite the seismic changes brought about by the web and social media, mainstream news sources still play a crucial role in democratic societies, informing the public about the important issues of the day and providing a platform for democratic debate. In this context, assuming an appropriate regulatory regime is in place to guarantee a diverse media ownership, the acid test for a democratic media landscape is related to *viewpoint diversity* [1], namely the extent by which media sources provide citizens with a robust range of alternative interpretations on a given topic. Here we use the term 'topic' to refer to a particular issue being discussed in the media—e.g., "the appropriate level of income tax in the UK", while a viewpoint refers to a position relevant to this issue—e.g., a position advocating a low taxation regime. In particular, as explained later in the paper, the concept of 'viewpoint' does not indicate an individual statement reported in the news but defines a category that groups together all statements that subscribe to a common position—e.g., all the statements in favour of a low taxation regime.

Assessing viewpoint diversity accurately requires effective content analysis methods and tools, which constitutes a major challenge, given i) the sheer scale of news and information in the contemporary online environment[1]—an information overload that facilitates disinformation [2], and ii) the limited take-up of automated content analysis solutions in media research and related fields [3]. Arguably, this limited take-up reflects a disconnect between the needs of media scholars and the state of the art in computational techniques for news media analysis [4]. As a result, current studies by both researchers and practitioners—e.g., see [5] —tend to be small scale and, by and large, rely on a manual analysis of topic and viewpoint coverage.

To address this issue, we have developed a hybrid human-machine approach, which relies on a Large Language Model (LLM) to help analysts to identify the various viewpoints relevant to the debate around a given topic, and then classifies the claims expressed in a news corpus of interest with respect to the identified categories. For instance, if we take immigration as our topic of interest, through an analysis of a corpus of news items we can identify a number of relevant viewpoints, covering different perspectives on the debate. These may include the view of immigration as beneficial to the economy, the characterization of immigrants as a threat to a nation's culture and social fabric, a humanitarian perspective on refugees as vulnerable people, and several others. Once these viewpoints have been identified, we can then use a LLM to classify statements expressed in the media by political and other actors with respect to these viewpoints, thus generating a multi-dimensional representation of the debate on the immigration topic over time. Here, we use the term 'claim' to refer to a statement about a topic, which has been expressed by an actor and reported in a news item. Our approach has been evaluated on a corpus of news items drawn from British media sources, showing promising results. In particular, as discussed in Sect. 3, our evaluation shows that GPT4o outperforms the other alternatives and can provide good support for this classification task, even when run in a zero-shot learning modality.

The rest of the paper is structured as follows. In the next section we discuss the notion of viewpoint and provide a brief overview of computational approaches to viewpoint extraction. In Sect. 3 we present our approach, illustrating our methodology in a concrete scenario, where we model the viewpoint dynamics associated with the UK's immigration debate, as discussed in mainstream news media over a three month period. In Sect. 4 we briefly discuss the application-level perspective on this work, providing an example of the kind of user-centric visualizations that we can generate to provide insights to media scholars and practitioners. Finally, in Sect. 5 we reflect on the key results presented in this paper and discuss the outstanding issues that will be the focus of future research.

---

[1] Even if we restrict our analysis to UK's mainstream news outlets, we already ought to consider thousands of news items being published on a daily basis – see https://tinyurl.com/ycxwprjm.

2 Viewpoints in the Media and Computer Science Literature

2.1 Viewpoints as Collections of Coherent Claims

The notions of viewpoint and viewpoint diversity in media sciences are discussed by Baden and Springer [1], who emphasise that differences in the ways arguments are *framed* [6] do not necessarily indicate different viewpoints. In particular, they point out that a specific position on an issue can be articulated through different framing devices—e.g., by presenting it in abstract terms through a thematic frame or alternatively by illustrating concrete examples, therefore making use of an episodic frame. According to Baden and Springer, in order to be distinct, viewpoints need instead to "construct different meaning in a consequential sense"—i.e., they need to open up perspectives that are semantically diverse. An important consequence of this approach is that viewpoints are therefore not atomic concepts but are instead constructed out of a number of individual positions expressed in the media. While these positions may be articulated using different terms and framing devices, they can be grouped around "commensurable interpretations" [1]—i.e., they subscribe to the same viewpoint. Building on these ideas, in a previous paper [7] we formalised the notion of viewpoint as a coherent collection of *claims*—i.e., one that only includes claims that satisfy the membership criterion associated with the viewpoint in question. In turn, a claim is defined in our model as a statement reported in a news item, which is associated to an agent and concerns a topic. This formalization has been realised in a formal OWL model, the News Classification Ontology (NCO), which can be found at http://data.open.ac.uk/ontology/newsclassification.

An example of instantiating the notion of viewpoint as a device for bringing together commensurable claims can be found in [5], where the authors analyse the immigration debate in mainstream media and identify four different viewpoints (*"Negative", "Administrative burden", "Positive", "Victimization"*), which abstract from the variety of individual positions about immigration that are reported in the news.

2.2 Computational Approaches to Viewpoint Extraction

The survey by Doan and Gulla [8] reviews the state of the art concerning automated approaches to identifying political viewpoints and concludes that it "falls somewhat short of our goal with automatic political viewpoint identification". In their survey, Doan and Gulla adopt a definition of political viewpoint that is broadly consistent with the one used in this paper. Essentially, they emphasise that, for a given topic, T, it is possible to identify a set of viewpoints, {p1, ...pn}, which summarise the contrastive positions that can be expressed about T^2. The task of a system for political viewpoint identification is then to classify a piece of text (what we refer to as a claim in our terminology) in terms of the given set of viewpoints.

The approach by Trabelsi and Zaïane [9] proposes a generative Topic-Viewpoint model, which uses unigrams to characterise probabilistically the vocabulary that is used

---

[2] Actually, the definition given in [8] does not explicitly includes the notion of topic, however we assume that this is simply an oversight, given that the notion of topic is informally discussed elsewhere in the paper and implicitly assumed throughout this work.

by an author expressing a particular viewpoint about a topic. A particular strength of this approach is that it is unsupervised and is generic with respect to specific topics or viewpoints. However, from the examples shown in the paper, it appears that the approach is limited to a binary classification, in favour or against a particular position related to a topic. This limits the approach's flexibility, in particular its ability to identify the variety of viewpoints relevant to the debate about a topic and then to classify relevant statements in terms of these various viewpoints. Another key aspect of this approach is that it takes advantage of interactions between different post creators on social media to identify contrastive opinions. While this is of course an interesting and clever heuristic in the context of social media, it also means that this approach is not directly applicable to our news scenario.

The work by Quraishi et al. [10] also focuses on debates in social media to identify contrastive viewpoints on a topic. Like Trabelsi and Zaïane, they take advantage of interactions in social media to identify different communities that in turn can be associated with different viewpoints. This work uses a graph partitioning method and improves on the approach by Trabelsi and Zaïane because it is able to discover several viewpoints associated with a topic. They also focus on explaining the discovered viewpoints by means of Iterative Rank Difference, a technique that identifies the descriptive keywords associated with a viewpoint. However, judging from the examples shown in the paper, the resulting descriptions appear to be both opaque and rather noisy. Moreover, in contrast with the aim of our work, there is no attempt here to try and provide a comprehensive account of the debate around a topic by identifying all the relevant viewpoints and using the resulting classification to structure all relevant claims.

The paper by Hada et al. [11] focuses instead on measuring the viewpoint diversity which emerges from interactions on X (formerly known as Twitter). In particular, they use the Fragmentation metric [12], which aims to measure the degree of exposure to alternative viewpoints for each user. Their analysis of conversations about the immigration topic shows that nearly 70% of users have a very low Fragmentation score, indicating virtually no exposure to alternative viewpoints. However, in contrast with our work, there is no attempt in this paper to model the various relevant viewpoints and the focus is purely on measuring diversity. In addition, as is the case with research on stance detection [13], the various positions around a topic are clustered around two broad categories, in favour or against a particular position, in contrast with the more fine-grained differentiation proposed in this paper.

Finally, Chen et al. [14] test different solutions on a variety of tasks concerned with extracting and classifying perspectives. A combination of Information Retrieval and a BERT model obtains a 50.8% F1 measure in perspective extraction, while BERT alone obtains a 63.7% F1 measure on the task of clustering equivalent perspectives—in our terminology, positions that belong to same viewpoint. This work is relevant to our research, as it focuses on both identifying and grouping together individual positions about a topic. However, they extract their data from debating websites, such as idebate.com, which naturally provide more structure to the debate than what is available from the broader news domain, which is instead the focus of our research. In addition, in contrast with our methodology, there is no attempt at classifying individual claims with respect to the set of identified viewpoints. Instead, the authors focus on classifying the stance of each perspective, which relies on a rather reductive assumption that any perspective on a topic can be classified in terms of binary support/oppose stance, which is not necessarily the case in the context of analysing complex issues[3].

3 Approach

3.1 Generating a Corpus of News Items Covering the UK Immigration Debate

As already pointed out, our goal is to develop an approach that can identify the range of viewpoints relevant to a topic and then automatically classify all the claims in a news corpus with respect to the identified viewpoints. The starting point for our experiments was a news corpus comprising 603 news items, published between 1 June and 31 August 2023. The corpus was generated by accessing the Aylien news service (www.aylien. com), which provides an API to access information from about 80,000 news sources from around the world. In particular, we retrieved news items that had already been tagged with the keyword 'immigration' by the Aylien service and/or contained such a keyword (or related ones, such as 'immigrants') in the title or body. The search was restricted to the following mainstream news sources: The Guardian, BBC, ITV, Daily Mail, The Sun, The Daily Express, The Daily Telegraph, The Evening Standard, The Independent, Reuters and Associated Press.

Having done this, the next step was to extract the claims reported in the news corpus. A claim is formally defined as a sextuple <*utterance, actor, news item, news source, date, topic*>, where the utterance is a statement made by an actor, about a particular topic, which is reported in a news item on a certain date[4]. The field 'news source' indicates the outlet that published the news item in question and, in our case study, can only be one of the mainstream media sources listed above. In the experiments reported in this paper the field 'topic' is fixed, given that we are only concerned with statements made about the topic 'Immigration'. Finally, we should also point out that we only consider claims that consist of utterances that can be attributed to an actor explicitly mentioned in the associated news item. Such an association between utterance and actor may be expressed directly in the news item, by reporting a statement by the actor using quotation marks, or indirectly—e.g., by using an expression, such as, "[actor] said that…"). However, while both direct and indirect quotations are considered, in either case the actor must be explicitly mentioned in the news item. This means that opinions expressed implicitly— e.g., an opinion expressed by a journalist authoring a news item—are considered as out of scope for this study.

---

[3] For example, our analysis of the immigration debate in UK includes a viewpoint "Immigration as a Management Issue", which cannot be reduced to a stance pro or against immigration.

[4] Needless to say, a specific utterance expressed by a specific actor may be reported by multiple news sources, often on the same day, leading to a number of distinct claims that share a common utterance and actor. Analogously, it is also possible for multiple actors to repeat the same utterance. Indeed, this is not uncommon in the UK's political debate, where politicians belonging to the same party may be required to put forward a specific agreed line.

3.2 Extracting Claims

To extract the set of claims we run GPT-4[5] on the news corpus, generating 3455 claims. However, once we analysed the extracted claims, we realised that they covered a variety of immigration contexts, not just in UK but also in other countries. Hence, we decided to create a more coherent corpus by restricting the analysis to UK immigration. This was achieved by considering only claims made by actors based in Britain[6] and, as a result, we ended up with 766 claims. These included both 'atomic' utterances, such as *"Britain has returned 1,800 migrants to Albania in just six months"*, as well as more articulated positions, such as *"While the principle of reducing net migration was right, there was a shortage of care workers in the UK. It's not as simple as just putting the salary thresholds up as well, there's quite a lot of skilled but lower paid people that we need coming into this country"*.

3.3 Identifying the Range of Viewpoints

A hybrid human-machine approach was used to generate a comprehensive set of viewpoints, using GPT-4 Turbo to produce an initial list that was then finalised by a human expert. In particular, we fed the utterances to the LLM in six batches and the model was asked to produce 5 viewpoints that could be used to classify the utterances. This produced a total of 30 viewpoints. Next, the LLM was asked to extract and synthesise the most significant and frequent viewpoints from all the ones generated by the various runs. The output was an initial set of viewpoints that was then checked by a human expert, to ensure that this set could provide a comprehensive range of dimensions to cover the UK immigration debate. In particular, the key role of the human expert was to provide clear distinct definitions for the various viewpoints and ensure complete coverage of the relevant perspectives. To this purpose, he also added an additional viewpoint, 'Immigration as a management issue', which had not been picked up by the LLM. This human-in-the-loop element of the methodology is in our view essential, given that our goal is to provide support to the various researchers and practitioners who engage in media analytics for regulatory or research purposes. Hence, it is crucial that the set of viewpoints used for classifying a debate is robust and consistent with the type of categories that expert analysts, such as media and political scientists, would be happy to consider.

The set of identified viewpoints is as follows:

1. **Immigration as a management issue**. This viewpoint characterises utterances that focus on the way immigration is managed, typically by the UK government. For example, criticisms of specific elements of immigration policy—e.g., the use of hotels to house immigrants—should be classified under this viewpoint, unless other factors, such as humanitarian considerations, are emphasised in the claim. A key aspect of this viewpoint is that it does not necessarily imply a stand in favour or against immigration.

---

[5] In the early phases of this work we used GPT-4 (for claim extraction) and GPT-4 Turbo (for viewpoint identification), while we later switched to GPT-4o, once this became available.

[6] These were identified by querying Wikidata [15] and, in the vast majority of cases, turned out to be British politicians.

2. **Immigrants as victims/Humanitarian emphasis**. This viewpoint is used to classify utterances that are sympathetic to the plight of immigrants—e.g., when a tragedy happens at sea.

3. **Immigrants as potential criminals or threat/National security emphasis**. This viewpoint classifies utterances that imply a view of migrants as criminals or the migration phenomenon as a threat to national security. This viewpoint also covers the rhetoric about "dodgy lawyers" —i.e., lawyers who instruct their refugee clients to lie in order to get asylum in UK. In this case immigrants are criminals by association. In addition, utterances that advocate the use of restraining measures that are normally used for criminals—e.g., security tags, also fall under this category.

4. **Enhancing/Maintaining immigration pathways**. This viewpoint is used to classify utterances which either advocate for measures that would make it easier to come to UK or alternatively criticise the introduction of new restrictions to immigrations. Interestingly, in the context of an immigration debate, statements that criticise a relaxation of immigration rules should not be classified under 'Maintaining immigration pathways'. In other words, 'Maintaining immigration pathways' is not a neutral category, but implies a (mild) positive attitude towards immigration. In addition, this viewpoint is not necessarily mutually exclusive with the 'Restricting immigration pathways' one, because an utterance may advocate for more legal migrants coming to UK while supporting stricter measures against illegal migrants.

5. **Restricting immigration pathways**. This viewpoint is used to characterise utterances that refer to measures that would make it more difficult to come to UK. It covers both legal and illegal immigration pathways. Furthermore, attempts to remove migrants from the UK and 'success stories' about sending migrants back to their country also fall under this category.

6. **Economic benefits of immigration**. This viewpoint is used to classify utterances that refer to the economic value of immigration. Note that this viewpoint is not mutually exclusive with the one labelled 'Economic cost of immigration', because an utterance may consider certain migrants as economically beneficial while maintaining that others introduce a financial burden for the country.

7. **Economic cost of immigration**. This viewpoint is used to classify utterances that refer to the economic cost of immigration—e.g., when talking about the cost of accommodation for illegal migrants.

8. **Integration policies/Multiculturalism as a positive force**. This viewpoint is used to classify utterances that propose practical measures for integrating migrants in UK society, emphasise the need to support the integration of migrants, or otherwise highlight the value of cultural diversity. This viewpoint is not mutually exclusive with the one labelled 'Anti-integration policies/Cultural identity preservation', because an utterance may express a favorable opinion about multiculturalism while advocating placing tracking tags on illegal migrants.

9. **Anti-integration policies/Cultural identity preservation**. This viewpoint is used to classify utterances that emphasise the cultural differences between UK people and foreign migrants as well statements that advocate separating migrants from the rest of the UK population. For instance, the use of tracking tags on immigrants implies both a view of immigrants as criminals and also enforces an anti-integration policy.

3.4   Generating a Gold Standard for Viewpoint Classification

In order to construct a gold standard on which to benchmark a variety of LLMs, 402 claims were randomly selected from the corpus of claims relevant to the UK immigration debate and five human annotators were given the task to classify each of them in terms of the nine relevant viewpoints—as already mentioned, an individual claim can indeed instantiate more than one viewpoint. Following an initial standardization phase on a small subset of claims, which were classified by all five annotators, each claim was rated by exactly three annotators and we then used majority voting to generate the gold standard. To facilitate the annotation task, a customised spreadsheet was provided to each annotator, which, among other things, made it easy for them to quickly interpret and classify an utterance in the context of the associated news item—i.e., they could access a customised rendering of the news item in which the relevant utterance had been highlighted. To produce a baseline gold standard, comprising the entire corpus of 402 claims, a simple majority rule was used, where an utterance, say u_i, would be classified (or not) under a viewpoint, say v_j, if and only if at least two annotators agreed that u_i should go under v_j (or not). However, as shown in the second column of Table 1, only moderate agreement was achieved on average between the human annotators on the corpus of claims[7]. The reason for this is that the classification of political statement is a rather contested task, even for humans. Statements by politicians can be ambiguous and difficult to interpret, and therefore, despite putting significant effort in calibrating and harmonizing the scores from different annotators, only a moderate level of agreement could be achieved. For this reason, we also produced a restricted version of the gold standard, by only keeping claims for which, for a given viewpoint, say v_i, either at least two annotators agreed that v_i was relevant to the claim in question or alternatively all annotators agreed that it was not relevant. In other words, all utterances for which one and only one annotator flagged any viewpoint as relevant were discarded. This restricted version provided us with a more robust basis for evaluating the performance of different LLMs on the claim classification task.

Our analysis also showed that viewpoint 8 (Integration policies/Multiculturalism as a positive force) was particularly problematic, with its opposite, viewpoint 9 (Anti-integration policies/Cultural identity preservation) also exhibiting a low agreement score. Therefore we also produced agreement scores with only seven viewpoints, removing viewpoints 8 and 9. As shown in Table 1 and Table 2, if limit ourselves to only seven viewpoints, the level of agreement increases and in particular we reach a substantial level of agreement on the restricted dataset. Both datasets are freely available and can be accessed at https://doi.org/https://doi.org/10.21954/ou.rd.26268025.

---

[7] Jacob Cohen himself has suggested that a score of 0.41 (moderate agreement) may be acceptable. However this position has been criticised [16] and in general a score denoting at least substantial agreement is expected (≥ 0.61), with almost perfect agreement (≥ 0.81) the recommended norm for critical domains, such as medical studies – see [16] for a discussion on this issue.

Table 1. Agreement between annotators on the different datasets.

Pair	Cohen's kappa on full dataset (402 claims) (9 viewpoints)	Cohen's kappa on full dataset (402 claims) (7 viewpoints)	Cohen's kappa on restricted dataset (219 claims) (7 viewpoints)
1–2	0.46	0.49	0.70
1–3	0.56	0.66	0.87
1–4	0.38	0.39	0.69
1–5	0.49	0.55	0.80
2–3	0.33	0.36	0.58
2–4	0.43	0.49	0.68
2–5	0.44	0.54	0.84
3–4	0.30	0.35	0.60
3–5	0.56	0.60	0.73
4–5	0.37	0.35	0.68
Average	*0.43*	*0.48*	*0.71*

Table 2. Average annotator agreement by viewpoint on the different datasets.

Viewpoint	Cohen's kappa on full dataset (402 claims)	Cohen's kappa on restricted dataset (219 claims) (7 viewpoints)
1	0.55	0.80
2	0.67	0.82
3	0.35	0.51
4	0.36	0.69
5	0.47	0.72
6	0.43	0.81
7	0.52	0.69
8	0.11	N/A
9	0.35	N/A

3.5 Using LLMs to Classify Claims in Terms of the Relevant Viewpoints

Having produced both a baseline and a restricted gold standard, we then tested a variety of LLMs on both datasets, to assess to what extent they can effectively support the task of classifying claims with respect to the relevant viewpoints. All the LLMs were tested in a zero-shot learning modality. This approach was chosen to establish an initial baseline for future developments and also because zero-shot learning is well-suited for supporting discourse analysis across any domain, a feature that is very important for our target users. Since most of the LLMs do not provide a 'clean' output—i.e., a yes/no binary classification result, regex patterns were employed to post-process the verbose output. For each of the experimental setups, standard binary classification metrics were computed for every model, including those of a random classifier, which provides a baseline reference for the performance of the other models. As is the norm for classification tasks, the performance of the LLMs was assessed in terms of precision, recall, and F1 scores.

In particular, we evaluated both open-source models (Llama 2, Llama 3, Mistral 8x7B) and commercial closed models accessible via API (GPT-4, GPT-4o, Titan Text Premier, Mistral Large). While the latter typically perform better in zero-shot settings, usually they do not disclose the number of parameters and other key characteristics, behaving as black boxes. Below, we briefly summarise the published characteristics of these models.

GPT-4 Turbo, developed by OpenAI, is a large commercial multimodal model featuring a context window of 128K tokens [17]. Originally, this was OpenAI's flagship model, however it has now been surpassed by GPT-4o. The training data extends up to December 2023.

GPT-4o is OpenAI's most advanced model[8], offering text generation that is twice as fast and 50% cheaper than GPT-4. It shares the same 128K-token context window as GPT-4 Turbo and uses training data up to October 2023.

Titan Text Premier is the latest addition to Amazon's Titan family of LLMs. It is designed for enterprise-grade text generation applications and was optimised for retrieval-augmented generation[9]. It has a context length of 32K tokens.

Mistral Large is Mistral's flagship language model [18], featuring a 32K-token context window. As with all the previous commercial models, the number of parameters has not been disclosed.

Mistral 8x7B is an open sparse mixture-of-experts network, consisting of 8 models, each with 7 billion parameters [18]. It uses a context length of 32K tokens. Specifically, we used the Mistral 8x7B Instruct, which has been fine-tuned through supervised learning and direct preference optimization for precise instruction following. It is regarded as one of the strongest open-weight models.

Llama 2 70B is the largest member of Meta's Llama 2 family, featuring a 4K-token context length [19]. For this study, we adopted Llama 2 70B Chat, a fine-tuned version of Llama 2 70B optimised for dialogue use cases. All Llama 2 models employ supervised fine-tuning and reinforcement learning with human feedback.

Llama 3 70B is the largest model in Meta's recently released Llama 3 family [20]. It offers a context window of 8.2K tokens and utilises Grouped-Query Attention to improve inference efficiency. It includes training data up to December 2023.

---

[8] https://openai.com/index/hello-gpt-4o/.

[9] https://aws.amazon.com/it/about-aws/whats-new/2024/05/amazon-titan-text-premier-amazon-bedrock/.

Table 3. Scores for each model on the full dataset (9 viewpoints).

Model	Precision	Recall	F1
Random	0.11	0.50	0.15
GPT-4 Turbo	0.36	0.69	0.46
GPT-4o	**0.49**	0.65	**0.52**
Llama 2 70B	0.20	0.66	0.30
Llama 3 70B	0.36	0.71	0.45
Titan Premier	0.42	0.48	0.43
Mixtral 8x7B	0.36	0.63	0.40
Mistral Large	0.36	**0.75**	0.46

Table 4. Scores for each model on the full dataset (7 viewpoints).

Model	Precision	Recall	F1
Random	0.12	0.50	0.17
GPT-4 Turbo	0.40	0.75	0.51
GPT-4o	**0.53**	0.80	**0.62**
Llama 2 70B	0.23	0.69	0.34
Llama 3 70B	0.41	0.84	0.52
Titan Premier	0.50	0.61	0.53
Mixtral 8x7B	0.39	0.75	0.48
Mistral Large	0.38	**0.86**	0.51

Table 5. Scores for each model on the restricted dataset (7 viewpoints).

Model	Precision	Recall	F1
Random	0.15	0.50	0.20
GPT-4 Turbo	0.51	0.75	0.60
GPT-4o	**0.71**	0.82	**0.73**
Llama 2 70B	0.27	0.65	0.37
Llama 3 70B	0.49	0.83	0.59
Titan Premier	0.64	0.58	0.58
Mixtral 8x7B	0.49	0.75	0.55
Mistral Large	0.47	**0.85**	0.58

As shown in Table 3, Table 4 and Table 5, GPT-4o outperforms all other models in all three test configurations. As expected, the performance of all LLMs improves monotonically in the three configurations, as we remove the most problematic viewpoints and claims. However, it is important to emphasise that GPT-4o exhibits a decent performance even in the most challenging scenario (Table 3), with pretty good performance in both datasets once viewpoints 8 and 9 have been removed. In particular, its performance on the restricted dataset (Table 5) is arguably good enough to provide useful insights in a media analytics application scenario. In addition, regardless of the model in question, it can also be seen that all LLMs significantly outperform the random classifier. Given the strong dataset imbalance, this observation is crucial to paint a clearer picture of the performance of the models, whose ability may be underestimated without comparing it to a baseline. In Table 6 and Table 7 we provide more details about the performance of the best model (GPT-4o) with respect to the individual viewpoints on both the full and restricted datasets. Apart from the poor performance on the problematic viewpoints, 8 and 9, it also possible to see that there are a couple of viewpoints (1 and 7) with very low precision and very high recall, for both the full and restricted datasets. The reason for this behaviour becomes clear if we analyse Fig. 1 and Fig. 2, which report the frequency of positives across the viewpoints for the two datasets and GPT-4o. In particular, viewpoints 1 and 7 are the ones for which the gap between the gold standard positives and the LLM positives is the largest. In addition, viewpoint 1 was the only one that was not picked up by the LLM in the various runs to identify potential viewpoints but was instead added by the domain expert. Hence, it is not surprising that, in a zero-shot setting, this is the one where GPT-4o exhibits the weakest performance.

Table 6. GPT-4o scores on the full dataset—all viewpoints

Viewpoint	Precision	Recall	F1
1	0.28	0.94	0.44
2	0.68	0.66	0.67
3	0.51	0.65	0.57
4	0.50	0.90	0.64
5	**0.71**	0.67	0.69
6	0.60	0.86	**0.71**
7	0.44	**0.96**	0.60
8	0.00	0.00	0.00
9	0.67	0.23	0.34

4 Application-Level Considerations

Despite the preliminary nature of the experiments reported in this paper, our results are already interesting from an application-level perspective and they appear to confirm the concerns expressed by scholars and commentators—see e.g., [21] and [22], who have

Table 7. GPT-4o scores on restricted dataset—7 viewpoints.

Viewpoint	Precision	Recall	F1
1	0.40	0.96	0.56
2	0.83	0.75	0.79
3	0.75	0.65	0.70
4	**0.88**	0.88	**0.88**
5	0.87	0.70	0.77
6	0.80	0.80	0.80
7	0.48	**1.00**	0.65

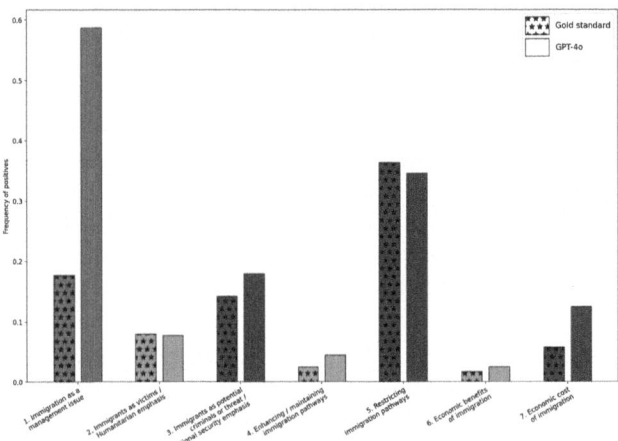

Fig. 1. Viewpoint percentages in full dataset and associated GPT-4o classification.

highlighted distortions and lack of balance in UK's mainstream media coverage of key issues. In particular, Fig. 1 and Fig. 2 show that very little media coverage is allocated to pro-immigration viewpoints (shown in red in the figures), compared to anti-immigration ones (shown in blue in the figures). Indeed, we believe that a particular strength of our approach is the focus on modelling the news dynamics in a way that is consistent with the analyses carried out by media and political scientists—e.g., see [5], who try to understand the dynamics of the debate on a particular issue. As shown in Fig. 3, our approach can also support granular visualizations of the viewpoint dynamics over time, a feature that is particular useful when analysing the debate about a particular topic over a crucial time period—e.g., during the run up to an election.

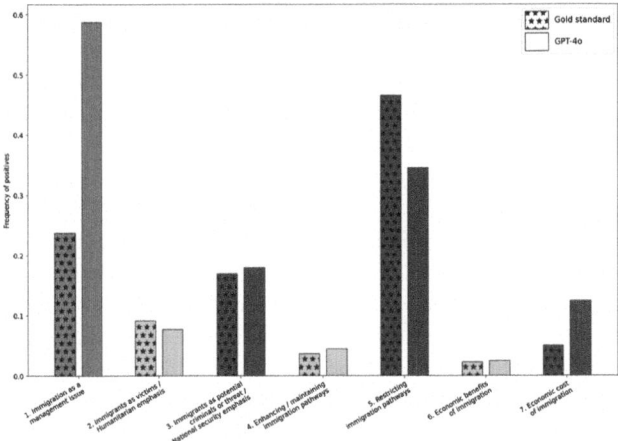

Fig. 2. Viewpoint percentages in restricted dataset and associated GPT-4o classification.

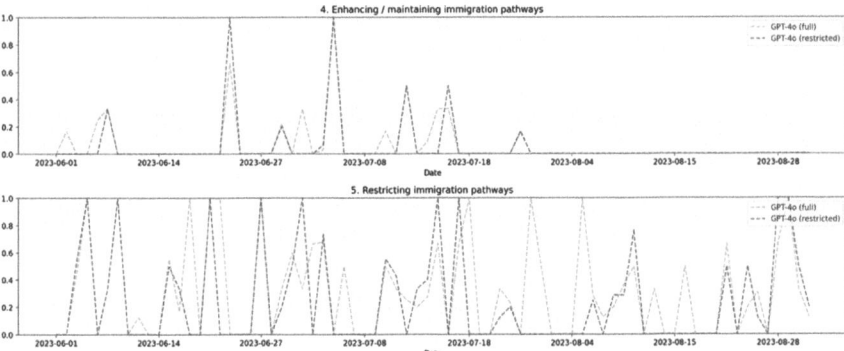

Fig. 3. Visualizing the viewpoint dynamics over time.

5 Conclusions

In this paper we have illustrated an approach and a set of initial experiments that use LLMs to model the viewpoint dynamics in the news. Our results indicate that, even in a zero-shot learning modality, the larger models, such as GPT-4o, already exhibit an acceptable level of performance. A key strength of our approach is that it goes beyond much computer science literature on capturing political opinions that, as discussed in Sect. 2.2, tends to adopt rather coarse-grained classifications—e.g., in favour or against a particular position—and fails to capture the variety of viewpoints that characterise the debate in the media. Having said so, we should also stress that this research is at a rather early stage and several issues still need to be tackled. These are discussed in what follows.

In the approach presented in this paper the viewpoints relevant to a debate are generated in advance of the claim classification task. Indeed, we believe that our hybrid human-machine approach worked well and the set of viewpoints presented in the paper provides a comprehensive and correct framework to analyse the immigration debate, as reported in the news during the June-August 2023 period. Nonetheless, this solution is structured in a rather waterfall fashion, where a corpus of news is considered, the viewpoints are generated and then the claims are classified. This works well for a retrospective analysis—e.g., a study on the election debate over the past few months, but would be unable to support the modelling of a live, ongoing debate, where new viewpoints may emerge dynamically. Hence, additional experiments will need to be carried out to test whether effective pipelines can be devised, which are able to handle the emergence of novel viewpoints in a debate. Such a solution would also make it possible to analyse the viewpoint dimensions themselves as an object of discourse—e.g., modelling the rate of change in viewpoint dimensions over time and across different topics.

Another issue concerns the typology of LLMs used in our study. Given the emphasis on measuring performance in a zero-shot setting, it is not surprising that the large, expensive and proprietary models produced the best results. Hence, more work is needed with smaller, cheaper and open models, first to test whether application-specific fine-tuning can significantly improve their performance in this task and secondly to develop practical pipelines where such fine-tuning can be effectively applied in user-centric solutions.

The claim extraction process also requires more work. While this aspect was not a priority in the context of the research reported here, more effort is needed to better characterise what is a claim and to develop highly performant techniques to extract them from a news corpus. Tackling these issues is essential to reduce the noise in the extracted corpus of claims, which ought to facilitate the manual construction of improved gold standards by human annotators and (most likely) lead to improvements in the performance of the LLMs on the claim classification task.

Another important challenge requires dealing with the contested nature of the domain. While we are happy with the robustness of the viewpoint generation process, it is clear that the claim classification task is a challenging one, given the inherent ambiguity of many claims expressed in the media. Indeed, as reported in Sect. 3.4, the level of agreement between human annotators was only moderate on the full dataset, thus affecting the robustness of our initial gold standard. While we were able to address the issue by generating a restricted version of the gold standard and by reducing the number of viewpoints, additional work is needed to give more robust foundations to the claim classification task, in particular, in terms of providing robust guidelines to human annotators. This aspect is closely related to the task discussed earlier: that is, it is our view that, if we are able to improve the claim extraction process and characterise more precisely what is a claim, then it will become easier to support human annotators in developing robust gold standards.

Finally, it is also important to emphasise that claim classification is a knowledge-intensive task. Statements by politicians can be ambiguous and in many cases can only be understood if the observer has enough background knowledge about their track record, political affiliation, key beliefs, etc. Hence, in the future we plan to extend our architecture by developing a *knowledge graph* [23] capturing key information about the actors expressing opinions in the media and by using it both to try and improve the performance of the system in the claim classification task and also to provide a wider range of domain analytics—e.g., by modelling the type and evolution of the positions expressed by individual actors or broader groups, such as political parties.

Acknowledgments. This work was partially supported by a grant from the Open Societal Challenges research programme of The Open University. The authors would also like to thank three anonymous EKAW reviewers for their insightful comments and criticisms.

Disclosure of Interests. The authors have no competing interests to declare, which are relevant to the content of this article.

References

1. Baden, C., Springer, N.: Conceptualizing viewpoint diversity in news discourse. Journalism **18**, 176–194 (2017). https://doi.org/10.1177/1464884915605028
2. Bermes, A.: Information overload and fake news sharing: a transactional stress perspective exploring the mitigating role of consumers' resilience during COVID-19. J. Retail. Consum. Serv. **61** (2021)
3. Boumans, J.W., Trilling, D.: Taking stock of the toolkit: an overview of relevant automated content analysis approaches and techniques for digital journalism scholars. Digit. J. **4**, 8–23 (2016). https://doi.org/10.1080/21670811.2015.1096598
4. Hamborg, F., Donnay, K., Gipp, B.: Automated identification of media bias in news articles: an interdisciplinary literature review. Int. J. Digit. Libr. **20**, 391–415 (2019). https://doi.org/10.1007/s00799-018-0261-y
5. Masini, A., et al.: Measuring and explaining the diversity of voices and viewpoints in the news: a comparative study on the determinants of content diversity of immigration news. J. Stud. **19**, 2324–2343 (2018)
6. Vreese, C.H.: News framing: theory and typology. Inf. Des. J. **13**, 51–62 (2005). https://doi.org/10.1075/idjdd.13.1.06vre
7. Motta, E., Daga, E., Gangemi, A., Gjelsvik, M.L., Osborne, F., Salatino, A.: The epistemology of fine-grained news classification. Semant. Web J. (2024). https://www.semantic-web-journal.net/system/files/swj3659.pdf
8. Doan, T.M., Gulla, J.A.: A survey on political viewpoints identification. Online Soc. Netw. Media **30** (2022). https://doi.org/10.1016/j.osnem.2022.100208
9. Trabelsi, A., Zaiane, O.: Unsupervised model for topic viewpoint discovery in online debates leveraging author interactions. In: Proceedings of the International AAAI Conference on Web and Social Media, vol. 12 (2018). https://doi.org/10.1609/icwsm.v12i1.15021
10. Quraishi, M., Fafalios, P., Herder, E.: Viewpoint discovery and understanding in social networks. In: Proceedings of the 10th ACM Conference on Web Science, pp. 47–56. Association for Computing Machinery (2018). https://doi.org/10.1145/3201064.3201076
11. Hada, R., et al.: Beyond digital "echo chambers": the role of viewpoint diversity in political discussion. IN: Proceedings of the Sixteenth ACM International Conference on Web Search and Data Mining, pp. 33–41. Association for Computing Machinery, New York, NY, USA (2023). https://doi.org/10.1145/3539597.3570487
12. Vrijenhoek, S., Kaya, M., Metoui, N., Möller, J., Odijk, D., Helberger, N.: Recommenders with a mission: assessing diversity in news recommendations. In: Proceedings of the 2021 Conference on Human Information Interaction and Retrieval, pp. 173–183. ACM (2021)

13. Küçük, D., Can, F.: Stance detection: a survey. ACM Comput. Surv. **53**, 12:1–12:37 (2020). https://doi.org/10.1145/3369026
14. Chen, S., Khashabi, D., Yin, W., Callison-Burch, C., Roth, D.: Seeing things from a different angle: discovering diverse perspectives about claims. In: Burstein, J., Doran, C., Solorio, T. (eds.) Proceedings of the 2019 Conference of the North American Chapter of the Association for Computational Linguistics: Human Language Technologies, Volume 1 (Long and Short Papers), pp. 542–557 (2019). https://doi.org/10.18653/v1/N19-1053
15. Vrandečić, D., Krötzsch, M.: Wikidata: a free collaborative knowledgebase. Commun. ACM **57**(10), 78–85 (2014)
16. McHugh, M.L.: Interrater reliability: the kappa statistic. Biochemia medica **22**(3), 276–282 (2012)
17. Achiam, J., et al.: GPT-4 Technical report. arXiv:2303.08774 (2023)
18. Jiang, A.Q., et al.: Mixtral of experts. arXiv:2401.04088 (2024)
19. Touvron, H., et al.: Llama 2: Open foundation and fine-tuned chat models. arXiv preprint arXiv:2307.09288 (2023)
20. Meta LLaMA Team: Introducing Meta Llama 3: The most capable openly available LLM to date. https://ai.meta.com/blog/meta-llama-3/ (2024)
21. Deacon, D., Downey, J., Harmer, E., Stanyer, J., Wring, D.: The narrow agenda: how the news media covered the Referendum. In: Jackson, D., Thorsen, E., Wring, D. (eds.) EU Referendum Analysis 2016, pp. 34–35 (2016)
22. Taylor, R.: How well does the UK's media system support democratic politics and represent citizens' interests? Democratic Audit Blog. https://tinyurl.com/nz9rppfz (2018)
23. Peng, C., Xia, F., Naseriparsa, M., Osborne, F.: Knowledge graphs: opportunities and challenges. Artif. Intell. Rev. **56**(11), 13071–13102 (2023)

Influence Beyond Similarity: A Contrastive Learning Approach to Object Influence Retrieval

Teresa Liberatore[1]([⊠]), Paul Groth[1], Monika Kackovic[1],
and Nachoem Wijnberg[1,2]

[1] University of Amsterdam, Amsterdam, Netherlands
{t.liberatore,p.t.groth,m.kackovic,n.m.wijnberg}@uva.nl
[2] University of Johannesburg, Johannesburg, South Africa

Abstract. Innovative art or fashion trends do not spring out of nowhere: they are products of societal contexts, movements and economic turning points. To understand the dynamics of innovation, it is necessary to understand influence relations between agents (e.g. artists, designers, creatives) and between the objects (e.g. clothes, paintings) that these agents produce. However, acquiring knowledge about these connections is challenging given that they are frequently undocumented. Recent literature has focused on discovering influence relations between agents, utilizing either object similarity or social network information. However, these methods often overlook the importance of direct relations between objects or oversimplify the complex nature of influence by approximating it with similarity.

To overcome this gap, we introduce Object Influence Retrieval (OIR), a task aimed at retrieving objects that potentially influenced a given object. To measure task performance, we describe two datasets for OIR: WikiartINFL (paintings) and iDesignerINFL (fashion items), both enriched with agent influence information. Additionally, we present CLOIR, a Contrastive Learning approach leveraging transfer learning from a pre-trained model to represent objects, incorporating agent influence information through contrastive learning. CLOIR shows up to a 30% improvement in Precision@k and Mean Reciprocal Rank in the OIR task compared to a baseline based on similarity between objects.

Keywords: Creative Influence · Computational Creativity · Knowledge Discovery · Content Based Image Retrieval · Contrastive Learning

1 Introduction

Determining the technology upon which an innovation is built or identifying the inspirations behind a painting can help uncover patterns of influence [1,2]. The web of influence relations that shape innovations across various domains has long

© The Author(s), under exclusive license to Springer Nature Switzerland AG 2025
M. Alam et al. (Eds.): EKAW 2024, LNAI 15370, pp. 35–52, 2025.
https://doi.org/10.1007/978-3-031-77792-9_3

fascinated academics and practitioners alike. Being able to acquire knowledge about the underlying influence relations between objects and agents is crucial to understanding the complexity of the creative process [3] and multifaceted nature of innovation [4].

However, acquiring such knowledge is challenging because it is often undocumented. For example, while academics explicitly reference their peers' papers, this is not true in other domains. A case in point is that painters often find inspiration in their peers' content, style, or approach; nevertheless, they do not declare it explicitly. Similarly, there is no established convention for explicitly acknowledging sources of inspiration in other creative industries, like fashion, design, architecture, and literature.

To acquire such knowledge at scale, systems are needed to help discover such influence information. Prior work has focused on relations between agents (i.e. individual creators) [5,6]. This is likely due to the fact that ground truth data for developing models is available in specific domains such as fine arts and music, which have been extensively studied by domain experts [5]. On the other hand, the task of retrieving influence relations between objects has been largely overlooked.

Hence, in this study, we introduce a new task - Object Influence Retrieval (OIR) - aimed at retrieving objects that potentially influenced another object. Along with the task, we present two datasets to develop and evaluate approaches to perform OIR: WikiartINFL, a collection of paintings with metadata enriched by artist influence information, and iDesignerINFL, which includes images of fashion items created by renowned designers also with corresponding influence relations. We also introduce CLOIR, a Contrastive Learning approach to perform OIR on the presented datasets. In CLOIR, a contrastive learning model is trained to represent objects in an embedding space that accounts for both (i) the similarity between objects and (ii) the influence relation between creators.

CLOIR outperforms baselines where similarity between objects serves as a proxy for influence, suggesting that CLOIR is better suited for finding potential influence between objects compared to similarity alone.

Summarizing, the main contributions of this paper are as follows:

1. Object Influence Retrieval (OIR): a new task with the goal of, given an object, retrieving the objects that potentially influenced it;
2. Contrastive Learning Object Influence Retrieval (CLOIR): a Contrastive Learning approach to solve OIR;
3. WikiartINFL and iDesignerINFL: two datasets augmented with agent influence information for evaluating approaches for solving OIR.

2   Related Work

To the best of our knowledge, this is the first study focused on retrieving object influences. Previous research on influence detection has primarily addressed the reverse problem: identifying influences between agents using either object similarity or social network information. Content-based image retrieval (CBIR) has explored finding similar images given a query image, but the problem of influence retrieval has not been explored yet.

Object Similarity to Determine Agent Influence: The literature [6–9] adopting this approach focuses on the fine art domain, where agents are artists and objects are artworks. A key characteristic of artworks is that, like pictures, they can be fully represented through their visual depiction. Additionally, in the fine art domain, art experts have extensively documented the influence of artists, providing reliable ground truth. Although the methodologies vary, these studies share a common framework:

1. Artworks are represented as feature vectors.
2. A similarity score is computed between these feature vectors.
3. The similarity between artworks is used to infer similarity between artists and suggest influence among the artists.
4. The discovered influences are then evaluated against the established artistic influence ground truth.

Social Network Information and Agent Influence: Most works that use social interactions to find influence between agents focus on the music domain [5,10]. In the music domain, there is abundant knowledge about interactions between artists, and the influence between objects is often made explicit through samples or covers of existing songs. In particular, one paper focuses on modeling the interactions between agents within the music domain using Knowledge Graph and Semantic Web technologies [10], whilst other works analyze interactions in the music industry explicitly linked to influence, such as sampling and covering [11,12]. Another work uses graph theory on artists' social networks to predict the corresponding influences [5].

In contrast to prior work on influence detection, our paper focuses on the connections between objects, introducing the OIR task and the CLOIR approach that incorporates object similarity with information about agent influences.

Content Based Image Retrieval. Image retrieval is a well-studied problem where given query images, similar images are retrieved from a database [13]. Among works in CBIR, studies that focus on unsupervised or pseudo-supervised CBIR are of particular interest for the problem at hand given that there is a lack of ground truth object influence information. In particular, [14,15] overcome the lack of supervision using a triplet network, where pseudo-labels for positive and negative examples are based on image similarity.

Drawing from these works, in CLOIR, we use a triplet network to shape the embedding space for retrieval, but differently from them, we aim at influence rather than similarity-based retrieval. Thus, the sampling of positive and negative examples is based on agent influences rather than similarity. We hypothesize that this dual-faceted representation will enhance the object representation space for OIR, compared to similarity alone. Our work thus introduces an operationalization of influence between objects that goes beyond object similarity.

3 Contrastive Learning Object Influence Retrieval

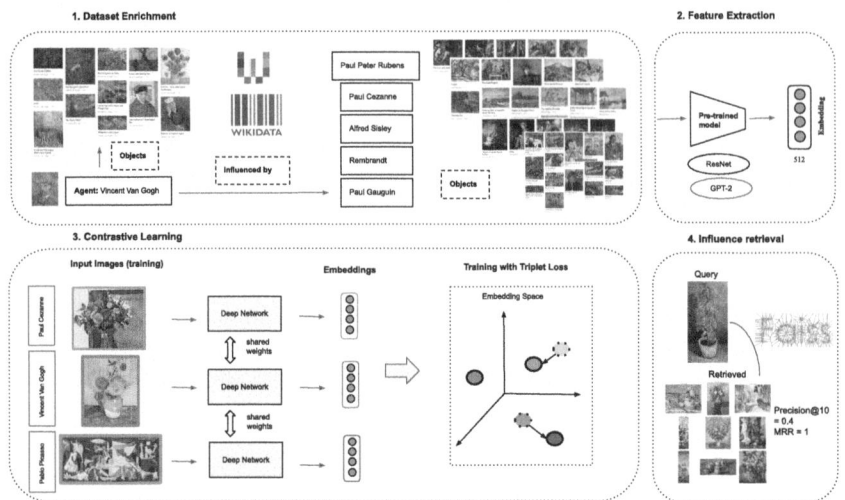

Fig. 1. Overview of CLOIR: Contrastive Learning approach to Object Influence Retrieval.

An overview of the approach is illustrated in Fig. 1 where the aim is, given an object, retrieve which objects potentially influenced its creation. We focus on objects whose main characteristics can be depicted through visual features, such as artworks and fashion items. However, our approach can be extended to domains where objects are represented in other modalities, such as text or audio, provided that (i) the modality, or a combination of modalities, can represent the fundamental characteristics of the objects, and (ii) pre-trained models can be used to extract features from the modalities of interest.

Specifically, with CLOIR, we aim to structure the object embedding space so that objects with visual similarities and which are produced by agents that are connected via their influence are positioned closely together.

Source code available at https://github.com/traopia/CLOIR .

3.1 Dataset Enrichment

The initial step involves sourcing information about influence between agents which is then mapped to the objects created by the agents.

Information about agent influence can be sourced in different ways. For WikiartINFl, for instance, it was sourced by querying Wikidata and Wikiart, as depicted in Fig. 1.

As a result, the enriched dataset for each observation should include (i) the object, (ii) the agent who produced it, (iii) the known influencers of the agent, and (iv) any additional metadata, if available.

3.2   Feature Extraction

The objects are represented with vector embeddings, leveraging vision pre-trained models, following prior works on fine-art object representations [16–18]. If metadata is available, text features are extracted from a language pre-trained model and concatenated to the visual embeddings.

We experiment with two different setups of pre-trained models to extract visual and textual features from data: a combination of ResNet34 and GPT2-small, and CLIP.

ResNet34 and GPT2. In the first setup we extract visual features with ResNet34 [19][1], a widely-used convolutional neural network architecture pre-trained on ImageNet; and text features with GPT2-small [20][2], a language model known for its proficiency in natural language processing tasks. Our choice of ResNet-34 and GPT-2 small, over other larger pre-trained models such as ResNet-50 or GoogleNet for vision, and BERT or GPT-4 for language, is motivated by the balance between model complexity and computational efficiency.

CLIP. In the second setup, we extract both visual and text features, using CLIP [21], a neural network trained to learn visual concepts from natural language. The multi-modal nature of the model allows us to extract both visual and text features from the same model, allowing for a smoother pipeline of feature extraction.

3.3   Contrastive Learning

The core of CLOIR revolves around fine-tuning the object representations obtained from pre-trained models according to the information about the influence between agents. In particular, we aim to shape the embedding space such that objects are proximal if (i) they are similar and (ii) the agents that produced them are linked through an influence relation.

We achieve this goal by combining approaches proposed in Multiple Instance Learning (MIL), fine-grained feature representation, and Content Based Image Retrieval (CBIR). MIL is a form of weakly supervised learning where training instances are arranged in sets, called bags, and a label is provided only for the entire bag [22]. In our case, the training instances are the objects arranged by agents, and the labels represent the influence relations. Specifically, from the MIL literature [23], we adopt the idea of mapping the class-level label (influence between agents) to all the objects and using an objective function based on instance-level similarity, which also respects the group-level label constraints. An objective function with these characteristics is the triplet constraint, which is also used in fine-grained feature representation tasks [24], where

---

[1] https://huggingface.co/microsoft/resnet-34.
[2] https://huggingface.co/openai-community/gpt2.

the goal is to distinguish subordinate classes by identifying instances with the same attributes. The triplet loss, learning from positive and negative examples for each learning anchor, encourages the model to identify objects sharing the same attributes, while preserving intra-class variation within the sub-classes. Triplet loss is widely used in CBIR too, because it directly optimizes the embedding space for similarity-based retrieval, ensuring that similar images are close together and dissimilar ones are far apart, thus improving retrieval performance [13].

To suggest influence between objects, we thus train a pseudo-supervised contrastive model with triplet loss, with pseudo-supervision coming from labels about influence between agents, as we don't have labels about object influences.

Triplet Construction. For an anchor object, we consider as positive examples the objects made by agents influential for the agent who produced the anchor object. On the other hand, negative examples are objects made by agents not considered influential. Both the pools of potential positive and negative examples are extensive and exhibit a high degree of internal variability. Therefore, samples of positives and negatives are considered for each anchor object, and in CLOIR we experiment with different sample sizes and sample strategies for triplet construction.

Triplet Network. The triplet network aims to maximize the discrimination of image representations, and it consists of three same networks that share weights. The triplet loss minimizes the distance between an anchor and positive examples and maximizes the distance between the anchor and negative examples.

Loss: In particular the triplet loss is defined as:

$$L(a, p, n) = \max\{d(a_i, p_i) - d(a_i, n_i) + \text{margin}, 0\}$$

where a_i represents the embedding of the anchor sample, p_i of the positive sample, n_i of the negative sample, d is the function measuring the distance between the samples, and the margin is a hyperparameter defining the minimum margin between positive and negative distances.

Model: The contrastive model trained with triplet loss is designed as a feedforward neural network composed of three linear layers with ReLU activation functions and dropout layers in between. The model reduces the dimensionality of the input features coming from the pre-trained models, to learn the patterns within the feature vectors. Training batches of anchor, positive, and negative examples are thus passed through the same network, whose weights are updated according to the triplet loss.

Implementation Details. The models are trained for 30 epochs with early stopping with patience set to 10 epochs, and batches of size 32. The optimization of the loss function is done with Adam optimizer with learning rate set to 0.0005, as suggested in [25]. To allow reproducibility and consistency among experiments, thus removing the randomness involved, a random seed is set to 42. All experiments have been performed on a GPU partition on an NVIDIA A100 GPU node.

3.4   Object Influence Retrieval

The final stage of our approach involves retrieving influential objects for a given query object via a vector search within the fine-tuned embedding space. To evaluate this, objects in the test set - those not previously encountered by the model as anchors, positive, or negative examples during training - are used as query objects. For each query object, we retrieve k-nearest objects in the trained embedding space. The FAISS library [26], which specializes in embedding similarity search tasks, is employed for this purpose, using Euclidean distance as the metric for vector closeness.

Evaluation. As is standard in information retrieval systems, performance is assessed using metrics such as Precision at K and Mean Reciprocal Rank (MRR). Given that there is no existing ground truth for this task, we make use of agent influence to compare performance. Specifically, retrieval is deemed correct if the retrieved object was created by an agent recognized as influential to the agent who created the query object. Additionally, to account for chains of agent influences, we extend these metrics to a second degree, where retrieval is also considered correct if the retrieved object was created by a direct influencer or an influencer of an influencer of the query object's agent.

4   Experiments

To evaluate the performance of our proposed approach, we conduct a series of experiments comparing our results against a baseline model. The baseline model retrieves potential influence objects, based solely on object similarity. We investigate various configurations within our approach, focusing on different sampling strategies, for example, selection in the contrastive model, and varying training/test splits.

Sampling Strategies. The core of our method is a contrastive model that requires both positive and negative examples for each query. We explore the impact of different sample sizes and sampling strategies on model performance. Specifically, we use sample sizes of 10 and 100, adhering to the minimum object requirement per artist. For positive examples, we compare random sampling against similarity-based sampling. We hypothesize that similarity-based sampling will improve performance by selecting examples that are more likely to be semantically related to the anchor objects, thereby reducing variability within the samples.

Training/Test Split. Additionally, we experiment with different training/test splits of the dataset. The first split is stratified, with the training set containing 70% of the objects from all agents and the remaining 30% in the test set. The second split, the Leave-out Agents split, is designed to evaluate the model's ability to retrieve object influences for agents not seen during training. In this configuration, the training set includes objects from 70% of the agents, while the remaining 30% of agents are excluded from the training set and reserved for the test set.

4.1   Data

We introduce two distinct datasets to evaluate our approach: WikiartINFL and iDesignerINFL. Both datasets contain objects, whose visual representations capture their main characteristics, along with the names of the agents who created them. In particular, WikiartINFL includes images of artworks made by artists, and iDesignerINFL images of fashion items made by fashion designers.

WikiartINFL Dataset. The Wikiart dataset[3] is a comprehensive collection of paintings and their associated metadata. It is one of the largest online repositories of digitized paintings and is frequently utilized to develop computational approaches to study fine arts. We use previously curated Wikiart data presented in prior studies [27]. This original dataset includes 75,921 artworks encompassing paintings, drawings, and illustrations.

We extended the Wikiart dataset with artist influence relations to create WikiartINFL. This information was gathered by querying Wikidata and Wikiart, specifically utilizing the "influenced by" property (P737)[4] to capture influence connections between artists within the dataset. We considered only artists with over 100 artworks in the painting collection and retained only those artists whose influencers were also present in the dataset. This selection ensures access to the artworks created by influential artists, which is essential for gathering positive examples to train a contrastive model.

iDesignerINFL Dataset. The iDesigner dataset[5] contains images of fashion items captured during runway shows of various designers. This dataset, introduced on Kaggle by Hearst magazine, was used in a challenge to predict which fashion designer created each item. The dataset includes multiple images of the same items taken from different angles during runway shows. To prevent data leakage, we ensured that images of the same item were assigned to the same split during the training and test phases. We considered images to refer to the same item if they exceeded a 95% similarity threshold and were made by the same designer.

---

[3] https://www.wikiart.org.
[4] https://www.wikidata.org/wiki/Property:P737.
[5] https://paperswithcode.com/dataset/idesigner.

To source information about the influences between fashion designers for the iDesigner dataset, we utilized a Large Language Model (LLM). This decision was driven by the scarcity of accessible information on designer influences and the potential of LLMs to provide labels when they are otherwise unavailable. Demonstrating that meaningful results can be achieved using influence data sourced via a LLM suggests that this approach could be applied in other domains with similarly scarce information. In particular, we prompted GPT-3 with the following query to gather information on designer influences: "Can you help me find the fashion designers that influenced the designers in this list? Specifically, can you create a dictionary where the keys are designers from the list and the values are their influencers, chosen from the same list of designers?".

Table 1. Descriptive statistics for WikiartINFL and iDesigner.

	WikiartINFL	iDesignerINFL
number of objects	39815	44204
number of agents	154	49
mean objects per agent	258	902
mean influencers	2.8	1.5
min,max influencers	1, 10	1, 3

Table 1 presents statistical summaries of the final versions of the Wikiart-INFL and iDesignerINFL datasets, both enriched with influence information. These statistics offer insights into the characteristics of the datasets after pre-processing and agent influences incorporation. Notably, while the two datasets are similar in size, WikiartINFL includes roughly three times the number of agents compared to iDesignerINFL. This implies that, on average, more objects are available per agent in WikiartINFL. However, it is important to note that for iDesignerINFL, these numbers may not reflect distinct objects, as multiple images can depict the same runway fashion item. Conversely, WikiartINFL reports nearly double the number of influencer agents per agent on average, resulting in greater variability across positive examples.

5   Results

5.1   Stratified Training/Test Split

Here we report the results from experiments conducted using a stratified training/test split, where: 70% of the objects for each agent are included in the training set, while the remaining 30% of objects for each agent are used for the test set.

Table 2. Results for the WikiartINFL dataset, stratified split. In green the highest values for the metrics. P@10 refers to Precision at 10 and MRR to Mean Reciprocal Rank, and their version with (2) represent the metric considering the second degree of chain of influence.

	Sampling	Size	Feature	Model	P@10	P@10(2)	MRR	MRR(2)
Baseline	–	–	Image	ResNet	0.108	0.162	0.188	0.283
				CLIP	0.128	0.172	0.199	0.275
			Image-Text	ResNet+GPT2	0.104	0.148	0.181	0.27
				CLIP	0.132	0.148	0.199	0.299
CLOIR	Random	10	Image	ResNet	0.17	0.217	0.238	0.339
				CLIP	0.169	0.203	0.215	0.198
			Image-Text	ResNet+GPT2	0.306	0.323	0.302	0.388
				CLIP	0.266	0.28	0.291	0.333
CLOIR	Similarity	10	Image	ResNet	0.125	0.181	0.217	0.315
				CLIP	0.133	0.175	0.215	0.288
			Image-Text	ResNet+GPT2	0.125	0.165	0.201	0.284
				CLIP	0.299	0.254	0.282	0.352
CLOIR	Random	100	Image	ResNet	0.193	0.238	0.25	0.351
				CLIP	0.18	0.212	0.236	0.31
			Image-Text	ResNet+GPT2	0.391	0.406	0.38	0.434
				CLIP	0.308	0.318	0.321	0.353
CLOIR	Similarity	100	Image	ResNet	0.201	0.255	0.276	0.371
				CLIP	0.165	0.199	0.24	0.303
			Image-Text	ResNet+GPT2	0.417	0.442	0.419	0.513
				CLIP	0.385	0.402	0.408	0.447

Table 3. Results for the iDesignerINFL dataset, stratified split. In green the highest values for the metrics.

	Sampling	Size	Model	P@10	P@10(2)	MRR	MRR(2)
Baseline	–	–	ResNet	0.083	0.093	0.218	0.24
			CLIP	0.068	0.077	0.224	0.247
CLOIR	Random	10	ResNet	0.094	0.108	0.258	0.285
			CLIP	0.072	0.984	0.25	0.273
CLOIR	Similarity	10	ResNet	0.072	0.082	0.232	0.255
			CLIP	0.051	0.061	0.226	0.249
CLOIR	Random	100	ResNet	0.126	0.144	0.273	0.304
			CLIP	0.086	0.1	0.251	0.278
CLOIR	Similarity	100	ResNet	0.129	0.146	0.294	0.324
			CLIP	0.085	0.099	0.261	0.288

In Table 2, we present the results for the WikiartINFL dataset. CLOIR outperforms the baselines across all experiments. Both when considering visual features only, or in combination with text features, the most significant improvement over the baseline is observed in the model where positive examples are sampled based on similarity with a sample size of 100. For visual features only, the improvement to the baseline reaches 10% across all metrics, which reaches up to 30% when considering text features too. Across metrics, the pre-trained model setup that leads to better performance is the combination of ResNet+GPT2. Furthermore, it is interesting to observe that sampling based on similarity leads to better performance when 100 examples per anchor are used, this trend does not hold when only 10 examples are considered. This result indicates that with fewer examples, greater variability ensures better generalization of the model. Conversely, when more examples are available, the increased sample size introduces variability, and a similarity-based sampling constraint helps the model better generalize the concept of influence.

A similar pattern can be observed in the experiments performed on the iDesignerINFL dataset, whose results are reported in Table 3. Namely, with a sample size of 10, random sampling performs better than similarity sampling - which in this case performs even worse than the baseline - but this trend reverses when 100 examples are considered, and the improvement is between 5% and 10% across metrics. The main conclusion from these experiments is that with a smaller sample size, introducing more variability through random sampling leads to better performance in OIR. Conversely, with a larger sample size, similarity-based sampling yields better results.

5.2   Leave-Out-Agents Training/Test Split

Here we report the results of experiments conducted using the leave-out-agents training/test split. In this setup, objects from 70% of the agents are included in the training set, while the remaining 30% of agents are left out for testing. This experimental design allows us to further study the generalization abilities of our approach, as it must suggest influences for objects of agents not seen during training.

In Table 4, the results for the WikiartINFL dataset are reported. We can observe that with this data split, CLIP leads to a decrease in performance with respect to the baseline. In general, the same pattern observed in the stratified data split can be observed: when only 10 examples per anchor are considered, random sampling leads to better performance, whilst similarity-based sampling is better when 100 examples are considered. In particular, the best-performing setup, leading to an improvement of up to 8% with respect to the baseline, can be observed with random sampling and sample size equal to 10.

For what concerns the iDesignerINFL dataset, the baseline performs similarly to CLOIR across experiments, as it can be observed in Table 5. This is a symptom that for this dataset the influence signal is not clear enough for CLOIR to capture and generalize over unseen agents.

Table 4. Results for the WikiartINFL dataset, leave-out-agents split. In green the highest values for the metrics.

	Sampling	Size	Feature	Model	P@10	P@10(2)	MRR	MRR(2)
Baseline	–	–	Image	ResNet	0.116	0.149	0.19	0.272
				CLIP	0.142	0.173	0.203	0.289
			Image-Text	ResNet+GPT2	0.114	0.143	0.188	0.27
				CLIP	0.163	0.187	0.213	0.294
CLOIR	Random	10	Image	ResNet	0.125	0.156	0.189	0.283
				CLIP	0.123	0.152	0.184	0.277
			Image-Text	ResNet+GPT2	0.201	0.226	0.197	0.314
				CLIP	0.163	0.182	0.192	0.283
CLOIR	Similarity	10	Image	ResNet	0.12	0.155	0.19	0.287
				CLIP	0.117	0.146	0.188	0.269
			Image-Text	ResNet+GPT2	0.15	0.183	0.218	0.318
				CLIP	0.153	0.18	0.211	0.289
CLOIR	Random	100	Image	ResNet	0.125	0.157	0.183	0.275
				CLIP	0.13	0.159	0.178	0.275
			Image-Text	ResNet+GPT2	0.178	0.2	0.197	0.301
				CLIP	0.165	0.187	0.185	0.281
CLOIR	Similarity	100	Image	ResNet	0.132	0.166	0.202	0.289
				CLIP	0.111	0.139	0.171	0.253
			Image-Text	ResNet+GPT2	0.191	0.217	0.21	0.326
				CLIP	0.15	0.169	0.187	0.252

Table 5. Results for the iDesignerINFL dataset, leave-out-agents split. In green the highest values for the metrics.

	Sampling	Size	Model	P@10	P@10(2)	MRR	MRR(2)
Baseline	–	–	ResNet	0.083	0.086	0.253	0.245
			CLIP	0.073	0.075	0.258	0.249
CLOIR	Random	10	ResNet	0.049	0.05	0.254	0.246
			CLIP	0.045	0.047	0.255	0.246
CLOIR	Similarity	10	ResNet	0.069	0.071	0.257	0.249
			CLIP	0.052	0.054	0.257	0.249
CLOIR	Random	100	ResNet	0.048	0.05	0.25	0.243
			CLIP	0.046	0.048	0.252	0.244
CLOIR	Similarity	100	ResNet	0.066	0.068	0.254	0.245
			CLIP	0.046	0.048	0.254	0.246

5.3   Embedding Space

The retrieval results show that our approach generally captures influence relations more effectively than the baseline. To further evaluate this, we compare the CLOIR embedding space with the baseline, quantifying their differences to demonstrate that our approach better supports object influence retrieval based on proximity. Additionally, we perform a qualitative analysis by visualizing the embedding space, focusing on a specific agent to illustrate how our approach captures influence through spatial proximity.

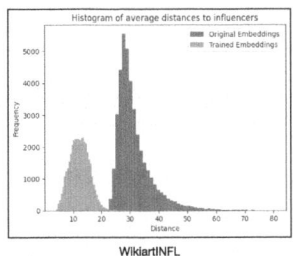 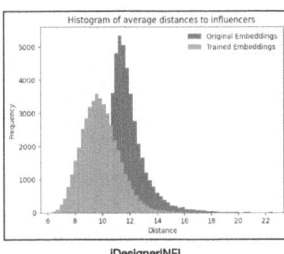

Fig. 2. Histogram of average distances between objects and their influencers. Comparison between baseline and CLOIR embedding space.

To quantitatively assess the difference between baseline and CLOIR object representations in the embedding space, we calculate the average Euclidean distance between each agent's objects and those of their respective influencers. This metric evaluates the effectiveness of our approach in capturing influence relations.

Figure 2 shows histograms of these average distances, indicating that CLOIR leads to a more compact embedding space where objects are closer to their respective influencers in both datasets.

Qualitative Embedding Space Evaluation. We qualitatively analyze the embedding space by visualizing it using the Uniform Manifold Approximation and Projection (UMAP) technique, which reduces dimensionality and computes Euclidean distances between objects. Our goal with CLOIR is to create an embedding space where object proximity indicates both similarity and influence.

To demonstrate this, we highlight objects from a specific agent and those influenced by them. In the WikiartINFL dataset, we use Vincent Van Gogh as an example (Fig. 3), and in the iDesigner dataset, we use Alexander McQueen (Fig. 4).

The visualizations show that in the CLOIR space, those two clusters overlap, whilst in the baseline embedding space, they do not.

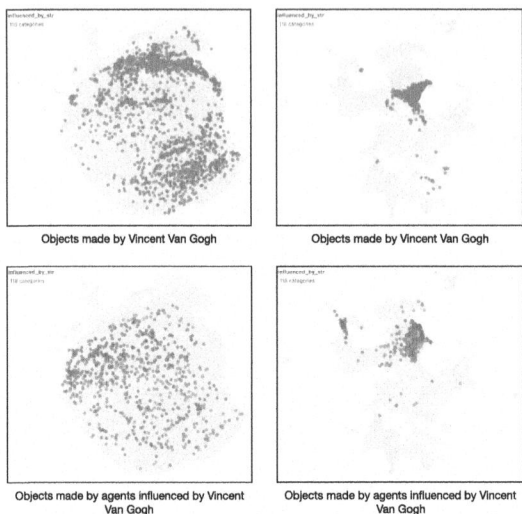

Fig. 3. WikiartINFL dataset embedding space: baseline on the left and CLOIR on the right. Top row: objects made by Vincent Van Gogh. Bottom row: objects made by agents influenced by Vincent Van Gogh.

5.4   Example of Influence-Based Retrieval

Lastly, we present example retrievals obtained using the baseline model and CLOIR. For each query object, we retrieve the 10 closest objects in both embedding spaces. Figures 5 and 6 show retrievals for the WikiartINFL and iDesignerINFL datasets, respectively. Objects created by influential agents are highlighted

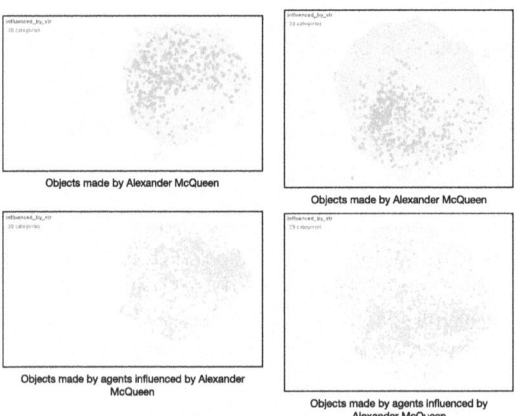

Fig. 4. iDesignerINFL dataset embedding space: baseline on the left and CLOIR on the right. Top row: objects made by Alexander McQueen. Bottom row: objects made by agents influenced by Alexander McQueen.

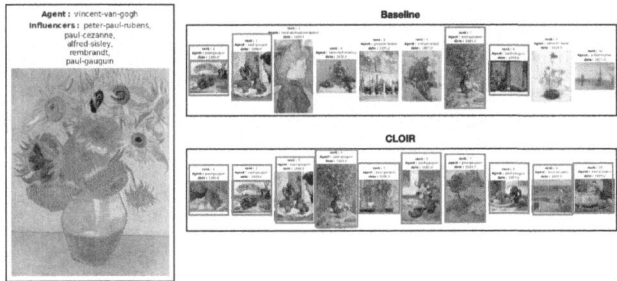

Fig. 5. WikiartINFL: Object Influence Retrieval comparing baseline with CLOIR retrieval.

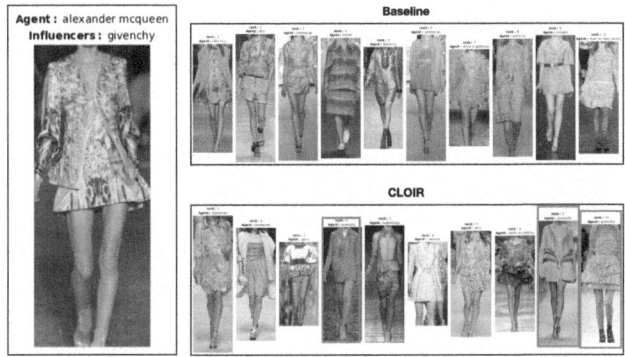

Fig. 6. iDesignerINFL: Object Influence Retrieval comparing baseline with CLOIR retrieval.

in green. We observe that CLOIR retrieves objects that not only share more characteristics with the query but also include a higher proportion of objects created by influential agents compared to the baseline.

6 Discussion

Our study demonstrates that CLOIR consistently outperforms traditional similarity-based methods in identifying potential influences between objects, establishing a foundation for performing Object Influence Retrieval (OIR).

While our results are promising, we acknowledge several limitations that also present opportunities for future research. A significant challenge is the lack of ground truth data on object influences to train and evaluate our approach robustly. To address this, we propose datasets incorporating agent influence as a proxy for object influence. However, this workaround, while innovative, would benefit from further ground truth data focused on object influence.

Data could be improved in other ways as well. Specifically, while the iDesigner dataset provides a collection of fashion item images and their creators, it lacks additional textual metadata and time information. Moreover, both datasets are unbalanced regarding the number of objects per agent. Future studies could thus focus on creating a new dataset that includes objects, corresponding agents, textual metadata (including time information), and an annotated test set with object influences.

Lastly, future research could explore other ways to integrate agent-influence information into object representation, leverage external domain knowledge to enhance representations and structure it in the form of knowledge graphs using neuro-symbolic approaches.

7   Conclusion

In this paper, we proposed a new task, Object Influence Retrieval (OIR), and CLOIR, an approach to solve it, by combining object similarity with influence knowledge between agents with a contrastive learning approach. We demonstrated the efficacy of CLOIR on two datasets, WikiartINFL and iDesignerINFL, achieving up to 30% improvement over a similarity-based baseline. These results suggest that combining similarity-based information with contextual influence knowledge can enhance the retrieval of objects that are influential for a query object. This work opens up new avenues for research in the automatic retrieval of influence, offering a framework that can be extended and refined in future studies. Research in this area has the potential to enable new applications in areas such as influence-based search engines, recommendation systems, historical media analysis, and the study of innovation dynamics. We hope that this work will pave the way for further studies in this direction, contributing to a deeper understanding of influence and innovation in various domains.

References

1. Park, D., Nam, J., Park, J.: Novelty and influence of creative works, and quantifying patterns of advances based on probabilistic references networks. EPJ Data Sci. **9**, 12 (2020)
2. Paulus, P.B., Dzindolet, M.: Social influence, creativity and innovation. Soc. Influ. **3**(4), 228–247 (2008)
3. Hermeren, G.: Influence in Art and Literature, ser. Princeton Legacy Library. Princeton University Press (2015). https://books.google.nl/books?id=SXt9BgAAQBAJ
4. Yoo, Y., Boland, R.J., Jr., Lyytinen, K., Majchrzak, A.: Organizing for innovation in the digitized world. Organ. Sci. **23**(5), 1398–1408 (2012)
5. Alfieri, F., Asprino, L., Lazzari, N., Presutti, V.: Creative influence prediction using graph theory. In: CREAI@ AI* IA, pp. 1–15 (2023)
6. Saleh, B., Abe, K., Arora, R.S., Elgammal, A.: Toward automated discovery of artistic influence. Multimedia Tools Appl. **75**, 3565–3591 (2016). https://link.springer.com/article/10.1007/s11042-014-2193-x
7. Shamir, L., Tarakhovsky, J.A.: Computer analysis of art. J. Comput. Cult. Herit. **5**(2), 1–11 (2012). https://doi.org/10.1145/2307723.2307726

8. Saleh, B., Abe, K., Elgammal, A.M.: Knowledge discovery of artistic influences: a metric learning approach. In: ICCC, pp. 163–172 (2014)
9. Dalmoro, B., Monteiro, C., Musse, S.R.: Measuring the influence of painters through artwork facial features. In: 2022 35th SIBGRAPI Conference on Graphics, Patterns and Images (SIBGRAPI), vol. 1, pp. 37–42 (2022)
10. Morales Tirado, A., et al.: Musical meetups knowledge graph (MMKG): a collection of evidence for historical social network analysis. In: Meroño Peñuela, A., et al. The Semantic Web. ESWC 2024. LNCS, vol. 14665, pp. 110–127. Springer, Cham (2024). https://doi.org/10.1007/978-3-031-60635-9_7
11. Bryan, N.J., Wang, G.: Musical influence network analysis and rank of sample-based music. In: ISMIR, pp. 329–334 (2011)
12. Kopel, M.: Analyzing music metadata on artist influence. In: Nguyen, N.T., Trawiński, B., Kosala, R. (eds.) ACIIDS 2015. LNCS (LNAI), vol. 9011, pp. 56–65. Springer, Cham (2015). https://doi.org/10.1007/978-3-319-15702-3_6
13. Dubey, S.R.: A decade survey of content based image retrieval using deep learning. IEEE Trans. Circuits Syst. Video Technol. **32**(5), 2687–2704 (2021)
14. Gu, Y., Zhang, H., Zhang, Z., Ye, Q.: Unsupervised deep triplet hashing with pseudo triplets for scalable image retrieval. Multimedia Tools Appl. **79**(47), 35:253–35:274 (2020)
15. Huang, S., Xiong, Y., Zhang, Y., Wang, J.: Unsupervised triplet hashing for fast image retrieval. Proc. Thematic Workshops ACM Multimedia **2017**, 84–92 (2017)
16. Banerjee, M., Cole, B.M., Ingram, P.: Distinctive from what? And for whom? Deep learning-based product distinctiveness, social structure, and third-party certifications. Acad. Manag. J. **66**, 1016–1041 (2023). https://journals.aom.org/doi/abs/10.5465/amj.2021.0175
17. Cetinic, E., Lipic, T., Grgic, S.: Fine-tuning convolutional neural networks for fine art classification. Expert Syst. Appl. **114**, 107–118 (2018)
18. Efthymiou, A., Rudinac, S., Kackovic, M., Worring, M., Wijnberg, N.: Graph neural networks for knowledge enhanced visual representation of paintings. MM 2021 - Proceedings of the 29th ACM International Conference on Multimedia, pp. 3710–3719 (2021). https://arxiv.org/abs/2105.08190v1
19. He, K., Zhang, X., Ren, S., Sun, J.: Deep residual learning for image recognition. In: Proceedings of the IEEE Conference on Computer Vision and Pattern Recognition, pp. 770–778 (2016)
20. Radford, A., Wu, J., Child, R., Luan, D., Amodei, D., Sutskever, I., et al.: Language models are unsupervised multitask learners. OpenAI blog **1**(8), 9 (2019)
21. Radford, A., et al.: Learning transferable visual models from natural language supervision. In: International Conference on Machine Learning, pp. 8748–8763. PMLR (2021)
22. Carbonneau, M.A., Cheplygina, V., Granger, E., Gagnon, G.: Multiple instance learning: a survey of problem characteristics and applications. Pattern Recogn. **77**, 329–353 (2018)
23. Kotzias, D., Denil, M., Freitas, N.D., Smyth, P.: From group to individual labels using deep features. In: Proceedings of the ACM SIGKDD International Conference on Knowledge Discovery and Data Mining, vol. 2015-August, pp. 597–606 (2015). https://doi.org/10.1145/2783258.2783380
24. Zhang, X., Zhou, F., Lin, Y., Zhang, S.: Embedding label structures for fine-grained feature representation. In: Proceedings of the IEEE Computer Society Conference on Computer Vision and Pattern Recognition, vol. 2016-December, pp. 1114–1123 (2015). https://arxiv.org/abs/1512.02895v2

25. Gondal, M.W., Joshi, S., Rahaman, N., Bauer, S., Wuthrich, M., Schölkopf, B.: Function contrastive learning of transferable meta-representations. In: International Conference on Machine Learning. PMLR, pp. 3755–3765 (2021)
26. Douze, M., et al.: The faiss library (2024)
27. Tan, W.R., Chan, C.S., Aguirre, H., Tanaka, K.: Improved artGAN for conditional synthesis of natural image and artwork. IEEE Trans. Image Process. **28**(1), 394–409 (2019). https://doi.org/10.1109/TIP.2018.2866698

Discovering a Representative Set of Link Keys in RDF Datasets

Nacira Abbas[1], Alexandre Bazin[2], Jérŏme David[1], and Amedeo Napoli[3(✉)]

[1] Université Grenoble Alpes, Inria, CNRS, Grenoble INP, LIG, 38000 Grenoble,
France
nacira.abbas@univ-grenoble-alpes.fr, jerome.david@inria.fr
[2] Université de Montpellier, CNRS, LIRMM, 34095 Montpellier, France
alexandre.bazin@lirmm.fr
[3] Université de Lorraine, CNRS, Loria, 54000 Nancy, France
amedeo.napoli@loria.fr

Abstract. A link key is based on a set of property pairs and can be used
to identify pairs of individuals representing the same real-world entity in
two different RDF datasets. Various algorithms are aimed at discovering
link keys which usually output a large number of candidates, making link
key selection and validation a challenging task. In this paper, we propose
an approach combining Formal Concept Analysis (FCA) for discovering
link key candidates and building a link key lattice, and then hierarchical
clustering over a given set of candidates for building a representative set
of link keys. Such a link key set should minimize the number of candidates
to be validated while preserving a maximal number of links between
individuals. The paper also provides a series of experiments which are
performed over different RDF datasets, showing the effectiveness of the
approach and the ability of hierarchical clustering to return a concise
and meaningful set of candidates while preserving the ordinal structure
of the link key lattice.

Keywords: Link Key Discovery · Formal Concept Analysis ·
Hierarchical Clustering · RDF Dataset · Representative Set

1 Introduction

The present research work relies on the discovery of *link keys*, which are expressions composed of sets of property pairs and a class pair, allowing to identify two individuals lying in different RDF datasets. For example, let us consider the link key $(\{(dsg, tit)\}, \{(dsg, tit), (cre, aut)\}, (book, novel))$. Whenever an instance a of class *book* in dataset D_1 has the same values for dsg (designation) as an instance b of class *novel* in dataset D_2 for tit (title), and in addition a and b share at least one value for cre (creator) and aut (author), then an identity link $(a, \texttt{owl:sameAs}, b)$ can be generated between a and b, and it can be inferred that a and b denote the same entity. More formally, a link

M. Alam et al. (Eds.): EKAW 2024, LNAI 15370, pp. 53–68, 2025.
https://doi.org/10.1007/978-3-031-77792-9_4

key is composed of two sets of pairs of properties and a pair of classes, i.e., $(\{(p_i, q_j)_{i \in I_1, j \in J_1}\}, \{(p_i, q_j)_{i \in I_2, j \in J_2}\}, (c_1, c_2))$. The first set of properties corresponds to a *universal quantification*, i.e., the sets of attached values should be equal, and the second set to an *existential quantification*, i.e., the sets of attached values should have a non-empty intersection. The application of link keys across two RDF datasets can be viewed as a data cleaning task, allowing duplicate identification and thus improving data quality.

Given two RDF datasets D_1 and D_2, every combination of property pairs and class pairs can be potentially considered as a link key expression. Then, link key discovery can be considered as a knowledge discovery problem, and efficient data mining algorithms should be designed for reducing the search space and mining interesting and useful link keys. Several dimensions should be taken into account to guide the mining process, among which *maximality, discriminability*, and *coverage*. Accordingly, a link key should generate a maximal link set, while the mapping between the pairs of instances in D_1 and D_2 should be close to a one-to-one mapping. In this paper, link key discovery relies on a specific algorithm based on Formal Concept Analysis (FCA) [10]. The algorithm takes as input two RDF datasets and returns a concept lattice, called the LK-lattice , where each concept encapsulates a so-called *link key candidate* generating a maximal set of potential identity links [2,7]. Then, discriminability and coverage are controlled thanks to adapted quality measures that enable the ranking and the validation of link keys.

Meanwhile, the number of discovered link key candidates can still be (very) large and some candidates may be preferred, raising a representation problem: is it possible to design a *compact* and *representative set* of candidates which preserves a maximal number of identity links and which can be navigated by a domain analyst? Finding such a set amounts to designing an algorithm capable of selecting a set of representative candidates preserving the largest part of the identity links. The size of the subset of candidates and the number of preserved links are suitable characteristics to be considered for computing such a representative subset of candidates. While optimizing at the same time the size of the set of candidates and the number of preserved links is not an easy task, an acceptable compromise can be achieved.

Accordingly, our objective is to propose the LKCLUST algorithm that builds a representative set of link keys, abbreviated as REPLKSET , including a minimal number of link keys and preserving a maximal number of identity links. Such a REPLKSET can be presented to domain analysts for validation in data interlinking [13], e.g., for detecting duplicates in library management or cleaning hand-made datasets. The algorithm LKCLUST is based on agglomerative hierarchical clustering (AHC [11]). As input, it takes an LK-lattice , a dissimilarity measure defined w.r.t. the links generated by the candidates, a cutting level, and a linkage criterion. As output it returns a REPLKSET . LKCLUST combines both FCA and hierarchical clustering for reducing a large search space of candidates in an original way, as the clustering process respects the LK-lattice ordering and every cluster corresponds to a concept interval in the LK-lattice . The present work proposes a follow-up to [1] where the reduction of the link key candidate sets is based on crisp set equality in the framework of FCA and partition pattern structures. Here we propose an alternative and extend the preceding purpose by considering set similarity and clustering.

There are several close approaches in data interlinking which are based on keys [3,14–16] and on link keys [6–8]. In both cases, keys and link keys can be seen as rules allowing to infer links between individuals, and can be used for checking data consistency as this is performed with keys and functional dependencies in database management systems. However, while keys are attached to only one dataset[1], link keys are involving two different datasets [6].

The summary of the paper is as follows. In Sect. 2 we recall useful basics about link keys, FCA, and hierarchical clustering. Then, in Sects. 3 and 4, we introduce the characteristics of a REPLKSET and then the LKCLUST algorithm for building such a REPLKSET . In Sect. 5, we report experiments performed on different RDF datasets that demonstrate the effectiveness of the present approach, before concluding.

2   Background

2.1   Link Key Candidates and Link Sets

In the following, an RDF dataset D is composed of a set of triples $(s, p, o) \in (U \cup B) \times U \times (U \cup B \cup L)$, where U is a set of IRIs[2], B is a set of blank nodes, and L is a set of literals. In (s, p, o), s denotes the subject, p the property or predicate, and o the object or value. Moreover, $C(D) = \{c \mid \exists s \ (s, \texttt{rdf:type}, c) \in D\}$ denotes the set of class identifiers in D, $I(c) = \{s \mid \exists s \ (s, \texttt{rdf:type}, c) \in D\}$ the set of instances of class $c \in C(D)$, $P(D) = \{p \mid \exists s, o \ (s, p, o) \in D\}$ the set of property identifiers in D, and $p(s) = \{o \mid (s, p, o) \in D\}$ the set of objects –or values– related to s through property p.

Given two RDF datasets D_1 and D_2, $k = (Eq, In, (c_1, c_2))$ is a link key expression composed of two sets of property pairs Eq and $In \subseteq P(D_1) \times P(D_2)$, with $Eq \subseteq In$, $c_1 \in C(D_1)$, and $c_2 \in C(D_2)$. For all for all $(p, q) \in Eq$, $p(a) = q(b)$ and $p(a) \neq \emptyset$, where $a \in c_1$ and $b \in c_2$, i.e., Eq is based on equality and corresponds to a $\forall$ quantifier. Moreover, for all $(p, q) \in In, p(a) \cap q(b) \neq \emptyset$, where $a \in c_1$ and $b \in c_2$, i.e., In is based on non-empty intersection and corresponds to an $\exists$ quantifier.

When such a link key expression is verified for an instance $a \in c_1$ and an instance $b \in c_2$, an identity link of the form $(a, \texttt{owl:sameAs}, b)$, actually an RDF triple, can be generated. For example, in Fig. 1, subjects a_3 and b_3 are sharing v_4 through (p_1, q_1) and have the same value v_5 for (p_2, q_2). Then the link

---

[1] The OWL2 construction `HasKey` allows keys to be defined for a given class, stating that each named instance of a class is uniquely identified by a property or a set of properties, as keys in a database system (see https://www.w3.org/TR/owl2-syntax/).

[2] Internationalized Resource Identifier.

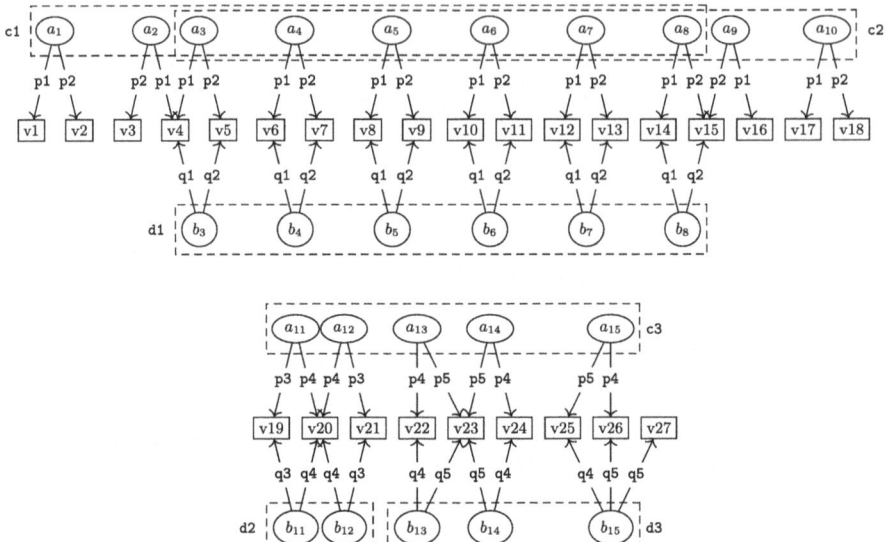

Fig. 1. Two examples of RDF datasets. Dataset D_1 includes instances prefixed by a and dataset D_2 those prefixed by b. D_1 includes classes $c1$, $c2$, $c3$, and D_2 contains classes $d1$, $d2$, and $d3$.

key $(\{(p_2, q_2)\}, \{(p_1, q_1), (p_2, q_2)\}, (c_1, d_1))$ generates the potential identity link (a_3, b_3), with $Eq = \{(p_2, q_2)\}$ and $In = \{(p_1, q_1), (p_2, q_2)\}$.

The expression $k_1 = (Eq_1, In_1, (c_1, c_2))$ over D_1 and D_2 denotes a *link key candidate* if the set of potential links generated by k_1 is not empty, i.e., $L(k_1) \neq \emptyset$, and k_1 is *maximal*. The latter means that there should not exist an expression $k_2 = (Eq_2, In_2, (c_1, c_2))$ such that $In_1 \subset In_2$, $Eq_1 \subset Eq_2$, and $L(k_1) = L(k_2)$. In the following, we will simply write "candidate" if there is no ambiguity. In addition and for the sake of simplicity, we will only consider the In part in a link key expression as $Eq \subseteq In$, i.e., $k = (In, (c_1, c_2))$.

The objective of link key discovery is to mine candidates in the possibly very large power set $P(D_1) \times P(D_2)$. The notion of candidate involves maximality, meaning that a candidate should be maximal and unique among a set of link key expressions generating the same link set. Maximal sets attached to a given relation, here inclusion, are usually related to a closure operator. This was one main reason for mining link key candidates thanks to Formal Concept Analysis (FCA [10]), as introduced in [5] and then revisited in [7]. The next section makes this approach more precise.

2.2   Link Key Discovery Based on FCA

We recall hereafter basics of FCA for allowing a good understanding of the paper. FCA [9,10] is a mathematical framework based on lattice theory and aimed at data analysis and classification. The basic data structure in FCA is a context

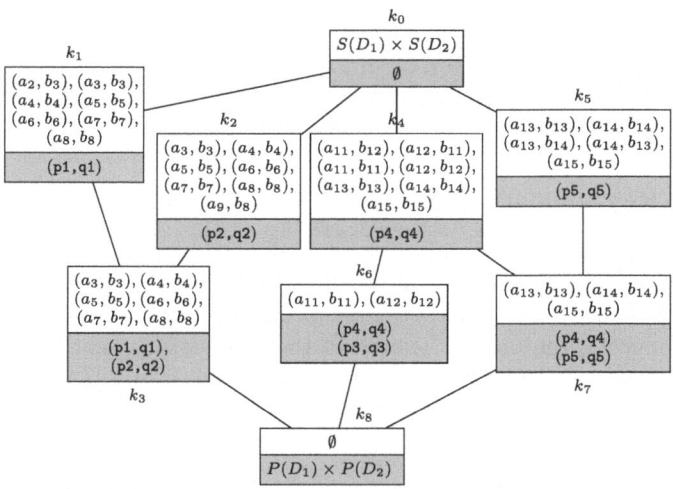

Fig. 2. The LK-lattice based on the datasets given in Fig. 1.

$K = (G, M, I)$ where G denotes a set of objects, M a set of attributes, and $I \subseteq G \times M$ a binary relation indicating that object g has attribute m.

Two derivation operators both denoted by $\cdot'$ are working in a dual way, (i) $A' = \{m \in M \mid \forall g \in A, (g, m) \in I\}$ with $\cdot' : 2^G \mapsto 2^M$, and (ii) $B' = \{g \in G \mid \forall m \in B, (g, m) \in I\}$ with $\cdot' : 2^M \mapsto 2^G$. Intuitively, A' is the set of all attributes common to objects in A while, dually, B' is the set of all objects having all attributes in B. The composition of the two operators $\cdot'$ is denoted by $\cdot''$ and corresponds to a *closure operator* (i.e., extensive, increasing, and idempotent).

A pair (A, B) is a concept in K iff $A = B'$ and $B = A'$. Then A is called the "extent" and B the "intent" of concept (A, B). In particular, A and B are closed sets, i.e., $A'' = A$ and $B'' = B$, where A is the largest subset of objects such that $A' = B$ and B the largest subset of attributes such that $B' = A$. The extent A of a concept (A, B) can be considered as a class of objects or instances, while the intent B corresponds to the description of the class. The set of concepts can be ordered by inclusion w.r.t. extents or intents. A concept (A_1, B_1) is subsumed by a concept (A_2, B_2) whenever $A_1 \subseteq A_2$ or dually $B_2 \subseteq B_1$. The set of all concepts is partially ordered thanks to this subsumption relation within a complete lattice named the *concept lattice* and including a to $(\top)$ and a bottom $(\bot)$ element.

For example, given the two RDF datasets D_1 and D_2 in Fig. 1, a context $K = (G, M, I)$ can be built, where the set G of objects contains pairs of individuals, say (a_i, b_j), and the set of attributes M includes pairs of properties (p_i, q_j) quantified by $\exists$. When a pair of individuals (a_i, b_j) verifies a such property pair (p_i, q_j), a cross fills the corresponding cell in the context. Then the concepts of K are calculated and a concept lattice called LK-lattice is built thanks to FCA algorithms, as shown in Fig. 2. Here it is not possible to show all the details of the construction of the LK-lattice , but the reader may check [2,7]. In the LK-lattice , the intent of k_3, i.e., $\{(p_1, q_1), (p_2, q_2)\}$, represents a link key candidate whose associated pair of classes is (c_1, d_1). The extent of k_3, namely $\{(a_3, b_3), (a_4, b_4), (a_5, b_5), (a_6, b_6), (a_7, b_7), (a_8, b_8)\}$, corresponds to the set of generated links $L(k_3)$. Then, for all $(a_i, b_j) \in L(k_3)$, $p_1(a_i)$ and $q_1(b_j)$, and as well $p_2(a_i)$ and $q_2(b_j)$, are sharing at least one value.

2.3   Link Key Validation Based on Discriminability and Coverage

The validation of link key candidates is usually performed within an unsupervised setting and is based on two quality measures, namely *discriminability* and *coverage*, which are defined w.r.t the set of links generated by a candidate [5]. Let us introduce L the link set related to the candidate k, and the two sets of individuals, $\pi_1(L) = \{a | (a, b) \in L\}$ and $\pi_2(L) = \{b | (a, b) \in L\}$. The *coverage* and the *discriminability* of a set of links L over classes c_1 and c_2 are defined as follows:

$$cov(L, c_1, c_2) = \frac{|\pi_1(L) \cup \pi_2(L)|}{|I(c_1) \cup I(c_2)|}, \quad dis(L, c_1, c_2) = \frac{min(|\pi_1(L)|, |\pi_2(L)|)}{|L|}.$$

Coverage and discriminability evaluate how close a set of links is to a one-to-one mapping between individuals of both datasets. Coverage is maximum when every instance of class c_1 is linked to at least one instance of class c_2, while discriminability is maximum when every instance of c_1 is linked to at most one instance of c_2. The *harmonic mean* of coverage and discriminability may be used to estimate the global quality of a candidate:

$$hmean(L, c_1, c_2) = \frac{2\, cov(L, c_1, c_2).dis(L, c_1, c_2)}{cov(L, c_1, c_2) + dis(L, c_1, c_2)}.$$

3   Characteristics of a Representative Set of Link Keys

The number of link key candidates in an LK-lattice can be very large and it can be very convenient, e.g., for an interactive analysis, to build a *representative set* of candidates denoted as REPLKSET , i.e., a subset of candidates which can be proposed as a summary capturing the significant elements of the original set of candidates. Three desirable characteristics should be verified by a REPLKSET , namely the *compression rate*, the *proportion of preserved identity links*, and the *preservation of the* LK-lattice *ordering*. Accordingly, the number of candidates which are retained in a REPLKSET should be low while the compression rate should be high. For example, the size of a REPLKSET associated with the LK-lattice given in Fig. 2 could be equal to the number of branches (4) in the LK-lattice , while the preserved identity links could be those lying in the extents of the candidates k_3, k_6, and k_7.

The candidates are partially ordered within the LK-lattice , where some candidates are too general while some others are too specific. Candidates lying in the upper levels of the LK-lattice have larger extents, i.e., large sets of potential identity links, than candidates lying in lower levels of the lattice. Then a particular form of *redundancy* can be observed where the same set of identity links can be generated by several candidates, some of which being more general and thus less easily interpretable than the others. For example, again the set of candidates $\{k_3, k_6, k_7\}$ is a good potential representative set of the LK-lattice ordering as all these candidates are lower bounds of concept intervals in the LK-lattice branches, i.e., $[k_1, k_3]$ or $[k_2, k_3]$, $[k_4, k_6]$, and $[k_5, k_7]$. Below, we formally define the three main features characterizing a REPLKSET .

- *The compression rate of a* REPLKSET *w.r.t. the* LK-lattice . A REPLK-SET should be minimal in size, i.e., the smaller the number of candidates in REPLKSET the better is the compression rate. Then the compression rate CRof a REPLKSET w.r.t. an LK-lattice can be calculated as follows:

  CR(REPLKSET ,LK-lattice) $= 1 - \frac{|\text{REPLKSET }|}{|\text{LK}-lattice |-2}$

  where $\top$ and $\bot$ are excluded in the LK-lattice . The compression rate is ranging from from 0 to 1, where the best values are close to 1. It is equal to 0 when —REPLKSET — = —LK-lattice —-2, i.e., there is no compression at all. The compression rate cannot be equal to 1 as the REPLKSET cannot be empty, i.e., it is assumed that there exists at least one candidate whose link set cannot be empty.

  For example, if REPLKSET $= \{k_3, k_6, k_7\}$ for the LK-lattice in Fig. 2, the compression rate is CR$= 1 - 3/7 \simeq 0.57$.

- *The proportion of preserved links (*PPL*).* A REPLKSET should preserve a maximal number of identity links, i.e., the higher the number of identity links preserved in REPLKSET the better the representativeness of REPLKSET . The proportion of preserved links PPLis evaluated thanks to the formula:

  PPL(REPLKSET ,LK-lattice) $= \frac{|\bigcup_{k_i \in \text{REPLKSET}} L(k_i)|}{|\bigcup_{k_j \in \text{LK}-lattice} L(k_j)|}$.

  It should be noticed that LK-lattice and as well $L(k_j)$ cannot be empty sets (thanks to the definition of a link key candidate). The proportion of preserved links PPLranges in $]0, 1]$, cannot be equal to zero, and is equal to 1 when all identity links are preserved, i.e., REPLKSET = LK-lattice .

- *The preservation of the* LK-lattice *ordering.* One requirement in building REPLKSET is to preserve in the clustering the ordering of the LK-lattice . Then a cluster should include a set of candidates such that the lower bound $x \wedge y$ and the upper bound $x \vee y$ of any pair of candidates x and y are also included in the cluster (recall that $x \wedge y$ and $x \vee y$ always exist in a lattice and are unique). There can be two main options: (i) a cluster is based either on a concept interval and includes candidates which are forming a chain, (ii) a cluster is based on a sublattice of the LK-lattice .

These three characteristics cannot be simultaneously optimized but a good compromise can be achieved, i.e., discovering a small set of candidates forming the REPLKSET , which maximizes the compression rate and the number of preserved identity links, and respects the LK-lattice ordering. It should be noticed that, given an LK-lattice , the REPLKSET is not unique. In the following, we make precise the construction of a REPLKSET , starting from an LK-lattice and using agglomerative hierarchical clustering.

4   The LKCLUST Algorithm for Building a REPLKSET

The LKCLUST algorithm constructs a representative set of link keys, namely REPLKSET , thanks to *agglomerative hierarchical clustering* (AHC) [11]. In LKCLUST (see Function 1), AHC takes as input the set of candidates in LK-lattice and builds a hierarchical partition of clusters. One main requirement is that the cluster hierarchy preserves –as much as possible– the partial ordering of concepts in the LK-lattice . In LKCLUST , AHC is based on a bottom-up strategy, starting with each candidate as a singleton cluster and then successively merging pairs of clusters based on their distance or dissimilarity. AHC stops when all clusters are merged into a top cluster.

At initialization, the dissimilarity between candidates c_1 and c_2 is defined as $\delta(c_1, c_2) = 1 - \frac{val(c_1 \wedge c_2)}{val(c_1 \vee c_2)}$, where $c_1 \vee c_2 \neq \bot$, i.e., c_1 and c_2 cannot be $\bot$ at the same time, and $val(c)$ is a function returning the size of the extent of c. The dissimilarity δ ranges from 0 to 1 and has the following properties:

(i) $\delta(c_1, c_2) = 0$ iff $c_1 = c_2$,
(ii) $\delta(c_1, c_2) = \delta(c_2, c_1)$ (symmetry),
(iii) $\delta(c_1, c_2) = 1$ iff $val(c_1 \wedge c_2) = 0$, then the extents of c_1 and c_2 are disjoint and c_1 and c_2 are not comparable, i.e., $c_1 \nleq c_2$ and $c_2 \nleq c_1$, c_1, and c_2 are lying in two different chains in the LK-lattice .

By contrast, when c_1 and c_2 are comparable, say $c_1 \leq c_2$, then $\delta(c_1, c_2) = 1 - val(c_1)/val(c_2)$ as $c_1 \wedge c_2 = c_1$ and $c_1 \vee c_2 = c_2$. The more c_1 is close to c_2 the less is $\delta(c_1, c_2)$. Accordingly the less dissimilar concept c_2 from a given concept c_1 is lying among the immediate lower or upper neighbor concepts in the LK-lattice , i.e., respectively the *lower cover* or the *upper cover* (more about this subject in [12]). More precisely:

Proposition 1. *If $c_1 \leq c_2 \leq c_3$ in the LK-lattice then $\delta(c_1, c_2) \leq \delta(c_1, c_3)$.*

Proof. Since $c_1 \leq c_2$, $\delta(c_1, c_2) = 1 - val(c_1 \wedge c_2)/val(c_1 \vee c_2) = 1 - val(c_1)/val(c_2)$, with $val(c_1) \leq val(c_2)$. In the same way, $c_1 \leq c_3$, $\delta(c_1, c_3) = 1 - val(c_1)/val(c_3)$, with $val(c_1) \leq val(c_3)$. Since $c_2 \leq c_3$, $val(c_2) \leq val(c_3)$, and it comes that $\delta(c_1, c_2) \leq \delta(c_1, c_3)$. In particular, if $c_1 \leq c_2$, then $c_1 \wedge c_2 \leq c_1 \leq c_2$, and it comes that $\delta(c_1, c_1 \wedge c_2) \leq \delta(c_1, c_2)$. □

Another parameter of AHC is the distance $dclust(X, Y)$ between two clusters X and Y, actually between a cluster and a singleton cluster. LKCLUST relies on the so-called *complete linkage*, i.e., $dclust(X, Y) = \max_{x \in X, y \in Y} \delta(x, y)$. Single linkage based on "min" and average linkage based on "mean distance" are other alternatives. However, if all pairs of points from two clusters X and Y are connected with complete linkage, this yields a "complete linkage" where all possible pairs are connected. Then, this ensures that dissimilarity between any two concepts in a cluster is bound by the dissimilarity of the infimum and supremum of the two concepts in the LK-lattice .

Table 1. Dissimilarity table between concepts lying in the LK-lattice in Fig. 2. Only values in bold are needed by the LKCLUST algorithm.

	k_0	k_1	k_2	k_3	k_4	k_5	k_6	k_7	k_8
k_0									
k_1	**0.97**								
k_2	**0.97**	**0.97**							
k_3	0.97	**0.14**	**0.14**						
k_4	**0.97**	1	1	1					
k_5	**0.98**	1	1	1	0.99				
k_6	0.99	1	1	1	**0.71**	1			
k_7	0.99	1	1	1	**0.57**	**0.4**	1		
k_8	1	1	1	1	1	1	1	1	

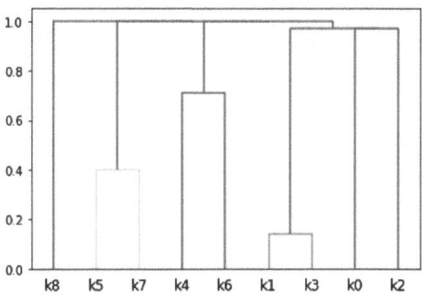

Fig. 3. Dendrogram built by AHC set up based on δ and complete linkage.

Let us now explain how the LKCLUST algorithm works. At initialization, in AHC , all $\delta(x,y)$ between singleton clusters x and y should be computed and recorded in a dissimilarity table, as shown in Table 1. However, in LKCLUST , thanks to proposition 1, only dissimilarities between neighboring candidates are computed, marked in bold in Table 1. Instead of computing $n.(n-1)/2$ dissimilarity values where n is the number of clusters as in AHC , a much smaller number should be computed in LKCLUST , depending on the number of neighbors for a given concept.

Moreover, at each step of AHC , the dissimilarity table should be updated, requiring the computing of $n-2$ dissimilarity values, i.e., the number of clusters minus the two clusters that are merged. Again, in LKCLUST , this number is much smaller as it only involves neighbor concepts. This is illustrated in the experiments, when comparing the running times of AHC and LKCLUST are compared.

For example, the LK-lattice in Fig. 2 contains 9 concepts and thus there are 36 dissimilarity values to compute at initialization. However, as shown in the dissimilarity table in Table 1, the computation is restricted to pairs of neighbor concepts, and only 12 values are calculated in fact. At first iteration, either concepts k_1 and k_3 or concepts k_2 and k_3 can be merged as their dissimilarity value is .14, the merging involving the minimal dissimilarity. In both cases, only 3 values need to be updated among the 7 possibilities at the next step.

The LKCLUST stops when all clusters are merged into one final cluster corresponding to the whole set of candidates. The output can be visualized as a *dendrogram*, i.e., a tree-based representation where the clusters are determined thanks to a *cutting level* (see Fig. 3). Setting a cutting level mainly depends on the objectives and on the characteristics of the application, and this is discussed in the next section about experiments.

5   Experiments

The discovery of representative link keys returned by the LKCLUST algorithm is evaluated thanks to two series of experiments. Firstly, we demonstrate that LKCLUST builds an REPLKSET with good characteristics and with a shorter runtime than a classical AHC algorithm. We also discuss the effects of selecting a cutting level in the dendrogram. Secondly, the experiments evaluates the characteristics of the representatives in a cluster, and as well the behavior of LKCLUST w.r.t. precision and recall in a supervised setting. The results confirm the high level capabilities of the LKCLUST algorithm in link key discovery.

Algorithm 1 The function LKCLUST iteratively builds the hierarchy of clusters, minimizing the number of dissimilarity values to be computed.

> **function** LKCLUST(L : a lattice)
> $values \leftarrow createheap()$
>   ▷ *Computing the dissimilarity values between a concept and its upper cover* ◁
>   **for** $x \in L$ **do**
>   **for** $y \in cover(x, L)$ **do**
> $d \leftarrow \delta(x, y)$
> $add(values, \langle x, y \rangle, d)$
>   **while** $size(values) > 1$ **do**
>   ▷ *Select and merge the clusters with the smallest dissimilarity* ◁
> $\langle x, y \rangle, d \leftarrow extractmin(values)$
> $clust \leftarrow createclust(x, y, d)$
>   ▷ *Update the dissimilarity values based on complete linkage* ◁
>   **for all** $(\langle x', y' \rangle, d') \in values$ **do**
>   **if** ($x = x'$ or $y = x'$) **then**
> $add(values, \langle clust, y' \rangle, dclust(clust, y'))$
> $remove(values, \langle x', y' \rangle)$
>   **else if** ($x = y'$ or $y = y'$) **then**
> $add(values, \langle clust, x' \rangle, dclust(clust, x')$
> $remove(values, \langle x', y' \rangle)$
>   **return** $extractmin(values)$

5.1   Datasets and Protocol

The experiments are performed over ten different *tasks*, where a task consists in considering two RDF datasets with a set of reference links (`a,owl:sameAs,b`), and then to discover the candidates. Seven of these tasks are based on synthetic datasets proposed by the "Ontology Alignment Initiative" (OAEI)[3]: (1) Restaurants, Person1, and Person2 tasks are taken from OAEI 2010; (2) Doremus tasks (1–3) about cultural institutions are taken from OAEI 2016; and (3) the SPIM-Bench task is taken from OAEI 2018. Any pair of the OAEI datasets is based on the same ontology/schema.

The three remaining tasks represent real-world cases of data interlinking, where the datasets are based on different ontologies. The "Libraries" task relies

---

[3] http://oaei.ontologymatching.org/.

Table 2. Statistics about the datasets considered in the experiments.

Task	datasets	#inst.	#prop.	#cl.	#LKC
Restaurants	Restaurant1	339	7	1	13
	Restaurant2	2,256	7	1	
Person1	Person11	2,000	14	1	537
	Person12	1,000	13	1	
Person2	Person21	2,400	14	1	471
	Person22	800	13	1	
Doremus1	PP-1	797	52	1	22
	BnF-1	692	48	1	
Doremus2	PP-2	4,053	52	1	74
	BnF-2	3,384	54	1	
Doremus3	PP-3	940	52	1	26
	BnF-3	822	53	1	
SPIMBench	Abox1	1,126	47	3	1,398
	Abox2	1,130	67	3	
Libraries	BnF	78,076	414	1	1,594
	Abes	290,247	128	1	
wiki-random	Wikidata	1,195	1,531	382	691
	DBPedia	1,184	413	175	
wiki-persons	Wikidata	8,314	4,092	65	3,788
	DBPedia	7,297	332	81	

on a sample of datasets provided by two French libraries, namely the "Bibliothèque nationale de France" (BnF)[4] and the "Agence bibliographique de l'enseignement supérieur" (Abes)[5]. They consist in a selection of the most frequent homonyms. The two last tasks are based on DBPedia and Wikidata samples. The wiki-random task contains randomly selected data while the wiki-person task consists in instances of persons sharing a name and a place of birth. Statistics about all datasets are provided in Table 2.

The LK-lattices are generated thanks to an FCA-based tool (not detailed here) which performs a basic normalization of data values and deals with property composition when possible (the datasets about "wiki" tasks only contain direct data property values). Moreover, as the size of "Libraries" datasets is very large, only properties instantiating 15% of subjects are considered. Finally, the column #LKCin Table 2 represents the number of link key candidates discovered in each task.

---

[4] https://data.bnf.fr/.

[5] https://www.idref.fr/.

All experiments have been conducted on a laptop equipped with an Intel(R) Core(TM) i7-10875H CPU @ 2.30GHz and 8GB of memory dedicated to the JVM. Clustering procedures have been applied to the sets of candidates excepting the top and bottom concepts.

5.2 Evaluation of the LKCLUST Algorithm

Below, the two main aspects which are evaluated are the performance of the LKCLUST algorithm compared to a classical AHC algorithm and the quality of the partitioning. The experiments were carried out with four configurations based on different variations of the two following parameters: (i) the clustering method, i.e., LKCLUST or AHC algorithms, (ii) whether the convexity of a cluster is forced or not. A convex cluster corresponds either to a concept interval (i.e., a chain) or a sublattice. To ensure the convexity of a cluster, the LKCLUST algorithm is modified and skip a candidate group which is not convex.

For every task and every configuration, the running time spent for clustering and the number of required dissimilarities are recorded. The results are presented in Fig. 4 and in Fig. 5. These two figures only show the four tasks in which the clustering procedure takes at least 2 s.

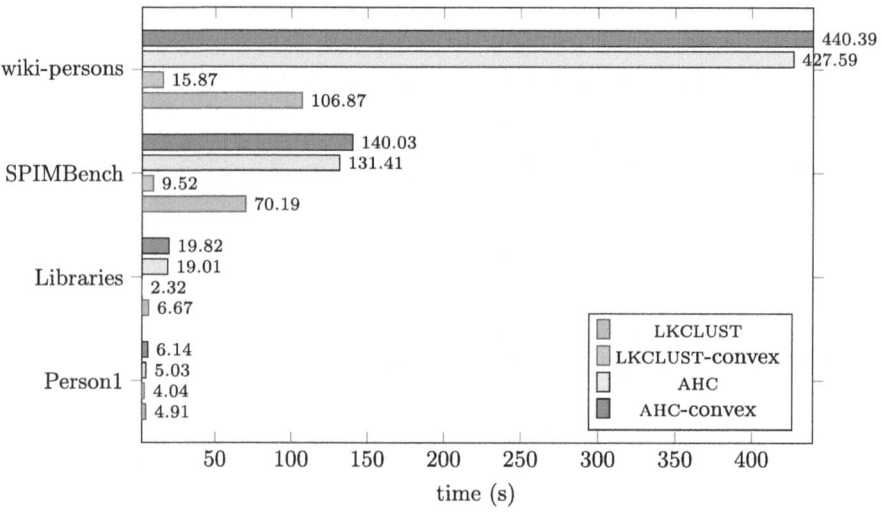

Fig. 4. Four clustering running times in different configurations and interlinking tasks.

It was expected that configurations based on LKCLUST algorithm are faster than those based on a classical AHC algorithm, and this is indeed the case as shown in Fig. 4. AHC works faster when we constrain the algorithm to only form convex sublattices. Actually, the convexity requirement reduces the number of possible groupings and therefore the number of dissimilarities to be computed as shown in Fig. 5. In line with these observed runtimes, Fig. 5 shows that the LKCLUST algorithm significantly reduces the number of calculated dissimilarities. This reduction is even more radical when the convexity constraint is enforced.

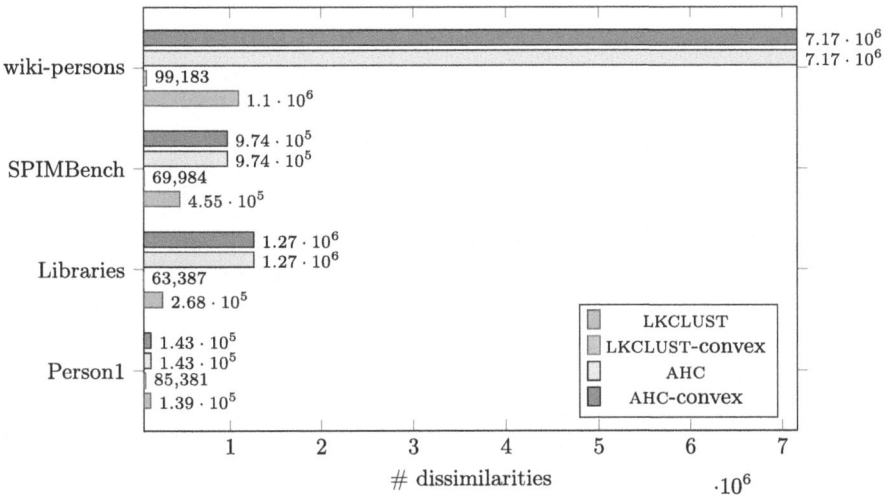

Fig. 5. The numbers of dissimilarities required in the different configurations and interlinking tasks.

To study the quality of partitions built by LKCLUST , the compression, link preservation, and convexity criteria are used. In the four above configurations, the cutting level is changed from 0 to 1 -excluded- in steps of length 0.1. For every cutting level, the compression rate, preservation rate, and convexity rate are monitored. Figure 6 shows the results in the four configurations for the "SPIMBench" task. The observations that can be made on this task also apply to the other tasks.

AHC and LKCLUST show similar results in term of compression and convexity. When convexity is required, all configurations behave in the same way and almost all links are preserved, independently of the cutting level. In term of preservation, AHC preserves more links than LKCLUST on this task. However, this observation does not apply necessarily to the other tasks where the preservation is roughly the same. This difference in preservation is mainly due to the presence of ties in the dissimilarities. In the event of a tie, a pairing is arbitrarily chosen, which leads to different partitioning. In this case, AHC tends to favor clusters including general concepts which are not necessarily neighbors, because the top concept is omitted.

To sum up, these results show that LKCLUST is much faster than a classical version of AHC . In addition, combining LKCLUST with the convexity constraint provides the best computation times, preserves almost all the links, while the compression ratio is slightly lower.

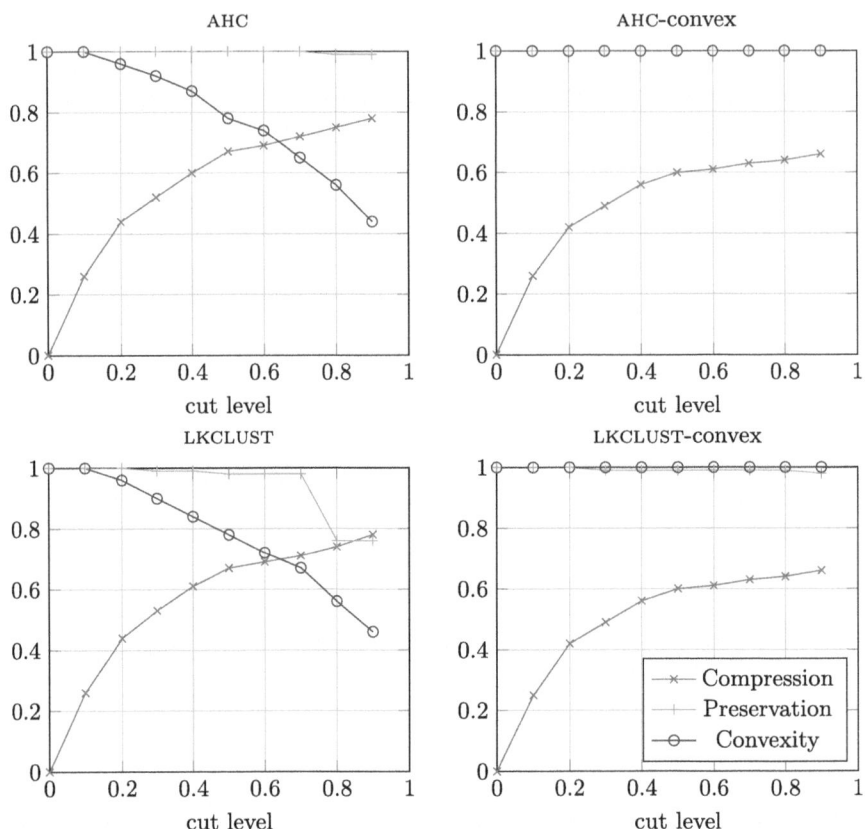

Fig. 6. The variations of compression, link preservation, and convexity w.r.t the cutting level in the task SPIMBench.

5.3 The Evaluation of the Candidates Lying in a REPLKSET

This second series of experimental results aims at analyzing the impact of LKCLUST on the selection of the best link key candidates, materialized by the *medoids*. The medoid in a cluster is the link key candidate whose sum of dissimilarities to all other candidates in the cluster is minimal. Then, while the cutting level is varied by step of 0.1, the medoids having a discriminability greater than or equal to 0.9 are selected at each step. The union of the links generated by the selected medoids is evaluated against a gold standard to assess precision and recall.

Figure 7 shows the precision, recall, and compression ratio that are measured for the tasks "wiki-random" and "Doremus1". These two tasks are illustrative of the observed trends. The compression ratio is computed over the selection of medoids whose discriminability is higher than 0.9.

On "wiki-random", the F-measure is almost stable whatever the cutting level. This shows that the medoids selected thanks to LKCLUST contain good link key candidates. We observe the same trends for the tasks "Person1", "Restaurants", "SPIMBench", and "Libraries".

Regarding "Doremus1", there is a drop in recall, while precision remains stable. For all "Doremus" tasks, this drop appears when the cutting level is high (> 0.8). For the tasks "Persons2" and "wiki-person", the recall break is less noticeable, i.e., -0.12 and -0.13 respectively, but occurs earlier, around 0.3. This drop becomes visible during an acceleration of the compression ratio. However an acceleration of the compression ratio does not necessarily imply a drop in recall in the other tasks. These sudden drops in recall could probably be due to a "discriminability threshold" effect. Indeed, the threshold was arbitrarily set at 0.9 in all the experiments, but should probably be adapted on a case-by-case basis.

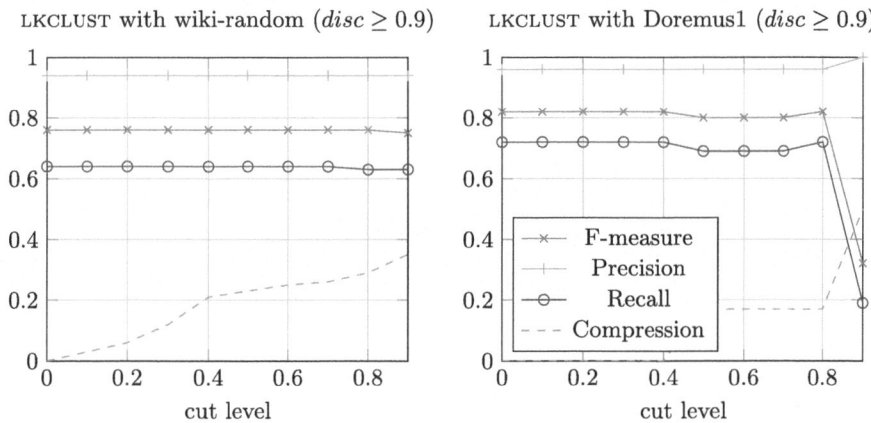

Fig. 7. The variation of quality measures and compression over a selection of link key candidates having a discriminability greater than 0.9 w.r.t the cutting level.

6 Conclusions and Perspectives

In this paper, we are interested in data interlinking based on the discovery of link keys generating identity links between individuals lying in two RDF datasets. Link keys are discovered thanks to FCA algorithms and are organized within a concept lattice called LK-lattice that can be very large in size, making it difficult to visualize and to interpret. Thus we introduce a representative set of link key candidates, i.e., the REPLKSET , which satisfies good properties, namely a high degree of compression, a high preservation of identity link, and the respect of the LK-lattice ordering. The REPLKSET is built thanks to an adapted algorithm named LKCLUST based on agglomerative hierarchical clustering. Experiments show that LKCLUST returns a REPLKSET satisfying the three constraints where the representatives in each cluster show good recall and precision.

As future work, we would like to study other clustering algorithms and distance measures to build a REPLKSET , and to characterize the mapping between an LK-lattice and a REPLKSET . In particular, this approach could also be well-suited to the difficult task of lattice reduction [4].

References

1. Abbas, N., Bazin, A., David, J., Napoli, A.: Discovery of link keys in resource description framework datasets based on pattern structures. Int. J. Approximate Reasoning **161**, 108978 (2023)
2. Abbas, N., David, J., Napoli, A.: Discovery of link keys in RDF data based on pattern structures: preliminary steps. In: Proceedings of CLA, pp. 235–246. CEUR Workshop Proceedings 2668 (2020)
3. Al-Bakri, M., Atencia, M., David, J., Lalande, S., Rousset, M.C.: Uncertainty-sensitive reasoning for inferring sameAs facts in linked data. In: Proceedings of ECAI, pp. 698–706 (2016)
4. Aragón, R.G., Medina, J., Ramírez-Poussa, E.: Reducing concept lattices by means of a weaker notion of congruence. Fuzzy Sets Syst. **418**, 153–169 (2021)
5. Atencia, M., David, J., Euzenat, J.: Data interlinking through robust linkkey extraction. In: Proceedings of ECAI, pp. 15–20 (2014)
6. Atencia, M., David, J., Euzenat, J.: On the relation between keys and link keys for data interlinking. Semantic Web J. **12**(4), 547–567 (2021)
7. Atencia, M., David, J., Euzenat, J., Napoli, A., Vizzini, J.: Link key candidate extraction with relational concept analysis. Discret. Appl. Math. **273**, 2–20 (2020)
8. Atencia, M., David, J., Scharffe, F.: Keys and pseudo-keys detection for web datasets cleansing and interlinking. In: ten Teije, A., et al. (eds.) EKAW 2012. LNCS (LNAI), vol. 7603, pp. 144–153. Springer, Heidelberg (2012). https://doi.org/10.1007/978-3-642-33876-2_14
9. Ganter, B., Obiedkov, S.A.: Conceptual Exploration. Springer, Heidelberg (2016). https://doi.org/10.1007/978-3-662-49291-8
10. Ganter, B., Wille, R.: Formal Concept Analysis. Springer, Heidelberg (1999). https://doi.org/10.1007/978-3-642-59830-2
11. Kaufman, L., Rousseeuw, P.J.: Finding Groups in Data: an Introduction to Cluster Analysis. Wiley (2009)
12. Monjardet, B.: Metrics on partially ordered sets-a survey. Discret. Math. **35**(1), 173–184 (1981)
13. Nentwig, M., Hartung, M., Ngonga Ngomo, A.C., Rahm, E.: A survey of current Link Discovery frameworks. Semantic Web J. **8**(3), 419–436 (2017)
14. Pernelle, N., Saïs, F., Symeonidou, D.: An automatic key discovery approach for data linking. J. Web Semantics **23**, 16–30 (2013)
15. Symeonidou, D., Armant, V., Pernelle, N., Saïs, F.: SAKey: scalable almost key discovery in RDF data. In: Mika, P., Tudorache, T., Bernstein, A., Welty, C., Knoblock, C., Vrandečić, D., Groth, P., Noy, N., Janowicz, K., Goble, C. (eds.) ISWC 2014. LNCS, vol. 8796, pp. 33–49. Springer, Cham (2014). https://doi.org/10.1007/978-3-319-11964-9_3
16. Symeonidou, D., Galárraga, L., Pernelle, N., Saïs, F., Suchanek, F.: VICKEY: mining conditional keys on knowledge bases. In: d'Amato, C., et al. (eds.) ISWC 2017. LNCS, vol. 10587, pp. 661–677. Springer, Cham (2017). https://doi.org/10.1007/978-3-319-68288-4_39

Understanding the Impact of Entity Linking on the Topology of Entity Co-occurrence Networks for Social Media Analysis

James Nevin[1][(✉)] ⓘ, Pengyu Zhang[1] ⓘ, Dimitar Dimitrov[2] ⓘ, Michael Lees[1] ⓘ,
Paul Groth[1] ⓘ, and Stefan Dietze[2] ⓘ

[1] University of Amsterdam, Amsterdam, Netherlands
{j.g.nevin,p.zhang,m.h.lees,p.t.groth}@uva.nl
[2] GESIS - Leibniz Institute for the Social Sciences, Cologne, Germany
{dimitar.dimitrov,stefan.dietze}@gesis.org

Abstract. A common form of analysis of textual data is entity co-occurrence, where networks of entities and their connections within the text are constructed and their topology analysed. As the analysis is focused on the entities and their relations, the tools used to extract them can have a potentially large effect on the results. A frequently used method as part of these analyses is entity linking, where extracted entities are mapped to a knowledge graph. Many established entity linking tools have been created for long text following standard spelling and grammar rules. As a result, the tools struggle on short, unstructured text such as tweets. On such text, it can be difficult to choose between tools and parameter settings, especially since ground truth is often unavailable. Given these challenges in entity linking on text and the direct influence of extracted entities on subsequent network analysis, we propose the need to apply multiple tools to create a more holistic set of results. We verify this assertion through a set of experiments. Using a dataset of approximately 21 million English-language tweets, we construct multiple entity co-occurrence networks using two tools (Fast Entity Linker and DBpedia Spotlight) and numerous confidence thresholds for each. We find that standard network analysis metrics, such as size, connectivity, and centrality are all heavily influenced by the choice of entity linking tool.

Keywords: Entity linking · Co-occurrence networks · Network analysis · Social media

1 Introduction

Large amounts of text are generated on the internet everyday. This text offers many opportunities for identifying trends and understanding online discourse [6]. However, many of these text data are unstructured. Given this lack of structure and the large size of these data, approaches that abstract the data

M. Alam et al. (Eds.): EKAW 2024, LNAI 15370, pp. 69–85, 2025.
https://doi.org/10.1007/978-3-031-77792-9_5

into interpretable forms can be powerful ways to analyse them [5,25,35]. One such abstract form is entity co-occurrence networks [3]. These co-occurrence networks are graphs[1] with entities being represented by vertices and weighted edges between them showing how frequently the entities occur together in pieces of text. The resultant graph can be analysed using standard network analysis metrics, including number of vertices or degree distributions [7]. Analysing these can give valuable insight into the text, such as, for example, topics of discussion through the entities identified [1,6]. The structure of the network can further show which entities play prominent roles (based on their centrality), or how well connected they are to each other [29,30]. Large scale corpuses of annotated text, such as TweetsKB [11], provide opportunities for creating these entity co-occurrence graphs and discovering novel links/relations between entities. The topology of such co-occurrence networks determines the results of the analysis. Hence, results and subsequent conclusions are solely dependent on the approach taken to identify the entities within the text.

In order to gain additional insights, entities can be not only identified, but also linked to a knowledge base (KB). These knowledge bases are carefully constructed to contain accurate, unambiguous information. By linking the entities to such a knowledge base, ambiguities in the identified entities can be removed, which can be important for the construction of accurate co-occurrence networks. However, this process of entity linking is a difficult task and an active area of research [24,27,31]. Because of this, the choice of entity linking algorithm and setting of parameters is not always straightforward. As the downstream co-occurrence network is heavily influenced by the algorithm, care needs to be taken in its selection.

One popular source of large scale text data is Twitter/X. Twitter is one of the most popular online social media websites, and is regularly used for information dissemination and discourse. However, the aforementioned challenges of lack of structure apply especially to tweets for a number of reasons: there is frequent use of acronyms; highly irregular grammar; and brevity plus little context. The usefulness of entities in analysing such short text has already been shown [2], but performing entity linking on text with these properties poses extra challenges [14]. Even state-of-the-art entity linkers achieve F1 scores of around 0.3 on a recent benchmark dataset [23]. Furthermore, it can be unfeasible to manually label additional datasets to determine which algorithms are best.

Given these challenges and the low accuracy scores, an alternative approach is needed for the selection and parameter setting of entity linking algorithms in the construction of co-occurrence networks. In the literature, it is common to test on a benchmark dataset and select only a single entity linker. However, since accuracy scores are so low and benchmark datasets relatively small, this chosen linker likely includes some bias and thus potentially overlooks different possibilities for the created network. Instead, we argue for application of multiple different entity linkers and settings on the full dataset. By analysing the

---

[1] Throughout this paper, we generally use the terms 'graph' and 'network' interchangeably.

multiple different constructed networks, one can identify the variance that can arise through changing linking algorithms and their parameters. In the case that downstream network results show large variance for different algorithms, it is prudent to report results based on the multiple algorithms, hence offering a more holistic picture. On the other hand, if results are robust, simply choosing the 'best' algorithm/parameters based on some standard or benchmark is reasonable.

We illustrate this approach through a set of experiments in constructing entity co-occurrence networks from tweets. We test two entity linking algorithms (Fast Entity Linker [4] and DBpedia Spotlight [22]) on a set of 21 million English-language tweets over a three year period, for which there is no entity linking ground truth. For each algorithm, we apply 5 different confidence thresholds. Networks are compared based on standard network analysis metrics, showing high sensitivity to the different possible algorithms and parameter settings. In some cases, these differences could influence the interpretation of the text, such as which topics are most important. This emphasises the need to report results across multiple algorithms.

The contributions of this paper are:

1. entity linking of a dataset of 21 million tweets with two different entity linking algorithms and multiple confidence thresholds;
2. comparison of entity co-occurrence networks created using these linked entities, showing the high sensitivity of standard network analysis metrics to algorithms and parameters.

The rest of the paper is structured as follows: Sect. 2 covers the related works; Sect. 3 introduces the data and their collection process; Sect. 4 describes the approach of the experiments; Sect. 5 analyses the results; and Sect. 6 offers conclusions.

2   Related Work

Network analysis is a powerful tool for identifying structure and importance of entities through their position in the network. For example, Named Entity Recognition (NER) has been applied to books to create an entity co-occurrence network, which was used to identify interesting topological features, such as the small world property, of the corpus [3]. In a similar fashion, Manaskasemsak et al. also used NER on a collection of tweets to identify clusters in entity co-occurrence networks [21]. The authors related these clusters to real-world events, and found this to be an effective approach for connecting tweets to the events. However, the use of NER poses potential risks. Fegley and Torvik investigated the effects of non-disambiguated or incorrectly combined entities in entity networks. They found that certain network measures, such as clustering and assortivity, can be strongly affected by ambiguity [13]. Because of this, going a step beyond entity recognition can improve the accuracy of network analysis results.

Entity linking aims to accomplish this disambiguation by linking keywords in text to entities in a knowledge graph, with various tools such as Fast Entity Linker (FEL) [4], DBpedia Spotlight [22], REL [34], and GENRE [8] available. However, there are a number of challenges to this linking task. There are frequent problems in benchmark datasets used for evaluating entity linking models [33]. Many of these datasets are long-tailed, making it sometimes difficult to evaluate linker performance [18]. Additionally, entities within the knowledge base can be a simplification of those appearing in the text [32], or be missing entirely [8,12]. Various approaches such as entity spaces [32], autoregressive language models [8], dense representations from Transformers [16], and entity co-occurrence [19] have been suggested for creating accurate models that address these problems.

Several studies have tackled the further challenges of entity linking in short, unstructured text like tweets. Numerous Twitter entity disambiguation datasets have been introduced for analysing method robustness on noisy texts and identifying key error sources [9,14]. The largest of these, TweetNERD [23], addresses previous limitations by providing a broad time window, consistent annotations, and splits for assessing out-of-domain and temporal generalization performance. Various approaches to disambiguation on such data have also been tested, including unsupervised systems [14] and hybrid approaches that combine dense retrieval with long contextual representations from Wikipedia [15]. Collective inference methods that integrate mention-entry, entry-entry, and mention-mention similarities have been shown to improve performance [20].

However, the studies above mainly focus on technical performance and evaluation metrics. It overlooks how the choice of entity linking tools affects downstream results – in this case, network analysis metrics like size, connectivity, and centrality.

3   Data and Data Collection

We employed the long term Twitter archive that is the foundation of TweetsKB [11]. The archive is based on continuously capturing a data stream of 1% randomly sampled tweets from Twitter/X [17]. The archive contains more than 14 billion tweets collected over the past 10 years. To reduce the computational time for the purposes of our analysis, we use a subset from the archive for a socially relevant topic, namely, nationalism during the COVID-19 pandemic. The dataset has been extracted using a list consisting of 73 keywords[2] about COVID-19 and nationalism and contains English-language tweets from 08/2019 to 08/2022. To extract tweets, we performed an exact string match in case sensitive fashion for the keywords "ppe" and "PCR". For the remaining keywords, a tweet is extracted from the archive if a keyword matches a sub-string of the tweet in a case insensitive manner. The dataset[3] contains approximately 21 million tweets.

---

[2] Keywords: https://github.com/jim-g-n/Tweet-Linked-Entity-Co-occurrence/tree/main/dataset.

[3] The dataset extraction is part of a bigger initiative to create a comprehensive dataset about nationalism during COVID-19 in English, German, and Italian by Mark Dang-Anh, Giorgia Riboni and Dimitar Dimitrov.

This dataset is large and carefully curated, exactly the kind to which entity co-occurrence analysis might be applied.

4   Approach

We perform a set of experiments to highlight the effect that choice of entity linking algorithm can have on co-occurrence network analysis.[4] As the dataset lacks a ground truth for entity linking, no conclusions can be drawn as to the quality of identified links. Hence, algorithms can only be compared based on differences in the downstream application. For each of the algorithm and parameter settings, we create a network by linking entities from the tweet dataset. Nodes and edges in the network are defined by the linked entities and how frequently they occur together in the same tweet.

4.1   Entity Linking Algorithms

We test two different entity linking algorithms, Fast Entity Linker (FEL) [4] and DBpedia Spotlight [22]. We choose these two as FEL is intended for short text such as tweets while DBpedia Spotlight is multilingual and highly configurable. Both tools are well-established.

FEL works in a unsupervised fashion and does not require any parameterisation while allowing changing the knowledge base and the entity embedding used for linking. The algorithm was initially developed to work on short text (i.e. search queries) but found adoption in the creation of TweetsKB [11] and TweetsCOV19 [10], two large Twitter corpuses containing annotated tweets. The version of FEL used for this paper links to a Wikipedia dump from November 2023. Each linked entity from a piece of text is given a confidence score from -3 to 0, where a higher score is more strict. We apply different confidence score cutoffs to exclude entities with low confidence.

DBpedia Spotlight is an entity linker that links to DBpedia [22]. As stated before, the main advantage of DBPedia Spotlight is that it is highly configurable, as it allows users to compute scores such as prominence (reflected by the number of times a resource is mentioned in Wikipedia), topical relevance (a similarity score between the paragraph containing the candidate resource and DBpedia resource's context) and contextual ambiguity (representing the relative difference in topic score between the first and the second ranked candidate resource, if more than one candidate is available), and confidence scores. We used the March 2022 version of the Spotlight model from the DBpedia Databus repository[5] and focus only on the confidence scores. Linking can be performed with confidence scores between 0 and 1, with higher being more strict. DBpedia Spotlight also returns a support score per entity, but this is not relevant for our study since we base entity prominence on how frequently they occur in the tweets.

---

[4] Created networks and analysis code are available here: https://github.com/jim-g-n/Tweet-Linked-Entity-Co-occurrence.

[5] https://databus.dbpedia.org/dbpedia/spotlight/spotlight-model/2022.03.01.

For the FEL algorithm, we test confidence thresholds of {-3.00, -2.75, -2.5, -2.25, -2.00}. For DBpedia Spotlight, we test confidence scores of {0.5, 0.6, 0.7, 0.8, 0.9}. These scores are presented as they cover a wide range of confidences. Analysis of other thresholds showed that they were too strict (very few entities), too noisy (an excessive number of entities), or did not change the main conclusions of this paper, so we exclude them in the interest of space. Networks based on these additional thresholds are included in the associated repository.

4.2   Network Construction and Analysis

For a given algorithm and confidence threshold, we create a set of linked entities for each tweet. The nodes in the network are the total collection of unique entities from all tweets. The edges between pairs of nodes are weighted based on the count of how many tweets the respective linked entities occur together in. There is a total of 10 different graphs – 5 created using FEL and 5 using DBpedia Spotlight.

We compare the different created networks using standard network analysis metrics [13]. These include: number of nodes (equivalent to the number of unique linked entities), number of edges, diameter (the shortest distance between the most distant nodes), average degrees (average number of edges per node, which can be weighted or scaled), and component sizes (number of nodes in connected components in the network). These metrics generally show how large and well-connected the network is, which has many downstream implications and is thus of frequent concern to those analysing such networks. For all of the constructed networks, we present these metrics and highlight how they compare across different algorithms and parameter settings. For much of the analysis, we restrict ourselves to the largest connected component within each network, as per standard analysis practices. There are two differences we wish to draw attention to: changes in network properties for different confidence thresholds for a given algorithm, and differences between the two algorithms for comparable confidence thresholds. In this case, by comparable confidence thresholds, we mean thresholds that identify a similar number of unique entities in the dataset.

We also consider the differences in top nodes in the connected networks. We use the degree centrality (scaled number of edges per node) to rank the nodes and compare the top 10 entities in all networks. Top nodes in networks give an indication of the topics being discussed and the roles entities play in connecting these. To emphasise the differences, we also show the top weighted edges in two of the graphs for the different algorithms.

Networks were created and analysed using the graph-tool python library [26].

5 Results

For the analysis, we first cover the overlap in linked entities of the two algorithms. Following this, we analyse the topology of the full graphs and their largest connected component. Lastly, we highlight prominent nodes in the connected components. Our analysis is broad but not in-depth, as our emphasis is on the differences that can arise rather than the precise results themselves.

5.1 Linked Entities Overlap

Figure 1 shows the similarity of the unique entities linked to with the different algorithms and confidence thresholds. The x-axis shows the DBpedia Spotlight confidence threshold, while different coloured and shaped dots represent different FEL confidence thresholds; the y-axis shows the Jaccard similarity of the set of unique entities created using the given DBpedia Spotlight and FEL confidence thresholds.

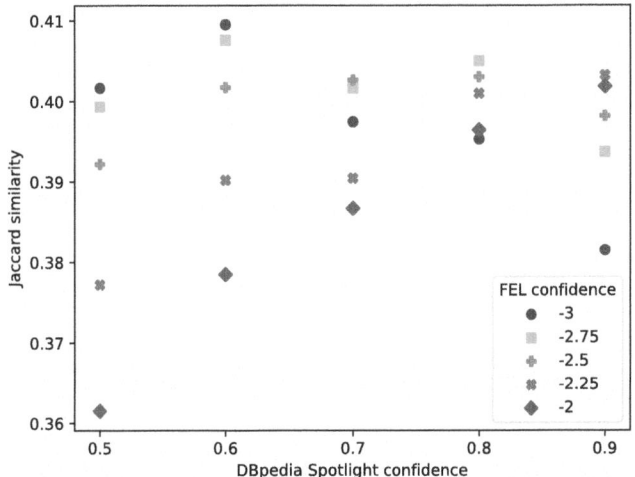

Fig. 1. Jaccard similarity (y-axis) of sets of unique entities linked using different DBpedia Spotlight (x-axis) and FEL (colours and shapes) confidence thresholds. (Color figure online)

The Jaccard similarity generally falls between 0.38 and 0.41. This is a relatively narrow range, given that the sizes of the sets can vary significantly for different confidence thresholds. That being said, the maximum value of 0.41 is fairly low, even when the size of the sets are similar, meaning the entities found by the two algorithms can differ substantially.

5.2 Full Graphs Topology

Tables 1 and 2 show, respectively, properties of the various co-occurrence networks created using FEL and DBpedia Spotlight. For both algorithms, the number of nodes (unique entities discovered) varies between approximately 535,000 and 355,000. The connectivity in the DBpedia Spotlight networks are relatively higher than the FEL networks for comparable confidence thresholds, with a larger number of edges and lower pseudodiameter (an estimate of the diameter).

Table 1. Graph properties of full co-occurrence networks created using FEL algorithm

Confidence	Num. nodes	Num. edges	Pseudodiameter	Average degree	Average weighted degree	Largest component
−3.00	535572	10019969	9	18.71	93.09	488506
−2.75	496670	7539889	10	15.18	75.77	442142
−2.50	456268	5496269	10	12.05	60.69	392744
−2.25	409826	3846504	11	9.39	46.83	336089
−2.00	357385	2594176	12	7.26	36.58	277843

Table 2. Graph properties of full co-occurrence networks created using DBpedia Spotlight algorithm

Confidence	Num. nodes	Num. edges	Pseudodiameter	Average degree	Average weighted degree	Largest component
0.5	515022	14113950	6	27.40	240.85	505326
0.6	473225	8975858	6	18.97	175.57	459781
0.7	435409	5855384	9	13.45	76.20	385827
0.8	401286	4176344	9	10.41	57.62	342799
0.9	358632	2920492	10	8.14	44.25	292811

Figure 2 shows the number of nodes (left) and percentage of nodes in the largest connected component (right) for different confidence thresholds of entity co-occurrence networks created using FEL (blue circles) and DBpedia Spotlight (orange diamonds). As seen in the tables, the number of nodes decreases for both algorithms as the confidence threshold is increased, and this decrease is roughly linear.

For both algorithms, the percentage of nodes in the largest connected component also decreases as the confidence threshold is increased. For the FEL algorithm, this decrease is slightly non-linear, reducing from 0.91 to around 0.78. For the DBpedia Spotlight algorithm, the decrease is not regular, with most nodes (¿0.95) remaining in the largest component with a confidence threshold of 0.5 or 0.6. There is a sharp drop from a confidence threshold of 0.6 to 0.7, followed by a roughly linear decrease. Overall, the DBpedia Spotlight graphs have generally higher connectivity than the FEL graphs, also relating back to the larger number of edges mentioned previously.

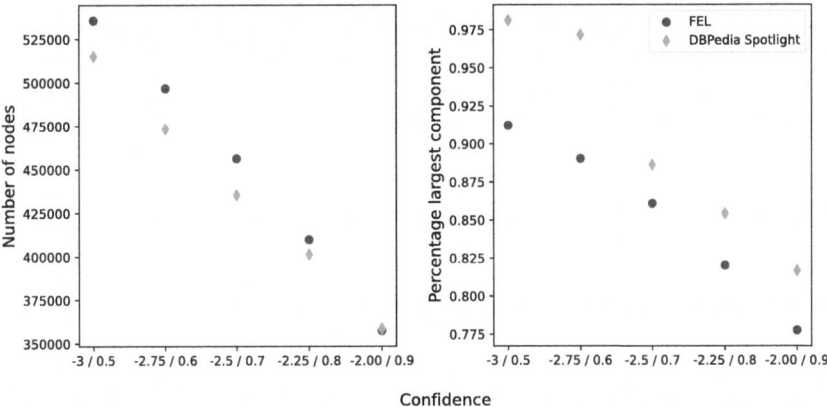

Fig. 2. Number of nodes (left) and percentage of nodes in largest connected component (right) for different confidence thresholds using FEL (blue circles) and DBpedia
Spotlight (orange diamonds) (Color figure online)

5.3 Connected Graphs Topology

We restrict the rest of the analysis to the largest connected component of each
graph, as this generally includes the majority of the nodes. Tables 3 and 4 show,
respectively, the properties of the connected graphs created using the FEL and
DBpedia Spotlight algorithms.

Table 3. Graph properties of connected co-occurrence networks created using FEL
algorithm

Confidence	Num. nodes	Num. edges	Pseudodiameter	Average degree	Average weighted degree
−3.00	488506	10017512	9	20.51	102.05
−2.75	442142	7536960	10	17.05	85.11
−2.50	392744	5492703	10	13.99	70.48
−2.25	336089	3842187	11	11.43	57.07
−2.00	277843	2589306	12	9.32	47.02

Table 4. Graph properties of connected co-occurrence networks created using DBpedia
Spotlight algorithm

Confidence	Num. nodes	Num. edges	Pseudodiameter	Average degree	Average weighted degree
0.5	505326	14113714	6	27.93	245.47
0.6	459781	8975501	6	19.52	180.70
0.7	385827	5853024	9	15.17	85.98
0.8	342799	4173397	9	12.17	67.43
0.9	292811	2917002	10	9.96	54.17

Figure 3 shows a number of different network metrics for the connected graphs created using both algorithms. Values for FEL are shown in blue circles, while DBpedia Spotlight are shown in orange diamonds. The average number of nodes (top left) are once again comparable between the two algorithms, with both showing a linear decrease as the confidence threshold is increased. However, as in the full graphs, other metrics show some large differences in behaviour.

The pseudodiameter, an approximation of the diameter of the graphs, is shown in the top right of the figure. In both cases, despite the number of nodes being smaller with a higher confidence threshold, increasing the confidence threshold increases the pseudodiameter, in a semi-stepwise fashion. Between the two algorithms, the pseudodiameter is much higher in the FEL graphs than the DBpedia Spotlight graphs, showing lower connectivity, even when the number of nodes is similar.

The middle row of the figure shows the (unscaled) average and weighted average degrees in the networks. We first note that both the average degree and average weighted degree are higher in the DBpedia Spotlight networks, but this difference becomes less pronounced at higher confidence levels. As the confidence threshold is increased, both the weighted and unweighted average degrees decrease for both algorithms, and this change is less linear in the DBpedia Spotlight networks than in the FEL networks. In particular, the weighted average degree has an inflection point at a confidence score of 0.7, but this behaviour is not observed in the FEL networks.

Similar behaviour can be observed in the scaled average weighted and unweighted degrees (bottom row of the figure). In this row, the average (weighted) degrees have been scaled based on the number of nodes in the networks. Differences between the two algorithms are slightly more pronounced, and the inflection point is still present in the weighted version for DBpedia Spotlight.

Figure 4 shows some calculated properties of the different connected networks. The top row shows the average local (left) and global (right) clustering, while the bottom row shows the unweighted (left) and weighted (right) assortivity, for different confidence thresholds. As before, FEL values are in blue circles and DBpedia Spotlight values in orange diamonds.

We again see differences between the two algorithms and different confidence thresholds for each. The average local clustering decreases as the confidence threshold is increased for both algorithms; however, this decrease is non-linear for DBpedia Spotlight, with a large drop going from 0.6 to 0.7. On the other hand, the global clustering values do not change monotonically, sometimes increasing and sometimes decreasing as the confidence threshold is increased. For the DBpedia Spotlight graphs, there is once again a large change when increasing the confidence threshold from 0.6 to 0.7. For all graphs, the average local clustering is fairly high, always above 0.4, indicating small-world properties regardless of choice of algorithm/confidence threshold.

The assortivity values are generally low for all networks. However, we still see trends, such as increasing unweighted assortivity for increasing confidence threshold and more sporadic metric behaviour in the DBPedia Spotlight graphs.

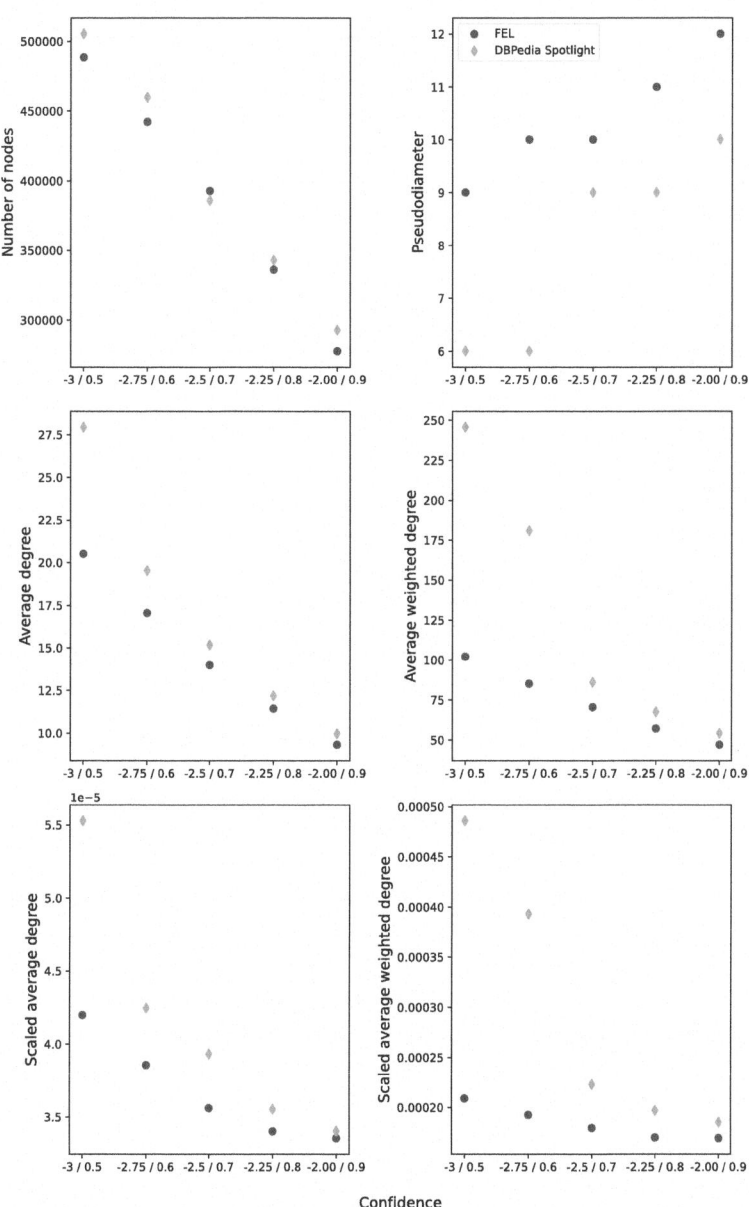

Fig. 3. Network metrics of largest connected components for different confidence thresholds (x-axes). Values for FEL and DBpedia Spotlight are in blue circles and orange diamonds, respectively. The first row shows the number of nodes (left) and pseudodiameters (right); the middle row shows the average degree (left) and average weighted degree (right); the bottom row shows the scaled average degree (left) and scaled average weighted degree (right). (Color figure online)

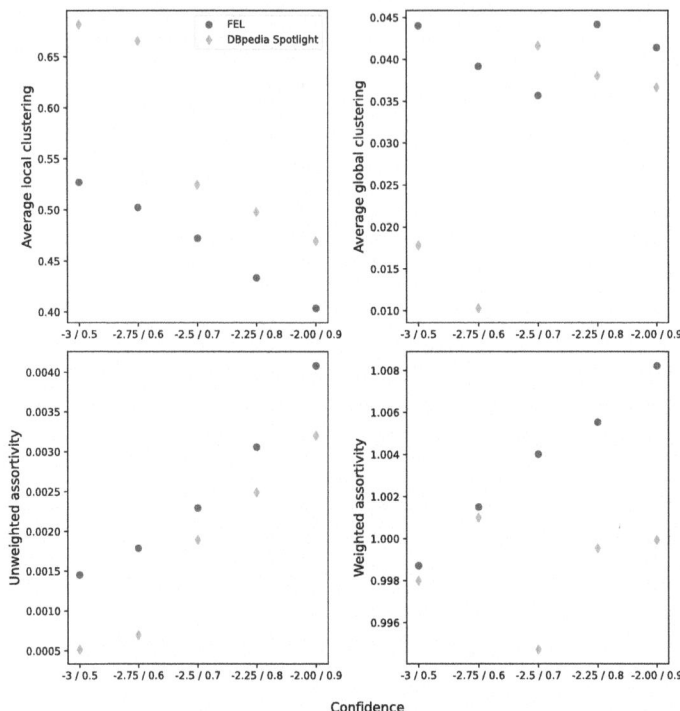

Fig. 4. Average network properties of largest connected components for different confidence thresholds (x-axes). Values for FEL and DBpedia Spotlight are in blue circles and orange diamonds, respectively. The top row shows the average local clustering (left) and average global clustering (right); the bottom row shows the unweighted assortivity (left) and weighted assortivity (right). (Color figure online)

5.4   Prominent Nodes

Tables 5 and 6 show the top 10 nodes by degree centrality in the FEL and DBpedia Spotlight networks, respectively. Top ranked nodes are generally more robust to changes in confidence thresholds than the other network properties, but there are some changes worth noting.

In all of the DBpedia Spotlight graphs, COVID-19 and related terms play a key role, regardless of confidence threshold. In the FEL graphs, however, making the threshold too strict (>-2.5) means that COVID-19 is no longer identified. We also see that the DBpedia Spotlight knowledge base has multiple entries relating to different aspects of COVID-19, but this is not the case with FEL. Besides for COVID-19, the top nodes in the two algorithms can be quite different, with the only other real overlap being 'China' and 'India'. Overall, the terms found by DBpedia Spotlight appear more reasonable.

We also note that applying stricter confidence thresholds can result in certain 'noisy' terms becoming more prominent. For example, 'Test_cricket' in DBpedia

Table 5. Top 10 degree nodes of connected co-occurrence networks created using FEL algorithm

−3	−2.75	−2.5	−2.25	−2
COVID-19	COVID-19	COVID-19	YouTube	YouTube
Twitter	Quarantine	India	India	India
Quarantine	India	YouTube	BTS	BTS
India	Midfielder	Twitter	China	Midfielder
Midfielder	YouTube	BTS	Midfielder	Twitter
Vaccination	Twitter	China	Twitter	Spotify
YouTube	Non-fungible_token	Midfielder	Non-fungible_token	Non-fungible_token
Non-fungible_token	China	Non-fungible_token	Spotify	Netflix
China	BTS	Spotify	Reblogging	Pfizer
BTS	Bachelor_of_Science	Bachelor_of_Science	Pfizer	MTV

Table 6. Top 10 degree nodes of connected co-occurrence networks created using DBpedia Spotlight Algorithm

0.5	0.6	0.7	0.8	0.9
Twitter	Twitter	COVID-19_vaccine	COVID-19_vaccine	COVID-19_vaccine
COVID-19_pandemic	COVID-19_pandemic	COVID-19_pandemic	COVID-19_pandemic	Test_cricket
COVID-19_vaccine	COVID-19_vaccine	COVID-19	Test_cricket	COVID-19
COVID-19	Coronavirus	Virus	COVID-19	Pfizer
Coronavirus	COVID-19	Test_cricket	Vaccine	COVID-19_pandemic
COVID-19_lockdowns	Virus	Vaccine	Pfizer	Vaccine
Virus	President_of_the_United_States	India	India	BTS
United_Kingdom	Test_cricket	Pfizer	China	Twitter
Quarantine	Vaccine	China	Twitter	YouTube
India	India	Radiotelephone	YouTube	India

Spotlight or 'Spotify' in FEL. This perhaps highlights potential issues with the algorithms, as it goes against the assumption that stricter confidence thresholds reduce noise.

Finally, Table 7 shows the top 10 weighted edges in the FEL -3 confidence threshold graph and the DBpedia Spotlight 0.5 confidence threshold graph. We see that the top edges with the two algorithms are completely different. The top edges in the DBpedia Spotlight graph are all related to the top nodes. This is not the case in the FEL graph, where many of the top edges connect to nodes not in the top 10. This once again highlights the differences in behaviour of the two algorithms.

5.5 Discussion and Possible Extensions

The results of these experiments highlight the sensitivity of network metrics to the choice of entity linker. Even when using confidence scores that identify a similar number of unique entities, DBpedia Spotlight finds these entities more regularly. The consequences of this are seen especially in the levels of network connectivity, with higher average degrees and largest components.

Table 7. Top 10 weighted edges in FEL graph with -3 confidence threshold and DBpedia Spotlight graph with 0.5 confidence threshold

FEL (-3 conf.)	DBpedia Spotlight (0.5 conf.)
(MTV, BTS)	(COVID-19_pandemic, Twitter)
(Pfizer, Moderna)	(Twitter, COVID-19_vaccine)
(Ministry_of_Health_and_Family_Welfare, India)	(Twitter, Coronavirus)
(Bachelor_of_Science, D%C3%A9FI)	(Twitter, COVID-19_lockdowns)
(Bachelor_of_Science, Whitelist)	(Twitter, COVID-19)
(Reblogging, Twitter)	(Quarantine, Twitter)
(D%C3%A9FI, Launchpad_%28website%29)	(Twitter, Test_cricket)
(Bachelor_of_Science, Launchpad_%28website%29)	(United_Kingdom, Twitter)
(D%C3%A9FI, Whitelist)	(Twitter, Virus)
(Umar, Riaz_%28actor%29)	(Twitter, India)

There are two clear avenues for further experiments. First, the already-created networks could be analysed in more sophisticated ways. There are a number of other metrics of interest, such as clustering and community analysis. These are common within the domain. However, they tend to be computationally expensive, and would need to be performed for all created networks. Another option would be running and analysing models on these networks. There are models for information spread or navigability, or sophisticated models for identifying important nodes in the network. These are also fairly expensive to run.

A second extension possibility is in the application of additional entity linking algorithms or network construction rules. Algorithms such as REL [34], GENRE [8], or Falcon2 [28] offer comparable performance to DBpedia Spotlight on tweets [23]. However, published versions of these link to an old KB, and thus miss crucial entities for this dataset such as COVID-19. Applying them would require refitting to a more recent KB. It would also be possible to add extra rules for including or excluding entities. For example, one could exclude entities that occur infrequently, allowing for less strict confidence thresholds. This would correspond to something like a global versus local significance criterion.

Finally, results of the analysis could be used to identify biases in the algorithms. Certain algorithms may perform better at finding long-tailed entities, while others might be better at identifying frequent entities. Comparing the networks constructed with different algorithms could be useful for seeing how such biases play out.

6   Conclusions

Analysing large scale text presents many challenges. The creation and interpretation of co-occurrence networks can offer useful insight into entities and their relationships in textual data. Extraction of these entities, and thus the construction of the networks, is not an easy task. This is especially the case for loosely structured text, like in tweets. In the literature, usually only a single entity linking algorithm and parameter set are used. Hence, we highlighted some of the changes that can be seen in the co-occurrence networks when changing algorithms and their settings.

We found that all regularly reported network metrics can be affected. These effects can differ between the algorithms tested, sometimes showing non-linear behaviour with respect to parameter settings. Things like identified entities, relative importance of entities, and level of connectivity can all be affected. Given the sensitivities observed, it would be prudent to report results for this kind of network analysis using multiple entity linking approaches. This offers a wider range of possibilities in interpretation, and can thus increase confidence in the robustness of results.

Acknowledgments. This publication has benefited from an invited research stay at GESIS - Leibniz Institute for the Social Sciences.

Disclosure of Interest. The authors have no competing interests to declare that are relevant to the content of this article.

References

1. Al-Moslmi, T., Ocaña, M.G., Opdahl, A.L., Veres, C.: Named entity extraction for knowledge graphs: a literature overview. IEEE Access **8**, 32862–32881 (2020)
2. Alam, M., Bie, Q., Türker, R., Sack, H.: Entity-based short text classification using convolutional neural networks. In: Keet, C.M., Dumontier, M. (eds.) EKAW 2020. LNCS (LNAI), vol. 12387, pp. 136–146. Springer, Cham (2020). https://doi.org/10.1007/978-3-030-61244-3_9
3. Amancio, D.R.: Network analysis of named entity co-occurrences in written texts. Europhys. Lett. **114**(5), 58005 (2016)
4. Blanco, R., Ottaviano, G., Meij, E.: Fast and space-efficient entity linking for queries. In: Proceedings of the Eighth ACM International Conference on Web Search and Data Mining, pp. 179–188 (2015)
5. Bono, C.A., Cappiello, C., Pernici, B., Ramalli, E., Vitali, M.: Pipeline design for data preparation for social media analysis. ACM J. Data Inf. Qual. **15**(4), 1–25 (2023)
6. Botzer, N., Weninger, T.: Entity graphs for exploring online discourse. Knowl. Inf. Syst. **65**(9), 3591–3609 (2023)
7. Cohen, R., Havlin, S.: Complex Networks: Structure, Robustness and Function. Cambridge University Press (2010)
8. De Cao, N., Izacard, G., Riedel, S., Petroni, F.: Autoregressive entity retrieval. arXiv preprint arXiv:2010.00904 (2020)
9. Derczynski, L., et al.: Analysis of named entity recognition and linking for tweets. Inf. Process. Manage. **51**(2), 32–49 (2015)
10. Dimitrov, D., et al.: TweetsCOV19-a knowledge base of semantically annotated tweets about the COVID-19 pandemic. In: Proceedings of the 29th ACM International Conference on Information & Knowledge Management, pp. 2991–2998 (2020)

11. Fafalios, P., Iosifidis, V., Ntoutsi, E., Dietze, S.: TweetsKB: a public and large-scale RDF corpus of annotated tweets. In: Gangemi, A., et al. (eds.) ESWC 2018. LNCS, vol. 10843, pp. 177–190. Springer, Cham (2018). https://doi.org/10.1007/978-3-319-93417-4_12

12. Färber, M., Rettinger, A., El Asmar, B.: On emerging entity detection. In: Blomqvist, E., Ciancarini, P., Poggi, F., Vitali, F. (eds.) Knowledge Engineering and Knowledge Management: 20th International Conference, EKAW 2016, Bologna, Italy, November 19-23, 2016, Proceedings, pp. 223–238. Springer International Publishing, Cham (2016). https://doi.org/10.1007/978-3-319-49004-5_15

13. Fegley, B.D., Torvik, V.I.: Has large-scale named-entity network analysis been resting on a flawed assumption? PLoS ONE **8**(7), e70299 (2013)

14. Harandizadeh, B., Singh, S.: Tweeki: linking named entities on twitter to a knowledge graph. In: Proceedings of the Sixth Workshop on Noisy User-generated Text (W-NUT 2020), pp. 222–231 (2020)

15. Hebert, L., Makki, R., Mishra, S., Saghir, H., Kamath, A., Merhav, Y.: Robust candidate generation for entity linking on short social media texts. arXiv preprint arXiv:2210.07472 (2022)

16. Heist, N., Paulheim, H.: Nastylinker: nil-aware scalable transformer-based entity linker. In: The Semantic Web: 20th International Conference, ESWC 2023, Hersonissos, Crete, Greece, May 28-June 1, 2023, Proceedings, pp. 174–191. Springer-Verlag, Berlin, Heidelberg (2023). https://doi.org/10.1007/978-3-031-33455-9_11

17. Twitter 1 stream. Accessed 20 Aug 2023

18. Ilievski, F., Vossen, P., Schlobach, S.: Systematic study of long tail phenomena in entity linking. In: Proceedings of the 27th International Conference on Computational Linguistics, pp. 664–674 (2018)

19. Ju, Y., Adams, B., Janowicz, K., Hu, Y., Yan, B., McKenzie, G.: Things and strings: improving place name disambiguation from short texts by combining entity co-occurrence with topic modeling. In: Blomqvist, E., Ciancarini, P., Poggi, F., Vitali, F. (eds.) EKAW 2016. LNCS (LNAI), vol. 10024, pp. 353–367. Springer, Cham (2016). https://doi.org/10.1007/978-3-319-49004-5_23

20. Liu, X., et al.: Entity linking for tweets. In: Proceedings of the 51st Annual Meeting of the Association for Computational Linguistics (Volume 1: Long Papers), pp. 1304–1311 (2013)

21. Manaskasemsak, B., Netsiwawichian, N., Rungsawang, A.: Entity co-occurrence graph-based clustering for twitter event detection. In: Barolli, L. (ed.) Advanced Information Networking and Applications: Proceedings of the 38th International Conference on Advanced Information Networking and Applications (AINA-2024), Volume 2, pp. 344–355. Springer Nature Switzerland, Cham (2024). https://doi.org/10.1007/978-3-031-57853-3_29

22. Mendes, P.N., Jakob, M., García-Silva, A., Bizer, C.: Dbpedia spotlight: shedding light on the web of documents. In: Proceedings of the 7th International Conference on Semantic Systems, pp. 1–8 (2011)

23. Mishra, S., Saini, A., Makki, R., Mehta, S., Haghighi, A., Mollahosseini, A.: TweetNERD-end to end entity linking benchmark for tweets. Adv. Neural. Inf. Process. Syst. **35**, 1419–1433 (2022)

24. Noullet, K., Ourgani, A., Färber, M.: A full-fledged framework for combining entity linking systems and components. In: Proceedings of the 12th Knowledge Capture Conference 2023, pp. 148–156 (2023)

25. Pastrav, C., Dignum, F.: Norms in social simulation: balancing between realism and scalability. In: Verhagen, H., Borit, M., Bravo, G., Wijermans, N. (eds.) Advances

in Social Simulation: Looking in the Mirror, pp. 329–342. Springer International Publishing, Cham (2020). https://doi.org/10.1007/978-3-030-34127-5_32

26. Peixoto, T.P.: The graph-tool python library. figshare (2014). https://doi.org/10.6084/m9.figshare.1164194, http://figshare.com/articles/graph_tool/1164194

27. Ristoski, P., Lin, Z., Zhou, Q.: KG-ZESHEL: knowledge graph-enhanced zero-shot entity linking. In: Proceedings of the 11th Knowledge Capture Conference, pp. 49–56 (2021)

28. Sakor, A., Singh, K., Patel, A., Vidal, M.E.: FALCON 2.0: an entity and relation linking tool over Wikidata. In: Proceedings of the 29th ACM International Conference on Information & Knowledge Management, pp. 3141–3148 (2020)

29. Salavati, C., Abdollahpouri, A., Manbari, Z.: Ranking nodes in complex networks based on local structure and improving closeness centrality. Neurocomputing **336**, 36–45 (2019)

30. Sciarra, C., Chiarotti, G., Laio, F., Ridolfi, L.: A change of perspective in network centrality. Sci. Rep. **8**(1), 15269 (2018)

31. Sevgili, Ö., Shelmanov, A., Arkhipov, M., Panchenko, A., Biemann, C.: Neural entity linking: a survey of models based on deep learning. Semant. Web **13**(3), 527–570 (2022)

32. Van Erp, M., Groth, P.: Towards entity spaces. In: Proceedings of the Twelfth Language Resources and Evaluation Conference, pp. 2129–2137 (2020)

33. Van Erp, M., et al.: Evaluating entity linking: an analysis of current benchmark datasets and a roadmap for doing a better job. In: Proceedings of the Tenth International Conference on Language Resources and Evaluation (LREC'16), pp. 4373–4379 (2016)

34. Van Hulst, J.M., Hasibi, F., Dercksen, K., Balog, K., de Vries, A.P.: REL: an entity linker standing on the shoulders of giants. In: Proceedings of the 43rd International ACM SIGIR Conference on Research and Development in Information Retrieval, pp. 2197–2200 (2020)

35. Zadgaonkar, A., Agrawal, A.J.: An approach for analyzing unstructured text data using topic modeling techniques for efficient information extraction. New Gener. Comput., 1–26 (2023)

Empowering CamemBERT Legal Entity Extraction With LLM Boostrapping

Julien Breton[1,2]([⊠]) [iD], Mokhtar Boumedyen Billami[2] [iD], Max Chevalier[1] [iD], and Cassia Trojahn[1] [iD]

[1] Informatics Research Institute of Toulouse (IRIT), Toulouse, France
{julien.breton,max.chevalier,cassia.trojahn}@irit.fr
[2] Berger-Levrault, Toulouse, France
{julien.breton,mb.billami}@berger-levrault.com

Abstract. The legal industry is characterized by the presence of large volumes and complex documents. Given the continuous evolution of these documents, there is a growing interest in automating the processing of legal texts to streamline compliance. One key step of this process is the extraction of legal entities. State-of-the-art methods for legal entity extraction, including rule-based systems, Bi-LSTM, and BERT, require substantial annotated data to be effective, a task that is time-intensive for domain experts. With the rise of Large Language Models (LLMs), research has increasingly focused on leveraging their capabilities and exploring zero-shot approaches. In this paper, we present a hybrid system that distils GPT-4 knowledge through rule-based methods into a CamemBERT model. This approach not only reduces the need for expert involvement compared to the standard CamemBERT system but also outperforms the GPT-4-only system, enhancing the F1 score for legal entities by 9–24% points.

Keywords: Legal Entity Extraction · Limited Annotated Data · CamemBERT · Large Language Models (LLMs) · Knowledge Distilation

1 Introduction

The legal industry is characterized by an extensive volume of evolving documents, such as contracts, legislation, court rulings, and regulatory filings. These documents are dense, complex, and rich in specialized language, making their analysis and processing both time-consuming and prone to human error. Automatic processing of such documents is essential for several applications, including legal research, compliance monitoring and contract analysis. It not only accelerates their analysis but also improves compliance, accuracy, and accessibility of legal information. Moreover, as laws and regulations frequently change, automated systems ensure that analyses are up-to-date with the latest legal standards. Companies are required to comply with the law and, if they fail to do so,

M. Alam et al. (Eds.): EKAW 2024, LNAI 15370, pp. 86–101, 2025.
https://doi.org/10.1007/978-3-031-77792-9_6

they will face severe penalties. As highlighted by Sassier et al. [19], in France, there are "more than 10,500 laws, 120,000 decrees, 7,400 treaties, 17,000 community's texts, tens of thousands of pages in 62 different codes. Some are constantly being modified: 6 modifications per working day for the 2006 Tax Code".

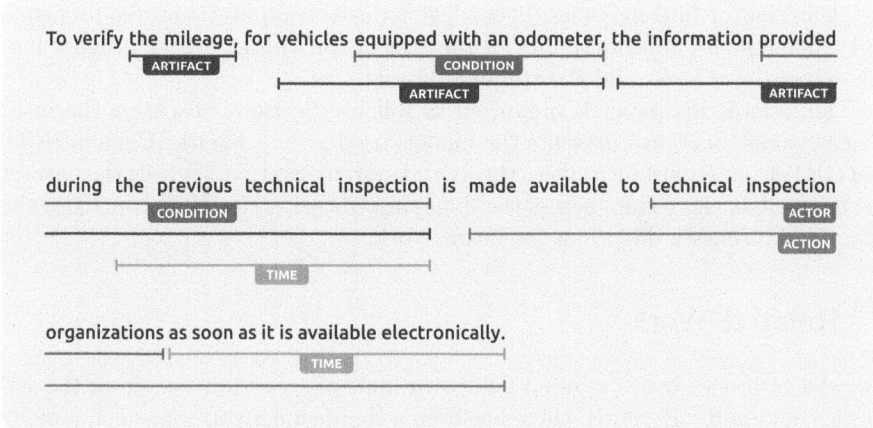

Fig. 1. Legal entity extraction from the following sentence: "To verify the mileage, for vehicles equipped with an odometer, the information provided during the previous technical inspection is made available to technical inspection organizations as soon as it is available electronically".

The backbone task for the automation of legal rules relies on formalizing them in a structured manner. Therefore, two primary tasks can be identified: the extraction of legal entities and the extraction of relations between these entities. This paper focus on the extraction of legal entities, as illustrated in Fig. 1, which is a sample of the dataset used in this paper. The original sentence example is in French and has been translated into English for the purpose of this paper.

The dataset utilized in our paper was introduced by Sleimi et al. [22], who based it on Luxembourg law documents. In their study, they developed a rule-based system to extract legal entities. The authors achieved a significant precision score of 0.874 and a recall score of 0.855. However, reaching this performance level requires a significant time investment from experts for annotation and rule-based pattern creation. Alternative methods, like LSTM (Long short-term memory) [20] or BERT (Bidirectional Encoder Representations from Transformers) [13], faced with the similar limitations in terms of annotated data. However, with the rise of large language models (LLMs) [30], new strategies have been emerged allowing for minimizing the involvement of experts in the extraction process. By leveraging foundational knowledge, LLMs are able, for example, to deal with the task of extracting named entities in biological documents [25].

This paper proposes a hybrid approach combining LLMs, rule-based methods, and BERT-based models. The objective is to train the CamemBERT model using a dataset generated by LLMs refined with rule-based methods. Distilling the knowledge from an LLM into a CamemBERT model with rule-based filtering serves several key purposes. First, this approach reduces the need for expert involvement, limiting their intervention to LLM input instructions and the definition of filtering rules. Second, it reduces computational requirements and enhances the interpretability of the results, providing clearer insights into the extraction process and the training dataset.

The rest of this paper is organized as follows. Section 2 discusses the main related works. Section 3 presents the models used as benchmark (CamemBERT and GPT-4). Section 4 introduces the hybrid system. Section 5 details the dataset used as well as the results across the different strategies. Section 6 concludes the paper and discusses directions for future work.

2   Related Work

Knowledge processing is a broad field with multiple contributions from the scientific community. Recently, there has been a significant trend toward employing neural networks for analysis [24] or classification tasks [5,18]. For example, David B. et al. [2] developed a system to anonymize named entities in German financial documents. The system identifies entities such as first and last names, postal and email addresses, and locations. The study evaluated various architectures, including RNN, LSTM, and Conditional Random Fields (CRF) [11]. The evaluation showed that RNN+CRF architectures achieved the highest performance, with over 97% recall without post-processing and around 99% recall after post-processing, while maintaining a precision of over 90%. Despite the effectiveness of the proposed approach, there are drawbacks concerning the involvement of experts and the time-consuming nature of the tasks. The method relies heavily on the manual annotation of documents by domain experts. Annotating a large corpus of 407 published German financial documents, which comprises a total of 189k tokens, is both labor-intensive and time-consuming.

BERT, introduced by Vaswani et al. [26], tends to mitigate this problem by conducting initial training on a large corpus of data, comprising nearly 3.3 million words sourced from the Toronto BookCorpus and English Wikipedia. This pre-training provides foundational knowledge that can be fine-tuned with smaller datasets. The legal community has adopted this architecture for Named Entity Recognition (NER) [14], as evidenced in Italian judgments [17] and Brazilian legal texts by Wang et al. [27]. These implementations achieved these results by combining BERT, Bi-LSTM, and CRF. However, all these methods still necessitate a substantial amount of data to train these architectures and time investment from experts, a challenge we address in this paper.

With the recent rise of LLMs, numerous studies have explored their capabilities in information extraction, yielding these results [1,6,8,28]. Additionally, there has been significant progress in distilling this performance into smaller models [10]. Yuxian et al. [9] concluded that distillation is an effective method for transferring knowledge from larger generative language models to smaller ones. This process results in more precise responses, reduced exposure bias, improved calibration, and enhanced long-text generation performance.

On the legal side, as described in the survey by Solihin et al. [23], the legal community has primarily focused on NER, resulting in limited research and datasets dedicated to the extraction of legal entities. While these two tasks might appear similar at first glance, they differ significantly. Named entities, such as instances of persons or organizations, are typically represented as single entities, like *"Stephen Hawking"*, and are usually short, comprising a few words. In contrast, legal entities can include broader concepts, named entities, or even complete phrases. For example, one of the legal concepts is Actor and can include an entity like *"the driver"*, which is not a named entity. Another legal concept is Condition, which may contain content like *"if the coupled set of vehicles consists of two self-propelled vehicles"*. These examples illustrate that our task extends beyond simple NER; it can vary from entity linking to span extraction, depending on the concept. Therefore, there is reasonable doubt that approaches effective for named entities [15,16,29,31] will be equally effective for legal entity extraction.

For legal entity extraction, studies such as those by Sleimi et al. [21], have developed their own datasets and employed a rule-based approach for this task. Recent research by Castano et al. [4] explores a similar process, focusing on the extraction of both concepts and entities from European legal documents. These extracted elements are then integrated and maintained within a knowledge management system. Other works, such as that by Dragoni et al. [7], demonstrate the feasibility and effectiveness of combining multiple Natural Language Processing (NLP) approaches for extracting legal rules from documents. Their findings suggest that this combined approach can significantly enhance the accuracy and efficiency of legal rule extraction.

Building on these conclusions, our paper aims to compare our hybrid system with benchmark systems, such as CamemBERT models and GPT-4, which are detailed in the following section.

3   Benchmark Systems

To compare our approach with state-of-the-art systems, we introduce Camem-BERT and GPT-4 models for the legal entity extraction task. The following sections provide a description of their implementation, highlighting the advantages and disadvantages of these architectures.

3.1   CamemBERT-Based Extraction

Due to the French nature of the dataset (detailed in Sect. 5.1), we employed a CamemBERT model, instead of the traditional BERT. Legal-CamemBERT-base [12] has been fine-tuned on over 22,000 legal articles from Belgian legislation in French, making it better suited for processing and understanding French legal texts. This choice was driven by accurately capturing the nuances and intricacies of the French language present in our dataset, thereby enhancing overall performance and reliability.

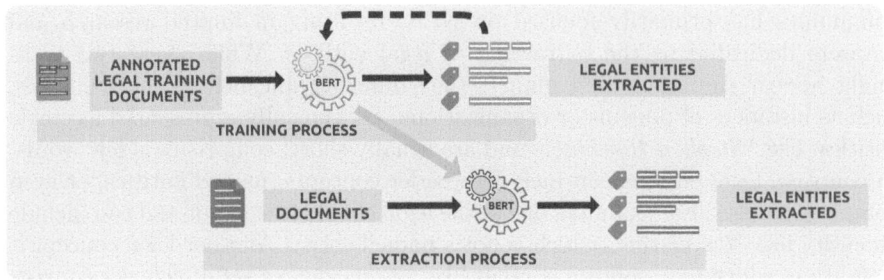

Fig. 2. Overall CamemBERT process by (i) fine-tuning the model on the training dataset and (ii) using this fine-tuned model to perform the evaluation.

Legal extraction based on CamemBERT necessitates data corpus produced by experts. The dataset should be divided into two parts: one for training and the other one for the evaluation, as illustrated in Fig. 2. Traditional entity extraction with CamemBERT relies on Inside-Outside-Beginning tagging, a common method in natural language processing for labelling sequences, particularly in tasks such as NER. However, in the legal entity extraction task, annotation overlaps exist in the document, illustrated in the example Fig. 1. Therefore, it necessitates to modify the internal architecture of the CamemBERT model.

	for	vehicles	equipped	with	an	odometer	...
CONDITION	0	0	1	1	1	1	
ARTIFACT	0	1	1	1	1	1	
...							

Fig. 3. Tokenizer matrix based on the short sentence: "for vehicles equipped with an odometer". Words activate legal concepts like Condition or Artifact with binary values (0 or 1).

The first modification pertains to the input matrix, used inside the PyTorch model, to allow overlapping classification. Figure 3 illustrate this modification, by allowing multiple tokens in a sentence to be activated simultaneously, such as the word "odometer". The second modification involves the custom metric used to reward the model during the training process. We compute a macro F1 score, which is the average F1 score across all concepts. The final modification involves creating a custom Model Class using the Transformer library from Huggingface. This modification allows us to override the loss function, switching from CrossEntropyLoss[1] to BCEWithLogitsLoss[2], which is a binary cross-entropy combined with a sigmoid function suitable for multi-label use cases. This custom architecture, along with the other experiments in this paper, is available in our GitLab repository[3].

As discussed, utilizing CamemBERT models necessitates a sufficient amount of training data, 200 statements in this dataset [22], produced by experts through time-consuming processes. However, the recent advent of powerful generative AI models, such as ChatGPT, has introduced new research avenues in zero-shot and few-shot learning. These models can perform legal entity extraction without the need for extensive data and expert intervention. This approach is examined in detail in the following section.

3.2 LLM Based Extraction

Figure 4 outlines the overall process of using GPT-4 to extract knowledge from legal texts. By combining legal statements with suitable prompts, we perform legal entity extraction. The advantage of this approach lies in the foundational training the model has received, enabling it to achieve high performance with minimal input information.

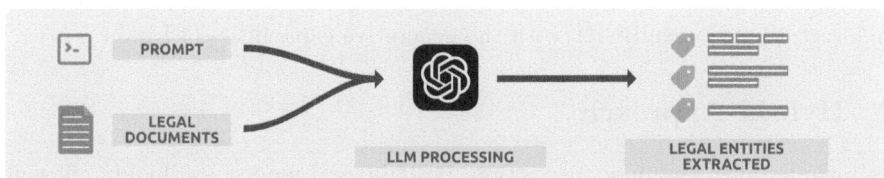

Fig. 4. Overall GPT-4 process, taking in input prompt and legal documents, to extract and produce structured legal entities as output.

The adoption of the prompt structure described below emerged from an empirical process of iterative experimentation, aimed at pinpointing the configuration that consistently delivers better outputs. This approach was informed by an adherence to best practices and guidelines recommended by OpenAI[4]. Through successive trials, adjustments were made to refine the prompt's components, including the precise definition of the model's role, the task description, and the format of the expected output.

---

[1] https://pytorch.org/docs/stable/generated/torch.nn.CrossEntropyLoss.html.
[2] https://pytorch.org/docs/stable/generated/torch.nn.BCEWithLogitsLoss.html.
[3] https://gitlab.irit.fr/ala/legal-concepts-extraction.
[4] https://platform.openai.com/docs/guides/prompt-engineering.

The prompt, available in our GitLab repository[5], is structured as follows: initially, a role is assigned to the LLM: "NLP expert", along with the task: "extracting entities from sentences". Indeed, the guidance provided in the OpenAI documentation suggests that clearly defining the role and capabilities of a model, can significantly enhance the quality and relevance of the outputs generated. The second part entails a description of the legal concepts and their definitions. These definitions represent the only necessary introduction of external knowledge, necessitating the involvement of an expert. Additionally, a statement example along with the desired output in JSON format is included at the end of the prompt, offering a direct illustration of the task to be performed by the model, increasing the performance. We finally incorporate the statement that we want to extract legal entities. This final prompt is then provided to the large language model for processing.

While LLMs can successfully extract domain-specific information, leveraging their zero-shot capability involves certain trade-offs. One significant drawback is their inconsistent accuracy in extracting legal entities. LLMs can often produce annotations that either include extraneous words or omit essential components of the legal entities. We named this problem the *boundary issue*. These inaccuracies can hinder the reliability of the extracted information. This limitation will be discussed in the results Sect. 5.3, where we will introduce the specific challenges and performance issues observed in our experiments.

Having explored the application of prompt engineering for information extraction using LLMs, we now turn our attention to our hybrid approach that leverages both CamemBERT models and LLMs for bootstrapping training data. The next section delves into the potential of combining the robust, pre-trained architecture of CamemBERT with the generative capabilities of LLMs.

4 Hybrid Approach

As introduced in the previous section, BERT-based models are highly efficient but require a substantial quantity of data. In contrast, LLMs leverage their foundational training to operate independently of specific training data, but introduce the boundary issue. To address the limitations of these existing strategies, we developed a hybrid architecture, as illustrated in Fig. 5, which integrates LLMs, the CamemBERT model, and rule-based systems.

The initial step of our pipeline involves conducting the first entity extraction from our legal documents using GPT-4. This procedure mirrors the one outlined in Sect. 3.2. Initially, experts craft a prompt that includes concepts, their definitions, and an example of the expected outcome from a sentence. Subsequently, the LLM performs the extraction using both the provided prompt and the legal documents. The output consists of the legal concepts extracted by the LLM.

Following the extraction with GPT-4, we refined the extracted concepts using a rule-based strategy. We aim to enhance the precision of the LLM by incorporating 15 syntactic rules developed by experts [21]. For instance, in the sentences

---

[5] https://gitlab.irit.fr/ala/legal-entity-extraction/-/raw/main/modules/llm/utils.py.

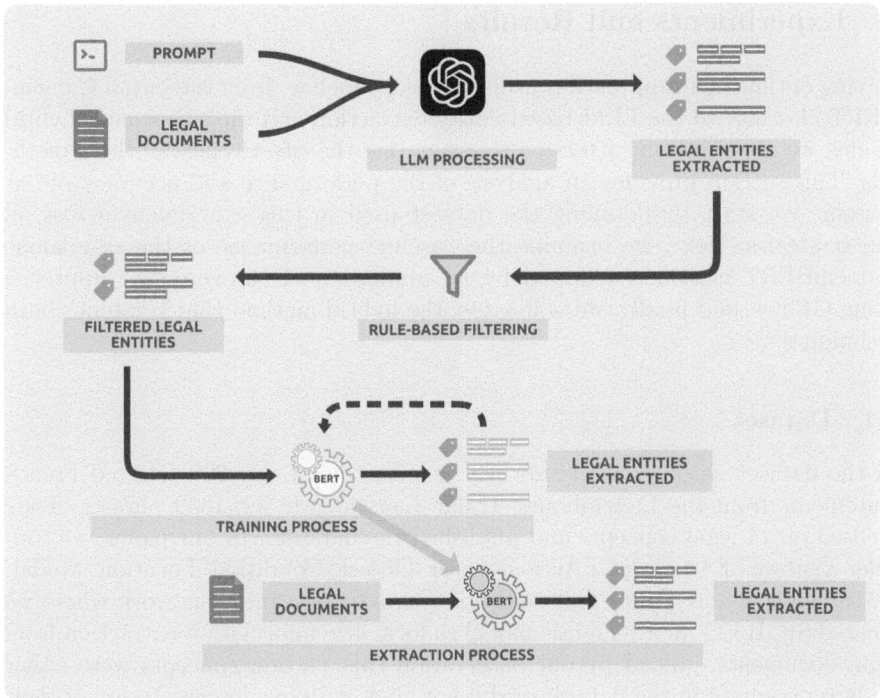

Fig. 5. Overall hybrid process including three main steps: (i) bootstrapping legal entity extraction with GPT-4, (ii) filtering the extraction using syntactic rules crafted by experts, and (iii) fine-tuning the Legal-CamemBERT model to perform the distillation process.

illustrated in Fig. 1, all artifact concepts can be identified as Noun Phrases (NP), while actions can be identified as Verb Phrases (VP). By applying these syntactic rules, we can tackle the boundary issue and ensure that the extracted entities are both precise and complete.

After refining the legal entities with the rule-based approach, we use this legal entities to train the CamemBERT classifier. We follow the methodology detailed in Sect. 3.1. The training process involved the same modification of adding a linear classifier at the output of the CamemBERT model and adjusting the loss function to suit the specific task. Upon completing this training phase, the model is ready to perform extractions on incoming documents.

By distilling the knowledge from an LLM into a CamemBERT model, we achieve a balance between leveraging advanced AI capabilities and ensuring practical, interpretable, and scalable solutions for industrial applications. This hybrid approach allows us to harness the strengths of both models, resulting in an efficient and effective system for legal entity extraction.

5   Experiments and Results

Having outlined the approaches utilized in our pipeline, from the initial Camem-BERT classifier to the LLM-based entity extraction and the subsequent hybrid model, we now turn our attention to evaluating the effectiveness of these methods. This section provides an analysis of the performance and accuracy of our system. We start by detailing the dataset used in this experiment across all the strategies. Next, we examine the baseline performance of the standalone CamemBERT classifier, followed by an evaluation of the zero-shot approach using GPT-4, and finally, we will assess the hybrid method that combines both techniques.

5.1   Dataset

In the dataset used in this study [22], experts annotated 200 selected French statements from the Luxembourg Traffic Law, identifying 1339 phrases. They focused on 14 legal concepts and published the dataset[6]. In our paper, we consider a subset of 8 concepts: Action, Actor, Object, Condition, Location, Modality, Reference, and Time. This selection was part of a previous work where we created SEMLEG [3] a semantic model to formalize information extraction from legal documents, guided by our observation that certain concepts were either underrepresented in the dataset or did not align with our interpretation of their definitions. To facilitate further study with broader applicability across various fields, we have chosen to focus on these concepts, which are described in Table 1.

Table 1. Concept definitions from Sleimi et al. [22] for the eight concepts used in our study.

Concept	Definition
Action	the process of doing something
Actor	an entity that has the capability to act
Artifact	a human-made object involved in an action
Condition	a constraint stating the properties that must be met
Location	a place where an action is performed
Modality	a verb indicating the modality of the action (e.g may, must, shall)
Reference	a mention of other legal provision(s) or legal text(s) affecting the current provision
Time	the moment or duration associated with the occurrence of an action

Figure 6 details the distribution of the concepts. An important point to note is the poor data balancing for the Artifact concept. While this imbalance may be acceptable for a rule-based approach by the authors, it will have significant implications for our CamemBERT model, detailed in the next section.

---

[6] https://sites.google.com/view/metax-re2018/.

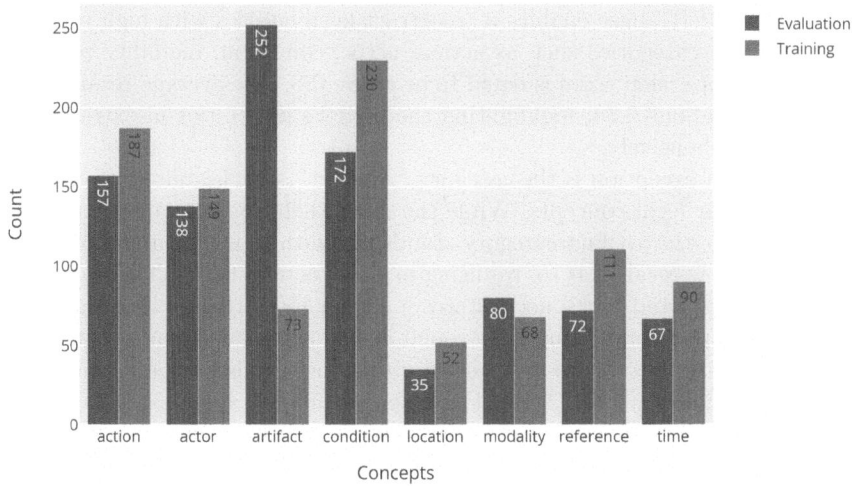

Fig. 6. Concepts distribution from the training and evaluation datasets [22]. Our paper employs eight legal concepts: Action, Actor, Artifact, Condition, Location, Modality, Reference, and Time.

5.2   Fine-Tuned CamemBERT Results

The results of our fine-tuned CamemBERT-based entity extraction model are presented in Table 2, which includes precision, recall, and F1 scores for the following legal entities: Action, Actor, Artifact, Condition, Location, Modality, Reference, and Time.

Table 2. Precision, Recall and F1 results using standalone Legal-CamemBERT-base [12] for legal entity extraction.

	Concepts							
	Action	Actor	Artifact	Condition	Location	Modality	Reference	Time
Precision	0.93	0.84	0.69	0.80	0.65	0.91	0.88	0.85
Recall	0.41	0.60	0.16	0.83	0.53	0.48	0.67	0.66
F1	0.57	0.70	0.26	0.82	0.59	0.63	0.76	0.74

The CamemBERT-based entity extraction performs with an average F1 score exceeding 0.63. The average precision is 0.81, indicating that the fine-tuning of

Legal-CamemBERT-base enables it to extract legal entities with high precision. Specifically, for categories such as action, actor, condition, modality reference, and time, the precision score is equal to or above 0.8. The average recall, on the other hand, is about 0.54, highlighting the model's difficulties in extracting all entities comprehensively.

However, an exception is the category "Artifact", which underperforms relative to the other legal concepts. While the model exhibits high precision across most categories, the artifact category stands out with a precision of 0.69 but an exceptionally low recall of 0.16, resulting in a F1 score of 0.26. This discrepancy between precision and recall necessitates a deeper investigation into the underlying causes. Upon conducting a thorough investigation, we identified that this discrepancy is attributable to imbalances within the training dataset from Sleimi et al. [22]. Specifically, the "Artifact" concept exhibits a significant disparity in the distribution of instances between the training and evaluation datasets. While most legal concepts maintain an approximate 50/50 split between training and evaluation, the "Artifact" concept is markedly skewed, with only 74 instances in the training set compared to 252 instances in the evaluation set. This imbalance impairs the model's ability to generalize effectively for the "Artifact" category, resulting in a poor recall score. To solve this issue, we should consider readjusting the balance of training and evaluation of this dataset.

Summing up, the fine-tuning of the Legal-CamemBERT-base model has demonstrated a good performance, achieving an average F1 score of 0.69 after removing the artifact concept. The artifact result clearly illustrates the need for substantial quantities of annotated data to enhance the recall. The creation of such datasets is inherently time-consuming and demands significant expertise.

5.3   GPT-4 Results

In this section, we detailed the results of prompt engineering for information extraction using GPT-4 across the 8 legal concepts. The detailed results are detailed in Table 3.

Table 3. Precision, Recall and F1 results using GPT-4 for legal entity extraction.

	Concepts							
	Action	Actor	Artifact	Condition	Location	Modality	Reference	Time
Precision	0.65	0.67	0.58	0.76	0.53	0.54	0.71	0.81
Recall	0.27	0.58	0.33	0.53	0.36	0.54	0.53	0.34
F1	0.38	0.63	0.42	0.63	0.42	0.54	0.61	0.48

The precision scores indicate the precision of the entity extraction performed by the model. The highest precision is observed for the Time concept (0.81), demonstrating the model's strong ability to correctly identify temporal entities within the text. Other categories, such as Condition (0.76) and Reference (0.71), also exhibit high precision, reflecting the model's effectiveness in these areas. Recall scores highlight the model's capability to identify all relevant instances of each entity. The Actor concept shows a recall of 0.58, signifying the model's competence in recognizing actors within the text. Conversely, the Action concept has the lower recall (0.27), suggesting difficulties in consistently identifying all actions in legal documents.

The model demonstrates precision, indicating its ability to accurately identify relevant entities. However, the recall scores are more variable, suggesting that while the model is precise, it may miss a significant number of relevant entities, particularly in categories like Action and Artifact. As introduced in Sect. 3.2, the LLM approach suffers from boundary issues, characterized by an inability to correctly delimit the boundaries of the extraction, often selecting only a subset of the expected annotation.

However, the boundary issue in legal entity extraction is significantly mitigated by the advantages of employing zero-shot learning approaches. This technique reduces the need for extensive, time-consuming data annotation processes, allowing models to leverage their pre-trained capabilities to perform tasks without specific task-oriented training. Consequently, while there are trade-offs in terms of boundary accuracy, the overall efficiency, scalability, and reduced need for expert involvement in data preparation present a compelling advantage, making zero-shot learning a highly practical solution for legal entity extraction. The next section attempts to improve the results by employing our hybrid approach.

5.4   Hybrid System Results

The results of our hybrid approach, which integrates large language models GPT-4, CamemBERT, and rule-based systems, demonstrate significant improvements in legal entity extraction across various concepts. Table 4 provides detailed results of precision, recall, and F1 for each concept.

Table 4. Precision, Recall and F1 results using our hybrid framework for legal entity extraction.

	Concepts							
	Action	Actor	Artifact	Condition	Location	Modality	Reference	Time
Precision	0.36	0.80	0.58	0.68	0.35	0.73	0.66	0.64
Recall	0.78	0.52	0.54	0.73	0.47	0.57	0.75	0.81
F1	0.50	0.63	0.56	0.70	0.41	0.64	0.70	0.72

As illustrated in Table 5, the hybrid architecture demonstrates substantial improvements over previous strategy by leveraging the strengths of GPT-4 for initial extraction, rule-based systems for refinement, and CamemBERT for efficient and scalable entity extraction. It also outperforms standalone GPT-4 model on almost every legal concept, increasing the F1 score of the legal concepts by

Table 5. Overall performance on F1 score with the three approaches: CamemBERT as baseline, GPT-4, and Hybrid as low involvement strategies. Bold values indicate the best between zero-shot strategies.

Category	Baseline CamemBERT	Zero-shot Benchmark GPT-4	Hybrid
Action	0.57	0.38	**0.50 (+12%)**
Actor	0.70	0.63 (0%)	0.63 (0%)
Artifact	0.26	0.42	**0.56 (+14%)**
Condition	0.82	0.63	**0.70 (+7%)**
Location	0.59	**0.42 (+1%)**	0.41
Modality	0.63	0.54	**0.64 (+10%)**
Reference	0.76	0.61	**0.70 (+9%)**
Time	0.74	0.48	**0.72 (+24%)**

9–24% points. The high precision scores across most categories underscore the model's precision, while the balanced F1 scores for several concepts indicate effective handling of both precision and recall. The enhanced performance can be attributed to the initial extraction capabilities of the LLM, combined with expert-crafted syntactic rules that enhance the precision of the extracted entities. This hybrid approach effectively addresses the trade-offs observed in LLM strategy and reduces the boundary issue.

It is important to note that the hybrid approach did not outperform the standalone CamemBERT model trained on a dedicated dataset produced by experts. While the hybrid architecture offers significant improvements, the precision and recall achieved by a dedicated CamemBERT model are superior due to the highly specialized nature of the training data. A standalone CamemBERT model benefits from being trained on meticulously curated datasets, which encapsulate domain-specific nuances and complexities that are often lost in generalized datasets produced by LLMs. Therefore, while the hybrid approach presents a practical and efficient solution with commendable performance metrics, it does not fully replicate the specialized accuracy and depth of a CamemBERT model trained on an expert-produced dataset. This behaviour has also been observed in a similar work [25] from the scientific community.

However, the detailed results demonstrate the potential of this methodology to enhance the efficiency and effectiveness of processing legal documents, while minimizing the need for expert intervention. This advancement paves the way for more sophisticated applications in the legal domain. By training the CamemBERT model on a refined dataset generated by GPT-4, we achieve a scalable solution that maintains high precision without the computational burden typically associated with continuous large language model usage.

6   Conclusion

This paper has presented an approach to enhance the extraction of legal entities from complex legal texts, addressing the challenges posed by the volume and complexity of legal documents while reducing expert involvement. We have evaluated three distinct strategies for legal entity extraction: fine-tuning the Legal-CamemBERT-base model, leveraging LLMs through prompt engineering, and developing a hybrid approach that combines LLMs, rule-based methods, and the CamemBERT model. All these strategies have been evaluated on the same dataset from the Luxembourg law. We demonstrated the effectiveness of the Legal-CamemBERT-base model by fine-tuning it on the Luxembourg law dataset. We explored the potential of GPT-4 to reduce expert involvement in the extraction process, emphasizing the role of prompt engineering in generating precise outputs. And finally, the hybrid approach aimed to distil the knowledge of a large language model into the CamemBERT model through rule-based filtering, achieving performance close to that of the fine-tuned Legal-CamemBERT-base model. This hybrid strategy notably outperformed the standalone GPT-4 model in almost every legal concept and significantly reduced the need for expert intervention to only: the creation of concepts, their definitions, and the formulation of syntactic rules to refine the LLM results.

In terms of generalizability, the approach offers significant advantages over traditional state-of-the-art methods, particularly by leveraging GPT-4 to bootstrap data. As long as GPT-4 is capable of understanding and managing a given domain, this approach should effectively reduce the need for expert involvement. Moreover, the SEMLEG model is domain-independent and, while it can be customized with domain-specific addons, it remains a highly suitable option for structuring legal rules across different domains.

Regarding multilingual limitations, although GPT-4 demonstrates strong capabilities in handling multilingual documents, subsequent components of the pipeline present challenges. First, the rule-based filtering relies on syntactic parsing, which can vary significantly between languages. As a result, syntactic rules developed for French may not be applicable to English, necessitating expert involvement to regenerate language-specific filtering rules. Additionally, while CamemBERT is well-suited for processing French texts, it would need to be replaced with an alternative, such as the original BERT model, when applied to English texts or other languages.

Future work will focus on extending these methods to the extraction of relationships between legal entities. Exploring further optimizations in hybrid approaches is also considered by improving the rule-based filtering. Instead of relying solely on experts to generate rules, we can consider methods to assist or even automate the generation of rules through unsupervised learning techniques, thereby reducing the need for manual intervention and enhancing the scalability of the extraction process.

Acknowledgments. This work was granted access to the HPC resources of IDRIS under the allocation 2024-AD011014922 made by GENCI.

References

1. Bellan, P., Dragoni, M., Ghidini, C.: Extracting business process entities and relations from text using pre-trained language models and in-context learning. In: Almeida, J.P.A., Karastoyanova, D., Guizzardi, G., Montali, M., Maggi, F.M., Fonseca, C.M. (eds.) Enterprise Design, Operations, and Computing: 26th International Conference, EDOC 2022, Bozen-Bolzano, Italy, October 3–7, 2022, Proceedings, pp. 182–199. Springer International Publishing, Cham (2022). https://doi.org/10.1007/978-3-031-17604-3_11

2. Biesner, D., et al.: Anonymization of German financial documents using neural network-based language models with contextual word representations. Int. J. Data Sci. Anal., 151–161 (2022)

3. Breton, J., Billami, M.B., Chevalier, M., Cassia, T.: Leveraging semantic model and LLM for bootstrapping a legal entity extraction: an industrial use case. In: 20th International Conference on Semantic Systems (SEMANTICS 2024) (2024)

4. Castano, S., Ferrara, A., Furiosi, E., Montanelli, S., Picascia, S., Riva, D., Stefanetti, C.: Enforcing legal information extraction through context-aware techniques: the ASKE approach. Comput. Law & Secur. Rev. **52**, 105903 (2024)

5. Chen, Y., Xiao, B., Lin, Z., Dai, C., Li, Z., Yan, L.: Multi-label text classification with deep neural networks. In: 2018 International Conference on Network Infrastructure and Digital Content (IC-NIDC), pp. 409–413. IEEE (2018)

6. Dagdelen, J., et al.: Structured information extraction from scientific text with large language models. Nat. Commun. **15**(1), 1418 (2024)

7. Dragoni, M., Villata, S., Rizzi, W., Governatori, G.: Combining NLP approaches for rule extraction from legal documents. In: 1st Workshop on MIning and REasoning with Legal texts (MIREL 2016)

8. Dunn, A., et al.: Structured information extraction from complex scientific text with fine-tuned large language models. ArXiv (2022)

9. Gu, Y., Dong, L., Wei, F., Huang, M.: MiniLLM: knowledge distillation of large language models (2024)

10. Hsieh, C.Y., et al.: Distilling step-by-step! outperforming larger language models with less training data and smaller model sizes. ArXiv (2023)

11. Lafferty, J., McCallum, A., Pereira, F., et al.: Conditional random fields: probabilistic models for segmenting and labeling sequence data. In: ICML, vol. 1, p. 3. Williamstown, MA (2001)

12. Louis, A., van Dijck, G., Spanakis, G.: Finding the law: enhancing statutory article retrieval via graph neural networks. In: Proceedings of the 17th Conference of the European Chapter of the Association for Computational Linguistics, pp. 2753–2768. Association for Computational Linguistics, Dubrovnik, Croatia (2023)

13. Medsker, L.R., Jain, L., et al.: Recurrent neural networks. Design Appl. **5**(64–67), 2 (2001)

14. Mohit, B.: Named entity recognition. In: Zitouni, I. (ed.) Natural Language Processing of Semitic Languages, pp. 221–245. Springer, Berlin, Heidelberg (2014). https://doi.org/10.1007/978-3-642-45358-8_7

15. Oliveira, V., Nogueira, G., Faleiros, T., Marcacini, R.: Combining prompt-based language models and weak supervision for labeling named entity recognition on legal documents. Artif. Intell. Law (2024). https://doi.org/10.1007/s10506-023-09388-1

16. Pakhale, K.: Comprehensive overview of named entity recognition: models, domain-specific applications and challenges. arXiv preprint arXiv:2309.14084 (2023)

17. Pozzi, R., Rubini, R., Bernasconi, C., Palmonari, M.: Named entity recognition and linking for entity extraction from Italian civil judgements. In: Basili, R., Lembo, D., Limongelli, C., Orlandini, A. (eds.) AIxIA 2023 – Advances in Artificial Intelligence: XXIInd International Conference of the Italian Association for Artificial Intelligence, AIxIA 2023, Rome, Italy, November 6–9, 2023, Proceedings, pp. 187–201. Springer Nature Switzerland, Cham (2023). https://doi.org/10.1007/978-3-031-47546-7_13

18. Prasanna, P.L., Rao, D.R.: Text classification using artificial neural networks. Int. J. Eng. Technol. **7**(1.1), 603–606 (2018)

19. Sassier, P., Lansoy, D.: Ubu Loi. Arthème Fayard, France (2008)

20. Sherstinsky, A.: Fundamentals of recurrent neural network (RNN) and long short-term memory (LSTM) network. ArXiv (2018)

21. Sleimi, A., Sannier, N., Sabetzadeh, M., Briand, L., Ceci, M., Dann, J.: An automated framework for the extraction of semantic legal metadata from legal texts. Empir. Softw. Eng. **26**, 1–50 (2021)

22. Sleimi, A., Sannier, N., Sabetzadeh, M., Briand, L., Dann, J.: Automated extraction of semantic legal metadata using natural language processing. In: 2018 IEEE 26th International Requirements Engineering Conference (RE), pp. 124–135 (2018)

23. Solihin, F., Budi, I., Aji, R.F., Makarim, E.: Advancement of information extraction use in legal documents. Int. Rev. Law, Comput. Technol. **35**(3), 322–351 (2021)

24. Suissa, O., Elmalech, A., Zhitomirsky-Geffet, M.: Text analysis using deep neural networks in digital humanities and information science. J. Am. Soc. Inf. Sci. **73**(2), 268–287 (2022)

25. Tang, R., Han, X., Jiang, X., Hu, X.: Does synthetic data generation of LLMs help clinical text mining? Arxiv (2023). arXiv preprint arXiv:2303.04360 (2023)

26. Vaswani, A., et al.: Attention is all you need. In: Advances in Neural Information Processing Systems, vol. **30** (2017)

27. Wang, Z., Wu, Y., Lei, P., Peng, C.: Named entity recognition method of Brazilian legal text based on pre-training model. J. Phys: Conf. Ser. **1550**, 032149 (2020)

28. Wei, X., et al.: Zero-shot information extraction via chatting with ChatGPT. arXiv preprint arXiv:2302.10205 (2023)

29. Zaratiana, U., Tomeh, N., Holat, P., Charnois, T.: GLiNER: generalist model for named entity recognition using bidirectional transformer (2023)

30. Zhao, W.X., et al.: A survey of large language models. arXiv preprint arXiv:2303.18223 (2023)

31. Zhou, W., Zhang, S., Gu, Y., Chen, M., Poon, H.: UniversalNER: targeted distillation from large language models for open named entity recognition (2023)

Lexicalization Is All You Need: Examining the Impact of Lexical Knowledge in a Compositional QALD System

David Maria Schmidt$^{(\boxtimes)}$, Mohammad Fazleh Elahi , and Philipp Cimiano

Semantic Computing Group, CITEC, Technical Faculty, Bielefeld University,
Bielefeld, Germany
{daschmidt,melahi,cimiano}@techfak.uni-bielefeld.de

Abstract. In this paper, we examine the impact of lexicalization on
Question Answering over Linked Data (QALD). It is well known that
one of the key challenges in interpreting natural language questions with
respect to SPARQL lies in bridging the lexical gap, that is mapping
the words in the query to the correct vocabulary elements. We argue
in this paper that lexicalization, that is explicit knowledge about the
potential interpretations of a word with respect to the given vocabulary,
significantly eases the task and increases the performance of QA systems. Towards this goal, we present a compositional QA system that can
leverage explicit lexical knowledge in a compositional manner to infer the
meaning of a question in terms of a SPARQL query. We show that such
a system, given lexical knowledge, has a performance well beyond current QA systems, achieving up to a 35.8% increase in the micro F_1 score
compared to the best QA system on QALD-9. This shows the importance and potential of including explicit lexical knowledge. In contrast,
we show that LLMs have limited abilities to exploit lexical knowledge,
with only marginal improvements compared to a version without lexical knowledge. This shows that LLMs have no ability to compositionally
interpret a question on the basis of the meaning of its parts, a key feature of compositional approaches. Taken together, our work shows new
avenues for QALD research, emphasizing the importance of lexicalization
and compositionality.

Keywords: Semantic Composition · Question Answering over Linked
Data · Large Language Models · Lexical Knowledge

1 Introduction

Question Answering over Linked Data (QALD) [63] is the task of automatically
mapping a natural language question to an executable SPARQL query such that
relevant information can be retrieved from RDF data sources. One of the seven
challenges [69] identified by the authors for the development of QALD systems
is handling the lexical gap [69], which requires bridging the way users refer to

© The Author(s), under exclusive license to Springer Nature Switzerland AG 2025
M. Alam et al. (Eds.): EKAW 2024, LNAI 15370, pp. 102–122, 2025.
https://doi.org/10.1007/978-3-031-77792-9_7

certain natural language terms and the way they are modeled in a given knowledge base. Consider the question *"Who is the mayor of Moscow?"*. In this case, *"mayor"* needs to be interpreted with respect to DBpedia as `dbo:leaderName`[1] to map the question correctly to the following SPARQL query: `SELECT ?o WHERE { dbr:Moscow dbo:leaderName ?o }`

Another important aspect of QALD is the principle of compositionality. That is, the meaning of a complex expression is determined by the meanings of its parts and the way they are syntactically combined. In the context of QALD, a complex question is represented by a SPARQL query that involves more than one triple pattern, excluding the predicates *rdf:type* or *rdfs:label*. For example, the SPARQL query of the complex question *"Who is the mayor of the capital of Russia?"* is as follows: `SELECT ?uri WHERE { dbr:Russia dbo:capital ?o . ?o dbo:leaderName ?uri }`. To handle complex questions, the QALD system requires using compositional reasoning to obtain the answer, which includes multi-hop reasoning, set operations, and other forms of complex reasoning.

Recent approaches based on machine learning models (e.g., deep neural networks [34,45,51,66], Seq2Seq neural networks [52], transformers [43,44,81], subgraph embeddings [6], probabilistic graphical models [30], bi-directional LSTMs [31], and tree-LSTMs [2]) have achieved promising results, and are currently mostly limited to answering simple questions (i.e., only one triple excluding the predicates *rdf:type* and *rdfs:label*). To deal with complex queries, Hakimov et al. [32] have proposed an approach that uses *Combinatory Categorial Grammar (CCG)* [65] for syntactic representations and typed lambda calculus expressions [8] for semantic representations. Some approaches [68,70] strongly resemble ours, as the motivation is very similar: using explicit lexical information and *Dependency-based Underspecified Discourse Representation Structures (DUDES)* [10,14] for semantic composition. However, these approaches generate all possible combinations of SPARQL queries for a natural language sentence, providing no mechanism for disambiguation; therefore, they produce many logically incorrect SPARQL queries.

Some QALD approaches [7,19,32,60,79] have made only limited use of lexicalization, while others [21,22,68] have used lexical knowledge but have not systematically investigated its impact. Recently, LLM-based approaches [3,4, 28,29,41,53,55] have proven to be powerful tools for NLP tasks. In particular, ChatGPT [24,67,80] has been shown to be an alternative to traditional QALD approaches. To our knowledge, *Generative Pretrained Transformer (GPT)* models have not been tested for their ability to compositionally interpret a question based on the meaning of its parts or the impact of lexical knowledge on their performance. In this paper, we thus address three research questions and provide the corresponding contributions listed below:

---

[1] We use namespace prefixes that are defined as follows: `dbr`: http://dbpedia.org/resource/, `dbo`: http://dbpedia.org/ontology/, `dbp`: http://dbpedia.org/property/, `rdfs`: http://www.w3.org/2000/01/rdf-schema#, `rdf`: http://www.w3.org/1999/02/22-rdf-syntax-ns#.

RQ1 How can a QA system leverage explicit lexical knowledge? Towards this goal, we present a new compositional QA system that relies on a dependency parse and bottom-up semantic composition.

RQ2 What is the impact of explicitly given lexical knowledge? Our experimental results show that our compositional system reaches (micro) F_1 measures of 0.72 on the QALD-9 dataset, which outperforms existing state of the art systems on the task by far (+ 35%).

RQ3 Can Large Language Models also leverage explicit lexical knowledge? Our experiments show that, when encoding lexical knowledge explicitly in the prompt, state-of-the-art LLMs can benefit from such knowledge, improving results. However, they are far from reaching improvements that match the performance of our compositional approach.

2   System Architecture

In this section, we detail a compositional approach to QALD, using *Dependency-based Underspecified Discourse Representation Structures (DUDES)* [10,14] for meaning representation and composition behavior, as well as leveraging explicit lexical knowledge, thus answering RQ1. The overall architecture of the pipeline is illustrated in Fig. 1 and serves as a blueprint for this section. Although we used DBpedia as a reference, our approach can be adapted to any particular ontology and vocabulary by providing a corresponding lexicon.

2.1   Explicit Lexical Knowledge

A necessary prerequisite for our approach is the availability of a Lemon lexicon [46] that describes by which lexical entries the elements (classes, properties) of a particular knowledge base (KB) can be verbalized in a particular language. We rely on the Lemon lexicon format that contains lexical entries and defines how their meaning is captured with respect to a given ontological vocabulary. A *lexical entry* represents a unit of analysis of the lexicon that consists of a set of grammatically related forms and a set of base meanings that are associated with all of these forms.[2] The lexicon is context-free in the sense that the possible meanings of words are described independently from their context.

2.2   Dependency Parsing

Our approach relies on a syntactic analysis of an input question by a dependency parser. To increase the chance that at least one correct dependency tree is generated, which is vital for our approach, we use multiple dependency parsing frameworks (i.e., SpaCy[3] and Stanza/CoreNLP framework [56]), configurations, and models[4]. Furthermore, some questions contain textual representations of

---

[2] Exemplary lexical entries can be found at https://lemon-model.net/.

[3] https://spacy.io/.

[4] `en_core_web_trf` and `en_core_web_lg`.

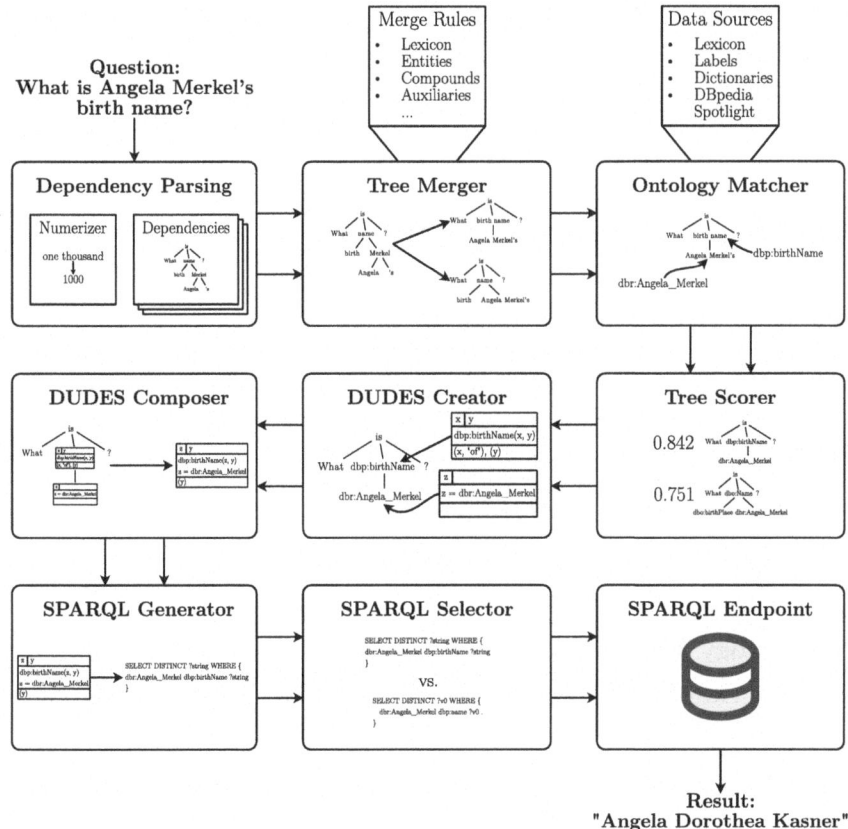

Fig. 1. Schema of the compositional question answering approach using DUDES

numbers that need to be translated into their numerical form to be used with, for example, `FILTER` expressions in SPARQL queries. However, for entities[5] that contain textual numbers (e.g., `dbr:One_Thousand_and_One_Nights`), this approach might be counterproductive with respect to the entity recognition process. Therefore, we consider both the converted and the original question in our approach if there is something to be converted. To do so, we use the `numerizer`[6] library.

2.3 Tree Merger

Matching nodes in the dependency tree to URIs representing entities and properties (a.k.a. *KB Linking*) is a central challenge in our approach. For this purpose,

---

[5] A wide range of subjects, including people, places, organizations, and various concepts, each identified by a unique URI (Uniform Resource Identifier).

[6] https://github.com/jaidevd/numerizer.

we introduce a number of merging rules over the dependency tree to yield phrases that facilitate matching to KB elements:

- *Generic Rules:* Several generic merge rules based on syntactic properties such as dependency tags[7] or part-of-speech (POS) tags[8] are applied, merging nodes based on, e.g., tags like `compound` or `det`, or comparative keywords like *"more"* or *"fewer"*.
- *Lexicon Marker Rules:* If a lexicon entry matches and includes a marker, the node of that marker (typically an `ADP` node) is merged into the node that bears the written representation of the corresponding lexical entry. For example, if the lexical entry contains a marker *"of"*, it is merged with the written representation *"birth name"*.
- *Entity Merging Rule:* The presented approach uses several methods for entity recognition (as discussed in Sect. 2.4), and these methods return several candidate entities for a given question. By this rule, the candidate entities, often found at different nodes of the dependency tree, are merged into one node, forming a merged candidate entity. As shown in the tree merger step in Fig. 1, the child node *"Angela"* is merged into its parent node *"Merkel"*, resulting in *"Angela Merkel"*.

2.4 Ontology Matcher

In this step, entities and properties are assigned to their respective tree nodes where possible. The matching methods are described below.

Property Matching: This matching method focuses on matching the nodes of the tree with DBpedia properties. First, each node of the tree is matched with the written representations of the lexical entries if possible. If there is no exact match for a node, the approach tries to find candidate lexical entries by applying several heuristics, such as omitting certain tokens from the node, e.g. by excluding trailing adpositions, which typically do not occur in the canonical forms of lexical entries. If the marker of a lexical entry matches with a token from a node, the corresponding candidate entry is prioritized. Finally, if there are remaining ambiguities, the candidate lexical entries are sorted in descending order using a Levenshtein distance-based similarity measure [42].

Entity Matching: To match tree nodes with entities from DBpedia, the approach uses all available `rdfs:label` information of entities, which are stored in a prefix trie [40] for efficient memory representation and lookup of similar labels. The similarities between the entity (e.g., *"Angela Merkel"*) in the tree and the entity labels in DBpedia are calculated using the Levenshtein similarity measure with a threshold set to 0.5. When both a shorter and a longer text span perfectly match certain labels, the longer match is generally prioritized higher. As an off-the-shelf solution, we also include the entity recognition results of DBpedia Spotlight [47] into the set of considered candidates to increase the chance of a correct match.

---

[7] https://universaldependencies.org/u/dep/.

[8] https://universaldependencies.org/u/pos/.

2.5   Tree Scorer

Each node of the tree is assigned a score based on matched properties or entities. Additionally, it considers the total number of tree nodes, as well as special terms and keywords which are neither properties nor entities. Two types of matching are taken into account: (i) exact matches, and (ii) matches under relaxed conditions. The scoring relies on a weighted average of three different scores: (i) fraction of nodes with exact matches (weight 3), (ii) fraction of nodes with matches under relaxed conditions (weight 1), and (iii) ratio of the number of nodes to the number of nodes in the dependency tree before merging nodes (weight 2). For (i) and (ii), single node weights (i.e., the number of tokens a node comprises) are multiplied with different multipliers based on whether the node has a matching lexical entry (multiplier 1.0), a matching entity or is a numeral (0.9) or is a special word like an ASK keyword, a comparative or *"in"* (0.8). Then, the weight is multiplied with that multiplier and added to a total sum. In the end, this sum is compared to the sum of all weights, forming the score value. The weighted average of these three scores forms the total score of a tree, according to which the trees are then prioritized in processing.

2.6   DUDES Creator

Now that we have a tree with KB elements (e.g., entities and properties) assigned to the nodes, the next task is to create *Dependency-based Underspecified Discourse Representation Structures (DUDES)* [10,14], which are used to compose the atomic meanings of the tree nodes. Our approach is slightly different from the latest version of DUDES [14] as we modify it for use with dependency trees instead of *Lexicalized Tree Adjoining Grammar (LTAG)* trees [36,62] and do not make use of subordination relations yet:

Definition 1. (Dependency-based Underspecified Discourse Representation Structures (DUDES) [14]) A DUDES is a triple (v, D, S) where:

- $v \in U \cup \{\epsilon\}$ is the main variable (also called referent marker or distinguished variable) where ϵ represents the absence of a main variable
- $D = (U, C)$ is a Discourse Representation Structure (DRS) [14,33,37] with
  - set of variables U (also called discourse universe or referent markers)
  - set of conditions C over variables U
- S is a set of selection pairs of the form $(v \in U, m)$ with v being a variable from U and m being a marker word for that variable with ϵ representing the empty marker, i.e. no marker being connected to that variable. Instead of writing ϵ, the second tuple component can also just be left out.

Entity DUDES: The simplest case of representing KB elements from the tree as a DUDES is representing entities (i.e., *Entity DUDES*). In Entity DUDES, an entity is assigned to a variable, for example, by adding a simple expression such as $z = dbr{:}Angela\_Merkel$. A full example for entity dbr:Angela_Merkel is illustrated in Fig. 2a.

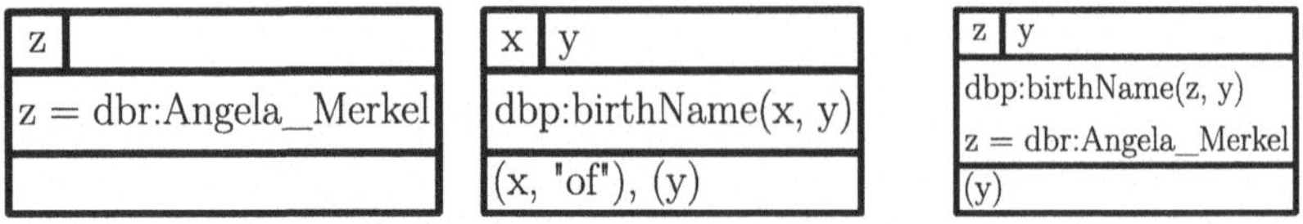

z	
z = dbr:Angela_Merkel	

x	y
dbp:birthName(x, y)	
(x, 'of'), (y)	

z	y
dbp:birthName(z, y)	
z = dbr:Angela_Merkel	
(y)	

(a) Entity DUDES **dbr:Angela_Merkel** with main variable z, condition z = *dbr:Angela_Merkel* and no selection pairs

(b) Property DUDES **dbo:birthName** with main variable x, condition *dbo:birthName(x, y)* and selection pairs $(x, "of")$ and (y, ϵ)

(c) Composition of Fig. 2a and 2b using selection pair $(x, "of")$. Results in condition pairs *dbo:birthName(z, y)* $\wedge\ z =$ *dbr:Angela_Merkel*

Fig. 2. Illustration of exemplary DUDES and their composition

Property DUDES: In contrast to entities, properties have variables that are intended to be replaced by entities or combined with other properties during DUDES composition. Additionally, variables can be restricted to certain markers that correspond to the subject or object position of the property. An example of the property **dbo:birthName** is shown in Fig. 2b. Note the variable x is associated with the marker *"of"* as another way of disambiguation which is however not used in this example. For, e.g., *"What is the birth name of Angela Merkel?"*, it would instead be used to determine the subject of the property.

2.7 DUDES Composer

The composition operation of DUDES [14] can be defined as follows:

Definition 2. (DUDES Composition) Let $d_1 = (v_1, D_1 = (U_1, C_1), S_1)$, $d_2 = (v_2, D_2 = (U_2, C_2), S_2)$ be two DUDES with disjoint variable sets, i.e. $U_1 \cap U_2 = \emptyset$. The DUDES composition operation $\odot$ for substituting d_1 into d_2 using selection pair $p = (x \in U_2, m) \in S_2$ and resulting in a composed DUDES $d_c = (v_c, D_c = (U_c, C_c), S_c)$, written $d_c = d_1 \overset{p}{\odot} d_2$, is defined as follows:

$$U_c = U_2[x{:=}v_1] \cup U_1 \qquad S_c = (S_2 \cup S_1) \setminus p \qquad v_c = \begin{cases} v_1 \text{ if } x = v_2 \\ v_2 \text{ else} \end{cases}$$
$$C_c = C_2[x{:=}v_1] \cup C_1$$

An example composition of the two DUDES (i.e., Fig. 2a and 2b) is shown in Fig. 2c. We apply a bottom-up composition strategy, merging child nodes into their parent nodes. DUDES compositions are performed until there is only a single composed DUDES (i.e., final DUDES) left at the root of the tree, representing the meaning of the whole question. For choosing a selection pair for composition, different heuristics and data sources are used. For example, in the case of modifier nodes, the parent DUDES is merged into the child DUDES and not the other way around. Additionally, the syntactic frames of the lexical entries, POS and dependency tags are used for selection pair determination. Instead of calculating all combinations at once, our approach composes one final DUDES at a time, limiting the effect of combinatorial explosion that would otherwise substantially increase the memory footprint.

2.8 SPARQL Generator

The logical expressions of the final DUDES represent the meaning of the given natural language question. Therefore, they are used to create a corresponding SPARQL query. For instance, the DUDES in Fig. 2c shows the triple pattern (i.e., dbr:Angela_Merkel dbo:birthName ?y) for the question *"What is Angela Merkel's birth name?"*. The triple patterns of DUDES contain entities (or literals), and variables. Our approach uses the Z3 SMT solver [49] to determine which variables in the final DUDES are bound to some values. Due to our combinatorial approach to dealing with ambiguities, multiple SPARQL queries are typically generated, from which one is selected by the SPARQL selector.

2.9 SPARQL Selector

For selecting the best SPARQL query, we use an LLM-based approach [58] trained to compare two queries, using the encoder of flan-t5-small [9] as a base model. As single LLM-based comparison results are still unreliable, various aggregation strategies are evaluated which make a final decision based on the pairwise comparisons of all candidate queries.

For each of the two candidate queries of a comparison, the model is given the input question, the candidate query, the number of its results, and the final DUDES. From this information, two output features are generated, representing the confidence in the respective queries. In order to reflect in the output how much better one query is than another, the model is trained to predict the F_1 scores of the respective queries.[9] We evaluate different strategies and configurations to select the final query:

- **BestScore:** Theoretically possible performance of our approach, selecting best queries based on their true F_1 score, clamped like in training data.
- **MostWins$_{p\%}^{top\ n}$:** Compares candidate queries pairwise per question and selects the query that "wins" the most comparisons (with margin of $> p\%$).
- **Accum$_{logits/sigmoid}^{top\ n}$:** For each candidate query, the model outputs are accumulated, either *logits* or *sigmoid* values, largest value is chosen.

If an exponent *top n* is given, the top n models (based on training micro F_1 score) are evaluated with their outputs summed together. Otherwise, single models are evaluated and presented with mean and standard deviation. Additionally, queries with no results or with too many results (threshold is the largest number of results for a train question $+ 10\%$) are discarded.

---

[9] To avoid high numbers of false positives affecting the micro F_1 score, queries with a false to true positive ratio of $10 : 1$ or worse are clamped to 0.0 in the training data.

3   Experimental Setup

The experiments were conducted using the well-known QALD-9 benchmark [73]. The dataset contains questions in multiple languages, along with corresponding SPARQL queries and answers from DBpedia. For the experiment, we followed Unger et al. [76] to manually create a lexicon covering the vocabulary elements in the training and test section of QALD-9. In particular, we created a total of 599 lexical entries for five syntactic frames [12]: 311 lexical entries for `NounPPFrame`, 96 lexical entries for `TransitiveFrame`, 143 lexical entries for `InTransitivePPFrame`, 28 lexical entries for `AdjectivePredicateFrame`, and 21 lexical entries for `AdjectiveSuperlativeFrame`. The time required for creating a lexical entry was approximately 2–5 min depending on the syntactic frame. The total time required to create our lexicon was approximately 16 h. All experiments[10] were conducted for the English part of QALD-9 only.

3.1   SPARQL Selection Model Training

In Sect. 2.8, we presented an LLM-based SPARQL selection approach for disambiguation of the generated SPARQL queries. In order to have training, validation and test data for the query selection model, we ran our approach for about 24 h on the QALD-9 benchmark and saved all generated candidate queries, separated by train and test questions. Afterwards, the training data was randomly split into 90% training questions and 10% validation questions.

The corresponding final data elements were generated in three steps for each part of the data (i.e., training, validation, and test). Each step involved generating 100 comparisons for each question and used for training in a symmetric way to avoid some general preference of the model for the first or second query. Comparisons between queries with (i) an F_1 score ≥ 0.01 and (ii) an F_1 score < 0.01 were added to the training data. Additionally, mixed comparisons with one query with an F_1 score ≥ 0.01 and one with an F_1 score < 0.01 were added.

In order to fine-tune the `google/flan-t5-small` model from the Huggingface transformers library [78], we first ran a hyperparameter optimization with 34 trials using Optuna [1], with an epoch search space between 1 and 5, an initial learning rate between $1e^{-5}$ and $1e^{-4}$ (logarithmic scale) and using a lambda learning rate scheduler with a lambda between 0.9 and 1.0 (logarithmic scale). As an optimizer, we used Adam [39] and trained 10 models using the parameters of the trial with the lowest validation loss discovered during the hyperparameter optimization. Each training was performed on a single Nvidia A40 with a batch size of 64. These 10 models were then used for evaluation.

---

[10] Software artifact: https://doi.org/10.5281/zenodo.12610054, System: AMD Ryzen 9 7900X3D, 96 GB RAM, NVIDIA GeForce RTX 4070, Arch Linux 6.9.2-arch1-1, Python 3.12.3, 12 parallel processes, total (elapsed real time) timeout of 10800 s for test benchmark, DBpedia version: 2016-10 https://downloads.dbpedia.org/2016-10/core-i18n/en/.

3.2   Experiments with GPT

We compared different GPT [54] models to our compositional approach. The previous research on QALD with GPT-3 [24] evaluated using the QALD-9 test dataset in three modes: zero-shot, few-shot, and fine-tuned model. In the zero-shot scenario, GPT-3 generated many invalid queries. Performance increased with the five-shot approach and even more with fine-tuning.

All of our experiments with GPT models are performed separately with 5 different prompts describing the task. The first prompt below has been hand-crafted. Afterwards, four additional prompts have been generated using Chat-GPT using the prompt *You are a world-class prompt engineer. Refine this prompt: <initial prompt>*. The resulting prompts used in our experiments are therefore:

1. You are a system which creates SPARQL queries for DBPEDIA from 2016-10 from natural language user questions. You answer just with SPARQL queries and nothing else.
2. Generate SPARQL queries from user questions for DBpedia from October 2016. Answer solely with SPARQL queries.
3. Develop a system capable of generating SPARQL queries for DBPedia based on user questions in natural language, with a knowledge base updated until October 2016. The system should exclusively respond with SPARQL queries and no additional information.
4. Craft SPARQL queries from October 2016 based on user questions in natural language, exclusively dedicated to extracting information from DBpedia. Your responses should consist solely of SPARQL queries.
5. Create SPARQL queries to generate responses to user questions by interpreting natural language queries, specifically targeting DBpedia, beginning from October 2016.

For our experiments we, prompted GPT-3.5-Turbo and GPT-4 in a zero-shot fashion and evaluated them on the entire training and test datasets of QALD-9. For the fine-tuned GPT-3.5-Turbo-0125 models, 10% of the original training dataset has been excluded and used as a validation dataset for fine-tuning. All experiments are executed with temperature 0 as well as both with and without lexical information in the prompt, i.e. lexical information was also present during fine-tuning. To the best of our knowledge, no work has investigated the impact of using lexical information (which is crucial for state-of-art performance for QALD) on the benchmark performance in this way yet. For the experiments with lexical information (as detailed in 2.1), we shorten the structure of lexical entries (the structure is detailed in previous work [5]) for prompting and training, as the lexical entries are not well-suited for direct usage.[11] This shortened representation consists primarily of pairs of field names and their values, e.g. *"Canonical form: birth name"* or *"Reference: dbp:birthName"*. To fit into the context window of the

---

[11] Datasets generated this way together with the shortened lexical entries can be found in our software artifact: https://doi.org/10.5281/zenodo.12610054.

used models, we restrict the entries appended to the prompt to entries which are relevant to the question.[12] Therefore, only the ability of GPT models to put the pieces together is tested, not whether they select the right entry from a much larger lexicon. The comparison is therefore not fair as our approach figures out the relevant lexical entries itself. The numbers presented in the evaluation section are therefore to be interpreted as "upper bounds" on the performance. The OpenAI API has been used for fine-tuning, as the model is not publicly available. The configurable hyperparameters *batch-size*, *learning rate multiplier* as well as *number of epochs* were optimized using the *auto* setting.

4   Evaluation

The evaluation of our approach is presented in three categories: *Single Model*, *Multi Model*, and *Upper Bounds*. The first category shows mean and standard deviation across the 10 trained models for SPARQL selection strategies using a single model. As these results show a high standard deviation for micro scores, we also evaluated the effect of bundling the outputs of multiple models. For bundling the model outputs (i.e., *top n* models), we focused on the strategies and models performing best on the training dataset of QALD-9. We also included *BestScore* strategies in the *Upper Bounds* category, demonstrating the highest achievable scores (i.e., upper bounds on the query selection model performances). Good scores in the *BestScore/Upper Bounds* category therefore indicate that the pipeline in principle generates the correct results, but those queries are not always identified during query selection. However, selecting the best query can be considered a much easier task than generating it from scratch, rendering these scores still reasonably realistic. Nevertheless, when comparing to other approaches, only the best performances achieved by a regular query selection strategy are used for fairness reasons.

During evaluation, all strategies with all 10 SPARQL selection models were tested simultaneously to ensure they were evaluated on the same generated queries. However, as we limited the elapsed real time to 3 h, this imposed a high overhead w.r.t. single strategies. Evaluating just one strategy and model at once would likely have achieved better results due to more tested candidates.[13] Illustrating the theoretical potential of the generated queries, a second evaluation with just *BestScore* being executed for 3 h was conducted (marked with *"(single)"* in Table 1), generating 815473 instead of 10552 queries and increasing scores from 0.37 to 0.51 (macro F_1).[14]

Table 1 shows that the multi-model strategies generally outperform single-model strategies, e.g., 0.43 vs. 0.72 vs. 0.85 for the micro F_1 scores of single-model, multi-model, and upper bound strategies, respectively. More precisely, single-model strategies on average achieve only about half of the upper bound

---

[12] This means where written representations occur in the question and ontology URIs in the gold standard query.

[13] Generated candidate queries per question: up to 5439, mean: 87.87 ± 495.02.

[14] Generated candidate queries per question: up to 59310, mean: 5660.98 ± 8268.85.

Table 1. Results for English QALD-9 test dataset after 3 h of elapsed real time. P refers to Precision, R to Recall, and F_1 to F_1 score. The best results of each category are marked in bold. *"(single)"* means running the benchmark without evaluating LLM strategy performance at the same time, reducing the corresponding overhead.

	Micro			Macro		
Strategy	$F_1 \pm \sigma$	$P \pm \sigma$	$R \pm \sigma$	$F_1 \pm \sigma$	$P \pm \sigma$	$R \pm \sigma$
Single Model						
Accum$_{logits}$	0.30 ± 0.10	0.24 ± 0.11	0.59 ± 0.25	0.31 ± 0.02	0.31 ± 0.02	0.34 ± 0.02
Accum$_{sigmoid}$	0.39 ± 0.13	0.31 ± 0.11	0.61 ± 0.20	$\mathbf{0.32 \pm 0.01}$	$\mathbf{0.32 \pm 0.01}$	$\mathbf{0.35 \pm 0.02}$
MostWins$_{0.0}$	0.29 ± 0.08	0.19 ± 0.06	$\mathbf{0.65 \pm 0.10}$	0.30 ± 0.02	0.30 ± 0.01	0.34 ± 0.01
MostWins$_{0.1}$	0.33 ± 0.08	0.23 ± 0.07	0.62 ± 0.14	0.31 ± 0.01	0.31 ± 0.01	0.34 ± 0.01
MostWins$_{0.25}$	0.40 ± 0.16	0.32 ± 0.22	0.65 ± 0.11	$\mathbf{0.32 \pm 0.01}$	$\mathbf{0.32 \pm 0.01}$	$\mathbf{0.35 \pm 0.02}$
MostWins$_{0.5}$	0.36 ± 0.13	0.28 ± 0.10	0.58 ± 0.21	0.31 ± 0.02	$\mathbf{0.32 \pm 0.02}$	0.34 ± 0.02
MostWins$_{0.75}$	$\mathbf{0.43 \pm 0.19}$	$\mathbf{0.38 \pm 0.23}$	0.56 ± 0.21	$\mathbf{0.32 \pm 0.02}$	$\mathbf{0.32 \pm 0.02}$	0.34 ± 0.02
MostWins$_{0.9}$	0.42 ± 0.18	0.36 ± 0.22	0.56 ± 0.21	$\mathbf{0.32 \pm 0.02}$	$\mathbf{0.32 \pm 0.02}$	0.34 ± 0.02
Multi Model						
MostWins$_{0.75}^{top\ 2}$	$\mathbf{0.72}$	$\mathbf{0.77}$	0.67	0.32	0.32	0.33
MostWins$_{0.9}^{top\ 2}$	0.64	0.61	0.67	0.32	0.32	0.34
MostWins$_{0.75}^{top\ 3}$	$\mathbf{0.72}$	$\mathbf{0.77}$	0.67	0.32	$\mathbf{0.33}$	0.33
MostWins$_{0.9}^{top\ 3}$	0.65	0.64	$\mathbf{0.68}$	0.32	$\mathbf{0.33}$	$\mathbf{0.35}$
MostWins$_{0.75}^{top\ 5}$	0.59	0.52	$\mathbf{0.68}$	$\mathbf{0.33}$	$\mathbf{0.33}$	$\mathbf{0.35}$
MostWins$_{0.9}^{top\ 5}$	0.59	0.53	$\mathbf{0.68}$	$\mathbf{0.33}$	$\mathbf{0.33}$	$\mathbf{0.35}$
MostWins$_{0.75}^{top\ 10}$	0.59	0.53	0.67	0.32	$\mathbf{0.33}$	0.34
MostWins$_{0.9}^{top\ 10}$	0.62	0.57	$\mathbf{0.68}$	$\mathbf{0.33}$	$\mathbf{0.33}$	0.34
Upper Bounds						
BestScore	0.81	$\mathbf{0.98}$	$\mathbf{0.69}$	0.37	0.38	0.38
BestScore (single)	$\mathbf{0.85}$	0.95	0.76	$\mathbf{0.51}$	$\mathbf{0.51}$	$\mathbf{0.54}$

performances, although they exhibit a high standard deviation, indicating their potential to yield substantially different results based on the specific trained model chosen for evaluation. However, aggregating the outputs of multiple models to select a query appears to combine the strengths of the bundled models without being affected by their weaknesses. This results in comparably stable performance across different numbers of bundled models (e.g., 0.59 to 0.72 for micro F_1 scores). In contrast, the macro F_1 scores are consistent across both single and multi-model strategies, indicating strong overall performance.

Comparison with GPT models: The results of our experiments (as shown in Table 2) with QALD-9 and GPT models [54] are examined with two different objectives: comparing the GPT performance with our approach, i.e. Table 1, (research question RQ2) and examining the effect of providing the lexical entries in the prompt (research question RQ3). Table 2 shows the F_1 scores of GPT-3.5-Turbo models with and without fine-tuning, as well as GPT-4 without fine-tuning, with and without a lexicon. Regarding the first objective (RQ2), our approach outperforms GPT models in terms of micro F_1 score, achieving 0.72 compared to 0.35 of the best-performing GPT model.

In contrast, the total best macro F_1 score of all evaluated GPT models outperforms the macro F_1 scores of our approach (0.33 vs. 0.42). However, this is only true for models that were provided with the correct lexical entries in the prompt, a substantial simplification compared to our approach which has

Table 2. GPT results for QALD-9 test dataset. P refers to Precision, R to Recall, F_1 to F_1 score, FT to fine-tuned, and Pr# to the prompt number. The best results of each experiment are marked in bold, total best scores of a category are underlined.

Model	FT	Pr#	Without Lexicon						With Lexicon					
			Micro			Macro			Micro			Macro		
			F_1	P	R	F_1	P	R	F_1	P	R	F_1	P	R
GPT-3.5-Turbo	✗	1	**0.15**	0.19	**0.12**	0.10	0.11	0.12	0.13	0.09	**0.21**	0.25	**0.27**	0.27
GPT-3.5-Turbo	✗	2	**0.15**	0.21	**0.12**	0.11	0.10	0.12	0.13	0.09	0.20	**0.26**	0.26	**0.31**
GPT-3.5-Turbo	✗	3	0.12	0.18	0.10	**0.13**	**0.13**	**0.15**	0.06	0.04	0.13	0.21	0.21	0.25
GPT-3.5-Turbo	✗	4	0.08	0.06	**0.12**	0.12	0.11	**0.15**	0.04	0.02	0.15	0.23	0.22	0.27
GPT-3.5-Turbo	✗	5	0.10	0.17	0.07	0.05	0.05	0.05	**0.16**	**0.45**	0.10	0.12	0.12	0.14
GPT-4	✗	1	0.22	0.29	0.18	0.26	0.27	0.28	<u>**0.35**</u>	0.81	0.22	**0.40**	**0.40**	**0.42**
GPT-4	✗	2	**0.34**	**0.68**	<u>**0.23**</u>	0.28	0.29	0.30	0.31	**0.87**	0.19	0.39	**0.40**	0.40
GPT-4	✗	3	0.22	0.25	0.20	0.26	0.27	0.28	0.12	0.09	**0.21**	0.39	**0.40**	**0.42**
GPT-4	✗	4	0.32	0.65	0.21	0.25	0.26	0.29	0.12	0.08	0.20	0.38	0.38	0.40
GPT-4	✗	5	0.19	0.17	0.21	0.20	0.19	0.25	0.24	0.32	0.20	0.26	0.26	0.29
GPT-3.5-Turbo	✓	1	0.28	0.81	0.17	0.24	0.24	0.25	0.22	0.28	0.18	<u>**0.42**</u>	<u>**0.43**</u>	0.42
GPT-3.5-Turbo	✓	2	**0.35**	0.88	**0.22**	0.24	0.25	0.25	0.11	0.08	0.16	0.40	0.42	0.42
GPT-3.5-Turbo	✓	3	0.26	0.50	0.18	**0.25**	**0.26**	**0.26**	0.09	0.06	0.15	0.37	0.37	0.40
GPT-3.5-Turbo	✓	4	**0.35**	0.91	**0.22**	0.23	0.24	0.24	0.10	0.07	**0.21**	0.41	0.41	<u>**0.44**</u>
GPT-3.5-Turbo	✓	5	0.34	<u>**0.92**</u>	0.21	0.24	0.25	0.25	**0.31**	**0.87**	0.19	0.38	0.38	0.40

to determine the relevant lexical entries from the whole lexicon. Without this advantage, all evaluated GPT models are outperformed by our approach, as the macro F_1 score does not exceed 0.28 then. Additionally, our upper bounds show scores up to 0.51. However, our current query selection models do not reach these scores, which remains to be solved in future work.

Regarding the second objective (RQ3), adding lexical entries to the prompt improves the macro scores in almost all cases, whereas the effect on the micro scores is mixed. For the non-fine-tuned GPT-3.5-Turbo model, the macro F_1 scores even doubles from 0.13 to 0.26. This effect is similarly large for GPT-4 (0.28 vs. 0.40) and fine-tuned GPT-3.5-Turbo (0.25 vs. 0.42).

Comparison with SOTA: In Table 3, we compare our approach with the most recent QA systems evaluated on QALD-9. TeBaQA [74] maps NL questions to SPARQL queries through learning templates from the QALD-9 dataset. However, the training dataset is small, consisting of only 403 questions, which limits the approach's ability to learn templates. gAnswer [83] and EDGQA [35] are graph-based approaches that interpret an NL question into a semantic query graph containing an edge for each relation mentioned in the question. SLING [48] and GenRL [59] are relationship linking frameworks developed for QALD. These approaches achieve the highest F_1 scores (ranging from 0.40 to 0.55) among all systems evaluated on the QALD-9 dataset. Our compositional approach outperforms all these methods, achieving an F_1 score of 0.72.

5 Related Work

Some QALD systems (such as ORAKEL [13], Pythia [68], QueGG [5,20,22], and LexExMachinaQA [21]) use Lemon lexica for lexicalization. For instance,

Table 3. Comparison with SOTA evaluated on the QALD-9 test dataset.

QALD System	Micro Precision	Micro Recall	Micro F_1 Score
Galactica [26]	0.14	0.02	0.03
Elon [72]	0.04	0.05	0.10
QASystem [72]	0.09	0.11	0.20
Falcon 1.0 [61]	0.23	0.23	0.23
WDAqua-core1 [16]	0.26	0.26	0.28
EDGQA [35]	0.31	0.40	0.32
TeBaQA [74]	0.24	0.24	0.37
gAnswer [83]	0.29	0.32	0.43
KGQAN [52]	0.49	0.39	0.43
SLING [48]	0.39	0.50	0.44
NSQA [38]	0.31	0.32	0.45
Zheng et al. [82]	0.45	0.47	0.46
GenRL [59]	0.49	0.61	0.53
Our Approach	**0.77**	**0.67**	**0.72**

Pythia [68] is built on Lexicalized Tree Adjoining Grammars [71] (LTAG) as a syntactic formalism and DUDES [11] for specifying semantic representations. The QueGG system [5,22] automatically generates a QA grammar from manually-created Lemon lexica. This grammar is then used to transform questions into SPARQL queries. However, the approach uses manually created sentence templates to cover syntactic variations and has very limited support for complex questions. The QueGG system [20] was compared with GPT-3.5 Turbo in a zero-shot scenario by prompting it with an instruction to generate a SPARQL query for a question related to DBpedia, without providing any lexical information. None of these approaches systematically evaluated the impact of lexical information on QALD performance. In contrast, our compositional approach uses dependency parsing, requiring no handwritten sentence templates. It addresses compositionality in a principled way using DUDES in combination with a tree merging and scoring component, covering a wide variety of complex questions.

WDAqua-core1 system [16] maps natural language sentences to KB elements by comparing an n-gram with the `rdfs:label` of an entity. For instance, the approach maps the natural language term *"writer"* to *dbo:writer* but fails to map it to other variations such as *dbo:creator* or *dbo:composer*. Some QALD systems (such as AskNow [19], DEANNA [79], SemQALD [32], QAKiS [7], QAnswer [60] etc.) use pattern dictionaries (e.g., BOA [25] or similar dictionaries) that map natural language terms to KB elements, while other approaches (Xser [19], gAnswer [19], CASIA [19]) use relational lexicalizations (e.g., PATTY [50]). The resources and dictionaries are very limited, and none of these approaches has investigated the impact of lexicalization on QALD.

One major limitation of state-of-the-art QALD systems is the lack of semantic compositionality for dealing with complex queries. To handle complex questions, Wang et al. [77] proposed a model that uses graph convolutional networks

(GCNs) [57] and performs reasoning over multiple KG triples. Similarly, Shekarpour et al. [64] use a combination of KB concepts with a HMM model. The approach first finds the segment (e.g., *"mayor"*, *"capital"*, *"Russia"*) of a query (e.g., *"Who is the mayor of the capital of Russia?"*) and then maps them to the appropriate resources. Other approaches (such as GETARUNS [15], IBM Watson [27] etc.) generate a logical form from a query. The approach generates triple patterns that are then split up again as properties are referenced by unions, resulting in many combinations of triples and wrong SPARQL queries. In contrast, our compositional approach selects the correct SPARQL query using a SPARQL selector (detailed in Sect. 2.9) from all possible combinations of SPARQL queries. There are rule-based architectures [17,18] to deal with complex questions, but the coverage of these approaches is completely limited to the rules added based on linguistic heuristics and observed patterns in the data.

6   Conclusion and Future Work

We have investigated the role and impact of explicitly given lexical knowledge in the context of QALD systems. We have presented a novel compositional system that uses this knowledge and demonstrated that it achieves performances in terms of micro F_1 scores well beyond the current state-of-the-art. In fact, our approach achieves a micro F_1 score of 0.72, which is 0.19 higher than the performance of the best state-of-the-art system on QALD (0.53). In this regard, our work has to be understood as providing a proof of concept that shows the impact of lexical knowledge and of a compositional approach.

Our approach handles complex queries using DUDES for semantic composition, combined with a tree merging and scoring component and a SPARQL selector, thereby covering a wide variety of complex questions. All we need is a lexicon in Lemon format. At the same time, our results show that LLMs are very limited in their ability to compose, as they cannot leverage provided lexical knowledge to the same extent as our proposed approach. Overall, our results suggest new avenues for QALD research by highlighting the role of explicit lexical knowledge and compositionality. However, there are also limitations.

First, a necessary prerequisite for our approach is the availability of a Lemon lexicon [46], which is manually created and takes approximately 16 h to produce for 599 lexical entries. Therefore, future work will focus on automating this process using one of the approaches, such as LexExMachina [23] and M-ATOLL [75], which automatically create a lexicon for the QA system. Another limitation is combinatorial explosion, i.e., the exponential growth of combinations when multiple candidate DUDES exist across multiple nodes of a tree, which increases response times considerably.

We provided a promising direction of QALD for future work consisting of the development of a hybrid system that combines the benefits of a compositional approach with the generalization abilities of large language models to bridge the lexical gap while leaving composition to a symbolic approach.

Acknowledgements and Funding. This work is partially funded by the Ministry of Culture and Science of the State of North Rhine-Westphalia under grant no NW21-059A (SAIL).

References

1. Akiba, T., Sano, S., Yanase, T., Ohta, T., Koyama, M.: Optuna: a next-generation hyperparameter optimization framework. In: Proceedings of the 25th ACM SIGKDD International Conference on Knowledge Discovery and Data Mining (2019)
2. Athreya, R.G., Bansal, S.K., Ngomo, A.C.N., Usbeck, R.: Template-based question answering using recursive neural networks. In: Proceedings of the 15th International Conference on Semantic Computing (ICSC), pp. 195–198 (2021)
3. Bai, Y., et al.: Benchmarking foundation models with language-model-as-an-examiner. In: Thirty-seventh Conference on Neural Information Processing Systems Datasets and Benchmarks Track (2023)
4. Bang, Y., et al.: A multitask, multilingual, multimodal evaluation of ChatGPT on reasoning, hallucination, and interactivity. In: Proceedings of the 13th International Joint Conference on Natural Language Processing and the 3rd Conference of the Asia-Pacific Chapter of the Association for Computational Linguistics (Volume 1: Long Papers), pp. 675–718. Association for Computational Linguistics (2023)
5. Benz, V., Cimiano, P., Elahi, M.F., Ell, B.: Generating grammars from lemon lexica for questions answering over linked data: a preliminary analysis. In: Proceedings of the 6th Natural Language Interfaces for the Web of Data Workshop (NLIWOD) co-located with the 19th International Semantic Web Conference (ISWC). CEUR Workshop Proceedings, vol. 2722, pp. 40–55 (2020)
6. Bordes, A., Chopra, S., Weston, J.: Question answering with subgraph embeddings. In: Proceedings of the 2014 Conference on Empirical Methods in Natural Language Processing (EMNLP) (2014)
7. Cabrio, E., et al.: Qakis: an open domain QA system based on relational patterns. In: International Semantic Web Conference, ISWC 2012 (2012)
8. Carpenter, B.: Type-Logical Semantics. MIT Press, Cambridge (1997)
9. Chung, H.W., et al.: Scaling instruction-finetuned language models. J. Mach. Learn. Res. **25**(70), 1–53 (2024)
10. Cimiano, P.: Flexible semantic composition with DUDES (short paper). In: Bunt, H., Petukhova, V., Wubben, S. (eds.) Proceedings of the Eight International Conference on Computational Semantics, IWCS 2009, Tilburg, The Netherlands, January 7–9, 2009. pp. 272–276. Association for Computational Linguistics (2009)
11. Cimiano, P.: Flexible semantic composition with DUDES (short paper). In: Proceedings of the 8th International Conference on Computational Semantics (IWCS), Tilburg, The Netherlands, pp. 272–276. Association for Computational Linguistics (2009)
12. Cimiano, P., Buitelaar, P., McCrae, J.P., Sintek, M.: Lexinfo: a declarative model for the lexicon-ontology interface. J. Web Semant. **9**(1), 29–51 (2011)
13. Cimiano, P., Haase, P., Heizmann, J., Mantel, M., Studer, R.: Towards portable natural language interfaces to knowledge bases the case of the orakel system. Data Knowl. Eng. **65**, 325–354 (2008)
14. Cimiano, P., Unger, C., McCrae, J.P.: Ontology-Based Interpretation of Natural Language. Synthesis Lectures on Human Language Technologies. Morgan & Claypool Publishers (2014)

15. Delmonte, R., et al.: Computational Linguistic Text Processing-Lexicon, Grammar, Parsing and Anaphora Resolution. Nova Science Publishers (2008)
16. Diefenbach, D., Both, A., Singh, K., Maret, P.: Towards a question answering system over the semantic web. Semantic Web **11**(3), 421–439 (2020)
17. Ding, Z., et al.: FORECASTTKGQUESTIONS: a benchmark for temporal question answering and forecasting over temporal knowledge graphs. In: Payne, T.R., et al. (eds.) ISWC 2023. LNCS, vol. 14265, pp. 541–560. Springer, Cham (2023). https://doi.org/10.1007/978-3-031-47240-4_29
18. Dubey, M.: Towards Complex Question Answering over Knowledge Graphs. Ph.D. thesis, University of Bonn, Germany (2021)
19. Dubey, M., Dasgupta, S., Sharma, A., Höffner, K., Lehmann, J.: AskNow: a framework for natural language query formalization in SPARQL. In: Sack, H., Blomqvist, E., d'Aquin, M., Ghidini, C., Ponzetto, S.P., Lange, C. (eds.) ESWC 2016. LNCS, vol. 9678, pp. 300–316. Springer, Cham (2016). https://doi.org/10.1007/978-3-319-34129-3_19
20. Elahi, M.F.: Multilingual Question Answering over Knowledge Graphs building on a Model of the Lexicon-ontology Interface. Ph.D. thesis, Bielefeld University, Bielefeld, Germany (2024)
21. Elahi, M.F., Ell, B., Cimiano, P.: Bridging the gap between ontology and lexicon via class-specific association rules mined from a loosely-parallel text-data corpus. In: Proceedings of the 4th Conference on Language, Data and Knowledge (LDK) (2023)
22. Elahi, M.F., Ell, B., Grimm, F., Cimiano, P.: Question answering on RDF data based on grammars automatically generated from lemon models. In: Proceedings of the 17th International Conference on Semantic Systems (SEMANTiCS) (2021)
23. Ell, B., Elahi, M.F., Cimiano, P.: Bridging the gap between ontology and lexicon via class-specific association rules mined from a loosely-parallel text-data corpus. In: Proceedings of the 3rd Conference on Language, Data and Knowledge (LDK) (2021)
24. Faria, B., Perdigão, D., Gonçalo Oliveira, H.: Question Answering over Linked Data with GPT-3. In: 12th Symposium on Languages, Applications and Technologies (SLATE 2023). Open Access Series in Informatics (OASIcs), vol. 113, pp. 1:1–1:15. Dagstuhl, Germany (2023)
25. Gerber, D., Ngonga Ngomo, A.C.: Bootstrapping the linked data web. In: 1st Workshop on Web Scale Knowledge Extraction@ ISWC (2011)
26. Glaese, A., et al.: Improving alignment of dialogue agents via targeted human judgements. arXiv preprint arXiv:2209.14375 (2022)
27. Gliozzo, A.M., Kalyanpur, A.: Predicting lexical answer types in open domain QA. Int. J. Semant. Web Inf. Syst. **8**(3), 74–88 (2012)
28. Gu, Y., et al.: Beyond I.I.D.: three levels of generalization for question answering on knowledge bases. In: Proceedings of the Web Conference 2021. WWW '21, New York, NY, USA, pp. 3477–3488. Association for Computing Machinery (2021)
29. Gu, Y., Su, Y.: ArcaneQA: dynamic program induction and contextualized encoding for knowledge base question answering. In: Proceedings of the 29th International Conference on Computational Linguistics, pp. 1718–1731. International Committee on Computational Linguistics, Gyeongju, Republic of Korea (2022)
30. Hakimov, S., Jebbara, S., Cimiano, P.: AMUSE: multilingual semantic parsing for question answering over linked data. In: Proceedings of the 16th International Semantic Web Conference (ISWC), pp. 329–346 (2017)

31. Hakimov, S., Jebbara, S., Cimiano, P.: Evaluating architectural choices for deep learning approaches for question answering over knowledge bases. In: Proceedings of the 13th IEEE International Conference on Semantic Computing (ICSC), pp. 110–113 (2019)

32. Hakimov, S., Unger, C., Cimiano, P.: Applying semantic parsing to question answering over linked data: addressing the lexical gap. In: Natural Language Processing and Information Systems (2015)

33. Hans, K.: A theory of truth and semantic representation. Formal Methods in the Study of language (1981)

34. Hao, Y., et al.: An end-to-end model for question answering over knowledge base with cross-attention combining global knowledge. In: Proceedings of the 55th Annual Meeting of the Association for Computational Linguistics ACL (2017)

35. Hu, X., Shu, Y., Huang, X., Qu, Y.: EDG-based question decomposition for complex question answering over knowledge bases. In: Hotho, A., et al. (eds.) ISWC 2021. LNCS, vol. 12922, pp. 128–145. Springer, Cham (2021). https://doi.org/10.1007/978-3-030-88361-4_8

36. Joshi, A.K., Schabes, Y.: Tree-adjoining grammars. In: Rozenberg, G., Salomaa, A. (eds.) Handbook of Formal Languages, pp. 69–123. Springer, Heidelberg (1997). https://doi.org/10.1007/978-3-642-59126-6_2

37. Kamp, H., Reyle, U.: From Discourse to Logic: Introduction to Modeltheoretic Semantics of Natural Language, Formal Logic and Discourse Representation Theory, vol. 42. Springer, Dordrecht (2013). https://doi.org/10.1007/978-94-017-1616-1

38. Kapanipathi, P., et al.: Leveraging abstract meaning representation for knowledge base question answering. In: Findings of the Association for Computational Linguistics: ACL-IJCNLP 2021, pp. 3884–3894. Association for Computational Linguistics (2021)

39. Kingma, D.P., Ba, J.: Adam: a method for stochastic optimization. In: Bengio, Y., LeCun, Y. (eds.) 3rd International Conference on Learning Representations, ICLR 2015, San Diego, CA, USA, May 7–9, 2015, Conference Track Proceedings (2015)

40. Knuth, D.: The Art of Computer Programming, vol. 3: Sorting And Searching. Addison-Wesley (1973)

41. Lehmann, J., et al.: Large Language Models for Scientific Question Answering: An Extensive Analysis of the SciQA Benchmark, pp. 199–217 (2024)

42. Levenshtein, V.I., et al.: Binary codes capable of correcting deletions, insertions, and reversals. In: Soviet Physics Doklady. vol. 10, pp. 707–710. Soviet Union (1966)

43. Liu, C., Liu, K., He, S., Nie, Z., Zhao, J.: Generating questions for knowledge bases via incorporating diversified contexts and answer-aware loss. In: Proceedings of the 2019 Conference on Empirical Methods in Natural Language Processing and the 9th International Joint Conference on Natural Language Processing (EMNLP-IJCNLP) (2019)

44. Lukovnikov, D., Fischer, A., Lehmann, J.: Pretrained transformers for simple question answering over knowledge graphs. In: Proceedings of the 18th International Semantic Web Conference (ISWC), pp. 470–486 (2019)

45. Lukovnikov, D., Fischer, A., Lehmann, J., Auer, S.: Neural network-based question answering over knowledge graphs on word and character level. In: Proceedings of the 26th International Conference on World Wide Web (WWW), pp. 1211–1220 (2017)

46. McCrae, J.P., Spohr, D., Cimiano, P.: Linking lexical resources and ontologies on the semantic web with lemon. In: Proceedings of the 8th Extended Semantic Web Conference on The Semantic Web: Research and Applications (ESWC), vol. 6643, pp. 245–259 (2011)

47. Mendes, P.N., Jakob, M., García-Silva, A., Bizer, C.: Dbpedia spotlight: shedding light on the web of documents. In: Ghidini, C., Ngomo, A.N., Lindstaedt, S.N., Pellegrini, T. (eds.) Proceedings the 7th International Conference on Semantic Systems, I-SEMANTICS 2011, Graz, Austria, September 7-9, 2011. pp. 1–8. ACM International Conference Proceeding Series, ACM (2011)

48. Mihindukulasooriya, N., et al.: Leveraging semantic parsing for relation linking over knowledge bases. In: Pan, J.Z., et al. (eds.) ISWC 2020. LNCS, vol. 12506, pp. 402–419. Springer, Cham (2020). https://doi.org/10.1007/978-3-030-62419-4_23

49. de Moura, L., Bjørner, N.: Z3: an efficient SMT solver. In: Ramakrishnan, C.R., Rehof, J. (eds.) Tools and Algorithms for the Construction and Analysis of Systems, pp. 337–340. Springer, Berlin Heidelberg, Berlin, Heidelberg (2008)

50. Nakashole, N., Weikum, G., Suchanek, F.: PATTY: a taxonomy of relational patterns with semantic types. In: Proceedings of the 2012 Joint Conference on Empirical Methods in Natural Language Processing and Computational Natural Language Learning, pp. 1135–1145, Jeju Island, Korea. Association for Computational Linguistics (2012)

51. Nikas, C., Fafalios, P., Tzitzikas, Y.: Open domain question answering over knowledge graphs using keyword search, answer type prediction, SPARQL and pre-trained neural models. In: Proceedings of the 20th International Semantic Web Conference (ISWC) (2021)

52. Omar, R., Dhall, I., Kalnis, P., Mansour, E.: A universal question-answering platform for knowledge graphs. In: Proceedings of the ACM SIGMOD/PODS International Conference of Management of Data (2023)

53. Omar, R., Mangukiya, O., Kalnis, P., Mansour, E.: ChatGPT versus traditional question answering for knowledge graphs: current status and future directions towards knowledge graph chatbots. CoRR abs/2302.06466 (2023)

54. Ouyang, L., e al.: Training language models to follow instructions with human feedback. In: Advances in Neural Information Processing Systems (2022)

55. Petroni, F., et al.: Language models as knowledge bases? In: Proceedings of the 2019 Conference on Empirical Methods in Natural Language Processing and the 9th International Joint Conference on Natural Language Processing (EMNLP-IJCNLP), Hong Kong, China, pp. 2463–2473. Association for Computational Linguistics (2019)

56. Qi, P., Zhang, Y., Zhang, Y., Bolton, J., Manning, C.D.: Stanza: a Python natural language processing toolkit for many human languages. In: Proceedings of the 58th Annual Meeting of the Association for Computational Linguistics: System Demonstrations (2020)

57. Ren, H., et al.: Graph convolutional networks in language and vision: a survey. Know.-Based Syst. **251**(C) (2022)

58. Rossiello, G., et al.: Generative relation linking for question answering over knowledge bases. In: Proceedings of the the 20th International Semantic Web Conference (ISWC) (2021)

59. Rossiello, G., et al.: Generative relation linking for question answering over knowledge bases. In: Hotho, A., et al. (eds.) ISWC 2021. LNCS, vol. 12922, pp. 321–337. Springer, Cham (2021)

60. Ruseti, S., Mirea, A., Rebedea, T., Trausan-Matu, S.: Qanswer-enhanced entity matching for question answering over linked data. In: CLEF (2015)

61. Sakor, A., et al.: Old is gold: linguistic driven approach for entity and relation linking of short text. In: Proceedings of the 2019 Conference of the North American Chapter of the Association for Computational Linguistics: Human Language Technologies, Minneapolis, Minnesota, pp. 2336–2346. Association for Computational Linguistics (2019)

62. Schabes, Y., Joshi, A.K.: An Earley-type parsing algorithm for Tree Adjoining Grammars. In: Proceedings of the 26th Annual Meeting of the Association for Computational Linguistics, Buffalo, New York, USA, pp. 258–269. Association for Computational Linguistics (1988)

63. Shekarpour, S., et al.: Question answering on linked data: challenges and future directions. In: Proceedings of the 25th International Conference Companion on World Wide Web (WWW), pp. 693–698 (2016)

64. Shekarpour, S., Ngonga Ngomo, A.C., Auer, S.: Query segmentation and resource disambiguation leveraging background knowledge. In: Proceedings of WoLE Workshop (2012)

65. Steedman, M.: Surface Structure and Interpretation, Linguistic Inquiry, vol. 30. MIT Press (1997)

66. Sun, H., Dhingra, B., Zaheer, M., Mazaitis, K., Salakhutdinov, R., Cohen, W.: Open domain question answering using early fusion of knowledge bases and text. In: Proceedings of the 2018 Conference on Empirical Methods in Natural Language Processing, pp. 4231–4242. Association for Computational Linguistics (2018)

67. Tan, Y., et al.: Can chatGPT replace traditional KBQA models? an in-depth analysis of the question answering performance of the GPT LLM family. In: Payne, T.R., et al. (eds.) ISWC 2023, pp. 348–367. Springer, Cham (2023). https://doi.org/10.1007/978-3-031-47240-4_19

68. Unger, C., Cimiano, P.: Pythia: compositional meaning construction for ontology-based question answering on the semantic web. In: International Conference on Applications of Natural Language to Data Bases (2011)

69. Unger, C., Freitas, A., Cimiano, P.: An introduction to question answering over linked data. In: Koubarakis, M., et al. (eds.) Reasoning Web 2014. LNCS, vol. 8714, pp. 100–140. Springer, Cham (2014). https://doi.org/10.1007/978-3-319-10587-1_2

70. Unger, C., Hieber, F., Cimiano, P.: Generating LTAG grammars from a lexicon/ontology interface. In: Proceedings of the 10th International Workshop on Tree Adjoining Grammar and Related Frameworks (TAG), pp. 61–68. Yale University (2010)

71. Unger, C., Hieber, F., Cimiano, P.: Generating LTAG grammars from a lexicon/ontology interface. In: Proceedings of the 10th International Workshop on Tree Adjoining Grammar and Related Frameworks (TAG), pp. 61–68. Yale University (2010)

72. Usbeck, R., Gusmita, R.H., Ngomo, A.N., Saleem, M.: 9th challenge on question answering over linked data (QALD-9). In: Joint proceedings of the 4th Workshop on Semantic Deep Learning (SemDeep-4) and NLIWoD4: Natural Language Interfaces for the Web of Data (NLIWOD-4) and 9th Question Answering over Linked Data challenge (QALD-9) co-located with 17th International Semantic Web Conference (ISWC), pp. 58–64. California, United States of America (2018)

73. Usbeck, R., et al.: Qald-10 - the 10th challenge on question answering over linked data. Semantic Web (2023)

74. Vollmers, D., Jalota, R., Moussallem, D., Topiwala, H., Ngomo, A.C.N., Usbeck, R.: Knowledge graph question answering using graph-pattern isomorphism. In: Proceedings of the 17th International Conference on Semantic Systems (SEMANTiCS) (2021)

75. Walter, S., Unger, C., Cimiano, P.: M-atoll: a framework for the lexicalization of ontologies in multiple languages. In: Proceedings of the 13th International Semantic Web Conference (ISWC) (2014)

76. Walter, S., Unger, C., Cimiano, P.: Dblexipedia: A nucleus for a multilingual lexical semantic web. In: Paulheim, H., van Erp, M., Filipowska, A., Mendes, P.N., Brümmer, M. (eds.) Proceedings of the Third NLP&DBpedia Workshop (NLP & DBpedia 2015) co-located with the 14th International Semantic Web Conference 2015 (ISWC 2015), Bethlehem, Pennsylvania, USA, October 11, 2015. CEUR Workshop Proceedings, vol. 1581, pp. 87–92. CEUR-WS.org (2015)

77. Wang, R., Rossetto, L., Cochez, M., Bernstein, A.: QaGCN: answering multi-relation questions via single-step implicit reasoning over knowledge graphs. In: Meroño Peñuela, A., et al. (eds.) The Semantic Web. LNCS (including subseries Lecture Notes in Artificial Intelligence and Lecture Notes in Bioinformatics), vol. 1, pp. 41–58. Springer, Germany (2024). https://doi.org/10.1007/978-3-031-60626-7_3

78. Wolf, T., et al.: Transformers: state-of-the-art natural language processing. In: Proceedings of the 2020 Conference on Empirical Methods in Natural Language Processing: System Demonstrations, pp. 38–45. Association for Computational Linguistics (2020)

79. Yahya, M., Berberich, K., Elbassuoni, S., Ramanath, M., Tresp, V., Weikum, G.: Natural language questions for the web of data. In: Proceedings of the 2012 Joint Conference on Empirical Methods in Natural Language Processing and Computational Natural Language Learning (EMNLP-CoNLL), pp. 379–390. Association for Computational Linguistics (2012)

80. Yih, W.T., Richardson, M., Meek, C., Chang, M.W., Suh, J.: The value of semantic parse labeling for knowledge base question answering. In: Proceedings of the 54th Annual Meeting of the Association for Computational Linguistics (ACL), Berlin, Germany, pp. 201–206. Association for Computational Linguistics (2016)

81. Yin, X., Gromann, D., Rudolph, S.: Neural machine translating from natural language to SPARQL. Futur. Gener. Comput. Syst. **117**, 510–519 (2021)

82. Zheng, W., Zhang, M.: Question answering over knowledge graphs via structural query patterns. arXiv preprint arXiv:1910.09760 (2019)

83. Zou, L., Huang, R., Wang, H., Yu, J.X., He, W., Zhao, D.: Natural language question answering over RDF: a graph data driven approach. In: 2014 ACM SIGMOD International Conference on Management of Data (2014)

On the Roles of Competency Questions in Ontology Engineering

C. Maria Keet[1]([⊠]) [iD] and Zubeida Casmod Khan[2] [iD]

[1] Department of Computer Science, University of Cape Town, Cape Town,
South Africa
mkeet@cs.uct.ac.za
[2] Council for Scientific and Industrial Research, Pretoria, South Africa
zdawood@csir.co.za

Abstract. Competency Questions (CQs) are not merely intended for
scoping the prospective content of an ontology and as information-
seeking queries posed over an ontology, but serve manifold purposes in
the ontology engineering processes. This position paper argues that CQs
should be viewed as complex acts with underlying motivations beyond
just eliciting facts. We explore the concept of questions in general, drawing on philosophical and logical perspectives to highlight the complexity
of questions and how they function beyond information seeking. This
understanding is applied to CQs for ontologies, revealing how they can
serve various purposes in ontology development, such as knowledge acquisition, knowledge organisation, and validation. The paper also introduces
the notion of types of CQs for ontology engineering and a first taxonomy of CQ types. Having identified different types of CQs, it may assist
research into devising more specific methods and tools to support the
development of CQs—be it manual authoring or automating it—and
their use at various stages and tasks in ontology engineering, as well as
contribute to a notion of quality of a CQ.

Keywords: Competency Question · Ontology Engineering · Ontology
Development · Foundational Ontology · Ontology methodology

1 Introduction

Ontology engineering concerns the development and structuring of knowledge in
various domains and applications. One of the key tools in the processes is the
use of Competency Questions (CQs). These questions are not just queries for
obtaining information but serve deeper purposes, playing a role also in defining the scope, requirements, and validation of ontologies, as demonstrated in,
among others, [1,4,6,7,9,15–17]. A systematic mapping of which CQs are to be
used where for which task in which ontology development methodology is still
outstanding.

This position paper argues that CQs should be viewed as complex acts with
underlying motivations beyond merely obtaining information. This requires a

© The Author(s), under exclusive license to Springer Nature Switzerland AG 2025
M. Alam et al. (Eds.): EKAW 2024, LNAI 15370, pp. 123–132, 2025.
https://doi.org/10.1007/978-3-031-77792-9_8

focused analysis on the technical functionality and limitations of CQs within ontologies. By exploring the concept of questions through philosophical and logical perspectives, we aim to highlight the complexity of questions and how they function beyond mere information seeking and scoping of a prospective ontology's content. This understanding is then applied to CQs, revealing how they can serve various purposes in ontology development, such as in knowledge acquisition, organisation, and validation.

There is an argument that philosophical and logical perspectives on questions might not be directly applicable to the specific context of CQs in ontology engineering. Notably, Sowa states that "the problem of matching language to logic is unsolvable if the two are considered totally different, irreconcilable systems" [14], which might also apply to CQs. This opposing viewpoint also aligns with the work of Bezerra and Santana, who present CQs as a method for evaluating ontologies [7]. In their approach, CQs function primarily as information-seeking queries within the domain. However, we argue that such philosophical and logical perspectives can provide context for the technical functionality and limitations of CQs. Analysing CQs based on these perspectives help to understand how to improve CQ formulation and interpretation, as well as their use, authoring, research into it, and development of tools for CQs. This is demonstrated by the development of a first taxonomy of CQ types, which provides a more informed understanding of their roles and functions.

The remainder of the paper is structured as follows. In Sect. 2 we examine CQs in ontologies, including a discussion of faulty CQs. Section 3 introduces pertinent philosophical and logical perspectives on questions, which inform clarifying types of CQs for ontologies. The taxonomy and a library of CQs as a direction towards a solution are presented in Sect. 4. Section 5 discusses and considers future research directions, and Sect. 6 concludes the paper.

2   Challenges with Determining the Quality of CQs

While the largest published dataset with CQs [11] may assist in understanding CQs, it was also found that not all questions were of exemplar quality [2,11], which is also an issue when trying to automatically generate them [1,3]. Therefore it is of use to further the understanding of CQs for ontologies to consider what a 'faulty' or 'bad' CQ is, and whether that would be in the absolute sense or relative to something.

Upfront, and applicable to all types of CQs, first, there are problems that all types of CQs may exhibit: syntax issues. The sentence may be a grammatically incomprehensible or an ambiguous question, or not be a question but a statement appended with a question mark. Second, there may be questions that no ontology will ever be able to answer, like "how do I apply for promotion to full professor?". Accordingly, we critically assessed the dataset of 234 CQs of [11]. Both authors evaluated each entry on grammar and semantic issues (including vagueness, ambiguity, answerability) to earn a Yes or a No, and then discussed to harmonise any differences. The outcome of the first round was 40 and 49 as problematic CQs of which 28 were initially judged differently. Discussion resolved each, mostly resulting in No (n=22), bringing the overall number of problematic questions to 53 (i.e., 23%; see the new ROCQS dataset in Sect. 4). Analysing those, 17 of the 53 are easily solvable grammar issues (e.g., 'what' versus 'which'), 9 were about 'can I do x'/'how to do x' rather than about content of the ontology, which are thus strictly unanswerable, and the rest had a range of issues, such as asking for the "fastest" software, imprecision with, e.g., "possibly problematic" behaviour, asking "where" to find something, and others.

There also exist questions that Wiśniewski calls "semantically faulty" [19]. He provides the example of "Which natural number is smaller than 0?". The answer is 'none', but that is presumably not intended because 'none' is not a natural number that was asked for in the question. Applying this notion to a CQ for an ontology, then it means that a CQ should not have always the empty set as answer, as a minimum, and maybe also that at least the category of the entity is among the intended one. For instance, if a physical object is expected as answer, to not have abstract objects in the answer. How this may be managed when querying an ontology is yet to be determined, especially regarding specifying upfront the category or upper-level entity or top-domain or domain-level class in the ontology, which may not be known upfront if the ontology is not aligned to a top-level ontology, let alone at the scoping stage of ontology development.

Then there are subtle issues that interfere with the goodness of a CQ, or its usability at least. We identify and discuss three. First, and especially for CQs used for validating the ontology: some questions cannot be converted into SPARQL or SPARQL-OWL (or a similar query language for another ontology language) to query the ontology due to the lack of expressiveness of the query language. One then either has to reformulate the question or check it manually. For instance, negation is not fully supported in SPARQL, and ranking is difficult. Example CQs that are at least ambiguous if not impossible to answer due to query language restrictions are, e.g.: "To what extent does [the software] support appropriate open standards?" due to the gradation inherent in 'to what extent' and the imprecision of what counts as 'appropriate'; "Which is the fastest software to read [this data]?" due to the comparator 'fastest'; and "Is there an animal that does not drink water?" due to the negation's intent. Of course, at least some of them can be tweaked to become answerable, such as "Which animal represented in the ontology is known not to drink water?" to limit it to declared knowledge in the ontology. Observe that this also shows why the set of CQs for scoping the ontology at the start of ontology development may differ from those CQs used to validate the ontology once built: a desired scope may be represented in more or less detail in the ontology eventually, be it due to scope creep or language limitations or rewording of an imprecise question.

Conversely, the query language may be expressive enough, but the ontology language it is used with is not. That is, second, there are CQs that cannot be answered by a particular ontology due to the restriction on the language that the ontology is represented in. For instance, a CQ "Does a narcissist love himself?" concerns the relational property of reflexivity, but if the ontology language prohibits it, it cannot possibly be answered as intended; likewise for a CQ inquiring about reflexivity of love. This does not make it a bad CQ of itself, however, just not a good one for the target ontology; or the CQ is good, but the ontology does not meet the user's requirements for their intended use. Thus, a CQ may be deemed good or bad within a specific context of use.

Third, there are questions that cannot be answered by a particular ontology because it lacks the coverage with respect to the content. For instance, with the "Does a narcissist love himself?" and where the ontology language allows reflexivity, but now it lacks the vocabulary, i.e., there is no Narcissist and/or no love in the ontology. The question is then not answerable for that version of the ontology. It could still be a good CQ. It would contextually turn into a 'not good' one only if the scope had been changed.

3   Philosophical and Logical Perspectives

As the previous section illustrated, all questions are not alike, and, in fact, already drew from the theories of questions, being that of Wiśniewski's notion of semantically faulty questions. Therefore, to arrive at a proposed solution direction, we first take a step back to consider theoretical works that focus on questions in general, which are more varied than CQs specifically for ontologies. Part of those insights are relevant, which we highlight here, where key aspects are italicised; their use will be demonstrated in the second paragraph of Sect. 4 and first paragraph of Sect. 5.

Cohen [8] reignited the 'question question' in research for the past century. He argues that questions are more than requests for information, and rather that they are complex entities that play a crucial role in shaping human thought and discourse. *Question complexity* concerns not only the number of variables involved, but also, according to Cohen, that questions may ask more than one thing at a time, that it is impossible to avoid ambiguous phrasing, and that question contain implicit assumptions. Most recently, Watson [18] highlights the limited attention given to questions from the philosophical point of view and argues that questions are an integral part of human life. She conducted a survey on questions and created a living question collection with as aim to construct a definition of questions. She found that defining a question as an interrogative sentence is insufficient and instead one must focus on its *function*, being an "information-seeking act". Ram [12] already refined the information-seeking aspect into underlying *knowledge acquisition* and *knowledge organisation* goals. Watson considers also *motivations* behind asking questions; e.g., information-seeking questions are asked with the motivation to "expose" a colleague for non-performance of duties [18].

Logics and linguistics-based approaches to questions are reviewed by Wiśniewski [19], who covers the history of research on questions, the different approaches and methods used by logicians and linguists, and the formal systems developed for representing questions. It covers various theories that aim to model natural language questions, including: questions as *sets of declaratives* (similarly advocated in [10]), as epistemic imperatives, as interrogative speech acts, as sentential functions, as *inquisitive semantics* etc. Lastly, types of questions (normal, regular, self-rhetorical, and proper) and answers (complete, partial, eliminative and corrective) are presented. Wiśniewski also noted that there is no universally accepted theory of questions yet.

Neither considers types of questions extensively, nor inventarises purposes, or goals for asking the questions. A practical example of a possible distinction between questions is demonstrated by Bertolazzi et al. [5]. Their work, they argue, provides a basis for understanding how questions evolve as cognitive abilities develop. They explore how ChatGPT builds and refines its hypothesis space through asking questions in the Twenty Questions Game. They use so-called *hypothesis-scanning* questions, which explicitly mention one of the candidate items (e.g., "is it a melon?"), and *constraint-seeking* questions, which do not (e.g., "can you eat it?"). Ruggeri and Lambrozo [13] refer to hypothesis-scanning questions are those that narrow down the search space by testing specific hypotheses or potential solutions, whereas constraint-seeking questions eliminate unlikely options by identifying features that many solutions have. From an ontological perspective, the former is to ascertain whether the fact holds and the latter to ascertain whether an entity has the property mentioned in the question. For instance, a hypothesis-scanning CQ from [20] may be: "Which software tool created [this data]?" and a constraint-seeking one "Does [this software] provide XML editing?", but extant CQs for ontologies have not yet been examined this way, to which we turn next.

4   Formulating CQs in Ontologies

Despite a recent increase in popularity in adoption of CQs, what they exactly are, or should be, or what 'exemplary' or 'good' CQs should look like, remains unclear, let alone the idea that there may be different types of CQs. The insights gleaned from philosophical and logical perspectives on questions can assist with this. By considering the complexity, function, motivation, and logical structure of questions, we can design CQs that not only seek information but also enhance the overall quality and usage of the ontology. We briefly describe a small taxonomy of types of CQs and a library of CQs in this section.

We first draw from the existing philosophical and logical works regarding questions. Applying the notions of hypothesis-scanning and constraint-seeking questions [13], we noted that these both may apply to ontology CQs; e.g., for a software ontology, the CQ "Who is the subject in the process of programming?", it is constraint-seeking, while for the CQ "Do I need a password to use the software?", it is hypothesis-scanning. Cohen's claim of complex and multi-faceted questions [8] holds for CQs whereby different terminologies are used (scoping vs. validation), each phrase chunk of a CQ may match vocabulary in the ontology, or a CQ may describe a property or characteristic of an entity, such as the 'unfolding in time' for perdurants. Cohen's claim on underlying assumptions ties in with Watson's motivations behind asking questions [18], which apply to CQs in that there are specific motivations, such as elicitating requirements, validating the content of an ontology, and obtaining a foundational ontology alignment. The 'questions as sets of declaratives' alludes to a tight relation between CQs and their formalisation in an ontology, whereas 'questions as inquisitive semantics' points to CQs' use to help with trying to determine the ontological nature of the entity that needs to be represented in the ontology.

Building on our analysis of what motivates asking questions and the components that make them up, we have identified five main types of CQs used in ontology development: scoping competency questions (SCQ), validating competency questions (VCQ), foundational competency questions (FCQ), metaproperty competency questions (MpCQ), and relationship competency questions (RCQ), and we describe them briefly here, alongside a small hierarchy, as shown in Fig. 1. For structuring purposes, they are divided into ontological CQs (OCQs) and domain CQs (DCQs), where the former focus on the ontological nature of the entity that is being interrogated and the latter on the entities with respect to the subject domain of the prospective ontology. The other ones are summarised as follows.

– Scoping CQ (SCQ): A question that mentions a domain entity and helps define the scope of an ontology for a specific subject domain. SCQs help establish what the ontology will be about.
– Validation CQ (VCQ): A question used to validate the content of an ontology by checking if the ontology adheres to its intended meaning and knowledge representation. VCQs ensure the ontology accurately reflects the domain it represents.
– Foundational CQ (FCQ): A question used to align a domain entity to an entity within a foundational ontology (a more general ontology). FCQs help ensure consistency between the domain ontology and a higher-level ontology.
– Relationship CQ (RCQ): A question that explores various characteristics of relationships within an ontology. There are four sub-types of RCQs:
  • Arity CQ (aRCQ): Determines the number of participants in a relationship.
  • Elementary Fact Type CQ (efRCQ): Asks whether a relationship can be further decomposed without losing information if split up (e.g., a ternary recast as two binary relationships).
  • Domain-Range CQ (drRCQ): Identifies the entities that participate in a relationship (domain and range); these may be either OCQs or DCQs.
  • Relational Property CQ (rpRCQ): Investigates specific properties of a relationship, such as transitivity.
– Metaproperty CQ (MpCQ): A question that classifies an entity according to a predefined set of metaproperties. Metaproperties are general characteristics that hold true across ontologies, such as being a sortal or whether it is telic (has a goal).

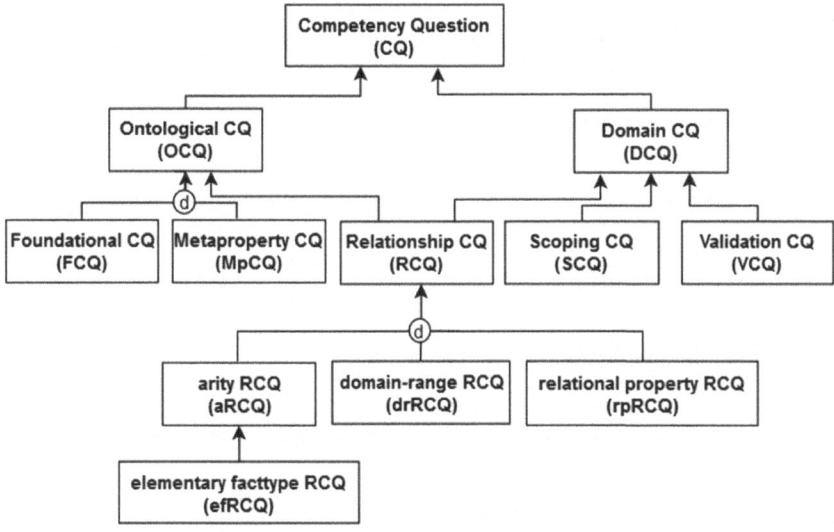

Fig. 1. Overview of the main types of CQs for ontologies, in EER diagram notation.

To foster further analysis of CQS, as well as their use and reuse at different stages of ontology development, we created a basic Repository of Ontology CQs, ROCQS[1]. It contains 38 FCQs, 33 VCQs, 323 SCQ, 27 RCQs, and 17 MpCQs, 48 of which were newly created by the authors. They are annotated with, including among others and depending on the type of CQ: source, whether they are templates or concrete CQs and if concrete which template they instantiate, which element of the question is the principal one (be it of the FO or the meta- or relational property), which ontology they were created for (if any), and it contains an illustrative table with examples, as illustrated in Fig. 2. The aforementioned SCQs evaluated as problematic (discussed in Sect. 2) are included in a separate tab and appended with a brief reason why.

5   Discussion

Our analysis of the philosophical and logical works reveals insights into the role of questions, particularly CQs, in ontology development. The insights from the philosophical and logical domains served as a foundation for analysing, identifying, and devising CQs. Recognising questions as complex acts for eliciting information, as in [12,18], informed by the logical and philosophical theories of questions [19], shed light on how to develop better CQs. For instance, FCQs fit the notion of Ram's [12] "knowledge organisation" goal—where precisely to link it to that ontology—whereas MpCQs and RCQs are information-seeking in the sense of knowledge acquisition, and VCQs align with Knuth's formalisation

---

[1] ROCQS is accessible from http://www.meteck.org/files/ROCQS/ROCQS.htm.

Type of OCQ	Purpose	Examples
Scoping CQ (SCQ)	Define the domain and scope of the ontology	Which predators eat rabbits? What information is clinically relevant for social interaction assessment?
Validation CQ (VCQ)	Verify the accuracy of the content	Is ruby a type of chocolate? What is the Base of ThinAndCrispyPizza?
Foundational CQ (FCQ)	Align entities with a foundational ontology	Is water bottle classified as a Material Entity in the BFO foundational ontology? Is coffee something that cannot be counted, or only in specific quantities?
Relationship CQ (RCQ)	Investigate the characteristics of relationships	What is the domain and range of the eating relationship? If a body contains a heart and a heart contains a cell, does the body contain the cell?
Metaproperty CQ (MpCQ)	Classify entities based on metaproperties	Is each instance of a coffee bean necessarily (at all times of its existence) an instance of a coffee bean? Does a thesis defense have a definite endpoint?

Properties recorded for each type of CQ:
SCQ: ID, SCQ, Ontology, Template (Y/N), Source, Link
VCQ: ID, VCQ, Ontology, Template (Y/N), Source, Link
FCQ: ID, FCQ, Foundational Ontology, Entity/ies described, Template (Y/N), Instantiates Template, Source, Link
RCQ: ID, RCQ, Entity/ies, Template (Y/N), Instantiates template, Source, Link, Sentence pattern
MpCQ: ID, MpCQ, Entity/ies, Template (Y/N), Instantiates template, Source, Link

Fig. 2. Illustrative table in ROCQS and the initial list of properties recorded for each type of CQ (screenshot from the ROCQS repository).

of Cox's definition of a question as "a system of assertions that answers that question" [10].

Also, by recognising ambiguity and various motivations behind CQs, we were able to elucidate types of CQs and catalog them in a repository for CQs, ROCQS. Both may motivate further research on multiple aspects, such as guiding automated CQ generation, which sentences structures work better than others, demonstrable effects of using questions of each type of CQ, quality metrics of CQs, and methods and techniques for more effective use of CQs in the various ontology development tasks.

Previous work regarding the generation of CQs using LLMs [1] did not include the distinction between hypothesis-scanning and constraint-seeking questions, but it is relevant given the examples provided in Sect. 3 and the LLM's demonstrated ability to deal with them [5]. Further investigation into such LLM-associated questions may benefit from the separation into different types of CQs and structure the training or output evaluation or filtering accordingly. One also might be able to take existing CQs, convert them to statements, use that to process a corpus in the domain of the prospective ontology, and from found matching sentences, generate new CQs.

We considered creating an ontology of CQs, but since the topic is not sufficiently stabilised yet and rather suggests more research avenues, we deemed it too premature to already develop the artefact. This is, perhaps, also due to, there being no universally accepted theory of questions, as already observed in [19], and possibly how to best structure the types of CQs we identified. Further, especially the collection of FCQs, RCQs, and MpCQs for ontologies would be helpful. To have guidelines and tools for that would assist the endeavour, yet, such research and development will also be assisted by having enough sample CQs to base it off.

Avenues for practical future work include, among others, integration of ROCQS with existing platforms, such as Protégé, to allow users to switch between formulating CQs and examining the relevant parts of the ontology. Another avenue is to conduct user studies with ontology developers to evaluate ROCQS to design it for both usability and CQ collection for analysis.

6 Conclusion

By recognising CQs as complex acts with diverse purposes that go further than information seeking, we can use them more effectively throughout ontology development. This paper explored questions through philosophical and logical perspective which revealed that CQs are used throughout the ontology engineering lifecycle—from scoping and validation to alignment and metaproperty analysis. This was captured in a taxonomy of CQ types that goes beyond previous work, providing a more comprehensive framework for understanding and applying CQs in ontology development. This, in turn, generated various new research and innovation directions to improve CQ quality and use.

Disclosure of Interests. The authors have no competing interests to declare that are relevant to the content of this article.

References

1. Alharbi, R., Tamma, V., Grasso, F., Payne, T.R.: An experiment in retrofitting competency questions for existing ontologies. ArXiv:abs/2311.05662 (2023)
2. Antia, M.J., Keet, C.M.: Assessing and enhancing bottom-up CNL design for competency questions for ontologies. In: Proceedings of the Seventh International Workshop on Controlled Natural Language (CNL 2020/21). ACL (2021). https:// aclanthology.org/2021.cnl-1.11
3. Antia, M.J., Keet, C.M.: Automating the generation of competency questions for ontologies with AgOCQs. In: Ortiz-Rodriguez, F., Villazón-Terrazas, B., Tiwari, S., Bobed, C. (eds.) KGSWC 2023. LNCS, vol. 14382, pp. 213–227. Springer, Cham (2023). https://doi.org/10.1007/978-3-031-47745-4_16
4. Bernabé, C.H., Keet, C.M., Khan, Z.C., Mahlaza, Z.: A method to improve alignments between domain and foundational ontologies. In: 14th International Conference on Formal Ontology in Information Systems 2023 (FOIS'23). FAIA, vol. 377, pp. 125–139. IOS Press (2023)
5. Bertolazzi, L., Mazzaccara, D., Merlo, F., Bernardi, R.: ChatGPT's information seeking strategy: insights from the 20-questions Game. In: Keet, C.M., Lee, H., Zarrieß, S. (eds.) Proceedings of the 16th International Natural Language Generation Conference, INLG 2023, Prague, Czechia, September 11–15, 2023. pp. 153–162. Association for Computational Linguistics (2023). https://aclanthology.org/2023. inlg-main.11
6. Bezerra, C., Freitas, F.: Verifying description logic ontologies based on competency questions and unit testing. In: ONTOBRAS'17, pp. 159–164 (2017)

7. Bezerra, C., Freitas, F., Santana, F.: Evaluating ontologies with competency questions. In: Proceedings of the 2013 IEEE/WIC/ACM International Joint Conferences on Web Intelligence (WI) and Intelligent Agent Technologies (IAT) - Volume 03, pp. 284–285. WI-IAT '13, IEEE Computer Society (2013). https://doi.org/10.1109/WI-IAT.2013.199
8. Cohen, F.S.: What is a question? The monist, pp. 350–364 (1929)
9. Keet, C.M., Ławrynowicz, A.: Test-driven development of ontologies. In: Sack, H., Blomqvist, E., d'Aquin, M., Ghidini, C., Ponzetto, S.P., Lange, C. (eds.) ESWC 2016. LNCS, vol. 9678, pp. 642–657. Springer, Cham (2016). https://doi.org/10.1007/978-3-319-34129-3_39
10. Knuth, K.H.: What is a question? In: AIP Conference Proceedings. vol. 659, pp. 227–242. AIP (2003). https://doi.org/10.1063/1.1570546
11. Potoniec, J., Wisniewski, D., Lawrynowicz, A., Keet, C.M.: Dataset of ontology competency questions to SPARQL-OWL queries translations. Data Brief **29**, 105098 (2020). https://doi.org/10.1016/j.dib.2019.105098
12. Ram, A.: A theory of questions and question asking. J. Learn. Sci. **1**(3/4), 273–318 (1991)
13. Ruggeri, A., Lombrozo, T.: Learning by asking: How children ask questions to achieve efficient search. In: Proceedings of the 36th Annual Meeting of the Cognitive Science Society, pp. 1335–1340 (2014). https://escholarship.org/uc/item/32c734r9
14. Sowa, J.F.: The Role of Logic and Ontology in Language and Reasoning, pp. 231–263. Springer, Dordrecht (2010). https://doi.org/10.1007/978-90-481-8845-1_11
15. Suarez-Figueroa, M.C., et al.: NeOn methodology for building contextualized ontology networks. NeOn Deliverable D5.4.1, NeOn Project (2008)
16. Thiéblin, E., Haemmerlé, O., Trojahn, C.: Complex matching based on competency questions for alignment: a first sketch. In: 13th International Workshop on Ontology Matching (OM@ISWC 2018), Monterey, US, pp. 66–70. CEUR-WS (2018)
17. Uschold, M., Gruninger, M.: Ontologies: principles, methods and applications. Knowl. Eng. Rev. **11**(2), 93–136 (1996). https://doi.org/10.1017/S0269888900007797
18. Watson, L.: What is a question. R. Inst. Philos. Suppl. **89**, 273–297 (2021). https://doi.org/10.1017/s1358246121000114
19. Wiśniewski, A.: Semantics of questions. The Handbook of Contemporary Semantic Theory, pp. 271–313 (2015)
20. Wiśniewski, D., Potoniec, J., Ławrynowicz, A., Keet, C.M.: Analysis of ontology competency questions and their formalizations in SPARQL-OWL. J. Web Semant. **59**, 100534 (2019). https://doi.org/10.1016/j.websem.2019.100534

Structured Representations for Narratives

Inès Blin[1,2]([⊠])[iD], Annette ten Teije[1][iD], Frank van Harmelen[1][iD],
and Ilaria Tiddi[1][iD]

[1] Vrije Universiteit Amsterdam, Amsterdam, The Netherlands
{i.blin,annette.ten.teije,frank.van.harmelen,i.tiddi}@vu.nl
[2] Sony Computer Science Laboratories-Paris, Paris, France

Abstract. Narratives are essential to the human sense-making process, and have sparked interest across multidisciplinary fields including Computer Science. However, efforts to unify terminology and identify core narrative requirements are limited, complicating the choice of a suitable representation. We first map identified narrative requirements to appropriate knowledge representations (1). We identify and organise narrative requirements from interviews we conducted (1a), and map these requirements to their most suited knowledge representation forms (1b). We then conduct a systematic survey to select candidate event and narrative-centric ontologies (2). We analyse how effectively these ontologies represent the main narrative requirements (2a), and rank them based on a five-star rating system (2b). We lastly provide recommendations on how to choose the best ontology for a narrative for a specific use case (3).

Keywords: Narratives · Ontologies · KGs · Event-centric KGs

1 Introduction

Narratives are fundamental for human understanding [13], offering invaluable insights into intricate events. They provide coherence to otherwise fragmented and unrelated data, building up coherent explanations from past data. Systems capable of constructing such narratives would be beneficial in developing truly "human-centric" systems. Such systems have already proved valuable in helping building daily narratives, e.g. in the legal [28] and historical [55] domains.

Throughout the literature, the term "narrative" is widely used but with varying interpretations and expectations, drawing from fields such as linguistics, philosophy, narratology, and artificial intelligence [16]. Narrative construction has gained prominence within Computer Science, as evidenced by workshops like the Computational Models for Narratives Workshop Series (2009–2016, [56]), the Narrative Understanding Workshop Series[1] (2019–2023), the Text2Story[2]

---

[1] https://sites.google.com/umass.edu/wnu2023/home.
[2] https://text2story24.inesctec.pt.

© The Author(s), under exclusive license to Springer Nature Switzerland AG 2025
M. Alam et al. (Eds.): EKAW 2024, LNAI 15370, pp. 133–154, 2025.
https://doi.org/10.1007/978-3-031-77792-9_9

(2018–2024), or the SEMMES[3] (2023–2024) workshops. Despite the importance of narratives, few efforts have been made to identify core elements common to each work such as events and agents, and to unify existing representations. A recurring simplification defines a narrative as a sequence of events.

The primary objective of this work is to identify the most suitable method, given a set of requirements, for representing a narrative in the form of a knowledge graph (KG) where an ontology is populated with data, that we call *narrative graphs*. The benefit of representing narratives in a structured format like KGs lies in their incorporation of a semantic layer, which allows for finer-grained representations. Designed to align with how humans naturally build and understand narratives [13], narrative graphs aim to facilitate understanding through improved querying and reasoning. KGs have been used as background knowledge to support narrative construction [35]. Debates persist at the syntactic level about optimal KG data models [83], encompassing options like simple RDF, Named Graphs[4] or RDF-star[5]. At the semantic level, the increasing number of narrative resources (ontologies, KGs, etc.) make it challenging to choose a suitable narrative representation for a given set of requirements or a use case.

In this paper, we propose to unify the terminology across the literature and to provide guidelines on which ontology model best to use for a specific use case. We divide our problem into three research questions: **RQ1.** What are the common requirements related to narratives? **RQ2.** Can existing ontologies represent these narrative requirements? **RQ3.** How to best choose a structured representation for a narrative? To answer **RQ1**, we conduct interviews to identify narrative requirements, and we organise these requirements into themes. We also map these requirements to their most suited knowledge representation forms. To answer **RQ2**, we conduct a systematic review on event-centric and narrative-centric ontologies, and evaluate the ontologies based on how effectively they can fullfill narrative requirements and on their vocabulary usage. To answer **RQ3**, we lastly provide recommendations on how to best identify a structured representation for a narrative, and apply them to the interviews use cases.

2   Narrative Requirements

We elicited narrative requirements from interviews to answer **RQ1**, and we organise these requirements into thematics. We also map these requirements to their most suited knowledge representation (KR) forms.

2.1   Identification of Narrative Requirements

We conducted 6 narrative interviews with 9 participants in the MUHAI[6] EU research consortium. MUHAI aims to advance AI systems by integrating narratives to enhance meaning and understanding. The project has two main use

---

[3] https://anr-kflow.github.io/semmes/.
[4] https://www.w3.org/2009/07/NamedGraph.html.
[5] https://w3c.github.io/rdf-star/cg-spec/editors_draft.html.
[6] Meaning and Understanding in Human-Centric AI: https://muhai.org.

cases: pragmatic everyday knowledge (e.g. households tasks such as cooking) and societal knowledge (historical episodes, socio-historical knowledge, and contemporary events). Since all participants were in MUHAI, this ensured consistency and relevance to the use cases, yielding focused insights; however, this approach possibly limits the generalisability of the findings to other domains, which could be explored in future work. The open unstructured interviews took 20 min on average. The starting point of the interviews were the following questions:

1. What elements are needed to represent your (domain-specific) narratives? Why and what do you need to represent with these?[7]
2. What kind of data structure is required to represent your narratives?
3. For which tasks do you use your narratives? E.g. querying, visualisation, etc.
4. Can you provide one example of a narrative in your use case?

Most participants had a computer science background. If not, we reformulated in non-technical terms. Table 1 provides a description of the interviews, with the domains, all related to MUHAI, the type of input data used to build the narrative, one example of a narrative and a practical application scenario for each narrative graph to be built. In the history and in the cognitive science domains (I1, I6), the main elements to be represented can be seen as extended version of the five Ws: Who, What, Where, When and Why. For the biomedical science (I2), the knowledge to be represented is much more structured and concise. The main difference for the social media analysis (I3) is to enable the representation of conflicting viewpoints, that is important when analysing discourse in social media. Arts & Literature (I5) additionally require context and background knowledge to enhance the comprehension as the text or the image is read. Lastly, the Robotic Kitchen (I4) involving a cooking robot in a virtual simulation needs to plan its actions to execute a recipe. In this context, each action is seen as an episode in a narrative. We detail hereafter the 6 main requirements we gathered and organized from these interviews.

- **Agents**[8]. These are animate entities like humans or animals.
  - *Types of agents* distinguish between humans, animals, animate objects, etc. Occasionally, people made the distinction between an active agent, who acts with a goal in mind, and a passive agent, who is more a spectator in a scene.
  - *Properties* describe agent attributes, such as physical characteristics.
  - *Roles* describe roles of agents in the events.
- **Objects.** These are physical or abstract objects that do not fall under the category of agents.
  - *Types of objects* distinguish between different types of objects.

---

[7] From a KR perspective, expressions that do not very easily fit into the regular <s,p,o> representation of RDF, like higher-order or n-ary relations.

[8] In these requirements, we differentiate agents from objects, with agents not necessarily participating in events directly, but their participation would be modeled through relationships.

Table 1. Interview Descriptions. #: number of people interviewed.

Id	#	Domain	Input Data	Narrative Example	Application scenario
I1	1	History	Historical events	Main **happenings** during the French Revolution [9]	**Query-answering** about historical events
I2	1	Biomedical Science	Scientific Claims	**PICOs** (Population, Intervention, Comparison and Outcomes) [81]	**Reasoning** and decision support in clinical research
I3	2	Social Media Analysis	User-generated content	**Claims** by users on Twitter regarding the Russo-Ukrainian conflict [74]	Analyzing and deriving **insights** from user-generated content and trends
I4	2	Robotic Kitchen	Recipes	**Robot execution** of a recipe [67]	Robotic **cooking task execution** based on recipe understanding
I5	2	Arts & Literature	Text (linguistics, semantics) & Images (paintings)	**Questions** arising when reading a text/painting, that form a narrative network together with the answers to these questions [80]	**Question-answering** to enable interpretive analysis from artistic or literary works
I6	1	Cognitive Science	Mental Models	Importance of **values** in our beliefs [89]	Exploring and **modeling** belief systems and cognitive processes

- *Properties* describe objects attributes, such as physical characteristics.
- *Roles* describe roles of objects in events.
- **Locations.** Places or settings in which events happens.
  - *Relations between locations* link locations between them.
- **Provenance.** This is a description of the origin of the data or statements.
  - *Narrative text* includes any original text narrative content.
- **Events and Actions.** Throughout the interviews, events appeared as the core element in the narrative. Similarly to agent, there is a distinction between an action and an event. An action happens because of the execution of a function by an agent, whereas an event simply happens. An action can be associated with pre-conditions and post-conditions, that enable the happening of the action, or result from the action.
  - *Types* distinguish between different types of events and actions.
  - *Properties* describe generic attributes, such as timestamps and locations.
  - *Relations* describe relation between events, including:
    * Mereological. Relations between parts and the wholes they form.
    * Temporal. Temporal relationships between events, such as the ones from Allen's Algebra [4].
    * Causal. Causal relationships between events.
- **Narrative Dynamics.** The higher-level elements that bind, influence, and drive the progression of the narrative. They play a critical role in shaping the overall story, providing the framework within which the narrative elements interact and evolve. They provide a type of narrative by bridging together elements from the above categories.
  - *Script* is a pattern of sequences of events that are related, like the restaurant scene [75]. This includes archetypes, plans, frames and processes.
  - *Storyline* is a coherent sequence of events, it is an instantiation of a script. This includes scenes, acts, episodes, sequences of events and situations.

- *State* is the particular condition that someone or something is at a specific time or during a specific time interval.
- *Goal* is a state that is desired by an agent.
- *Outcome/Effect* is a state that results from a causal relation, representing the new condition after an event.
- *Perspectives* are different interpretations of the same event.
- *Conditions* include conditions under which events or actions can happens. This includes pre-conditions, results or post-conditions, complete-condition, coherence-condition, and enablement.

2.2   Narrative Requirements Representation

In Sect. 2.1, we identified requirements for including semantic concepts in the representations of narratives across a broad range of domains. In this section, we map these representations to a variety of syntactic constructs for KGs. We first describe each KR formalism, ordered by representation complexity. The complexity level is based on the introduction of new abstraction levels, the ability to handle more entities, or the incorporation of more sophisticated reasoning.

- **basic:** Only binary relations without variables;
- **n-ary:** Relations involving more than two entities;
- **variables:** Placeholders that can be replaced with actual values or entities;
- **higher-order:** Representations that involve functions or predicates applied to other functions or predicates;
- **temporal relation:** Relationships that involve time aspects;
- **epistemic:** Relates to knowledge and beliefs about knowledge;
- **causal:** Cause-and-effect relationships between entities or events;
- **rules:** Logic-based statements that define conditions and implications.

The left part of Table 2 summarises the mapping of narrative requirements to the interviews. The most frequent requirements are events and actions, agents, location, and time. The frequency of (physical) objects is low, except in specific use cases such as Robotic Kitchen and Arts & Literature. The right side of Table 2 maps narrative requirements to their most suited KG representation forms. Objects, locations, and agents require only basic or n-ary representations. Provenance requires higher-order representations, while most events are simple except for temporal and causal relationships. Narrative dynamics demand the most complex representations, involving variables, higher-order, epistemic, or rules. Objects, locations, and agents are therefore the easiest to represent, while events and narrative dynamics require more advanced KR formalisms.

Lastly, Table 3 provides a mapping of KR formalisms (cf. Table 2) to various syntactic constructs (KG, query, rules). The left part of Table 3 describes the different existing types of KGs. The right part of Table 3 shows which KR formalisms can be used to represent each construct. Since some KGs are additive, e.g. blank nodes contain simple KG triples, we focus on the additional complexity introduced by the more advanced KG constructs described in the "Description"

Table 2. Mapping of narrative requirements to the interviews and their required KG representations. For a requirement r and interview I_i, a ✓ indicates that c was identified as a requirement in I_i. For a requirement r and a representation r, a ✓ indicates that c can be represented with r. Req: Requirements.

	Interviews							Syntactic Constructs								
Narrative Req.	I1	I2	I3	I4	I5	I6	Total	basic	n-ary	variables	higher-order	temporal	epistemic	causal	rules	Total
Agents	✓	✓	✓	✓	✓	✓	6	✓	✓							2
types	✓		✓				2	✓								1
properties	✓		✓				2	✓								1
roles	✓		✓	✓	✓		4	✓	✓							2
Objects				✓	✓		2	✓	✓							2
types				✓	✓		2	✓								1
properties				✓	✓		2	✓								1
roles				✓	✓		2	✓	✓							2
Locations	✓		✓	✓	✓		4	✓	✓							2
relations			✓	✓			2	✓	✓							2
Provenance	✓			✓	✓		3				✓					1
narrative text	✓			✓	✓		3				✓					1
Events/Actions	✓	✓	✓	✓	✓	✓	6	✓	✓			✓		✓		4
types	✓		✓				2	✓								1
properties	✓						1	✓								1
relations	✓	✓	✓	✓	✓	✓	6	✓	✓			✓		✓		4
Dynamics	✓	✓	✓	✓	✓	✓	6			✓	✓	✓		✓		4
script	✓		✓	✓	✓		4			✓					✓	2
storyline			✓				1				✓		✓			2
state		✓		✓	✓		3			✓						1
goal			✓	✓			2				✓		✓			2
outcome/effect						✓	1				✓		✓			2
perspective			✓	✓	✓		3				✓		✓			2
conditions	✓		✓	✓			3			✓					✓	2
Total	14	5	15	18	15	5		14	6	3	8	2	4	2	3	

column. The table highlights how different KR formalisms, ranging from basic structures like simple triples to more complex constructs such as rules and named graphs, require corresponding sophisticated KG representations.

Table 3. Mapping of KG/Query/Rules representations to KR formalisms. NG: Named Graph, LPG: Labelled Property Graph.

Name	Construct	Repr.	Description	basic	n-ary	variables	higher-order	temporal	epistemic	causal	rules
				KR formalisms							
Simple KG	KG	RDF	simple (s, p, o) triples	✓				✓		✓	
Blank Node			nodes with no IRI	✓	✓						
Singleton Property			triples annotated	✓							
Reification			statement nodes				✓	✓	✓		
NG		NG	triples grouped			✓					
LPG		LPG	key-value pairs				✓	✓	✓		
Hyperrelational Graph		RDF-Star	⟨s p1 o1⟩⟩ p2 o2				✓	✓	✓		
Query	Query	–	–			✓					
Rule	Rule	–	–					✓	✓		✓

3   Can Existing Ontologies Represent Narratives?

To answer **RQ2** on assessing whether ontologies can represent narrative requirements, we conduct a systematic review on event-centric and narrative-centric
ontologies. We then evaluate the ontologies based on how effectively they can
represent narrative requirements and based on their vocabulary usage.

3.1   Systematic Review

We conducted a systematic review to compare existing ontologies based on their
ability to represent narratives in a structured format. The process we used is
illustrated in Fig. 1. We used the query *"narrative OR event-centric"*, focusing
on abstracts and titles. We then incorporated references from survey papers
identified during the search or from other relevant sources, and included ontologies cited in the identified papers. Survey papers included surveys on narrative-
centric ontologies [88] and event-centric ontologies [66,76,78]. We included only
papers that presented a novel ontology relevant for representing a narrative
graph, with inclusion criteria determined through abstract review.

   This resulted in 61 papers: 45 were papers presenting an ontology, 1 presented both an ontology and a resource KG, and 15 were resources based on
these ontologies. Among the 46 (45 + 1) ontologies, we further analysed the
other ontologies on which they were built, such as DUL [31] or CIDOC-CRM [21].
This resulted in three categories of ontologies: (1) event-centric ontologies, which
includes (1a) top-level ontologies and (1b) ontologies extended from ontologies
from (1a),(2) narrative-centric ontologies, which includes (2a) narrative-centric
ontologies and (2b) ontologies extended from ontologies from (2a), and (3) media-
centric ontologies, which we discarded for our comparison since these were about
video content that we did not have in our interviews. Tables 4 and 5 describes
the 15 resources and 46 ontologies respectively. Figure 2 shows the visual dependencies on the ontologies from category (1b) and (2b).

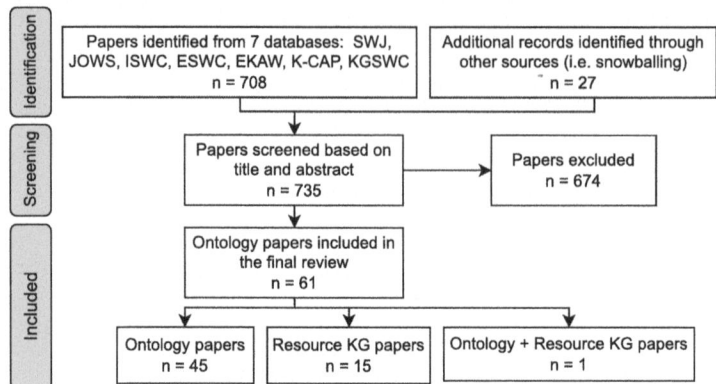

Fig. 1. Process used to retrieve papers for our comparison (PRISMA [63]). For the snowballing, we looked at paper references from the first iteration

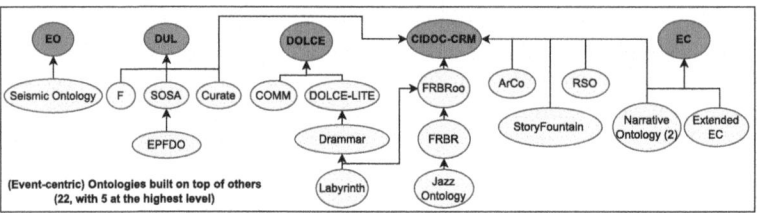

Fig. 2. The connections between the various event-centric ontologies (1). The ones coloured in blue (from 1a) are the ones we kept for our comparison: EO [71], DUL [31], DOLCE [30], CIDOC-CRM [21] and EC [45] (Color figure online)

We refer to our supplementary online material[9] for further details on the systematic review. The main document contains a description of our methodology and sources, and the folders contain the papers retrieved with the search engines. The ontologies highlighted in blue in Table 5 are the ones we retained for our comparison, i.e. the ontologies from categories (1a) and (2a). All the ontologies from (1b) and (2b) were extended from others from (1a) and (2a), so we focus on (1a) and (2a) ontologies instead. Table 6 describes the references found from the search engines: "all" refers to the number of results we got from the search queries, and "useful" refers to the number of references we kept.

---

[9] https://bit.ly/ekaw-2024-systematic-review.

Table 4. Resource Paper Description.

Id	Resource	Ontology
1	Structured Event Entity Resolution [42]	EO [71]
2	VLX-Stories [27]	VLX-Stories [27]
3,4	EventKG [33,34]	SEM [87]
5	DIVE [20]	SEM [87]
6	ECG [73]	SEM [87]
7	Wind in Our Sails [77]	CIDOC-CRM [21]
8	BiographySampo [38]	CIDOC-CRM [21]
9	ArCo [15]	CIDOC-CRM [21]
10	SCUPLTEUR [3]	CIDOC-CRM [21]
11	Using Event Spaces [59]	CIDOC-CRM [21], LODE [78]
12	EventMedia [44]	CIDOC-CRM [21], LODE [78]
13	OceanGraph [94]	SOSA [39] (extended from DUL [31])
14	3cixty [86]	DUL [31], Schema [2]
15	History of Vienna [46]	Schema [2]

3.2   Description of the Retained Ontologies

Some models we included come from a survey on narrative-centric ontologies [88], that reviewed 11 ontologies that explicitly mention the term "narrative" in their description. [60] proposes a storytelling ontology using RST [51] that describes events and objects, and how they are connected. Both [6,57] are extended from CIDOC-CRM [21], an ontology widely used in the cultural heritage domain. [6] focuses on representing core notions in narratology such as fabula, plot and narration, and [57] provides an interface to discover links between entities. A follow-up of [6] is the Narrative Ontology [54] to be used in Digital Libraries. The Mediation Ontology [41] is in the domain of international relations. The Labyrinth system [18] allows to discover narratives between entities in the domain of digital humanities and western archetypes. It re-uses ontologies like Drammar [50] and FRBRoo [22]. [84] presents a fabula model for emergent narratives, and re-uses the General Transition Network [85]. The main focus is to represent causal transitions between events. ODY-ONT [43] is used to query a literary text, and BKOnto [93] to represent biographical narratives. The CRBOnto ontology [65] relies upon Propp's morphology of folks stories [68]. [11] proposes an ontology to represent transmedia fiction such as the Marvel universe. [19] proposes a model based on Drammar [50] and DOLCE-LITE to represent actions motivated by heroes in the literature.

Event-centric ontologies encompass diverse domains such as: SEO [26] for scientific events, CIDOC-CRM [21] and EDM [23] for cultural heritage, SEM [87], LODE [78], ABC [48], EO [71], EC [45] and E [92] for more generic usages. CIDOC-CRM [21] is widely used for the cultural heritage domain, and several

Table 5. Extended from specifies the original ontology, if applicable, from which the listed *Ontologies* were derived

Type	Id	#	Ontologies	Extended from
Event	1a	16	EO [71], DUL [31], DOLCE [30], CIDOC-CRM [21], EC [45], SEM [87], LODE [78], the Enslaved Ontology [79], ABC [48], EDM [23], VLX-Stories [27], E [92], SEO [26], FARO [72], ProppOntology [64], BBO [47]	–
Event	1b	1	Seismic Ontology [52]	E [71]
		3	F [76], SOSA [39], EPFDO [53]	DUL [31]
		1	Curate [58]	DUL [31], CIDOC-CRM [21]
		2	COMM [5], DOLCE-LITE	DOLCE [30]
		1	FRBR [62]	FRBRoo [22]
		1	the Jazz Ontology [69]	FRBR [62]
		1	Extended EC [17]	EC [45]
Narrative	2a	7	RST [60], Mediation [41], Princess Lovely [84], StoryTeller [93], ODY-ONT [43], DL Ontology for Fairy Tale [65], Transmedia Fictional Worlds [11]	–
Narrative	2b	1	Drammar [19]	DOLCE-LITE
		1	AO [18]	Drammar [19], FRBRoo [22]
		4	FRBRoo [22], ArCo [15], StoryFoutain [57], RSO [61]	CIDOC-CRM [21]
		1	Narrative Ontology [6]	CIDOC-CRM [21], EC [45]
Media	3	6	OntoMedia [49], SsVM [25], Eventory [91], DISC [32], VERL [29], RDF-Events [82]	–

Table 6. Provenance description of papers from the systematic review. N: Narrative, E: Event-centric. For *K-CAP/N (useful)*, 5 papers were not accessible.

Provenance	N (all)	N (useful)	E (all)	E (useful)
SWJ	7	2	2	2
JOWS	21	2	147	7
ISWC	35	9	66	2
ESWC	32	2	106	1
EKAW	16	1	14	1
K-CAP	44	1	176	2
KGSWC	2	1	40	4

other ontologies have since been derived from, such as FRBR [62] for bibliographic records. DUL [31] is a combination and a simplification of DOLCE [30] and the DnS [31] ontologies. Several ontologies were extended from DUL, such as F [76], originally for emergency response and SOMA-NARR [67]. The Enslaved Ontology [79] integrates content about the historical slave trade from diverse sources, while VLX-Stories [27] structures text from media feeds into an event-centric ontology. ProppOntology [64] is built upon Propp's morphology of the folktale [68]. Lastly, BBO [47] represents dynamic BPMN process executions in structured format and FARO [72] focuses on event relations.

3.3 Evaluation of Narrative Ontologies

Background for Ontology Evaluation. Tim Berners-Lee defined a 5-star ranking schema to assess the quality of Linked Data [1]. This scoring system was based on four rules which did not include any considerations with respect to the vocabulary that includes broader categories such as ontologies and thesauri. [40] therefore proposed a new 5-star system to evaluate the quality of the vocabulary, displayed in Fig. 3. We re-use this 5-star system to compare the ontologies.

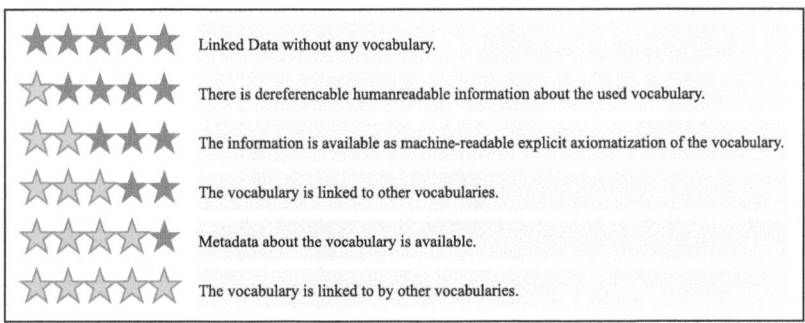

Fig. 3. The 5-star system to evaluate ontologies from [40]

Analysis of the Ontologies. Table 7 illustrates the coverage of our narrative requirements by the narrative and event-centric ontologies. The ontologies are listed in descending order of their 5-star scores [40], and subsequently in descending order of the number of requirements they cover. Each column in the table shows its 5-star score, how effectively an ontology fulfills the narrative requirements, whether the ontology is event-centric or narrative-centric, and whether it is domain-specific. We hereafter detail the main findings from Table 7.

Simple Requirements with Event-Centric Ontologies. For narratives with simple requirements such as event without intricate relationships or narrative dynamics, an event-centric ontology like SEM [87] may suffice. Event-centric ontologies generally offer higher quality in terms of 5-star score, but tend to represent less complex narrative requirements, including basic relationships between events and narrative dynamics.

The Case of DOLCE [30] and DUL [31]. They can represent the broadest range of requirements, albeit at the cost of complexity and readability in terms of representation [72], since these are upper-level ontologies. By design, they

Table 7. Coverage of narrative requirements for narrative and event-centric ontologies. For a requirement r and an ontology o, a ✓ indicates that r can be represented in o. The number of stars on top indicate the 5-star ranking from Sect. 3.3. E: event-centric, N: narrative-centric. Ontologies marked with an asterisk (*) are unofficial designations used for clarity in this table and the remainder of the paper, as the original papers did not provide official names.

Narrative Requirements	DOLCE [30]	DUL [31]	CIDOC-CRM [21]	EDM [23]	EO [71]	EC [45]	ProppOntology [64]	Enslaved [79]	SEO [26]	SEM [87]	Transmedia* [11]	FARO [72]	LODE [78]	ABC [48]	ODY-ONT [43]	Storytelling* [60]	BBO [47]	CRBOnto [65]	VLX-Stories [27]	Labyrinth* [18]	E [92]	Mediation [41]	Emergent* [84]	BKOnto [93]	Total
	★★★★★	★★★★★	★★★★★	★★★★★	★★★★★	★★★★★	★★★★☆	★★★★☆	★★★★☆	★★★★☆	★★★★☆	★★★★☆	★★★★☆	★★★☆☆	★★★☆☆	★★★☆☆	★★☆☆☆	★★☆☆☆	★☆☆☆☆	★☆☆☆☆	★☆☆☆☆	★☆☆☆☆	★☆☆☆☆	☆☆☆☆☆	
Agents	✓	✓	✓	✓	✓		✓	✓	✓	✓	✓		✓	✓	✓	✓	✓	✓	✓	✓	✓	✓	✓		21
types	✓	✓					✓		✓	✓		✓			✓		✓		✓	✓					10
properties	✓	✓							✓			✓			✓										5
roles	✓	✓					✓	✓		✓					✓	✓		✓							8
Objects	✓	✓	✓	✓			✓			✓				✓			✓		✓	✓					10
types	✓	✓	✓	✓			✓			✓				✓			✓		✓	✓					10
properties	✓	✓	✓	✓			✓			✓				✓				✓	✓	✓					10
roles	✓	✓													✓		✓	✓							5
Locations	✓	✓	✓	✓	✓		✓	✓	✓	✓	✓		✓	✓	✓		✓	✓	✓	✓	✓	✓		✓	20
relations			✓	✓	✓				✓											✓					5
Provenance	✓	✓	✓	✓				✓	✓								✓		✓		✓	✓			10
narrative text	✓	✓	✓					✓	✓								✓								6
Events/Actions	✓	✓	✓	✓	✓	✓	✓	✓	✓	✓	✓	✓	✓	✓	✓		✓	✓	✓	✓	✓	✓	✓	✓	23
types	✓	✓	✓				✓	✓	✓	✓		✓		✓			✓		✓				✓		12
properties									✓																1
relations	✓	✓	✓	✓	✓	✓	✓	✓	✓				✓	✓	✓	✓				✓	✓	✓		✓	17
Dynamics	✓	✓																							2
script	✓	✓																	✓				✓		4
storyline	✓	✓		✓			✓			✓					✓		✓								7
state	✓	✓	✓							✓									✓	✓					6
goal	✓	✓					✓			✓													✓		5
outcome/effect	✓	✓																					✓		3
perspective	✓	✓														✓							✓		4
conditions	✓	✓											✓										✓		4
Total	21	21	11	11	5	2	13	10	9	7	7	5	5	8	7	5	7	7	11	9	7	6	6	4	0
Event vs. Narrative-centric	E	E	E	E	E	E	E	E	E	E	N	E	E	E	N	N	E	N	E	N	E	N	N	N	
Domain-specific			✓	✓			✓	✓	✓	✓	✓		✓		✓	✓			✓	✓		✓	✓	✓	15

are made to represent any domain, but users would need to define their own templates for specific applications or requirements, such as perspectives, which could be time-consuming.

Domain-Specific Ontologies. Apart from DUL [31] and DOLCE [30], the ontologies that can represent the most requirements tend to be domain-specific ontologies, in domains such as the digital humanities [21,23] and literary texts [64]. This is good if one wants to build a narrative in this particular domain, but might be challenging to extend to other use cases.

Ontologies with Lower 5-Star Rating. Some ontologies such as Mediation [41] or Emergent*[10] [84] adhere less to established web standards, which

---

[10] Cf. Table 7 for the explanation on the asterisk.

not only lowers their 5-star scores but also significantly their interoperability and integration.

Narrative Dynamics. Fewer ontologies can represent the narrative dynamics, that are core to binding and driving the progression of the narrative elements. Among the 4-rated domain-specific ontologies, SEO [26] and Transmedia* [11] can represent scripts and storylines. Among the 4-rated generic ontologies, SEM [87] can represent perspectives through the `sem:View` class and blank nodes, and FARO [72] goals, states and conditions. The other ontologies that can represent narrative dynamics have lower 5-star ranking (cf. 1-score models in Table 7).

Other. The number of requirements an ontology can represent varies significantly, from 21 in DUL [31] and DOLCE [30] at the cost of complexity, to 2 in EC [45], although both with a high 5-star score. There is a positive correlation between key requirements, such as agent or event representation, and their presence in ontologies. It is important to note that the presence of a checkmark ($\checkmark$) indicates the representation of a requirement but does not reflect the number of such relations an ontology can represent. For instance, some ontologies only include basic relationships between events, while others offer a more extensive list. Lastly, the simpler narrative requirements such as agents and objects are well covered by event-centric ontologies, whereas the more complex ones like narrative dynamics tend to be less covered or less adhering to web standards, highlighting the ongoing challenges of narrative representation in structured format.

4   Recommendations

To answer **RQ3** on the best ontology to represent a narrative, we provide generic recommendations on how to best identify a structured representation for a narrative. We lastly provide tailored recommendations from the interviews use case (cf. Tables 1 and 2).

4.1   Generic Recommendations

Based on the previous analyses on the narrative requirements and their KR formalisms (Sect. 2), and on how effectively ontologies fullfill these narrative requirements (Sect. 3), we recommend using the following steps to effectively build a narrative in a structured format for a specific application:

1. Refer to Table 2 to **identify the narrative requirements** that are relevant to your project. These tables categorise narrative requirements by their complexity, aiding in the selection process.
2. Based on the chosen narrative requirements, identify the **types of KR formalisms** needed from Tables 2 and 3. This will help in understanding the level of complexity your model must support.

3. Choose **ontologies that can effectively represent** the identified narrative requirements. Use Table 7 to find ontologies that cover your narrative requirements and prioritise ontologies that have a high score on the 5-star schema ranking [40].
4. **Extend existing ontologies** by picking the one that fits best, and ensure it adheres to web standards to facilitate future integration, reuse and collaboration.
5. **Consider developing your own ontology** if existing ones are insufficient, and ensure it adheres to web standards to facilitate future integration, reuse and collaboration.

4.2   Tailored Recommendations from Interviews Use Cases

Table 8 provides an overview of the recommendations we give. **I1 and I3**, within the History and Social Media domains respectively, share numerous narrative requirements. The primary distinction is that I1 includes more generic event descriptions, while I3 encompasses more narrative dynamics such as goals and perspectives. Our recommendations are tailored to the type of KG employed, each with its own advantages and drawbacks. For example, reification is compliant with RDF standards but may be inefficient. If using reification exclusively, we suggest employing the core classes of DOLCE [30] or DUL [31], whose ontologies and design patterns are based on n-ary relationships. For I3 and the perspectives, SEM [87] is recommended due to its existing implementation. Other KGs, such as Named Graphs, are more compact but can become overloaded with numerous identifiers, while RDF-star is more efficient but still not widely adopted. In these cases, we recommend using DOLCE [30] or DUL [31] for simple narrative requirements and extending them to accommodate higher-order relationships. Additionally, we advise considering FARO [72], which, to our knowledge, can represent a wide range of event relations, including the goals pertinent to I3.

Table 8. Recommendations from interview use-cases.

Id	Domain	Recommendation
I1	History	DOLCE [30]/DUL [31] if reification, FARO [72]
I2	Biomedical Science	PICO with FARO [72]
I3	Social Media Analysis	Same as I1 + SEM [87] for perspectives
I4	Robotic Kitchen	DUL [31] or derived ontologies (as per the literature)
I5	Arts & Literature	ProppOntology [68], extending Emergent* [84] if time permits
I6	Cognitive Science	Extend or create an ontology

I2 and I6, within the Biomedical Science and Cognitive Science domains respectively, both have a limited number of narrative requirements. The primary difference is that the input data for I2 is already structured according to the PICO ontology[11], whereas the input data for I6 is unstructured. Given their limited conceptual needs, we recommend targeting ontologies that precisely represent these requirements rather than selecting the most comprehensive ones. For I2, which requires agents and relationships between events and states, we suggest combining PICO with FARO [72]. FARO is capable of representing numerous relationships between events and states, while PICO provides domain-specific vocabulary for agents. For I6, which requires agents, relationships between events, and outcomes/effects, Table 7 indicates that this last requirement is not well-represented across ontologies. The options available are the generic DOLCE [30]/DUL [31] or the domain-specific Emergent* [84]. Designing new patterns over DOLCE/DUL may be too complex for those not specialized in the semantic web. Therefore, we recommend either extending existing ontologies by creating new classes or properties, or building a new one tailored to these specific requirements.

I4 in the Robotic Kitchen domain has the most narrative requirements including objects and numerous narrative dynamics. From the literature, it is also much linked to sensory information from robots [7,8]. Most work in this area like SOSA [39] has used or built upon DUL [31]. In line with the existing work, we would recommend continuing using DUL [31] or derived ontologies.

I5 in the Arts and Literature domain overlaps with domains-specific ontologies in Table 7. We recommend starting with these ontologies. ProppOntology [68] covers most simple requirements and some complex ones like storyline. For other narrative dynamics such as outcome/effect or conditions, designing patterns over DOLCE [30]/DUL [31] is an option. However, given the domain-specific nature of this application, extending Emergent* [84] to cover more narrative dynamics and improve its sustainability is a preferable approach if time permits.

5   Discussion and Conclusion

In this paper, we map identified narrative requirements to appropriate knowledge representations. We conduct six interviews and identify core narrative requirements, as well as the KR formalisms they require. We find that requirements like objects or agents require simpler KR formalisms, while requirements such as events or narrative dynamics require more complex KR formalisms (RQ1). We only included domains relevant to MUHAI, limiting our findings to text-based or sensor-based narratives and potentially reducing applicability to other domains such as multimedia content, that could be explored in future work.

We then conduct a systematic review to extract event-centric and narrative-centric ontologies to represent narratives. We analyse the 24 retained ontologies with respect to how effectively they can represent narrative requirements and how well they are rated on the 5-star vocabulary system [40] (RQ2). We found that simple requirements are best represented by event-centric ontologies. Event-centric ontologies such as DUL [31] and DOLCE [30] cover a broad range of

---

[11] https://linkeddata.cochrane.org/pico-ontology.

requirements but this comes at the cost of complexity [72], since these are upper-level ontologies, which means that one might need additional work to define templates. Lower-rated ontologies struggle with interoperability due to poor web standards adherence. Fewer ontologies effectively represent narrative dynamics, with SEM [87] and FARO [72] being notable exceptions with high ratings for goals, perspectives and conditions. Coverage of complex requirements is more limited and often poorly standardized.

We lastly provide recommendations on how to best pick a model to represent a narrative, and apply them to the interview s (RQ3). The interview subjects and domain-specific ontologies cover key domains of narrative representation, but the list is not exhaustive, potentially limiting generalizability. Our semantic-web focus may narrow the scope of our recommendations to semantic web-related techniques, but broader exploration could lead to more comprehensive guidance across various KR paradigms.

In future work, we aim to enhance the evaluation of models by integrating other metrics for ontology evaluation, and to improve the generalisability of this study. Various works [12,14,24,36,37,70,90] have attempted to define criteria to evaluate an ontology from various perspectives such as Software Engineering, verification, validation or semiotics. Most importantly, we are actively working on automating the generation of structured narratives from KGs. Our initial efforts included a prototype for generating a narrative on the French Revolution [10] using the SEM ontology [87], where the data collection was performed manually. We are now developing a method to automatically retrieve relevant content to construct a narrative in the form of a KG.

Acknowledgments. This work was funded by the European MUHAI project (Horizon 2020 research and innovation program) under grant agreement number 951846 and the Sony Computer Science Laboratories - Paris. We thank our MUHAI collaborators for participating in the interviews, and Martina Galletti and the reviewers for their constructive feedback.

References

1. Linked Data - Design Issues. https://www.w3.org/DesignIssues/LinkedData.html
2. Schema.org (2024). https://schema.org. Accessed 17 Sep 2024
3. Addis, M., et al.: SCULPTEUR: towards a new paradigm for multimedia museum information handling. In: Fensel, D., Sycara, K., Mylopoulos, J. (eds.) ISWC 2003. LNCS, vol. 2870, pp. 582–596. Springer, Heidelberg (2003). https://doi.org/10.1007/978-3-540-39718-2_37
4. Allen, J.F.: Maintaining knowledge about temporal intervals. Commun. ACM **26**(11), 832–843 (1983)
5. Arndt, R., Troncy, R., Staab, S., Hardman, L., Vacura, M.: COMM: designing a well-founded multimedia ontology for the web. In: Aberer, K., et al. (eds.) ASWC/ISWC -2007. LNCS, vol. 4825, pp. 30–43. Springer, Heidelberg (2007). https://doi.org/10.1007/978-3-540-76298-0_3
6. Bartalesi, V., Meghini, C., Metilli, D.: Steps towards a formal ontology of narratives based on narratology. In: 7th Workshop on Computational Models of Narrative (CMN 2016). Schloss Dagstuhl-Leibniz-Zentrum fuer Informatik (2016)

7. Beßler, D., Porzel, R., Pomarlan, M., Beetz, M., Malaka, R., Bateman, J.: A formal model of affordances for flexible robotic task execution. In: ECAI 2020, pp. 2425–2432. IOS Press (2020)
8. Beßler, D., et al.: Foundations of the socio-physical model of activities (SOMA) for autonomous robotic agents 1. In: Formal ontology in information systems, pp. 159–174. IOS Press (2021)
9. Blin, I.: Building a French revolution narrative from wikidata. In: SEM4NBU@ IJCAI, pp. 1–5 (2022)
10. Blin, I.: Building narrative structures from knowledge graphs. In: Groth, P., et al. The Semantic Web: ESWC 2022 Satellite Events. ESWC 2022. LNCS, vol. 13384, pp. 234–251. Springer, Cham (2022). https://doi.org/10.1007/978-3-031-11609-4_38
11. Branch, F., Arias, T., Kennah, J., Phillips, R., Windleharth, T., Lee, J.H.: Representing transmedia fictional worlds through ontology. J. Am. Soc. Inf. Sci. 68(12), 2771–2782 (2017)
12. Brank, J., Grobelnik, M., Mladenic, D.: A survey of ontology evaluation techniques. In: Proceedings of the Conference on Data Mining and Data Warehouses (SiKDD 2005), pp. 166–170. Citeseer (2005)
13. Bruner, J.: The narrative construction of reality. Crit. Inq. 18(1), 1–21 (1991)
14. Burton-Jones, A., Storey, V.C., Sugumaran, V., Ahluwalia, P.: A semiotic metrics suite for assessing the quality of ontologies. Data Knowl. Eng. 55(1), 84–102 (2005). https://doi.org/10.1016/j.datak.2004.11.010. https://www.sciencedirect.com/science/article/pii/S0169023X0400223X
15. Carriero, V.A., et al.: ArCo: The Italian Cultural Heritage Knowledge Graph. In: Ghidini, C., et al. (eds.) ISWC 2019. LNCS, vol. 11779, pp. 36–52. Springer, Cham (2019). https://doi.org/10.1007/978-3-030-30796-7_3
16. Caselli, T., Hovy, E., Palmer, M., Vossen, P.: Computational Analysis of Storylines: Making Sense of Events. Cambridge University Press, Cambridge (2021)
17. Cervesato, I., Franceschet, M., Montanari, A.: A guided tour through some extensions of the event calculus. Comput. Intell. 16(2), 307–347 (2000)
18. Damiano, R., Lieto, A.: Ontological representations of narratives: a case study on stories and actions. In: 2013 Workshop on Computational Models of Narrative, pp. 76–93. Schloss Dagstuhl Leibniz-Zentrum für Informatik (2013)
19. Damiano, R., Lombardo, V., Pizzo, A.: The ontology of drama. Appl. Ontol. 14(1), 79–118 (2019)
20. De Boer, V.: Dive into the event-based browsing of linked historical media. J. Web Semant. 35, 152–158 (2015)
21. Doerr, M.: The CIDOC conceptual reference module: an ontological approach to semantic interoperability of metadata. AI Mag. 24(3), 75–75 (2003)
22. Doerr, M., Bekiari, C., LeBoeuf, P., nationale de France, B.: FRBROO, a conceptual model for performing arts. In: 2008 Annual Conference of CIDOC, Athens, pp. 15–18 (2008)
23. Doerr, M., Gradmann, S., Hennicke, S., Isaac, A., Meghini, C., Van de Sompel, H.: The Europeana data model (EDM). In: World Library and Information Congress: 76th IFLA General Conference and Assembly. vol. 10, p. 15. IFLA (2010)
24. Duque-Ramos, A., Fernández-Breis, J.T., Stevens, R., Aussenac-Gilles, N.: OQuaRE: A SQuaRE-based approach for evaluating the quality of ontologies. J. Res. Pract. Inf. Technol. 43(2), 159–176 (2011)
25. Ekin, A., Tekalp, A.M., Mehrotra, R.: Integrated semantic-syntactic video modeling for search and browsing. IEEE Trans. Multimedia 6(6), 839–851 (2004)

26. Fathalla, S., Vahdati, S., Lange, C., Auer, S.: SEO: A scientific events data model. In: Ghidini, C., et al. (eds.) ISWC 2019. LNCS, vol. 11779, pp. 79–95. Springer, Cham (2019). https://doi.org/10.1007/978-3-030-30796-7_6

27. Fernàndez-Cañellas, D., et al.: VLX-stories: building an online event knowledge base with emerging entity detection. In: Ghidini, C., et al. (eds.) ISWC 2019. LNCS, vol. 11779, pp. 382–399. Springer, Cham (2019). https://doi.org/10.1007/978-3-030-30796-7_24

28. Filtz, E., Navas-Loro, M., Santos, C., Polleres, A., Kirrane, S.: Events matter: Extraction of events from court decisions. Legal Knowledge and Information Systems, pp. 33–42 (2020)

29. Francois, A.R., Nevatia, R., Hobbs, J., Bolles, R.C., Smith, J.R.: VERL: an ontology framework for representing and annotating video events. IEEE Multimedia **12**(4), 76–86 (2005)

30. Gangemi, A., Guarino, N., Masolo, C., Oltramari, A., Schneider, L.: Sweetening ontologies with DOLCE. In: Gómez-Pérez, A., Benjamins, V.R. (eds.) EKAW 2002. LNCS (LNAI), vol. 2473, pp. 166–181. Springer, Heidelberg (2002). https://doi.org/10.1007/3-540-45810-7_18

31. Gangemi, A., Mika, P.: Understanding the semantic web through descriptions and situations. In: Meersman, R., Tari, Z., Schmidt, D.C. (eds.) OTM 2003. LNCS, vol. 2888, pp. 689–706. Springer, Heidelberg (2003). https://doi.org/10.1007/978-3-540-39964-3_44

32. Geurts, J., Bocconi, S., van Ossenbruggen, J., Hardman, L.: Towards ontology-driven discourse: from semantic graphs to multimedia presentations. In: Fensel, D., Sycara, K., Mylopoulos, J. (eds.) ISWC 2003. LNCS, vol. 2870, pp. 597–612. Springer, Heidelberg (2003). https://doi.org/10.1007/978-3-540-39718-2_38

33. Gottschalk, S., Demidova, E.: EventKG: a multilingual event-centric temporal knowledge graph. In: Gangemi, A., et al. (eds.) ESWC 2018. LNCS, vol. 10843, pp. 272–287. Springer, Cham (2018). https://doi.org/10.1007/978-3-319-93417-4_18

34. Gottschalk, S., Demidova, E.: EventKG - the Hub of Event Knowledge on the Web - and Biographical Timeline Generation. vol. 10, pp. 1039–1070. IOS Press (2019)

35. Guan, S., et al.: What is event knowledge graph: a survey. IEEE Trans. Knowl. Data Eng. **35**(7), 7569–7589 (2022)

36. Gómez-Pérez, A.: Ontology evaluation. In: Staab, S., Studer, R. (eds.) Handbook on Ontologies. International Handbooks on Information Systems, pp. 251–273. Springer, Berlin, Heidelberg (2004). https://doi.org/10.1007/978-3-540-24750-0_13

37. Hlomani, H., Stacey, D.: Approaches, methods, metrics, measures, and subjectivity in ontology evaluation: a survey. Semantic Web J. **1**(5), 1–11 (2014)

38. Hyvönen, E., et al.: BiographySampo – publishing and enriching biographies on the semantic web for digital humanities research. In: Hitzler, P., et al. (eds.) ESWC 2019. LNCS, vol. 11503, pp. 574–589. Springer, Cham (2019). https://doi.org/10.1007/978-3-030-21348-0_37

39. Janowicz, K., Haller, A., Cox, S.J., Le Phuoc, D., Lefrançois, M.: SOSA: a lightweight ontology for sensors, observations, samples, and actuators. J. Web Semantics **56**, 1–10 (2019)

40. Janowicz, K., Hitzler, P., Adams, B., Kolas, D., Vardeman, C., II.: Five stars of Linked Data vocabulary use. Semantic Web **5**(3), 173–176 (2014). https://doi.org/10.3233/SW-140135. https://content.iospress.com/articles/semantic-web/sw135, publisher: IOS Pressv

41. Kattagoni, V., Singh, N.: IREvent2Story: a novel mediation ontology and narrative generation. In: Text2Story@ ECIR, pp. 5–14 (2018)

42. Kejriwal, M., Peng, J., Zhang, H., Szekely, P.: Structured event entity resolution in humanitarian domains. In: Vrandečić, D., et al. (eds.) ISWC 2018. LNCS, vol. 11136, pp. 233–249. Springer, Cham (2018). https://doi.org/10.1007/978-3-030-00671-6_14

43. Khan, A.F., Bellandi, A., Benotto, G., Frontini, F., Giovannetti, E., Reboul, M.: Leveraging a narrative ontology to query a literary text. In: 7th Workshop on Computational Models of Narrative (CMN 2016). Schloss Dagstuhl-Leibniz-Zentrum fuer Informatik (2016)

44. Khrouf, H., Troncy, R.: EventMedia: a LOD dataset of events illustrated with media. Semantic Web **7**(2), 193–199 (2016). https://doi.org/10.3233/SW-150184. https://content.iospress.com/articles/semantic-web/sw184, publisher: IOS Press

45. Kowalski, R., Sergot, M.: A logic-based calculus of events. N. Gener. Comput. **4**, 67–95 (1986)

46. Krabina, B.: Building a knowledge graph for the history of Vienna with semantic MediaWiki. J. Web Semantics **76**, 100771 (2023). https://doi.org/10.1016/j.websem.2022.100771. https://www.sciencedirect.com/science/article/pii/S1570826822000555

47. Krause, F., Kurniawan, K., Kiesling, E., Paulheim, H., Polleres, A.: On the representation of dynamic BPMN process executions in knowledge graphs. In: Ortiz-Rodriguez, F., Villazón-Terrazas, B., Tiwari, S., Bobed, C. (eds.) Knowledge Graphs and Semantic Web. KGSWC 2023. LNCS, vol. 14382, pp. 97–105. Springer, Cham (2023). https://doi.org/10.1007/978-3-031-47745-4_8

48. Lagoze, C., Hunter, J.: The ABC ontology and model (2002)

49. Lawrence, K.F., et al.: OntoMedia-creating an ontology for marking up the contents of heterogeneous media (2005)

50. Lombardo, V., Damiano, R., Pizzo, A.: Drammar: a comprehensive ontological resource on drama. In: Vrandečić, D., et al. (eds.) ISWC 2018. LNCS, vol. 11137, pp. 103–118. Springer, Cham (2018). https://doi.org/10.1007/978-3-030-00668-6_7

51. Mann, W.C., Thompson, S.A.: Rhetorical Structure Theory: A Theory of Text Organization. University of Southern California, Information Sciences Institute Los Angeles (1987)

52. Mastoras, V., et al.: Towards a framework for seismic data. In: Ortiz-Rodriguez, F., Villazón-Terrazas, B., Tiwari, S., Bobed, C. (eds.) Knowledge Graphs and Semantic Web. KGSWC 2023. LNCS, vol. 14382, pp. 106–119. Springer, Cham (2023). https://doi.org/10.1007/978-3-031-47745-4_9

53. Mederos, A.L., García-Duarte, D., Lio, D.G., Hidalgo-Delgado, Y., Ruíz, J.A.S.: An ontological model for the failure detection in power electric systems. In: Villazón-Terrazas, B., Ortiz-Rodríguez, F., Tiwari, S.M., Shandilya, S.K. (eds.) KGSWC 2020. CCIS, vol. 1232, pp. 130–146. Springer, Cham (2020). https://doi.org/10.1007/978-3-030-65384-2_10

54. Meghini, C., Bartalesi, V., Metilli, D.: Representing narratives in digital libraries: the narrative ontology. Semantic Web **12**(2), 241–264 (2021)

55. Meroño-Peñuela, A., et al.: Semantic technologies for historical research: a survey. Semantic Web **6**(6), 539–564 (2015). https://doi.org/10.3233/SW-140158. https://content.iospress.com/articles/semantic-web/sw158, publisher: IOS Press

56. Miller, B., Lieto, A., Ronfard, R., Ware, S., Finlayson, M.: Proceedings of the 7th workshop on computational models of narrative. In: 7th Workshop on Computational Models of Narrative (CMN 2016). vol. 53 (2016)

57. Mulholland, P., Collins, T., Zdrahal, Z.: Story fountain: intelligent support for story research and exploration. In: Proceedings of the 9th International Conference on Intelligent User Interfaces, pp. 62–69 (2004)

58. Mulholland, P., Wolff, A., Collins, T.: Curate and storyspace: an ontology and web-based environment for describing curatorial narratives. In: Simperl, E., Cimiano, P., Polleres, A., Corcho, O., Presutti, V. (eds.) ESWC 2012. LNCS, vol. 7295, pp. 748–762. Springer, Heidelberg (2012). https://doi.org/10.1007/978-3-642-30284-8_57

59. Mulholland, P., Wolff, A., Kilfeather, E., McCarthy, E.: Using event spaces, setting and theme to assist the interpretation and development of museum stories. In: Janowicz, K., Schlobach, S., Lambrix, P., Hyvönen, E. (eds.) EKAW 2014. LNCS (LNAI), vol. 8876, pp. 320–332. Springer, Cham (2014). https://doi.org/10.1007/978-3-319-13704-9_25

60. Nakasone, A., Ishizuka, M.: Storytelling ontology model using RST. In: 2006 IEEE/WIC/ACM International Conference on Intelligent Agent Technology, pp. 163–169. IEEE (2006)

61. Oldman, D., Tanase, D.: Reshaping the knowledge graph by connecting researchers, data and practices in ResearchSpace. In: Vrandečić, D., et al. (eds.) ISWC 2018. LNCS, vol. 11137, pp. 325–340. Springer, Cham (2018). https://doi.org/10.1007/978-3-030-00668-6_20

62. O'Neill, E.T.: FRBR: functional requirements for bibliographic records. Libr. Resour. Tech. Serv. **46**(4), 150–159 (2011)

63. Page, M.J., et al.: PRISMA 2020 explanation and elaboration: updated guidance and exemplars for reporting systematic reviews. BMJ **372**, n160 (2021). https://doi.org/10.1136/bmj.n160D. https://www.bmj.com/content/372/bmj.n160, publisher: British Medical Journal Publishing Group Section: Research Methods & Reporting

64. Pannach, F., Sporleder, C., May, W., Krishnan, A., Sewchurran, A.: Of lions and Yakshis. Semantic Web **12**(2), 219–239 (2021). https://doi.org/10.3233/SW-200417. https://content.iospress.com/articles/semantic-web/sw200417, publisher: IOS Press

65. Peinado, F., Gervás, P., Díaz-Agudo, B.: A description logic ontology for fairy tale generation. In: Proceedings of the Workshop on Language Resources for Linguistic Creativity, LREC. vol. 4, pp. 56–61 (2004)

66. Piryani, R., Aussenac-Gilles, N., Hernandez, N.J.: Comprehensive survey on ontologies about event. In: ESWC Workshops on Semantic Methods for Events and Stories (SEMMES @ ESWC 2023). vol. 3443, pp. 1–15. ceur-ws.org, Hersonissos, Greece (2023). https://hal.science/hal-04283673, backup Publisher: Pasquale Lisena and Ilaria Tiddi and Simon Gottschalk and Luc Steels

67. Porzel, R.: On formalizing narratives. In: JOWO (2021)

68. Propp, V.: Morphology of the Folktale. University of Texas Press (1968)

69. Proutskova, P., et al.: The jazz ontology: a semantic model and large-scale RDF repositories for jazz. J. Web Semantics **74**, 100735 (2022). https://doi.org/10.1016/j.websem.2022.100735. https://www.sciencedirect.com/science/article/pii/S1570826822000245

70. Raad, J., Cruz, C.: A survey on ontology evaluation methods. In: Proceedings of the International Conference on Knowledge Engineering and Ontology Development, part of the 7th International Joint Conference on Knowledge Discovery, Knowledge Engineering and Knowledge Management. Lisbonne, Portugal (2015). https://doi.org/10.5220/0005591001790186, https://hal.science/hal-01274199

71. Raimond, Y., Abdallah, S.: The event ontology (2007)

72. Rebboud, Y., Lisena, P., Troncy, R.: Beyond causality: representing event relations in knowledge graphs. In: Corcho, O., Hollink, L., Kutz, O., Troquard, N., Ekaputra,

F.J. (eds.) Knowledge Engineering and Knowledge Management. EKAW 2022. LNCS(), vol. 13514, pp. 121–135. Springer, Cham (2022). https://doi.org/10.1007/978-3-031-17105-5_9

73. Rospocher, M., et al.: Building event-centric knowledge graphs from news. J. Web Semantics **37**, 132–151 (2016)

74. Santagiustina, C.R.M.A., Spillner, L., Mildner, T., Porzel, R., et al.: Towards conflictual narrative mechanics. In: CEUR WORKSHOP PROCEEDINGS, pp. 19–27. CEUR-WS (2022)

75. Schank, R.C., Abelson, R.P.: Scripts, plans, and knowledge. In: IJCAI. vol. 75, pp. 151–157. New York (1975)

76. Scherp, A., Franz, T., Saathoff, C., Staab, S.: F–a model of events based on the foundational ontology DOLCE+ DNS ultralight. In: Proceedings of the Fifth International Conference on Knowledge Capture, pp. 137–144 (2009)

77. Schouten, S., De Boer, V., Petram, L., Van Erp, M.: The wind in our sails: developing a reusable and maintainable Dutch maritime history knowledge graph. In: Proceedings of the 11th on Knowledge Capture Conference, pp. 97–104 (2021)

78. Shaw, R., Troncy, R., Hardman, L.: LODE: linking open descriptions of events. In: Gómez-Pérez, A., Yu, Y., Ding, Y. (eds.) ASWC 2009. LNCS, vol. 5926, pp. 153–167. Springer, Heidelberg (2009). https://doi.org/10.1007/978-3-642-10871-6_11

79. Shimizu, C., et al.: The enslaved ontology: peoples of the historic slave trade. J. Web Semantics **63**, 100567 (2020). https://doi.org/10.1016/j.websem.2020.100567. https://www.sciencedirect.com/science/article/pii/S1570826820300135

80. Steels, L., Verheyen, L., Van Trijp, R.: An experiment in measuring understanding. In: Schlobach, S., Pérez-Ortiz, M., Tielman, M. (eds.) Frontiers in Artificial Intelligence and Applications. IOS Press (2022). https://doi.org/10.3233/FAIA220203, https://ebooks.iospress.nl/doi/10.3233/FAIA220203

81. Stork, L., Tiddi, I., Spijker, R., ten Teije, A.: Explainable drug repurposing in context via deep reinforcement learning. In: Pesquita, C., et al. The Semantic Web. ESWC 2023. LNCS, vol. 13870, pp. 3–20. Springer, Cham (2023). https://doi.org/10.1007/978-3-031-33455-9_1

82. Stühmer, R., Anicic, D., Sen, S., Ma, J., Schmidt, K.-U., Stojanovic, N.: Lifting events in RDF from interactions with annotated web pages. In: Bernstein, A., et al. (eds.) ISWC 2009. LNCS, vol. 5823, pp. 893–908. Springer, Heidelberg (2009). https://doi.org/10.1007/978-3-642-04930-9_56

83. Suchanek, F.M.: The need to move beyond triples. In: Text2Story@ ECIR, pp. 95–104 (2020)

84. Swartjes, I., Theune, M.: A Fabula model for emergent narrative. In: Göbel, S., Malkewitz, R., Iurgel, I. (eds.) TIDSE 2006. LNCS, vol. 4326, pp. 49–60. Springer, Heidelberg (2006). https://doi.org/10.1007/11944577_5

85. Trabasso, T., Van den Broek, P., Suh, S.Y.: Logical necessity and transitivity of causal relations in stories. Discourse Process. **12**(1), 1–25 (1989)

86. Troncy, R., et al.: 3cixty: building comprehensive knowledge bases for city exploration. J. Web Semantics **46–47**, 2–13 (2017). https://doi.org/10.1016/j.websem.2017.07.002. https://www.sciencedirect.com/science/article/pii/S1570826817300318

87. Van Hage, W.R., Malaisé, V., Segers, R., Hollink, L., Schreiber, G.: Design and use of the simple event model (SEM). J. Web Semantics **9**(2), 128–136 (2011)

88. Varadarajan, U., Dutta, B.: Models for narrative information: a study. arXiv preprint arXiv:2110.02084 (2021)

89. Vilarroya, Ó.: Somos lo que nos contamos. Cómo los relatos construyen el mundo en que vivimos, Editorial Ariel, Barcelona (2019)
90. Vrandečić, D.: Ontology evaluation. In: Staab, S., Studer, R. (eds.) Handbook on Ontologies. IHIS, pp. 293–313. Springer, Heidelberg (2009). https://doi.org/10.1007/978-3-540-92673-3_13
91. Wang, X.j., Mamadgi, S., Thekdi, A., Kelliher, A., Sundaram, H.: Eventory–an event based media repository. In: International Conference on Semantic Computing (ICSC 2007), pp. 95–104. IEEE (2007)
92. Westermann, U., Jain, R.: Toward a common event model for multimedia applications. IEEE Multimedia **14**(1), 19–29 (2007)
93. Yeh, J.h.: StoryTeller: an event-based story ontology composition system for biographical history. In: 2017 International Conference on Applied System Innovation (ICASI), pp. 1934–1937. IEEE (2017)
94. Zárate, M., Rosales, P., Braun, G., Lewis, M., Fillottrani, P.R., Delrieux, C.: *Ocean-Graph*: some initial steps toward a oceanographic knowledge graph. In: Villazón-Terrazas, B., Hidalgo-Delgado, Y. (eds.) KGSWC 2019. CCIS, vol. 1029, pp. 33–40. Springer, Cham (2019). https://doi.org/10.1007/978-3-030-21395-4_3

Comparing Symbolic and Embedding-Based Approaches for Relational Blocking

Daniel Obraczka[1,2]([envelope])[iD] and Erhard Rahm[1,2][iD]

[1] Center for Scalable Data Analytics and Artificial Intelligence (ScaDS.AI)
Dresden/Leipzig, Leipzig, Germany
{obraczka,rahm}@informatik.uni-leipzig.de
[2] Department of Computer Science, Leipzig University, Leipzig, Germany
https://scads.ai

Abstract. Blocking aims to avoid unnecessary comparisons in a data matching pipeline. The heterogeneous nature of Knowledge Graphs (KG) is challenging for blocking approaches that traditionally were implemented for tabular data. While there is vast research on blocking approaches in general, the domain of KGs lacks systematic investigation, especially when comparing embedding-based and symbolic approaches. In this study, we generalize relational blocking, which incorporates neighborhood information of entities, to enable a variety of approaches across the neuro-symbolic spectrum. The results of our study are three-fold: (1) The relational enhancements to state-of-the-art approaches significantly improve their results. (2) (Neuro-)Symbolic approaches can outperform sophisticated deep-learning-based methods in terms of speed and quality. (3) Hybrid methods that combine symbolic and embedding-based techniques are promising avenues that have not been explored thoroughly yet. Our experiments were run on 16 real-world datasets of varying sizes with mono- and multi-lingual settings. We ensure statistical significance with a Bayesian analysis. We release our framework as open-source library.

Keywords: Entity Resolution · Knowledge Graphs · Blocking · Data Integration · Knowledge Graph Embedding · Entity Alignment

1 Introduction

Knowledge Graphs (KGs) have become a vital backbone for the information needs of the modern world. Complex tasks such as question answering [36] and recommendations [33] rely on high-quality data integration from multiple sources. The matching of KGs has seen wide research attention with approaches that use a variety of different methods [19,34,38]. A major performance bottleneck in the Entity Resolution (ER) step of the data integration

M. Alam et al. (Eds.): EKAW 2024, LNAI 15370, pp. 155–173, 2025.
https://doi.org/10.1007/978-3-031-77792-9_10

process is comparing entity pairs from all sources, which in cases of binary matching has an a-priori quadratic complexity. Blocking aims to decrease this complexity and reduce the number of unnecessary comparisons by assigning likely matches into separate blocks. When dealing with databases, blocking approaches can often rely on the (relative) homogeneity of the underlying schemata, especially since entity resolution on tables most often only considers a single entity type (e.g. persons). KGs, on the other hand, contain a plethora of different entity types with vast heterogeneity across the schemata. Therefore, schema-agnostic approaches such as Token Blocking [25] were needed to match heterogeneous data sources [28]. While approaches initially relied on symbolic overlap across data sources to create blocks (e.g. by using common tokens), newer approaches utilize pre-trained word embeddings and even deep learning-based methods [35]. Originally from the database area, most blocking approaches are not built to utilize the rich relational information present in KGs. While Knowledge Graph Embeddings (KGEs) have been used as a method to encode this data in a lower-dimensional embedding space [34] and find matches via nearest neighbors, they have not been used for blocking. To tackle the blocking task in the KG domain, an approach needs to be schema-agnostic and able to utilize the relational information present in the KG. While there is already work [24] comparing embedding-based with symbolic methods on tabular data and approaches like MinoanER [7] that utilize relational information, but consist solely of symbolic techniques, there is no systematic study comparing such relational blocking techniques across the spectrum of neuro-symbolic approaches.

The main contributions of our study are, therefore, the following:

- We generalize the composite blocking scheme of MinoanER [7] to utilize not only symbolic but all techniques across the neuro-symbolic spectrum ranging from the sophisticated DeepBlocker [35] variants to approaches that rely on symbolic overlap like Token Blocking.
- Furthermore, we present a novel hybrid blocking approach that combines Token Blocking with token/attribute embedding clustering, which shows a promising future research direction.
- Our results show that the relational enhancements of our generalized framework significantly outperform their non-relational counterparts. Furthermore, the hybrid and purely symbolic methods outperform the sophisticated DeepBlocker variants as well as a state-of-the-art KGE method.
- Our experiments were performed on 16 real-world datasets. We ensured the significance of our results with a Bayesian signed rank test, and we released our framework as an open-source Python library to enable reproducibility and simplify future research[1].

We begin by defining some preliminaries, followed by a discussion of related work in Sect. 3. In Sect. 4, we present our framework and the integrated methods. After we present our experimental results in Sect. 5, we end with a conclusion and future work.

---

[1] https://github.com/dobraczka/klinker.

2 Preliminaries

Knowledge Graphs (KG) consist of triples in the form of $(entity, property, value)$, where $property$ can be either an attribute property or a relationship and $value$ a literal or another entity, respectively. Therefore, a KG is a tuple $\mathcal{KG} = (\mathcal{E}, \mathcal{R}, \mathcal{A}, \mathcal{L}, \mathcal{T})$, where $\mathcal{E}$ is the set of entities, $\mathcal{A}$ the set of attribute properties, $\mathcal{R}$ the set of relationship properties, $\mathcal{L}$ the set of literals, and $\mathcal{T}$ is the set of triples. We distinguish attribute triples $\mathcal{T}_A$ and relationship triples $\mathcal{T}_R$, where $\mathcal{T}_A : \mathcal{E} \times \mathcal{A} \times \mathcal{L}$ are triples connecting entities and literals, e.g. (dbr:Emma_Caulfield, dbo:birthDate, ''1973-04-08'') and $\mathcal{T}_R : \mathcal{E} \times \mathcal{R} \times \mathcal{E}$ connect entities, e.g. (dbr:Buffy_the_Vampire_Slayer, dbo:starring, dbr:Sarah_Michelle_Geller) as seen in Fig. 1. For a relation triple (e_1, r, e_2), we will refer to e_1 as the *head* and e_2 as the tail entity. The task of Entity Resolution (ER) aims to find $\mathcal{M} = \{(e_1, e_2) \in \mathcal{E}_1 \times \mathcal{E}_2 | e_1 \equiv e_2\}$, where $\equiv$ refers to the equivalence relation. ER has an a-priori quadratic complexity because every entity pair $(e_1, e_2) \in \mathcal{E}_1 \times \mathcal{E}_2$ would have to be compared. Blocking aims to reduce this complexity by eliminating unnecessary comparisons by only clustering likely matches in a set of blocks $\mathcal{B}$. The number of comparisons in a set of blocks $\mathcal{B}$ is given by $||\mathcal{B}|| = \sum_{b_i \in \mathcal{B}} ||b_i||$, with $||b_i||$ denoting the number of comparisons in a single block b_i [28]. A blocking approach aims to optimize two measures: $Recall(\mathcal{B}, \mathcal{M}) = truePositive(\mathcal{B})/|\mathcal{M}|$ shows the ratio of true positives compared to all matches. The Reduction Ratio $RR(\mathcal{B}, \mathcal{E}_1, \mathcal{E}_2) = ||\mathcal{B}||/(|\mathcal{E}_1||\mathcal{E}_2|)$ signifies the relation between comparisons with blocking versus without. Since either one of these measures can be optimized by making sacrifices in the other, the harmonic mean $h3r(rec, rr) = 2(rr * rec)/(rr + rec)$ can be used as an aggregate measure.

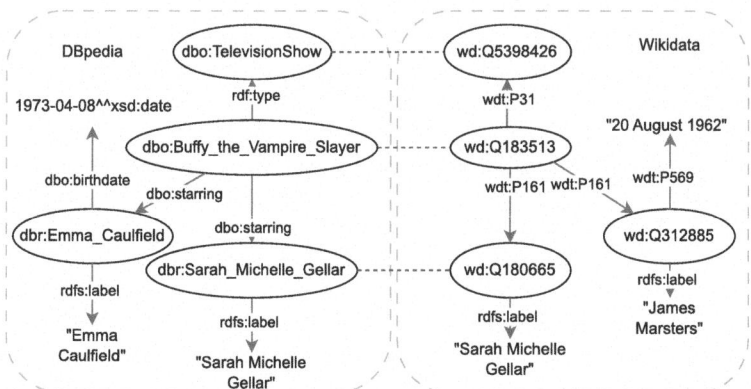

Fig. 1. Subgraphs of DBpedia and Wikidata. Blue dashed lines show entities that should be matched. Some URIs are shortened for brevity. (Color figure online)

3   Related Work

Entity Resolution is a crucial step in the construction of KGs [13] and has seen significant research attention. Approaches range from probabilistic [32], clustering-based [30,31] over methods relying on traditional machine learning [21] all the way to transformer-based [12] and neuro-symbolic methods [23]. In the last years, a large family of techniques has relied on Knowledge Graph Embeddings (KGEs). These approaches encode entities from the given KGs into dense vectors and minimize the distance in the embedding space between similar entities. While many approaches are supervised, there are also models like LightEA [16], which can be used in the unsupervised setting. This specific approach relies on a three-view label propagation scheme, making it time-efficient while still achieving state-of-the-art quality. An overview of such approaches is given in these surveys [34,38]. The usefulness of such KGEs for blocking is largely unexplored.

Numerous approaches have been developed to perform blocking in a schema-agnostic manner on tabular data. Token Blocking [25] places entities into the same block that share a common token. Attribute Clustering Blocking [27] first clusters attributes across data sources by their values and then only places entities into the same block if the common tokens they share also belong to the same attribute cluster. Another notable work is the neural approach DeepBlocker [35], which we will describe in detail in Sect. 4.

A variety of approaches deal specifically with blocking on KGs. Prefix-Infix-Suffix Blocking [26] exploits naming patterns in the entities URI's. Multi-Block [14] first builds indices over multiple similarity measures, preserving the distances between entities. In the next phase, these indices are aggregated into a single multi-dimensional index. MinoanER [7] uses a composite blocking scheme, where entities are assigned to the same blocks if they (1) have the same *name* that is not used by any other entity, (2) common tokens or (3) their top neighbors share a common token. The *name* properties are defined by top-k attributes whose literal values have high importance. Similarly, the top neighbors are connected to entities via relations that appear often but have many distinct values.

For a general overview of blocking methods, we recommend this survey [28]. Several benchmark studies also exist in the area of blocking. Efthymiou et al. [8] show that Token and Attribute Clustering Blocking work well if datasets are derived from common sources, while Prefix-Infix-Suffix was more promising for more diverse data sources. In their study, they mention that utilizing neighborhood information could be an important signal for blocking algorithms. Zeakis et al. [37] investigate the usefulness of pre-trained embeddings for Entity Resolution in the domain of tabular data. While their study is not focused on blocking, it is one aspect of their analysis. They use averaged token embeddings and language models like BERT [6] or Sentence-BERT [29]. Their results suggest that the Sentence-BERT models are best suited for blocking. Papadakis et al. [24] investigate a variety of blocking techniques on tabular data in the schema-based and schema-agnostic settings. Similar to Zeakis et al. [37], they investigate embedding-based methods.

While there is work [24] on comparing deep learning-based approaches and symbolic blocking on tabular data, there is, to our knowledge, no systematic investigation of blocking approaches that utilize neighborhood information comparing embedding-based and symbolic approaches in the domain of KGs.

4 Method

4.1 General Framework

To compare embedding-based and non-embedding-based approaches in a common framework, our investigation relies on a relaxed generalization of the blocking scheme described in [7]. Our open-source Python implementation is named `klinker`. We use a composite blocking scheme

$$\Theta(\mathcal{KG}_1, \mathcal{KG}_2) = \Theta_A(\mathcal{C}_{a_1}, \mathcal{C}_{a_2}) \vee \Theta_R(\mathcal{C}_{r_1}, \mathcal{C}_{r_2}) \tag{1}$$

consisting of the *attribute* blocking function Θ_A and the *relational* blocking function Θ_R. This blocking scheme is a relaxed variant of the one used in MinoanER [7] because it drops the unique name blocking scheme, and it is generalized because it allows any blocking function as Θ_A and Θ_R. The concatenated attribute values of the most important attributes are signified as $\mathcal{C}_a$, and the concatenated attribute values of the most important neighbors are denoted as $\mathcal{C}_r$. In order to determine the importance of a relation or an attribute, we use the respective support and discriminability described by Efthymiou et al. [7]. The **support of a relation** $r \in \mathcal{R}$ is defined as $support(r) = \frac{|instances(r)|}{|\mathcal{E}|^2}$, with $instances(r) = \{(h, t)|(h, r, t) \in \mathcal{T}_R\}$ being the set of head and tail entity tuples of all relation triples that contain r.

The **support of an attribute** $a \in \mathcal{A}$ is defined as $support(a) = \frac{|heads(a)|}{|\mathcal{E}|}$, with $heads(a) = \{h|(h, a, t) \in \mathcal{T}_A\}$ being the set of head entities of the attribute triples that contain a.

The **discrimininablity** of a relation $r \in \mathcal{R}$ is defined as $discriminability(r) = \frac{|tails(r)|}{|instances(r)|}$, with $tails(r) = \{t|(h, r, t) \in \mathcal{T}_R\}$. The formula is analogous for an attribute $a \in \mathcal{A}$.

The **importance** of a relation $r \in \mathcal{R}$ is then defined as the harmonic mean of support and discriminability: $importance(r) = 2 \cdot \frac{support(r) \cdot discriminability(r)}{support(r) + discriminability(r)}$. The formula for the importance of an attribute $a \in \mathcal{A}$ is again analogous. We can now formally define the aforementioned concatenated attribute values of the entity and its most important neighbors.

$$\mathcal{C}_a = \{(e, l_c)|l_c = \bigoplus_{(e, a, l_i) \in \mathcal{T}_A \wedge a \in top_{n_a}(e)} l_i\} \tag{2}$$

Here, $\oplus$ represents the concatenation operation, and $top_{n_a}(e)$ is the n_a attributes connected to entity e with the highest importance score.

$$\mathcal{C}_r = \{(e_i, l_{c_r})|l_{c_r} = \bigoplus_{r \in top_{n_r}(e_i) \wedge (e_i, r, e_j) \in \mathcal{T}_R \wedge (e_j, l_{c_j}) \in \mathcal{C}_a} l_{c_j}\} \tag{3}$$

with $top_{n_r}(e)$ being the n_r relations connected to entity e with the highest importance score.

4.2   Embedding-Based Blockers

Blocking schemes that rely on embeddings are made up of two basic building blocks: *Frame Encoders*, that encode entities into an embedding matrix, and an *Embedding Block Builder* that creates blocks from the given embeddings.

Frame Encoders. Each of the implemented Frame Encoders is an encoder function $\psi : \mathcal{C} \to \mathbb{R}^d$, which encodes each tuple (e, l_c) into an embedding vector $\mathbf{v_e}$ of dimensionality d.

Token Embedding Aggregation. Since entity attribute values often consist of a varying number of tokens, we need to aggregate token embeddings from pretrained word embeddings into a uniform shape. We investigate two variants for aggregation. The first is the SIF (smooth inverse frequency) approach [2], which has been used successfully in previous blocking studies [35]. The second relies on sentence embeddings using siamese BERT-networks [29].

DeepBlocker. [35] uses deep learning methods for blocking. In `klinker`, we adapted three methods from this framework: Autoencoder, Cross-Tuple Training (CTT), and hybrid. All of them rely on token embedding aggregation as the first step. The Autoencoder variant uses encoder and decoder layers in the following steps. The aggregated entity embedding $\mathbf{v_e}$ is used as input for the encoder, which outputs the hidden vector $\mathbf{u_e} \in \mathbb{R}^{d_u}$. Subsequently, the decoder uses $\mathbf{u_e}$ to produce the output $\mathbf{o_e} \in \mathbb{R}^{d_e}$. During training the model learns a concise representation for $\mathbf{u_e}$ by minimizing $||\mathbf{v_e} - \mathbf{o_e}||$. The Cross-Tuple Training variant learns to generate representations on entity pairs e_1, e_2 and associated match/non-match labels. Since no training data is provided, the pairs are generated synthetically. Given a tuple $(e, l_c) \in \mathcal{C}_a$ and it's tokens $w_e = tokens(l_c)$, a subset w'_e of tokens in w_e is selected in order the create a positive pair (w_e, w'_e). A negative pair can be created by sampling another entity $e_n \neq e$. Analogous to the Autoencoder method, the aggregated entity embedding is used. Here, the embedding pairs $(\mathbf{v_{e_1}}, \mathbf{v_{e_2}})$ are sent through a Siamese summarizer and classifier to learn a good representation utilizing the synthetically created match/non-match pairs and labels. The Siamese summarizer uses the same model parameters to generate the representations for e_1 and e_2. The third DeepBlocker variant is called *hybrid* because it combines the Autoencoder and CTT approach by first training the Autoencoder and then using the representations generated by the Autoencoder as input for the Siamese summarizer and classifier of the CTT model.

Embedding Block Builders. Creating entity embeddings does not give us blocks yet. We present two variants for creating blocks from embeddings. We start with k nearest neighbor search and then illustrate another method using clustering. Given an entity $e_a \in \mathcal{E}_a$ of the Knowledge Graph $\mathcal{KG}_a$ with the embedding $\mathbf{v}_{e_a}$ and the entity embeddings $\mathbb{E}_b = [\mathbf{v}_{e_{b_1}}, ..., \mathbf{v}_{e_{b_m}}]$ of the entities $\{e_{b_1}, ..., e_{b_m}\} \in \mathcal{E}_b$ belonging to $\mathcal{KG}_b$, we create the block $b_a = \{e_a, kNN(\mathbf{v}_{e_{a_i}}, \mathbb{E}_b)\}$. Where $kNN(\mathbf{v}_{e_{a_i}}, \mathbb{E}_b)$ returns the k nearest neighbors of $\mathbf{v}_{e_{a_i}}$ among $\mathbb{E}_b$ w.r.t a specific distance measure (e.g. euclidean distance). We create a set of n blocks $\mathcal{B} = \{b_1, ..., b_n\}$ for all $\{e_{a_1}, ..., e_{a_n}\} \in \mathcal{E}_a$. The number of comparisons in $\mathcal{B}$ is therefore $k|\mathcal{E}_a|$. To improve the speed, we use Faiss as the nearest neighbor library [15].

Alternatively, we can use a clustering function that assigns a cluster label $\omega_i \in \Omega$ to each embedding $\mathbf{v_e} \in \{\mathbb{E}_a, \mathbb{E}_b\}$. Then we can create a set of blocks $\mathcal{B} = \{b_1, ..., b_n\}$ for all $\{\omega_1, ..., \omega_n\} \in \Omega$, with all entities in b_i having the same cluster label ω_i for their embedding.

4.3   Generalized Relational Blocking

The most straightforward approach for relational blocking uses token blocking [25] for Θ_A and Θ_R. For each tuple $(e, l_c) \in \mathcal{C}$ with $\mathcal{C} \in \{\mathcal{C}_a, \mathcal{C}_r\}$, the concatenated attribute values l_c are tokenized and entities that share a token t are put into the same block:

$$\forall e_1, e_2 \in b_i : b_i \in \mathcal{B}_{tok} \wedge tokens(l_{c_1}) \cap tokens(l_{c_2}) \neq \emptyset \tag{4}$$

Here $tokens$ is the tokenization function, and $\mathcal{B}_{tok}$ is the set of generated blocks. Since we do not use the unique name blocking of MinoanER [7], this blocking scheme has at least the same recall but at most the same reduction ratio as theirs. We will refer to this blocking scheme as **RelTB**.

Hybrid Block Building via Embedding Clustering. To investigate different granularities for the embedding block builder's clustering variant, we use token or attribute embeddings as inputs for the clustering method. For example, for the token embeddings clustering approach, we take the concatenated literal values l_{c_i} of an entity e_i and use $token(l_{c_i})$ as input for an embedding function. This provides us with multiple embeddings $\{\mathbf{v}_{e_{i_1}}, ..., \mathbf{v}_{e_{i_n}}\}$ per entity e_i if the number of tokens is n. This can analogously be done by using the embeddings of literal values. In this case, the number of embeddings for each entity e_i is equal to the number of attribute triples $(e_i, a, l_j) \in \mathcal{T}_A$. Clustering methods like HDBSCAN [17] can cluster specific data points as noise. In the domain of blocking, we see this as a hint that there is insufficient semantic overlap to create a block. Instead, we perform token blocking for all token/attribute embeddings assigned to the noise cluster. We use the other cluster labels as described in Sect. 4.2. We will refer to the token/attribute embedding clustering-based variant as **RelTB**$_{TC}$ or **RelTB**$_{AC}$ respectively.

Enhanced DeepBlocker. Approaches that rely on DeepBlocker for Θ_A and Θ_R will be denoted as **RelDeepBlocker**. Furthermore, we also incorporate hybrid approaches, which use token blocking for Θ_A but use DeepBlocker as Frame Encoder and k nearest neighbor search as embedding block builder for Θ_R. We will denote these methods as **RelTB**$_{DeepBlocker}$. In a preliminary study, we investigated whether clustering could also be used with these embeddings, but HDBSCAN assigned most of the entities to the noise cluster.

5   Experiments

5.1   Datasets

Several benchmark datasets were considered for this study but ultimately were not deemed the right fit. The movie datasets used by Obraczka et al. [23] have a shallow graph structure but are relatively small and do not necessarily require blocking. The datasets used in the KG track of the Ontology Alignment Evaluation Initiative [1] contain millions of entities, making them challenging in terms of scalability and therefore a perfect candidate for our experiments. However, we found that Token Blocking (without involving relations) yields 100% recall on the available partial gold standard, making it unsuitable for this investigation. The datasets used by Efthymiou et al. [7] are also sufficiently large. Still, we could not load them properly due to inconsistent encodings across the gold standard and respective data sources, and we found some entities in the gold standard that do not show up in either data source. Ultimately, we chose the 16 matching tasks published by Sun et al. [34], which consist of matching samples of the KGs DBpedia, Wikidata, and YAGO. Half of the matching tasks consist of a multi-lingual setting, which aim to match entities from different DBpedia variants (English-German and English-French). The similar but larger DBP1M dataset [10] is unsuitable for our study because it does not contain attribute triples. Table 1 shows the statistics of the matching tasks. With up to 2.3 million triples, these datasets are challenging w.r.t scalability, but they still have problems. They rely on an unrealistic 1-to-1 matching scenario, where each dataset contains a respective counterpart for the other dataset. Furthermore, all the datasets are derived from Wikipedia in one way or the other, providing limited heterogeneity. More research is needed to create realistic benchmark datasets, but this is outside the scope of this study.

5.2   Setup

We tuned the hyperparameters[2] on S-DW1 where necessary and used these values for all other datasets. Using all attributes was the best setting for the top top_{n_a} attributes. Still, for the top_{n_r} relations, we used the 90th percentile of the number of distinct relations as the cutoff point to avoid highly connected nodes gathering a vast neighborhood. For all approaches, we cleaned the literals from

---

[2] Detailled information can be found here: https://github.com/dobraczka/klinker/tree/main/run_scripts/hyperparam_sweeps.

Table 1. Dataset statistics

Task Name	Dataset	$|\mathcal{E}|$	$|\mathcal{T}_R|$	$|\mathcal{T}_A|$	$|\mathcal{R}|$	$|\mathcal{A}|$	$|\mathcal{L}|$	$|\mathcal{M}|$
S-DW1	DBpedia	15000	38265	52134	248	341	28236	15000
	Wikidata	15000	42746	138246	169	649	118515	15000
S-DW2	DBpedia	15000	73983	51378	167	174	25690	15000
	Wikidata	15000	83365	175686	121	457	146977	15000
S-DY1	DBpedia	15000	30291	52093	165	256	25297	15000
	YAGO	15000	26638	117114	28	34	105710	15000
S-DY2	DBpedia	15000	68063	49602	72	89	22561	15000
	YAGO	15000	60970	116151	21	19	104546	15000
L-DW1	DBpedia	100000	293990	334911	413	492	133931	100000
	Wikidata	100000	251708	687860	261	874	542921	100000
L-DW2	DBpedia	100000	616457	360696	318	327	137483	100000
	Wikidata	100000	588203	878219	239	760	682367	100000
L-DY1	DBpedia	100000	294188	360415	287	378	101386	100000
	YAGO	100000	400518	649787	32	37	497633	100000
L-DY2	DBpedia	100000	576547	374785	230	276	97433	100000
	YAGO	100000	865265	755161	31	35	578596	100000
S-ED1	DBpedia-EN	15000	47676	62403	215	285	28973	15000
	DBpedia-DE	15000	50419	133776	131	193	35630	15000
S-ED2	DBpedia-EN	15000	84867	59511	169	170	23831	15000
	DBpedia-DE	15000	92632	161315	96	115	33185	15000
S-EF1	DBpedia-EN	15000	47334	57164	267	307	30281	15000
	DBpedia-FR	15000	40864	54401	210	403	28760	15000
S-EF2	DBpedia-EN	15000	96318	52396	193	188	22761	15000
	DBpedia-FR	15000	80112	56114	166	220	21645	15000
L-ED1	DBpedia-EN	100000	335359	423666	381	450	147142	100000
	DBpedia-DE	100000	336240	586207	196	251	199527	100000
L-ED2	DBpedia-EN	100000	622588	430752	323	325	139867	100000
	DBpedia-DE	100000	629395	656458	170	188	200356	100000
L-EF1	DBpedia-EN	100000	309607	384248	400	465	145103	100000
	DBpedia-FR	100000	258285	340725	300	518	157791	100000
L-EF2	DBpedia-EN	100000	649902	396150	379	363	145382	100000
	DBpedia-FR	100000	561391	342768	287	467	157564	100000

XSD datatype information. For Token Blocking, we removed stopwords[3] and tokens shorter than character length 3. Approaches that utilize nearest neighbor search use $k = 500$ on the small datasets and $k = 1000$ on the large datasets.

---

[3] Using NLTK's [5] English stopword list.

Table 2. Reduction Ratio (RR) and Recall (Rec) results in percent. The RR values of the respective variants are shown in a single column, since they are identical by design. Per Task the RR and Rec values for the three highest h3r values are colored with a darker color implying a better value. All results shown use SIF embeddings.

| | DeepBlocker | | | | RelDeepBlocker | | | | RelTB$_{DeepBlocker}$ | | | |
| | | AE | CTTs | hyb | | AE | CTT | hyb | | AE | CTT | hyb |
	RR	Rec	Rec	Rec	RR	Rec	Rec	Rec	RR	Rec	Rec	Rec
S-DW1	99.904	8.6	16.9	9.8	99.902	28.0	34.9	14.3	99.914	45.4	48.5	46.0
S-DW2	99.920	14.8	14.6	9.2	99.840	27.6	33.0	12.6	99.918	47.5	47.8	39.9
S-DY1	99.885	26.3	19.6	12.4	99.769	52.9	33.8	17.8	99.922	70.3	66.8	60.3
S-DY2	99.873	21.8	13.8	6.7	99.872	49.5	37.8	18.8	99.905	46.4	55.7	52.5
S-ED1	99.914	18.4	17.1	13.1	99.914	28.3	25.6	12.3	99.877	77.7	76.6	72.5
S-ED2	99.924	13.1	14.6	13.0	99.923	14.3	22.9	11.4	99.869	73.6	73.4	71.3
S-EF1	99.782	27.3	26.6	21.8	99.777	36.2	34.4	24.8	99.846	65.5	69.7	67.7
S-EF2	99.759	24.8	31.3	29.7	99.753	23.5	43.4	27.1	99.825	83.6	82.2	79.6

For RelDeepBlocker, the Θ_A and Θ_R components use half the k to ensure comparability. Our experiments include a comparison to LightEA, a state-of-the-art *unsupervised* KGE method, as an embedding method that incorporates the rich relational information present in the KG in a more elaborate manner. This approach was chosen for its speed and high-quality results [16]. As previously mentioned, we investigate two different embedding approaches: SIF aggregated token embeddings, which rely on fasttext word embeddings, and Sentence-BERT embeddings (ST), where we use GTR-T5 [20] for the mono-lingual datasets and LaBSE [9] for the multi-lingual tasks. On the large datasets, we reduce the dimensionality to fit the embeddings in one GPU. For the Sentence-BERT embeddings, we perform PCA on 30% of the data and use the first principal component as the final layer[4] of the Sentence-BERT model to reduce to dimensionality from 768 to 196 dimensions. The fasttext embeddings can be reduced using the dimensionality reduction provided by the fasttext library. Here, we reduce the dimensionality from 300 to 100 dimensions. For RelTB$_{AC}$ and RelTB$_{TC}$, we further reduce the dimensionality since HDBSCAN works best on lower dimensional embeddings. Here, we reduce the Sentence-BERT embeddings to 32, and the fasttext embeddings via UMAP [18] to 25 dimensions. In the future, we want to investigate whether hubness-reduction methods are viable alternatives here [22]. We use the RAPIDS[5] library for faster GPU implementations of HDBSCAN and UMAP. The experiments were run on a machine with an AMD EPYC 7551P CPU. For the small datasets, we used a GeForce RTX 2080 Ti GPU; for the large datasets,

---

[4] see   https://www.sbert.net/examples/training/distillation/README.html# dimensionality-reduction.

[5] https://rapids.ai/.

Table 3. Reduction Ratio (RR) and Recall (Rec) results in percent. The RR values of the RelTB$_{DeepBlocker}$ variants are shown in a single column, since they are identical by design. Per Task the RR and Rec values for the three highest h3r values are colored with a darker color implying a better value. Runs that did not complete after 10 h are marked with '-'.

		LightEA		RelTB$_{DeepBlocker}$				RelTB$_{AC}$		RelTB$_{TC}$		RelTB	
				AE		CTT	hyb						
	EM	RR	Rec	RR	Rec	Rec	Rec	RR	Rec	RR	Rec	RR	Rec
S-DW1	SIF	99.896	29.9	99.914	45.4	48.5	46.0	97.631	91.1	95.796	96.5		
	ST	99.896	33.8	99.914	49.3	42.6	47.4	99.414	89.8	99.014	90.5	98.886	90.1
S-DW2	SIF	99.917	23.9	99.918	47.5	47.8	39.9	96.397	99.9	94.866	100.0		
	ST	99.917	29.2	99.918	41.6	39.8	43.5	98.619	97.6	97.193	98.1	97.649	97.5
S-DY1	SIF	99.877	55.2	99.922	70.3	66.8	60.3	99.759	94.9	96.118	98.4		
	ST	99.877	36.8	99.922	67.2	59.7	56.8	99.835	95.1	99.617	96.1	99.625	95.4
S-DY2	SIF	99.870	35.4	99.905	46.4	55.7	52.5	99.561	99.0	93.730	100.0		
	ST	99.870	28.0	99.905	46.6	46.9	46.4	99.193	99.1	95.008	99.1	98.209	99.1
L-DW1	SIF	99.957	26.1	99.934	43.8	43.7	43.1	95.973	92.8	95.316	94.1		
	ST	99.957	29.6	99.934	46.1	44.9	46.2	99.312	83.7	99.031	84.6	98.815	84.2
L-DW2	SIF	99.968	20.4	99.929	44.3	45.2	-	98.245	97.3	91.984	99.8		
	ST	99.968	26.2	99.929	44.3	46.6	46.0	97.502	97.5	96.101	97.6	95.486	97.6
L-DY1	SIF	99.957	42.7	99.950	68.4	68.8	-	96.889	96.8	95.496	98.5		
	ST	99.957	26.4	99.950	68.7	69.3	70.4	99.600	96.4	99.316	96.9	99.258	96.5
L-DY2	SIF	99.965	38.1	99.938	74.9	-	-	96.878	99.3	93.387	99.9		
	ST	99.965	24.0	99.938	74.3	73.8	74.1	98.008	99.4	96.851	99.3	96.201	99.5
S-ED1	SIF	99.910	30.4	99.877	77.7	76.6	72.5	99.421	92.3	97.035	96.2		
	ST	99.910	41.5	99.877	84.0	72.5	70.2	99.246	92.5	99.233	92.6	99.155	92.1
S-ED2	SIF	99.922	26.4	99.869	73.6	73.4	71.3	98.822	95.9	96.682	99.0		
	ST	99.922	34.7	99.869	80.1	71.4	69.6	98.383	96.3	98.421	96.4	98.453	95.9
S-EF1	SIF	99.759	37.4	99.846	65.5	69.7	67.7	99.196	84.6	93.686	88.5		
	ST	99.759	63.8	99.846	76.3	71.0	66.0	98.758	85.0	98.664	84.5	98.947	84.2
S-EF2	SIF	99.745	40.5	99.825	83.6	82.2	79.6	96.935	98.2	86.246	98.8		
	ST	99.745	72.4	99.825	87.1	80.2	76.0	95.012	98.1	95.199	97.0	96.934	96.9
L-ED1	SIF	99.960	20.6	99.925	66.5	66.4	67.1	99.531	89.3	95.902	96.2		
	ST	99.960	38.8	99.925	66.3	68.4	68.6	99.274	89.7	99.331	89.8	99.323	89.3
L-ED2	SIF	99.965	22.8	99.907	63.6	63.6	63.6	98.289	92.0	98.071	92.6		
	ST	99.965	33.4	99.907	63.4	65.1	64.9	98.277	92.4	98.353	92.6	98.475	92.0
L-EF1	SIF	99.924	22.2	99.939	59.5	59.8	59.8	99.027	80.4	98.611	81.1		
	ST	99.924	45.5	99.939	59.1	63.0	62.5	99.247	80.6	99.136	80.7	99.082	80.5
L-EF2	SIF	99.926	25.8	99.928	68.4	68.5	68.4	96.861	95.6	93.910	95.8		
	ST	99.926	53.1	99.928	67.9	70.8	70.6	96.876	96.0	94.740	95.9	95.515	95.5

we used one Tesla V100. More detailed information on reproducibility can be found in our GitHub repository.

5.3   Results

We rely on the Bayesian analysis for comparing machine learning models proposed by Benavoli et al. [3] to make sound statistical performance comparisons. A Bayesian signed rank test [4] determines significant differences between the

two approaches. Compared to frequentist hypothesis testing, this allows reject-
ing *or* accepting a null hypothesis. Furthermore, a *region of practical equivalence*
(ROPE) can be defined for approaches with equally good performance. We rely
on the `Autorank` [11] package to automatically set the ROPE in relation to effect
size. The Bayesian analysis gives us a probability that one approach is better,
worse, or equal to another. In our study, we decide if one of these probabilities
is $\geq 95\%$, or else we give no verdict (i.e., see it as inconclusive).

We first investigate the performance of the different approaches that utilize DeepBlocker. Table 2 shows Recall and Reduction Ratio values on the
small datasets using the SIF embeddings. Since the number of neighbors con-
trols the reduction ratio, we show these values in a single column for each family of approaches. For space reasons, the other datasets and embedding variants are omitted here. Per matching task, the three best approaches w.r.t h3r
values are highlighted in decreasing strength of color. It is evident that the
RelTB$_{DeepBlocker}$ variants outperform their counterparts, and using our statistical comparison regime, we can say that this difference is significant. We can
also see that RelDeepBlocker is better than DeepBlocker, except for the hyb variant, which generally performs the worst. For AE and CTT, the RelDeepBlocker
variant is also significantly better than its non-relational counterpart.

In Table 3, we show the results on all datasets with both embedding variants. Per matching task, the three best approaches w.r.t h3r values are again
highlighted in decreasing strength of color. We can see that RelTB and the
attribute/token embedding clustering variants RelTB$_{AC}$ and RelTB$_{TC}$ perform
the best. In fact, on all 16 matching tasks, the best value is achieved either by
RelTB$_{AC}$ or RelTB$_{TC}$. Looking at the RR and Recall values, we can see that
RelTB$_{AC}$ generally has a higher RR than RelTB$_{TC}$, with the latter dominat-
ing in Recall, even achieving 100% on two datasets (S-DW2 and S-DY2). The
KGE-based approach LightEA performs the worst, not managing to outperform
RelTB$_{DeepBlocker}$ on a single dataset. In Fig. 2, we provide the result of the
Bayesian statistical analysis w.r.t h3r values. We also distinguished the different
embedding approaches to provide a more detailed analysis. All other approaches
outperform LightEA. It is also notable that RelToken and the RelTB variants
without DeepBlocker are significantly better than the other approaches. Among
these five dominant approaches, we cannot reach a conclusive answer as to which
of these performs the best. In Fig. 3, we take a more detailed analysis w.r.t Recall
and Reduction ratio on these five approaches. While RelTB$_{TC}$ with SIF embeddings is significantly better regarding Recall than all other approaches, it is also
significantly worse regarding Reduction Ratio. RelTB$_{AC}$ with SentenceTransformer embeddings strikes the best balance between Recall and Reduction Ratio
of these approaches. It performs similarly to, e.g., RelToken regarding Recall
but is significantly better in terms of Reduction Ratio. While it is inconclusive whether RelTB$_{AC}$ with SentenceTransformer embeddings is significantly
better than RelToken w.r.t h3r values, we see this nuanced analysis on Recall
and Reduction Ratio as a sign that hybrid approaches deserve more research
attention.

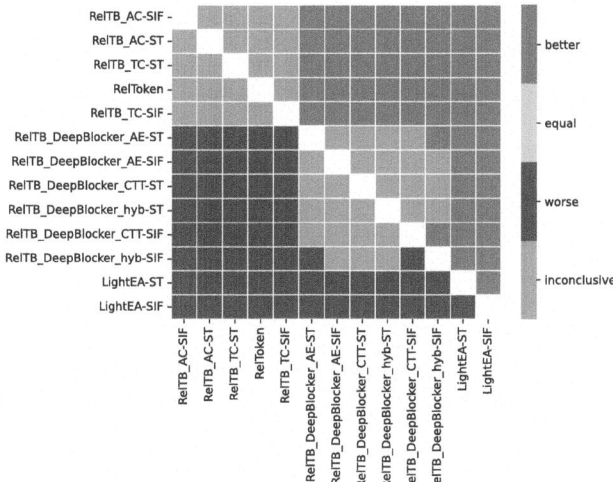

Fig. 2. Decision matrix comparing blocking approaches using Bayesian signed rank tests w.r.t h3r values. Each cell shows the decision when comparing the row approach to the column approach. A decision is reached if the posterior probability is $\geq 95\%$.

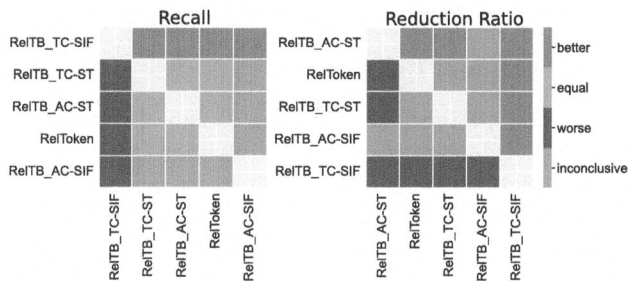

Fig. 3. Decision matrices comparing a subset of blocking approaches using Bayesian signed rank tests w.r.t Recall and Reduction Ratio. Each cell shows the decision when comparing the row approach to the column approach. A decision is reached if the posterior probability is $\geq 95\%$.

Since the main goal of employing blocking is to speed up the matching process, runtime is an important factor. In Fig. 4, we show the runtimes of the different approaches. The purely symbolic approach RelToken is the fastest of the high-quality methods. $RelTB_{TC}$ and $RelTB_{AC}$ are roughly on par with the AutoEncoder variant of $RelTB_{DeepBlocker}$. The more sophisticated DeepBlocker variants are one order of magnitude slower. While LightEA is the fastest (especially when using Sentence-BERT embeddings), it does not produce high-quality results.

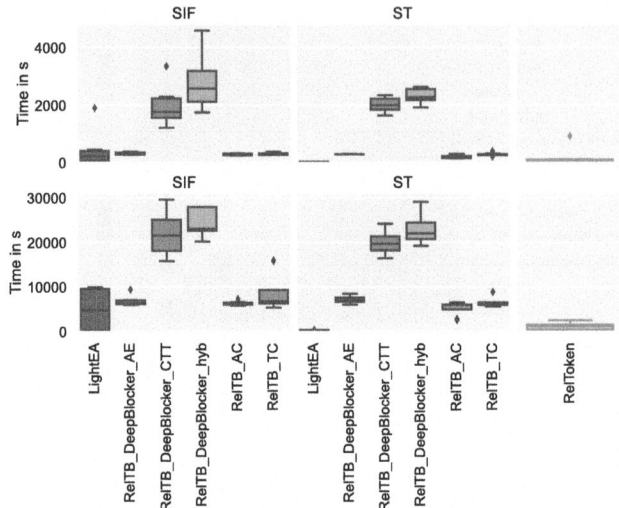

Fig. 4. Time in seconds as aggregated box plots, with the first row showing the 15k datasets and the second row showing the 100k datasets. Note the differing ranges of the y-axis per row.

5.4   Impact of Relational Enhancement

Since one of our contributions is enhancing some state-of-the-art blocking approaches for the KG domain, we will now examine the differences between relational and non-relational blocking. In Fig. 5, we compare relational blockers and their non-relational counterparts on S-DY2. Since LightEA is relational by design, we compare it with the SIF aggregated fasttext embeddings. For the RelTB$_{DeepBlocker}$ variants, we compare them with their DeepBlocker counterparts. In terms of Recall, we can, in many cases, more than double the values while retaining a high Reduction Ratio. The highest cost of relational enhancement is in terms of speed. For example, Token Blocking without rela-

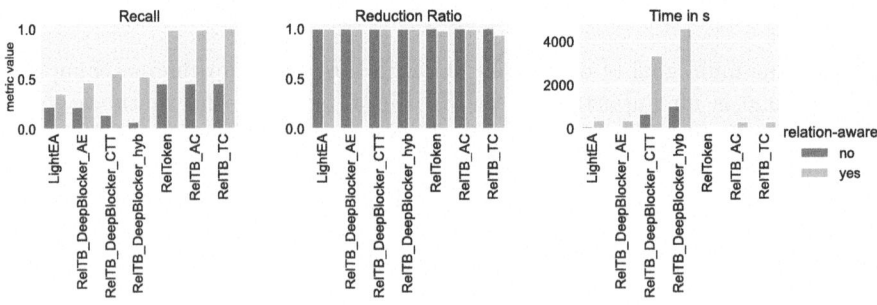

Fig. 5. Comparison between relational blockers with their non-relational counterparts on S-DY2, with SIF aggregated fasttext embeddings (where applicable).

tional enhancement takes 8 s, while RelToken takes 27. The speed penalty is higher for the attribute/token embedding clustering variants, where the additional steps of embedding and clustering take most of the time and up to 313 s. The dimensionality reduction needed for HDBSCAN is included in this time as well.

6   Conclusion and Future Work

In this work, we presented a unified framework that integrates embedding-based and symbolic approaches, enhances state-of-the-art methods to utilize relational information, and explores various hybrid methods that combine embedding-based and symbolic techniques. Our study showed that simple symbolic approaches can outperform sophisticated, state-of-the-art deep learning methods like DeepBlocker, even when adapted to incorporate neighborhood information. We showed that Relational Token Blocking and the hybrid variants that utilize clustering on token/attribute embeddings are significantly better than the other approaches we investigated. Token Blocking relies on symbolic overlap at a lower level of granularity than entity embeddings. Our study has shown that similarities at this level can be exploited by embedding-based methods as well. Subsequent studies should investigate how these findings can be incorporated into new hybrid methods, which are currently an underexplored field in (relational) blocking research. Future work must address the current limitations of existing benchmark datasets to enable more robust investigations. Even on some of the datasets used in this study, no approach managed blocking quality that would be deemed acceptable in a practical setting (e.g., on L-EF1 we achieved at best 81% Recall), emphasizing the importance of additional research.

Acknowledgments. The authors acknowledge the financial support by the BMBF and by the SMWK in the programme Center of Excellence for AI-research "Center for Scalable Data Analytics and Artificial Intelligence Dresden/Leipzig", project identification number: ScaDS.AI. The authors gratefully acknowledge the computing time made available to them on the high-performance computer at the NHR Center of TU Dresden. This center is jointly supported by the Federal Ministry of Education and Research and the state governments participating in the NHR (www.nhr-verein.de/unsere-partner). Computations for this work were done (in part) using resources of the Leipzig University Computing Center.

Disclosure of Interests. The authors have no competing interests to declare that are relevant to the content of this article.

References

1. Algergawy, A., et al.: Results of the ontology alignment evaluation initiative 2019. In: Shvaiko, P., Euzenat, J., Jiménez-Ruiz, E., Hassanzadeh, O., Trojahn, C. (eds.) Proceedings of the 14th International Workshop on Ontology Matching co-located with the 18th International Semantic Web Conference (ISWC 2019), Auckland, New Zealand, October 26, 2019. CEUR Workshop Proceedings, vol. 2536, pp. 46–85. CEUR-WS.org (2019). https://ceur-ws.org/Vol-2536/oaei19_paper0.pdf
2. Arora, S., Liang, Y., Ma, T.: A Simple but tough-to-beat baseline for sentence embeddings. In: 5th International Conference on Learning Representations, ICLR 2017, Toulon, France, April 24-26, 2017, Conference Track Proceedings. OpenReview.net (2017). https://openreview.net/forum?id=SyK00v5xx
3. Benavoli, A., Corani, G., Demsar, J., Zaffalon, M.: Time for a change: a tutorial for comparing multiple classifiers through bayesian analysis. J. Mach. Learn. Res. **18**, 77:1–77:36 (2017). http://jmlr.org/papers/v18/16-305.html
4. Benavoli, A., Corani, G., Mangili, F., Zaffalon, M., Ruggeri, F.: A Bayesian Wilcoxon signed-rank test based on the Dirichlet process. In: Proceedings of the 31th International Conference on Machine Learning, ICML 2014, Beijing, China, 21-26 June 2014. JMLR Workshop and Conference Proceedings, vol. 32, pp. 1026–1034. JMLR.org (2014). http://proceedings.mlr.press/v32/benavoli14.html
5. Bird, S., Klein, E., Loper, E.: Natural Language Processing with Python. O'Reilly (2009). http://www.oreilly.de/catalog/9780596516499/index.html
6. Devlin, J., Chang, M.W., Lee, K., Toutanova, K.: BERT: Pre-training of deep bidirectional transformers for language understanding. In: Burstein, J., Doran, C., Solorio, T. (eds.) Proceedings of the 2019 Conference of the North American Chapter of the Association for Computational Linguistics: Human Language Technologies, NAACL-HLT 2019, Minneapolis, MN, USA, June 2-7, 2019, Volume 1 (Long and Short Papers), pp. 4171–4186. Association for Computational Linguistics (2019). https://doi.org/10.18653/V1/N19-1423,
7. Efthymiou, V., Papadakis, G., Stefanidis, K., Christophides, V.: MinoanER: schema-agnostic, non-iterative, massively parallel resolution of web entities. In: Herschel, M., Galhardas, H., Reinwald, B., Fundulaki, I., Binnig, C., Kaoudi, Z. (eds.) Advances in Database Technology - 22nd International Conference on Extending Database Technology, EDBT 2019, Lisbon, Portugal, March 26-29, 2019, pp. 373–384. OpenProceedings.org (2019). https://doi.org/10.5441/002/EDBT.2019.33,
8. Efthymiou, V., Stefanidis, K., Christophides, V.: Benchmarking blocking algorithms for web entities. IEEE Trans. Big Data **6**(2), 382–395 (2020). https://doi.org/10.1109/TBDATA.2016.2576463. https://ieeexplore.ieee.org/document/7485873/
9. Feng, F., Yang, Y., Cer, D., Arivazhagan, N., Wang, W.: Language-agnostic BERT Sentence Embedding. In: Muresan, S., Nakov, P., Villavicencio, A. (eds.) Proceedings of the 60th Annual Meeting of the Association for Computational Linguistics (Volume 1: Long Papers), pp. 878–891. Association for Computational Linguistics, Dublin, Ireland (2022). https://doi.org/10.18653/v1/2022.acl-long.62, https://aclanthology.org/2022.acl-long.62

10. Ge, C., Liu, X., Chen, L., Zheng, B., Gao, Y.: LargeEA: aligning entities for large-scale knowledge graphs. Proc. VLDB Endow. **15**(2), 237–245 (2021). https://doi. org/10.14778/3489496.3489504, http://www.vldb.org/pvldb/vol15/p237-gao.pdf

11. Herbold, S.: Autorank: a python package for automated ranking of classifiers. J. Open Source Softw. **5**(48), 2173 (2020). https://doi.org/10.21105/JOSS.02173

12. Hertling, S., Portisch, J., Paulheim, H.: KERMIT - A Transformer-Based Approach for Knowledge Graph Matching. CoRR **abs/2204.13931** (2022). https://doi.org/ 10.48550/ARXIV.2204.13931, arXiv: 2204.13931

13. Hofer, M., Obraczka, D., Saeedi, A., Köpcke, H., Rahm, E.: Construction of knowledge graphs: current state and challenges. Inf. **15**(8), 509 (2024). https://doi.org/ 10.3390/INFO15080509

14. Isele, R., Jentzsch, A., Bizer, C.: Efficient multidimensional blocking for link discovery without losing recall. In: Marian, A., Vassalos, V. (eds.) Proceedings of the 14th International Workshop on the Web and Databases 2011, WebDB 2011, Athens, Greece, June 12, 2011 (2011), http://webdb2011.rutgers.edu/papers/Paper%2039/ silk.pdf

15. Johnson, J., Douze, M., Jégou, H.: Billion-scale similarity search with GPUs. IEEE Trans. Big Data **7**(3), 535–547 (2021). https://doi.org/10.1109/TBDATA.2019. 2921572

16. Mao, X., Wang, W., Wu, Y., Lan, M.: LightEA: a scalable, robust, and interpretable entity alignment framework via three-view label propagation. In: Proceedings of the 2022 Conference on Empirical Methods in Natural Language Processing, pp. 825–838. Association for Computational Linguistics, Abu Dhabi, United Arab Emirates (2022). https://doi.org/10.18653/v1/2022.emnlp-main.52, https:// aclanthology.org/2022.emnlp-main.52

17. McInnes, L., Healy, J.: Accelerated hierarchical density based clustering. In: Data Mining Workshops (ICDMW), 2017 IEEE International Conference on, pp. 33–42. IEEE (2017)

18. McInnes, L., Healy, J.: UMAP: uniform manifold approximation and projection for dimension reduction. CoRR **abs/1802.03426** (2018). http://arxiv.org/abs/1802. 03426

19. Nentwig, M., Hartung, M., Ngomo, A.C.N., Rahm, E.: A survey of current Link Discovery frameworks. Semantic Web **8**(3), 419–436 (2017). https://doi.org/10. 3233/SW-150210

20. Ni, J., et al.: Large dual encoders are generalizable retrievers. In: Goldberg, Y., Kozareva, Z., Zhang, Y. (eds.) Proceedings of the 2022 Conference on Empirical Methods in Natural Language Processing, pp. 9844–9855. Association for Computational Linguistics, Abu Dhabi, United Arab Emirates (2022). https://doi.org/10. 18653/v1/2022.emnlp-main.669, https://aclanthology.org/2022.emnlp-main.669

21. Obraczka, D., Ngomo, A.-C.N.: DRAGON: decision tree learning for link discovery. In: Bakaev, M., Frasincar, F., Ko, I.-Y. (eds.) ICWE 2019. LNCS, vol. 11496, pp. 441–456. Springer, Cham (2019). https://doi.org/10.1007/978-3-030-19274-7_31

22. Obraczka, D., Rahm, E.: Fast hubness-reduced nearest neighbor search for entity alignment in knowledge graphs. SN Comput. Sci. **3**(6), 501 (2022). https://doi. org/10.1007/S42979-022-01417-1

23. Obraczka, D., Schuchart, J., Rahm, E.: Embedding-assisted entity resolution for knowledge graphs. In: Chaves-Fraga, D., Dimou, A., Heyvaert, P., Priyatna, F., Sequeda, J.F. (eds.) Proceedings of the 2nd International Workshop on Knowledge Graph Construction co-located with 18th Extended Semantic Web Conference (ESWC 2021), Online, June 6, 2021. CEUR Workshop Proceedings, vol. 2873. CEUR-WS.org (2021). https://ceur-ws.org/Vol-2873/paper8.pdf

24. Papadakis, G., Fisichella, M., Schoger, F., Mandilaras, G., Augsten, N., Nejdl, W.: Benchmarking filtering techniques for entity resolution. 2023 IEEE 39th International Conference on Data Engineering (ICDE), pp. 653–666 (2023). https://doi.org/10.1109/ICDE55515.2023.00389, https://ieeexplore.ieee.org/document/10184692/, conference Name: 2023 IEEE 39th International Conference on Data Engineering (ICDE) ISBN: 9798350322279 Place: Anaheim, CA, USA Publisher: IEEE

25. Papadakis, G., Ioannou, E., Niederée, C., Fankhauser, P.: Efficient entity resolution for large heterogeneous information spaces. In: King, I., Nejdl, W., Li, H. (eds.) Proceedings of the Forth International Conference on Web Search and Web Data Mining, WSDM 2011, Hong Kong, China, February 9-12, 2011, pp. 535–544. ACM (2011). https://doi.org/10.1145/1935826.1935903

26. Papadakis, G., Ioannou, E., Niederée, C., Palpanas, T., Nejdl, W.: Beyond 100 million entities: large-scale blocking-based resolution for heterogeneous data. In: Adar, E., Teevan, J., Agichtein, E., Maarek, Y. (eds.) Proceedings of the Fifth International Conference on Web Search and Web Data Mining, WSDM 2012, Seattle, WA, USA, February 8-12, 2012, pp. 53–62. ACM (2012). https://doi.org/10.1145/2124295.2124305

27. Papadakis, G., Ioannou, E., Palpanas, T., Niederee, C., Nejdl, W.: A blocking framework for entity resolution in highly heterogeneous information spaces. IEEE Trans. Knowl. Data Eng. **25**(12), 2665–2682 (2013). https://doi.org/10.1109/TKDE.2012.150. http://ieeexplore.ieee.org/document/6255742/

28. Papadakis, G., Skoutas, D., Thanos, E., Palpanas, T.: Blocking and filtering techniques for entity resolution: a survey. ACM Comput. Surv. **53**(2), 1–42 (2020). https://doi.org/10.1145/3377455

29. Reimers, N., Gurevych, I.: Sentence-BERT: sentence embeddings using siamese BERT-networks. In: Inui, K., Jiang, J., Ng, V., Wan, X. (eds.) Proceedings of the 2019 Conference on Empirical Methods in Natural Language Processing and the 9th International Joint Conference on Natural Language Processing, EMNLP-IJCNLP 2019, Hong Kong, China, November 3-7, 2019, pp. 3980–3990. Association for Computational Linguistics (2019). https://doi.org/10.18653/V1/D19-1410,

30. Saeedi, A., David, L., Rahm, E.: Matching entities from multiple sources with hierarchical agglomerative clustering. In: Aveiro, D., Dietz, J.L.G., Filipe, J. (eds.) Proceedings of the 13th International Joint Conference on Knowledge Discovery, Knowledge Engineering and Knowledge Management, IC3K 2021, Volume 2: KEOD, Online Streaming, October 25-27, 2021, pp. 40–50. SCITEPRESS (2021). https://doi.org/10.5220/0010649600003064

31. Saeedi, A., Peukert, E., Rahm, E.: Using link features for entity clustering in knowledge graphs. In: Gangemi, A., et al. (eds.) ESWC 2018. LNCS, vol. 10843, pp. 576–592. Springer, Cham (2018). https://doi.org/10.1007/978-3-319-93417-4_37

32. Suchanek, F.M., Abiteboul, S., Senellart, P.: PARIS: probabilistic alignment of relations, instances, and schema. Proc. VLDB Endowment **5**(3), 157–168 (2011). https://doi.org/10.14778/2078331.2078332

33. Sun, R., et al.: Multi-modal knowledge graphs for recommender systems. In: d'Aquin, M., Dietze, S., Hauff, C., Curry, E., Cudré-Mauroux, P. (eds.) CIKM '20: The 29th ACM International Conference on Information and Knowledge Management, Virtual Event, Ireland, October 19-23, 2020, pp. 1405–1414. ACM (2020). https://doi.org/10.1145/3340531.3411947

34. Sun, Z., et al.: A benchmarking study of embedding-based entity alignment for knowledge graphs. Proc. VLDB Endowment **13**(12), 2326–2340 (2020). https://doi.org/10.14778/3407790.3407828

35. Thirumuruganathan, S., et al.: Deep learning for blocking in entity matching: a design space exploration. Proc. VLDB Endowment **14**(11), 2459–2472 (2021). https://doi.org/10.14778/3476249.3476294

36. Usbeck, R., et al.: Benchmarking question answering systems. Semantic Web **10**(2), 293–304 (2019). https://doi.org/10.3233/SW-180312

37. Zeakis, A., Papadakis, G., Skoutas, D., Koubarakis, M.: Pre-trained embeddings for entity resolution: an experimental analysis. Proc. VLDB Endow. **16**(9), 2225–2238 (2023). https://doi.org/10.14778/3598581.3598594, https://www.vldb.org/pvldb/vol16/p2225-skoutas.pdf

38. Zhang, R., Trisedya, B.D., Li, M., Jiang, Y., Qi, J.: A benchmark and comprehensive survey on knowledge graph entity alignment via representation learning. VLDB J. Int. J. Very Large Data Bases **31**(5), 1143–1168 (2022). https://doi.org/10.1007/s00778-022-00747-z

UniQ-Gen: Unified Query Generation Across Multiple Knowledge Graphs

Daniel Vollmers[(✉)] [ID], Nikit Srivastava [ID], Hamada M. Zahera [ID],
Diego Moussallem [ID], and Axel-Cyrille Ngonga Ngomo [ID]

Data Science Group, Paderborn University, Paderborn, Germany
{daniel.vollmers,nikit.srivastava,hamada.zahera,
diego.moussallem,axel.ngonga}@uni-paderborn.de

Abstract. Generating SPARQL queries is crucial for extracting relevant information from diverse knowledge graphs. However, the structural and semantic differences among these graphs necessitate training or fine-tuning a tailored model for each one. In this paper, we propose UniQ-Gen, a unified query generation approach to generate SPARQL queries across various knowledge graphs. UniQ-Gen integrates entity recognition, disambiguation, and linking through a BERT-NER model and employs cross-encoder ranking to align questions with the Freebase ontology. We conducted several experiments on different benchmark datasets such as LC-QuAD 2.0, GrailQA, and QALD-10. The evaluation results demonstrate that our approach achieves performance equivalent to or better than models fine-tuned for individual knowledge graphs. This finding suggests that fine-tuning a unified model on a heterogeneous dataset of SPARQL queries across different knowledge graphs eliminates the need for separate models for each graph, thereby reducing resource requirements.

Keywords: SPARQL Generation · Question Answering over Knowledge Graphs · Large Language Models · KGQA

1 Introduction

Large language models (LLMs) have recently shown significant performance in various NLP tasks, including answering questions on Knowledge Graphs (KGQA) [1]. These models are often fine-tuned on a domain-specific dataset (e.g., QALD-10) to convert natural text to corresponding logical forms like a SPARQL query [2] or S-expression [3]. However, training or fine-tuning LLMs is a resource-intensive process that require a lot of computational resources such as extensive GPU hours [4]. Current approaches typically train or fine-tune an LLM on a single domain-specific dataset or knowledge graph [2]. However, these methods require further tuning when applied to new domains or knowledge graphs. This is due to knowledge graphs (e.g.g, Freebase [5], Wikidata [6], and DBpedia [7]) have significant variances in data representation. For instance, Wikidata represents *'The Apls Mountains'* with semantic identifiers like *wd:Q1286*, while

M. Alam et al. (Eds.): EKAW 2024, LNAI 15370, pp. 174–189, 2025.
https://doi.org/10.1007/978-3-031-77792-9_11

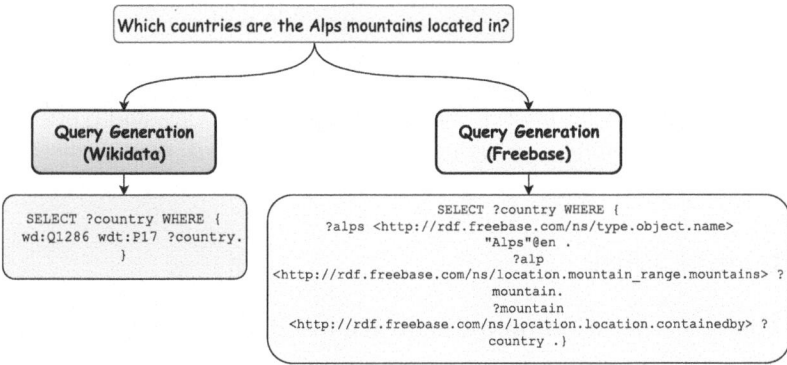

Fig. 1. An example of two individual models for generating SPARQL queries across different knowledge graphs.

Freebase uses encoded identifiers such as *?alps* <http://rdf.freebase.com/ns/type.object.name> *Alps @en* as shown in Fig. 1. Moreover, Freebase hierarchically organizes semantic relations between entities, which differs from the structures in other knowledge graphs. These differences pose challenges in developing systems that are compatible with multiple knowledge graphs. Each knowledge graph requires a different entity linking and query generation components. Consequently, adapting to these differences needs re-training or fine-tuning of LLMs to ensure effective performance.

In this paper, we propose a unified approach to fine-tune a single large language model for generating SPARQL queries across different knowledge graphs. Our approach involves fine-tuning one LLM on multiple knowledge graphs rather than training separate models for each graph. To achieve this, we combine training data from multiple knowledge graphs in one dataset. This joint fine-tuning of one LLM allows to better generalization across different data representations and structures. As result, a single model for multiple KGs significantly reduces the resource requirements in a productive environment, as only one model needs to be deployed rather multiple models for different knowledge graphs. To evaluate the performance of our approach, we perform a comparative analysis between models tailored to each knowledge graphs and our unified model. In our experiments, we use two different knowledge graphs (Wikidata and Freebase), which have significant different in their SPARQL queries. Furthermore, we extract relevant information (e.g., entities, relations, types) from the knowledge graph and investigate which information has the most impact of the performance of query generation. The evaluation results demonstrate that our approach achieves performance comparable to, or the same as, single KG models, reducing the need for training or fine-tuning the model separately for each knowledge graph. We summarize the main contributions of this paper as follows:

– We propose a unified approach for generating SPARQL queries for multiple knowledge graphs.

- Our approach achieves equivalent or comparable performances as individual models (which tailored for each KG), eliminating the need for separate training or fine-tuning an LLM for each KG.
- Incorporating relevant information (e.g., entities, relations, and types) within the LLM prompt improves the performance of SPARQL query generation.
- The source code and datasets used in our experiments are publicly available.[1]

2   Related Work

2.1   Query Generation on a Single Knowledge Base

Wikidata. Recent approaches [2,8] treat the query generation problem as a translation problem. The task is to translate a natural language question into a SPARQL query. As inputs these models use the natural language question itself, plus linked knowledge such as entities and relations from the knowledge base [2]. Applying fine-tuning techniques to language models also become increasingly popular over recent years. Other approaches use patterns for generating a set of candidate queries [9,10]. Afterward, a ranking approach is applied to computing the final prediction. Commonly all approaches apply entity linking, which usually consists of a span detection step and a disambiguation step. The output of the disambiguation is either a ranked list of entities per span or only one entity per span in the case of an end-to-end entity linking setup.

Freebase. On the Freebase knowledge base, semantic parsing is usually solved by iteratively predicting and ranking queries in the form of S-Expressions [2,3, 11]. For example, the RnG-KBQA [3] framework uses a combination of ranking and query generation for predicting queries. Different from the translation approaches used for Wikidata question answering, candidate queries are generated and introduced in the generation model. Other approaches such as Pangu [12], construct queries, by iteratively extending and ranking a set of query sub-plans. All models share the characteristic of computing a large number of sub-plans, which demand significant resources in terms of GPU memory and time, making an end-to-end implementation usually unavailable.

2.2   Query Generation on Multiple Knowledge Graphs

Since the Semantic Web's inception, various methods have been developed for semantic parsing over multiple or interlinked knowledge graphs. One early method, PowerAqua by Lopez et al. [13], uses semantic similarity between ontology terms and user queries to generate query triples, which an inference engine uses to retrieve answers. Its updated version [14] works on large KGs like DBpedia[2] but struggles with scalability, performance, and effectiveness, especially with

---

[1] https://github.com/dice-group/KATRINA.
[2] https://www.dbpedia.org/.

large-scale data, complex queries, and integrating data from different ontologies, requiring significant effort for updates and maintenance. SINA, introduced by Shekarpour and Auer [15], is a data-semantics-aware keyword search approach that converts natural language and keyword queries into SPARQL queries for accessing interlinked KGs within the Linked Open Data Cloud. It uses a hidden Markov model for query disambiguation and resource identification, leveraging Linked Data topology to construct federated SPARQL queries for information retrieval from multiple KGs. However, this process is computationally complex, and the keyword-based approach can overlook the question's syntax.

OQA by Fader et al. [16] enhances answer accuracy and coverage by integrating curated KBs like Freebase with automatically extracted KBs, employing NLP for query translation and paraphrase-driven learning for query variability. In a similar manner, MULTIQUE by Bhutani et al. [17] uses semantic parsing through neural networks to handle complex queries by combining curated and extracted KBs. These approaches rely heavily on textual data, making it challenging to manage and computationally expensive to extract relevant information on-demand. Zhang et al. [18] employed a rule-based method to handle queries across multiple KGs by identifying resources, forming triple patterns, aligning variables, and performing joint inference to create accurate SPARQL queries. However, string matching for entity linking can cause mismatches and missed entities due to its inability to disambiguate similar names and handle naming variations accurately. Neelam et al. [19] introduced SYGMA, which streamlines query generation through KB-agnostic "Question Understanding" and KB-specific "Question Mapping & Reasoning." It uses abstract meaning representation to create a KB-agnostic lambda expression, refined with specific KB details before being converted into a SPARQL query using a rule-based system. SYGMA's modular design aids generalization but is sensitive to individual module performance, particularly relation linking.

3   Approach

Figure 2 shows an overview of approach (UniQ-Gen) for generating SPARQL queries for multiple knowledge graphs (e.g., Wikidata and Freebase) using a single model. We achieve this by fine-tuning the T5 [20] model on a mixed dataset, containing training examples of (natural questions, SPARQL-for-Freebase) and (question, SPARQL-for-Wikidata). In this way, the fine-tuned T5 model learns how to generate SPARQL queries for both graphs, rather than fine-tuning two separate models for each graph. Accordingly, we reduce the computational and maintenance costs associated with fine-tuning and managing separate models for each knowledge graph. The following sections describes the details of each module in our approach, including *Knowledge Extraction* and *Query Generation*.

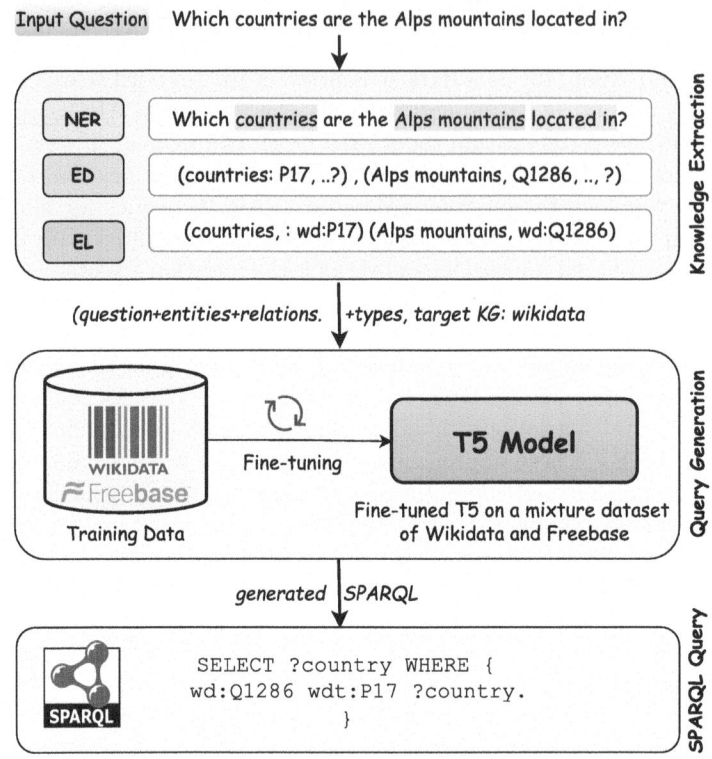

Fig. 2. Our approach (UniQ-Gen) for generating SPARQL queries using one model for multiple knowledge graphs.

3.1 Knowledge Extraction

The knowledge extraction process involves three main tasks: *named entity recognition* (NER), *entity disambiguation* (ED) and *entity linking* (EL). Named entity recognition identifies and classifies entity spans within the text [21], while entity disambiguation associates these spans with corresponding entities in the target knowledge graph. First, we extract relevant information (e.g., entities, relations, and types) from the input question. Noteworthy, This process varies between Wikidata and Freebase due to their different structures and representations of information.

Extracting Knowledge from Wikidata. Many question-answering approaches rely on pre-built frameworks, such as DBpedia Spotlight, which generally yield satisfactory results, due to the significant resources required to develop an efficient linking systems. However, these frameworks often struggle with identifying and categorizing relations and types. For instance, in a query like "Which mountains are located in the US?", a typical NER framework would only recognize "US" as an entity. For accurate query generation, it is essential to also link

Question: Which countries are the Alps mountains located in? [SEP]

Entities: Country: P17, Alps mountains: Q1286; ,[SEP]

Relations: located in: P138 [SEP]

Target Knowledge Graph: Wikidata.

Fig. 3. An example input for generating SPARQL query with Wikidata

the term "mountains" to the knowledge graph. Additionally, these frameworks usually fail to predict relationships between entities, a critical factor for enhancing QA system accuracy. To address these limitations, we employ the following methods:

- *Entity Recognition, Disambiguation, and Linking*: We use Flair [22], a state-of-the-art framework that employs an LSTM network with contextual string embeddings to accurately recognize and classify entity spans within text. For entity disambiguation, we use GENRE [23], which applies an autoregressive transformer architecture based on the pre-trained BART model with constrained decoding and beam search to predict Wikipedia titles. We link these titles with Wikidata using a dictionary, assuming each entity has a unique label mapping to a single Wikipedia title.
- *Extracting Relations and Types*: We use a fine-tuned T5 model to predict types and relations that are not directly mapped to spans in the input sequence. For example, in the question "Which is the highest mountain in the US?", the relation "is located in" is needed but not directly present in the text. We extend target labels to include relations, e.g., "Who is the spouse of Obama[SEP]entities: Barack Obama, relations: spouse."

We combine the outputs of these methods and use a dictionary to map Wikidata labels to their URIs, assuming each label is unique within the knowledge graph. These outputs are then used as input for the *Query Generation* module, as shown in Fig. 2. The input includes: the *question, entities, relations* and *a target knowledge graphs*. All are concatenated using a separation token, *[SEP]* as shown in Fig. 3.

Extracting Knowledge from Freebase. Many types and relations in Freebase share identical labels due to its hierarchical ontology, making pre-built methods (e.g., Flair) for linking entities, relations, and types in Wikidata inadequate. Therefore, we adapt existing methods for extracting knowledge from Freebase as follows:

- *Entity Recognition, Disambiguation, and Linking*: Our approach aligns with the RNG-KGQA framework [3], employing a BERT-NER model to accurately detect entity mentions within the text. For entity disambiguation, we use

Table 1. Training Samples

Question	Entities	Relations	Target	SPARQL
Is Kevin Costner owner of Fielders Stadium?	wd:Q11930 wd:Q5447154	wdt:P1830	wikidata	ASK...
how many hadrons are in the family meson?	physics.hadron, m.04_rh	physics.particle. family	freebase	SELECT (COUNT...
What periodical literature does Delta Air Lines use as a mouthpiece?	wd:Q188920 wd:Q1002697	wdt:P2813 wdt:P31	wikidata	SELECT...

a pre-trained BERT-based model that leverages relation information linked with each entity, thereby improving the ranking of the target entity. For entity linking, our approach matches these mentions with surface forms from the Freebase KG, ranks them using popularity scores, and retains the top 5 candidates. The ranking model employed is a cross-encoder model, as described in Eq. 1.

- *Types and Relations Linking*: Our approach follows the schema retrieval method from the TIARA Framework [11], using a cross-encoder ranker to rank relations and types from the Freebase Ontology. The score for question (x) and a schema (c) is computed as:

$$s(x, c) = \text{Linear}(\text{BertCls}([x; c])) \tag{1}$$

where BertCLS represents the CLS token from a BERT-encoder [11]. We use the top-5 relations and types as input for our query generation model.

3.2 Query Generation

We employ the T5 model in this module, which has demonstrated promising results in query generation task [2,3]. In particular, we fine-tune the T5 model on diverse training dataset containing SPARQL queries from multiple knowledge graphs (Wikidata and Freebase). Table 1 show show some training examples of the dataset used to fine-tune the T5 model, including examples for question-to-SPARQL$_{(wikidata)}$ and question-to-SPARQL$_{(freebase)}$. One example includes: an input question, the (entities relations, and types), a target KG (e.g., Wikidata), and the ground-truth SPARQL query. During the training phase, these samples are shuffled into one batch and the T5 model is trained to generate SPARQL queries based on the input questions and the target knowledge graph. This process involves fine-tuning the model to accurately map the linguistic structures of the questions to the target knowledge graph.

Language models often encounter challenges with special tokens such as { or }, which are integral to SPARQL queries. To address this issue, we replace these tokens in the target strings as follows: { with _cbo_ and } with _cbc_. We normalize variables by replacing the leading *?*-token. For example, a variable like

?uri is replaced with _var_<id>, or _result_<id> if it is a part of the result set. Here, <*id*> represents a number, as a query can contain multiple variables. During the T5 model's inference, we revert these substitutions to generate a valid SPARQL query. For variables, we only replace the leading underscore with the *?*-token. Note that types are included under the entity tag to shorten the input string, as entities and types are used similarly in SPARQL queries. For Freebase the approach is the same, except that the string *target:freebase* is appended at the end of the input string instead of the string

4   Experiments

We conducted our experiments to answer the following research questions:

- **RQ$_1$:** How well does our unified model perform compare to individual language models, trained on single knowledge graphs, for SPARQL query generation?
- **RQ$_2$:** How does our unified model perform compared to state-of-the-art baselines?
- **RQ$_3$:** What is the impact of incorporating knowledge such as entities, relations, and types on the performance of the query generation models?
- **RQ$_4$:** How does integrating knowledge from different resource extraction frameworks into the training datasets affect the performance of query generation models?

4.1   Datasets

In our experiments, we use different benchmark datasets, namely, LC-QuAD 2.0 [24], and QALD-10 [25] on Wikidata knowledge graph and GrailQA [26] on Freebase knowledge graph. We briefly describe these datasets as follows:

- **LC-QuAD 2.0** [24] is a large dataset with 30k questions in English, each paired with a corresponding Wikidata query. The dataset is divided into a training subset with around 24k questions and a test set with 6k questions.
- **QALD-10** [25] this dataset is manually annotated with 394 question-query pairs across different languages. It is an updated version of the QALD-9 dataset, referred to as QALD-9-plus. As the dataset is comparably small, we initially trained our model on the LC-QuAD 2.0 dataset as a pre-trained model (i.e., foundation model), then fine-tuned it on the QALD-9-plus dataset, following the same training strategy as [27].
- **GrailQA** [26] This dataset is a large, crowdsourced collection from Freebase KG, containing around 64k questions. The dataset provides not only SPARQL queries but also contains S-expressions as alternative logical representations. The dataset is divided into a train split with 44K questions a development with 6k and a test split with 13k.

4.2   Experiment Setup

In our experiments, we use Nvidia-H100 GPUs for efficient models training. The T5-base model is used as a foundation model,since it is widely used in query generation research and ensures comparability with other methodologies. Each model is trained for a maximum 50 epochs with an early stopping mechanism to mitigate overfitting. For our unified model, we combine the LC-QuAD 2.0 [26] and GrailQA [24] datasets. Despite GrailQA containing approximately $20k$ more questions than LC-QuAD 2.0, our experiments show no significant impact on performance, indicating that data balancing is unnecessary. For the QALD-9-plus dataset, we fine-tuned our pre-trained LC-QuAD and Freebase model and supplemented the training data with an equivalent number of randomly selected entries from the GrailQA dataset to match the volume of the QALD-9-plus dataset.

4.3   Evaluation

We evaluated the performance of SPARQL query generation using the GERBIL-QA framework [28]. This framework is well-established with different benchmark datasets and evaluation metrics, including Micro-F1, Macro-F1, and Macro-F1 QALD, which are used in the QALD challenge.[3]. We adopted the same evaluation setting of Usbeck et al. [29], and also included metrics such as Macro Precision, Macro Recall, Macro F1, and Macro F1-QALD. The Macro F1 score is calculated per question and uses the geometric mean for the final score. For clarity, we refer to Macro Precision, Macro Recall, Macro F1, and Macro F1-QALD as Precision, Recall, F1, and F1-QALD, respectively. We set up the Virtuoso Triple Store for Freebase following instructions from the GrailQA repository[4], and for Wikidata, we used the official Triple Store.[5] Each query is generated regardless of whether the triple store returns an empty result set. We did not verify the correctness of the generated queries; therefore, improperly formatted queries result in an empty set of results. Finally, we compiled all outputs into a QALD-formatted JSON file and submitted it using the *'upload result file'* function in the GERBIL-QA framework to calculate the final results.

4.4   Results and Discussion

Comparison of Unified Model and Single KG Models (RQ_1). To answer this question, we implemented two different variants of models: the first model is a unified model which is fined-tuned on a heterogeneous dataset of SPARQL queries for Wikidata and Freebase. The other variants are single models tailored for Freebase and Wikidata, which are trained only on the train subsets for the respective KG.

---

[3] https://www.nliwod.org/challenge.
[4] https://github.com/dki-lab/GrailQA.
[5] https://query.wikidata.org/.

Table 2. Comparison of joint and single KG models $\mathbf{RQ}_1$

Dataset	Experiment	Precision	Recall	F1	F1 QALD
LC-QuAD	Gold resources joint	0.88	0.87	0.88	**0.92**
	Gold resources only LC-QuAD data	0.88	0.88	0.89	**0.92**
	End-to-end joint	0.47	0.47	0.47	0.62
	End-to-end only LC-Quad data	0.42	0.43	0.42	0.59
GrailQA	Gold resources joint	0.51	0.59	0.54	0.67
	Gold resources only Grail QA data	0.54	0.62	0.57	**0.68**
	End-to-end joint	0.30	0.34	0.31	0.49
	End-to-end only Grail QA data	0.30	0.34	0.31	0.49
QALD-10	Joint gold resources	0.49	0.49	0.49	**0.64**
	Gold resource only QALD	0.47	0.47	0.47	0.62
	Joint end-to-end model	0.44	0.45	0.44	0.60
	End-to-end only QALD-10	0.45	0.45	0.45	0.61

We conducted two experiments using these models. The first one, referred as the *gold-resource experiment*, involved evaluating the models using high-quality input data derived from the test splits. This evaluation process involves extracting entities, types, and relations from the test split, then incorporated as additional information into the model's input. In contrast, the second experiment, referred to as the *end-to-end experiment*, used knowledge acquired from our *Knowledge Extraction* module approach as direct input for the model. Our analysis on the GrailQA dataset shows that in the end-to-end configuration, the unified model performs equivalently to the single KG models. Conversely, in the gold-resource setup, the performance disparity between the unified and single KG models is minimal. Similarly, on the LC-QuAD dataset, the end-to-end performance of the unified model surpasses the single KG model. For the GrailQA dataset, the difference in performance between models trained and evaluated using gold resources was negligible. On the QALD-10 dataset, the unified model's performance with gold-resource input slightly outperforms the single KG model. In the end-to-end experiment, the single KG model achieves a higher F-Measure by one percent compared to the unified model, though this difference is marginal, consistent with results from other datasets. Overall, the results show that the unified model achieves comparable or equivalent performance as single KG models across all experiments.

Comparison with State-of-the-Art Models ($\mathbf{RQ}_2$).

Results on Wikipedia Datasets. We conducted two experiments on the LC-QuAD 2.0 dataset to compare the performance with state-of-the-art baselines in SPARQL query generation. First, we evaluated the performance of our system using golden entities and relations as inputs. Remarkably, the performance of our approach is aligned with the performance of Banerjee et al. [2], which uses also the T5-small model. This similarity in performance can be attributed to

Table 3. Baseline comparison on Wikidata datasets RQ_2

(a) Results on the QALD-10 dataset

Approach	F1 QALD
Borroto et al. [8]	0.59
Diefenbach et al. [9]	0.58
Shivashanker et al. [30]	0.49
Baramiia et al. [31]	0.43
Joint model end-to-end	0.60
Single KG model end-to-end	**0.61**

(b) End-to-end Results on the LC-QuAD 2.0 datasets

Approach	F1
GPT 3.5. [1]	0.39
Chat GPT [1]	0.42
Joint model end-to-end	**0.46**
Single KG model end-to-end	0.42

(c) Baseline comparison on the LC-QuAD dataset with gold knowledge

Approach	F1 QALD
Banerjee et al. [2] (T5 base)	0.91
Banerjee et al. [2] (T5 small)	**0.92**
Banerjee et al. [2] (PGN-BERT)	0.86
Joint model gold knowledge	**0.92**
Single KG model gold knowledge	**0.92**

Table 4. Baseline comparison on the GrailQA dataset RQ_2

Approach	Precision	Recall	F1	F1 QALD
Shu et al. 2022 [11]	0.59	0.71	0.62	0.71
Shu et al. 2024 [32]	0.59	0.71	0.62	0.71
Yu et al. 2023 [12]	0.64	0.79	0.68	**0.72**
Yu et al. 2024 [33]	0.62	0.79	0.67	0.71
Joint model end-to-end	0.30	0.34	0.31	0.49
Joint model gold	0.51	0.59	0.54	**0.67**

the similar inputs, despite the fact that our study employs a unified model for multiple KGs. In an end-to-end setup, our unified model achieves outperforming results compared to the model by Tan et al. [1] in terms of Macro F1-measure. For additional comparison, we refer to the results from the KGQA-leaderboard[6]. Furthermore, on the QALD-10 dataset, our approach achieves state-of-the-art results in terms of F1-QALD measure. Across all Wikipedia datasets, our unified model achieves the same or comparable results as single models.

Results on Freebase (GrailQA) dataset. In the next step, we compare our results with the baseline models on the GrailQA dataset. It is important to note that the evaluation script used by GrailQA differs from ours, since it

---

[6] https://github.com/KGQA/leaderboard/.

Table 5. comparison of different input data RQ_3

Dataset	Experiment	Precision	Recall	F1	F1 QALD
LC-QuAD	Baseline	0.37	0.37	0.37	0.53
	Entities & types	0.70	0.71	0.71	0.81
	Golden resources	0.88	0.87	0.88	**0.92**
GrailQA	Baseline	0.20	0.24	0.21	0.39
	Golden entities & Golden types	0.33	0.40	0.35	0.54
	Golden resources	0.51	0.59	0.54	**0.67**

assesses queries in the form of S-expressions instead of SPARQL queries. To evaluate on the Gerbil-QA framework, we only included systems that provide their results in a format compatible with QALD. Our findings indicate that the results are not as strong as those achieved by the baseline models. This is mainly because the GrailQA dataset focuses on queries that require detailed knowledge of the Freebase structure, especially its hierarchical ontology. Existing methods usually generate and rank sub-queries, making it possible to learn the knowledge graph structure. However, these methods are resource intensive, as they need to generate and rank a large set of queries, resulting in a slow processing [11,33], which may not be suitable for use in a production environment.

Influence of KG Knowledge on the Model Performance (RQ_3). To answer this question, we conducted different experiments on extensive datasets, LC-QuAD 2.0 and GrailQA, to ensure that the models are trained on sufficient number of entities and relations. We carried out three experiments for each knowledge graph: i) The first experiment, referred to as a *baseline* experiment that used only the question as input. ii) The second experiment includes additional information such as entities and types as well as the question as an input. iii) The third experiment is referred a golden-resource experiment, where entities, types and relations are included with the question as an input.

Table 5 shows the evaluation results of all experiments, indicating that the model's performance improved with the additional information (entities, relations, and types). Specifically, adding entities and types led to a significant performance boost. While relations also improved performance with less noticeable compared to entities. Overall, the LC-QuAD dataset demonstrates better results (F-measure of 0.88), with the additional information. In contrast, the GrailQA dataset reaches an F-measure of only 0.54. Therefore, future research should focus on adding more detailed information, such as the structural aspects of knowledge graphs.

Experiments with Different Training Data (RQ_4). Typically, training data is enriched with additional information by adding golden-resource entities and relations from the target SPARQL queries. This method often causes the model to duplicate the input information without distinguishing between relevant and irrelevant data. We address this issue by extracting relevant information using

Table 6. Comparison of different training setups $\mathbf{RQ}_4$

Dataset	Experiment	Precision	Recall	F1	F1 QALD
LC-QuAD	Gold resources	0.88	0.87	0.88	**0.92**
	Gold resources inc. KE	0.84	0.85	0.84	0.90
	End-to-end	0.46	0.46	0.46	0.60
	End-to-end inc. KE	0.47	0.47	0.47	0.62
GrailQA	Gold resources	0.33	0.40	0.35	0.54
	Gold resources inc. KE	0.51	0.59	0.54	**0.67**
	End-to-end	0.12	0.15	0.12	0.26
	End-to-end inc. KE	0.30	0.34	0.31	0.49
QALD-10	Gold resources inc. KE	0.49	0.49	0.49	**0.64**
	Gold resources	0.48	0.49	0.48	**0.64**
	End-to-end inc. KE	0.44	0.45	0.44	0.60
	End-to-end model	0.27	0.28	0.28	0.43

our Knowledge Extraction module and include ine the training data. To achieve this, we carried out several experiments per dataset, by training the model only with the golden-resource information from the dataset and additional input from knowledge extraction. Afterward, we performed the same experiments as in Sect. 4.4, evaluating the models with both gold input and in an end-to-end setup.

Our findings indicate that including relevant information improves the model performance across all datasets in the end-to-end setup. However, on the LC-QuAD dataset, the performance improvement is minimal compared to training with gold-standard data. On the GrailQA dataset, we achieve a significant improvement, as the results improved from 0.26 to 0.49 in terms of the F1 QALD measure. Similarly, on the QALD dataset, the model's performance improved from 0.43 to 0.6. These variations in performance across datasets can be attributed to different Knowledge Extraction methods in linking data. For instance, entity mentions in LC-QuAD closely align with those in Wikidata knowledge graph, whereas QALD-10 presents greater ambiguity. For example, the question: *"Do the princes William and Harry share the same mother?"*, where the entities are referred to only by their first names. In the evaluation setup, with golden-resource information, we observe a performance improvement only on the GrailQA dataset. This is not surprising, as incorporating golden information can introduce noise into the model inputs.

5   Conclusion and Future Work

This paper presents UniQ-Gen, a unified approach for fine-tuning a single model to generate SPARQL queries across different knowledge graphs. Our results demonstrate that training a unified model on a heterogeneous dataset (e.g., including samples from Wikidata and Freebase) achieves comparable performance to single models for individual knowledge graphs, eliminating the need for separate models for each graph. Moreover, incorporating additional information such as entities, relations, and types, significantly enhances the performance of query generation models. While there are many effective solutions for entity linking, accurate and efficient relation linking remains a challenge in the field of knowledge graph question answering. However, our one-shot query generation approach lacks the incorporating of structural information about the knowledge graph. In our future work, we plan to address this limitation by including structural information (e.g., hierarchical relationships between entities) in our unified model. Furthermore, we will also adapt our approach to handle structural differences between knowledge graphs by integrating KG-specific structural knowledge. .

Acknowledgement. This work has been supported by the German Federal Ministry of Education and Research (BMBF) within the projects, COLIDE (grant no 01I521005D), KIAM (grant no 02L19C115), the European Union's Horizon Europe research and innovation programme (grant No 101070305), and the Deutsche Forschungsgemeinschaft (DFG, German Research Foundation): TRR 318/1 2021 - 438445824.

References

1. Tan, Y. et al.: Can ChatGPT replace traditional KBQA models? An in-depth analysis of the question answering performance of the GPT LLM family. In: Payne, T.R., et al. (eds.) The Semantic Web – ISWC 2023. ISWC 2023. LNCS, vol. 14265. Springer, Cham (2023). https://doi.org/10.1007/978-3-031-47240-4_19
2. Banerjee, D., Nair, P.A., Kaur, J.N., Usbeck, R., Biemann, C.: Modern baselines for sparql semantic parsing. In: Proceedings of the 45th International ACM SIGIR Conference on Research and Development in Information Retrieval, SIGIR '22. ACM (2022). https://doi.org/10.1145/3477495.3531841
3. Ye, X., Yavuz, S., Hashimoto, K., Zhou, Y., Xiong, C.: RnG-KBQA: generation augmented iterative ranking for knowledge base question answering (2022)
4. Yin, J., Dong, J., Wang, Y., De Sa, C., Kuleshov, V.: ModuLoRA: finetuning 2-bit LLMs on consumer GPUs by integrating with modular quantizers. **TMLR** (2024)
5. Bollacker, K., Evans, C., Paritosh, P., Sturge, T., Taylor, J.: Freebase: a collaboratively created graph database for structuring human knowledge. In: Proceedings of the 2008 ACM SIGMOD International Conference on Management of Data, SIGMOD '08, pp. 1247–1250, New York, NY, USA (2008). Association for Computing Machinery. https://doi.org/10.1145/1376616.1376746
6. Vrandečić, D., Krötzsch, M.: Wikidata: a free collaborative knowledgebase. Commun. ACM **57**(10), 78–85 (2014). https://doi.org/10.1145/2629489
7. Auer, S., Bizer, C., Kobilarov, G., Lehmann, J., Cyganiak, R., Ives, Z.: DBpedia: a nucleus for a web of open data. In: Aberer, K., et al. (eds.) ASWC/ISWC -2007. LNCS, vol. 4825, pp. 722–735. Springer, Heidelberg (2007). https://doi.org/10.1007/978-3-540-76298-0_52
8. Borroto Santana, M.A., Ricca, F., Cuteri, B.: A system for translating natural language questions into SPARQL queries with neural networks: preliminary results (discussion paper). In: SEBD 2021: Italian Symposium on Advanced Database Systems, pp. 226–234, Aachen, Germany (2021). RWTH Aachen

9. Diefenbach, D., Both, A., Singh, K., Maret, P.: Towards a question answering system over the semantic web. Semant. Web **11**, 421–439 (2020). https://doi.org/10.3233/SW-190343

10. Vollmers, D., et al.: Knowledge Graph Question Answering Using Graph-Pattern Isomorphism, pp. 103–117. IOS Press (2021). https://doi.org/10.3233/ssw210038

11. Shu, Y., Zhiwei, Yu., Li, Y., Karlsson, B.F., Ma, T., Yuzhong, Q., Lin, C.Y.: Tiara : multi-grained retrieval for robust question answering over large knowledge bases (2022)

12. Gu, Y., Deng, X., Su, Y.: Don't generate, discriminate: a proposal for grounding language models to real-world environments (2023)

13. Lopez, V., Motta, E., Uren, V.: PowerAqua: fishing the semantic web. In: Sure, Y., Domingue, J. (eds.) ESWC 2006. LNCS, vol. 4011, pp. 393–410. Springer, Heidelberg (2006). https://doi.org/10.1007/11762256_30

14. Lopez, V., Fernández, M., Motta, E., Stieler, N.: PowerAqua: supporting users in querying and exploring the semantic web. Semant. Web, **3**(3), 249–265 (2012)

15. Shekarpour, S., Marx, E., Ngomo, A.C.N., Auer, S.: Sina: semantic interpretation of user queries for question answering on interlinked data by Saeedeh Shekarpour with Prateek Jain as coordinator. In: SIGWEB Newsletter (2014). https://doi.org/10.1145/2641730.2641733

16. Fader, A., Zettlemoyer, L., Etzioni, O.: Open question answering over curated and extracted knowledge bases. In: Proceedings of the 20th ACM SIGKDD International Conference on Knowledge Discovery and Data Mining, KDD '14, pp. 1156–1165, New York, NY, USA (2014). Association for Computing Machinery. https://doi.org/10.1145/2623330.2623677

17. Bhutani, N., Zheng, X., Qian, K., Li, Y., Jagadish, H.: Answering complex questions by combining information from curated and extracted knowledge bases. In: Hassan Awadallah, A., Su, Y., Sun, H., Yih, S.W.T. (eds.) Proceedings of the First Workshop on Natural Language Interfaces, pp. 1–10 (2020). Association for Computational Linguistics. https://doi.org/10.18653/v1/2020.nli-1.1

18. Zhang, Y., He, S., Liu, K., Zhao, J.: A joint model for question answering over multiple knowledge bases. In: Proceedings of the AAAI Conference on Artificial Intelligence, vol. 30, no. 1 (2016). https://doi.org/10.1609/aaai.v30i1.10381

19. Neelam, S., et al.: SYGMA: a system for generalizable and modular question answering over knowledge bases. In: Goldberg, Y., Kozareva, Z., Zhang, Y. (eds.) Findings of the Association for Computational Linguistics: EMNLP 2022, pp. 3866–3879, Abu Dhabi, United Arab Emirates (2022). Association for Computational Linguistics. https://doi.org/10.18653/v1/2022.findings-emnlp.284

20. Raffel, C., et al.: Exploring the limits of transfer learning with a unified text-to-text transformer. J. Mach. Learn. Res. **21**(140), 1–67 (2020). http://jmlr.org/papers/v21/20-074.html

21. Nadeau, D., Sekine, S.: A survey of named entity recognition and classification. Lingvisticae Investigationes **30**, 3–26 (2007). https://api.semanticscholar.org/CorpusID:8310135

22. Akbik, A., Bergmann, T., Blythe, D., Rasul, K., Schweter, S., Vollgraf, R.: FLAIR: an easy-to-use framework for state-of-the-art NLP. In: NAACL 2019, 2019 Annual Conference of the North American Chapter of the Association for Computational Linguistics (Demonstrations), pp. 54–59 (2019)

23. De Cao, N., Izacard, G., Riedel, S., Petroni, F.: Autoregressive entity retrieval. In: 9th International Conference on Learning Representations, ICLR 2021, Virtual Event, Austria, May 3-7 (2021). https://openreview.net/forum?id=5k8F6UU39V

24. Dubey, M., Banerjee, D., Abdelkawi, A., Lehmann, J.: LC-QuAD 2.0: a large dataset for complex question answering over Wikidata and DBpedia. In: Ghidini, C., et al. (eds.) ISWC 2019. LNCS, vol. 11779, pp. 69–78. Springer, Cham (2019). https://doi.org/10.1007/978-3-030-30796-7_5

25. Usbeck, R., et al.: QALD-10 - the 10th challenge on question answering over linked data. Semant. Web J. (2023). https://www.semantic-web-journal.net/system/files/swj3357.pdf

26. Gu, Y., et al.: Beyond IID: three levels of generalization for question answering on knowledge bases. In: Proceedings of the Web Conference 2021, pp. 3477–3488. ACM (2021)

27. Zhou, Y., Geng, X., Shen, T., Zhang, W., Jiang, D.: Improving zero-shot cross-lingual transfer for multilingual question answering over knowledge graph. In: Toutanova, K., et al. (eds.) Proceedings of the 2021 Conference of the North American Chapter of the Association for Computational Linguistics: Human Language Technologies, pp. 5822–5834 (2021). Association for Computational Linguistics. https://doi.org/10.18653/v1/2021.naacl-main.465

28. Usbeck, R., et al.: Benchmarking question answering systems. Semant. Web **10**(2), 293–304 (2019). https://doi.org/10.3233/SW-180312

29. Usbeck, R., Röder, M., Hoffmann, M., Conrads, F., Huthmann, J.: Axel-Cyrille Ngonga Ngomo, Christian Demmler, and Christina Unger. Benchmarking Question Answering Systems. Semantic Web **10**(2), 293–304 (2019). https://doi.org/10.3233/SW-180312. URL http://www.semantic-web-journal.net/system/files/swj1578.pdf

30. Shivashankar, K., Benmaarouf, K., Steinmetz, N.: From graph to graph: AMR to SPARQL. In: Proceedings of the 7th Natural Language Interfaces for the Web of Data (NLIWoD) co-located with the 19th European Semantic Web Conference (ESWC 2022) (2022)

31. Baramiia, N., Rogulina, A., Petrakov, S., Kornilov, V., Razzhigaev, A.: Ranking approach to monolingual question answering over knowledge graphs. In: Proceedings of the 7th Natural Language Interfaces for the Web of Data (NLIWoD) co-located with the 19th European Semantic Web Conference (ESWC 2022) (2022)

32. Shu, Y., Yu, Z.: Distribution shifts are bottlenecks: extensive evaluation for grounding language models to knowledge bases. In: Falk, N., Papi, S., Zhang, M. (eds.) Proceedings of the 18th Conference of the European Chapter of the Association for Computational Linguistics: Student Research Workshop, pp. 71–88, St. Julian's, Malta (2024). Association for Computational Linguistics. https://aclanthology.org/2024.eacl-srw.7

33. Gu, Y., Su, Y.: ArcaneQA: dynamic program induction and contextualized encoding for knowledge base question answering. In: Calzolari, N., et al. (eds.) Proceedings of the 29th International Conference on Computational Linguistics, pp. 1718–1731, Gyeongju, Republic of Korea (2022). International Committee on Computational Linguistics. https://aclanthology.org/2022.coling-1.148

LLM-Driven Knowledge Extraction in Temporal and Description Logics

Damiano Duranti[1] , Paolo Giorgini[1] , Andrea Mazzullo[1,2(✉)] ,
Marco Robol[1] , and Marco Roveri[1]

[1] University of Trento, Trento, Italy
{damiano.duranti,paolo.giorgini,andrea.mazzullo,
marco.robol,marco.roveri}@unitn.it
[2] Free University of Bozen-Bolzano, Bolzano, Italy

Abstract. Trustworthy knowledge extraction represents a bottleneck in
the development of autonomous AI agents capable of integrating learning and reasoning capabilities. As a foundational framework of neuro-
symbolic knowledge acquisition systems from semi-structured data, we
introduce an approach that combines Large Language Model (LLM)
functionalities with symbolic verification modules. In a process mining context, we propose to leverage LLMs to generate linear temporal logic specifications starting from sets of finite traces that represent
event logs. In a knowledge representation setting, we focus instead on
LLM-based extraction of description logic concepts to obtain human-
readable conceptualizations that separate positive and negative labeled
data instances. We integrate chat interfaces based on state-of-the-art
LLMs with formal verification modules: in the process mining case, we
employ model checking tools for linear temporal logic on finite traces;
and, for description logic concept learning, we perform entailment checks
using dedicated reasoning engines. First, we conduct a proof-of-concept
evaluation of these architectures, comparing the performance of the
LLMs on each task. We then provide an implementation of a GPT-based
toolchain to automate the candidate generation and verification steps.

Keywords: LLM-Driven Knowledge Extraction · Linear Temporal
Logic · Process Mining · Description Logics · Concept Learning

1 Introduction

Exploiting generative capabilities of LLMs to address knowledge representation
problems is a subject of ongoing discussion in the literature [34,37,46,52,57].
While LLMs are known to perform well in natural language generation tasks,
several approaches currently investigate the possibility of complementing them
with *reasoning* modules [13,28,38,41,45,56], enhancing their ability to provide
trustworthy and explainable answers. When used with time-dependent data in
dynamic environments, or in knowledge-intensive application domains, LLM-
based systems also need to face significant *knowledge extraction* challenges,
involving the acquisition of structured information from raw or semistructured
data, in turn to enable temporal or domain-specific reasoning potential.

M. Alam et al. (Eds.): EKAW 2024, LNAI 15370, pp. 190–208, 2025.
https://doi.org/10.1007/978-3-031-77792-9_12

In temporal settings, a knowledge acquisition problem for *process mining* [1] requires to identify, from sets of event logs, a formal specification that captures the underlying process structure. *Linear-time temporal logic on finite traces* (LTL_f) [16], has been widely investigated as a foundational formalism for declarative approaches to the problem [11,12,14,47]. Strictly related is the task of finding a LTL_f specification that is capable of discerning a set of *positive* logs, i.e., examples of successful processes, from a set of *negative* ones, i.e., examples of undesired dynamics [19,40]. Such tasks are of paramount importance to Business Process Modeling applications, as well as in safety-critical domains, where the extraction of meaningful specifications from sets of positive/negative event traces is an important step in the development of trustworthy and explainable systems that are transparent to the human user.

In knowledge representation applications based on *description logics* (DLs) [5], and particularly in the field of ontology generation, the *concept learning* problem deals instead with the task of automatically generating, from a set of data instances, DL concept descriptions capable of correctly representing them under a given ontology [42]. A strictly related problem, known as *concept separability*, asks whether there exists a concept formulated in a given DL language that is capable of discerning *positive* labeled data instances from *negative* ones [31]. Applications of DL concept learning and separability can be found in ontology-based data management, particularly for the integration of heterogeneous datasets or incomplete knowledge bases, as well as in entity resolution problems, and in computational linguistics, when addressing referring expression generation tasks.

In this paper, we explore whether and to what extent LLMs can be used to support the automated generation of provably correct specifications for the knowledge extraction problems mentioned above. To this end, we propose an integrated approach that employs the generative capabilities of LLMs to address dedicated knowledge acquisition tasks, both in LTL_f and in DL settings, complementing them with reasoning modules for symbolic verification [25]. Our architecture consists of the following elements. First, an LLM-based interface is prompted with an instance of a knowledge extraction problem, formulated in the LTL_f or in the DL setting, with the request of providing a candidate specification capable of separating positive from negative instances given as part of the input. Such candidate is then submitted to a dedicated formal verification module. In the process mining case, we employ a model checker to verify whether the LLM generated LTL_f formula correctly separates the positive and negative finite traces given in the input. For the concept learning case, we perform instance checking using a dedicated reasoning engine, to control that the provided DL concept applies correctly to the positive data instances and discards the negative ones, as requested by the prompt. Whenever a specification is found to be correct after the formal verification step, the search is successfully completed. Otherwise, the prompt is re-submitted to the LLM for another generation attempt.

We investigate the viability of our approach to LLM-driven knowledge extraction by performing, over dedicated sets of problem instances for LTL$_f$ process mining and for learning of the DL concept, a proof-of-concept experimental evaluation that compares the performance of three chatbots based on state-of-the-art LLMs, namely GPT3.5, GPT4, Bard, and Gemini. In addition, we implement a dedicated GPT-based toolchain to automate candidate generation and verification cycles, testing its performances on a similar dataset.

Related Work. The theoretical motivation behind our approach lies in the field of *reverse engineering of formulas* for LTL$_f$ [19,20,24,40], as well as in *ontology* and *concept learning* [22,27,33,42,43] or *separability* [23,31] for DLs, where the problem of concept generation has been addressed also by concrete tool implementations [8,9,18,30]. Related problems are considered also in the fields of *inductive* and *abductive reasoning* [17,39]. Moreover, our LLM-driven knowledge extraction approach can be related to the model of *active learning with membership queries* in computational learning theory [3,6,7,49], as well as to the *query-by-example* approach from database theory [35]. Our framework shares connections also with *counterexample-guided inductive synthesis* [2]. Finally, related work is currently focusing on *LLM-based ontology construction* [21,36] and *planning* [29,44,50,51,54,55].

2   Preliminaries

Hereafter we provide the background required for linear-time temporal logic over finite traces (LTL$_f$) and for the description logics (DLs) considered in the paper.

Linear-time Temporal Logic on Finite Traces. Given a countable set Σ of *proposition letters*, an LTL$_f(\Sigma)$ *formula* φ is generated according to the grammar: $\varphi ::= p \mid \neg\varphi \mid \varphi \wedge \varphi \mid \mathsf{X}\varphi \mid \varphi\mathsf{U}\varphi$, where $p \in \Sigma$, X is the *strong next*, and U is the *until* operator. Standard abbreviations are employed for $\bot := p \wedge \neg p$ (with $p \in \Sigma$ fixed), $\top := \neg\bot$, other Boolean connectives ($\vee, \rightarrow, \leftrightarrow$), and temporal operators: *weak next*, $\mathsf{N}\varphi := \neg\mathsf{X}\neg\varphi$; *release*, $\varphi\mathsf{R}\psi := \neg(\neg\varphi\mathsf{U}\neg\psi)$; *eventually*, $\mathsf{F}\varphi := \top\mathsf{U}\varphi$; *globally*, $\mathsf{G}\varphi := \bot\mathsf{R}\varphi$; *weak until*, $\varphi\mathsf{W}\psi := (\varphi\mathsf{U}\psi) \vee \mathsf{G}\varphi$.

We call $\sigma \in (2^\Sigma)^+$ a *finite trace* over 2^Σ, and we set the *length* of $\sigma = (\mu_0, \ldots, \mu_{n-1}) \in (2^\Sigma)^+$ as $|\sigma| = n$. The *satisfaction* of an LTL$_f(\Sigma)$ formula φ in σ at instant $0 \leq i < |\sigma|$, denoted by $\sigma, i \models \varphi$, is defined as follows:

$$\sigma, i \models p \text{ iff } p \in \mu_i; \quad \sigma, i \models \neg\psi \text{ iff } \sigma, i \not\models \psi; \quad \sigma, i \models \psi \wedge \chi \text{ iff } \sigma, i \models \psi \text{ and } \sigma, i \models \chi;$$

$$\sigma, i \models \mathsf{X}\psi \text{ iff } i + 1 < |\sigma| \text{ and } \sigma, i + 1 \models \psi;$$

$$\sigma, i \models \psi\mathsf{U}\chi \text{ iff } \exists j, i \leq j < |\sigma|, \text{ s.t.} \sigma, j \models \chi, \text{ and } \sigma, k \models \psi, \forall k \text{s.t.} i \leq k < j;$$

We say that σ is a *model* of φ, written $\sigma \models \varphi$, iff $\sigma, 0 \models \varphi$.

Description Logics. Let $\mathsf{N_C}$, $\mathsf{N_R}$, and $\mathsf{N_I}$ be mutually disjoint and countably infinite sets of *concept*, *role*, and *individual* names. An $\mathcal{ALCO}$ concept is defined according to the following grammar: $C ::= A \mid \{a\} \mid \neg C \mid C \sqcap C \mid \exists r.C$, where $A \in \mathsf{N_C}$, $a \in \mathsf{N_I}$, and $r \in \mathsf{N_R}$. We also employ usual abbreviations: $\bot$ for $A \sqcap \neg A$, with $A \in N_C$ arbitrarily fixed; $\top$ for $\neg\bot$; $C \sqcup D$ for $\neg(\neg C \sqcap \neg D)$; $\{a_1, \ldots, a_n\}$ for $\{a_1\} \sqcup \ldots \sqcup \{a_n\}$; and $\forall r.C$ for $\neg\exists r.\neg C$. We denote by $\mathcal{ALC}$ the fragment obtained from $\mathcal{ALCO}$ by dropping concepts of the form $\{a\}$.

Given $\mathcal{L} \in \{\mathcal{ALCO}, \mathcal{ALC}\}$, an $\mathcal{L}$ *concept inclusion* ($\mathcal{L}$ *CI*) takes the form $C \sqsubseteq D$, with C and D $\mathcal{L}$ concepts, and an $\mathcal{L}$ *ontology* is a finite set of $\mathcal{L}$ CIs. A *dataset* $\mathcal{D}$ is a finite set of *assertions* of the form $A(a)$ or $r(a,b)$, with $a, b \in N_I$, $A \in N_C$, and $r \in N_R$. An $\mathcal{L}$ *axiom* is an $\mathcal{L}$ CI, an assertion, or an $\mathcal{L}$ *concept assertion* of the form $C(a)$, where C is an $\mathcal{L}$ concept and $a \in N_I$. An $\mathcal{L}$ *knowledge base* ($\mathcal{L}$ KB) is a pair $\mathcal{K} = (\mathcal{O}, \mathcal{D})$, where $\mathcal{O}$ is an $\mathcal{L}$ ontology and $\mathcal{D}$ is a dataset. For a syntactic object X, we denote by Σ_X the *signature of* X, that is, the set of concept, role, and individual names occurring in X. For $\mathcal{L} \in \{\mathcal{ALCO}, \mathcal{ALC}\}$ and an $\mathcal{L}$ KB $\mathcal{K}$, an $\mathcal{L}(\Sigma_\mathcal{K})$ concept C is an $\mathcal{L}$ concept such that $\Sigma_C \subseteq \Sigma_\mathcal{K}$.

The semantics is given in terms of *interpretations* $\mathcal{I} = (\Delta^\mathcal{I}, \cdot^\mathcal{I})$, where $\Delta^\mathcal{I}$ is a non-empty set, called *domain* of $\mathcal{I}$, and $\cdot^\mathcal{I}$ is a function mapping every $A \in N_C$ to a subset of $\Delta^\mathcal{I}$, every $r \in N_R$ to a subset of $\Delta^\mathcal{I} \times \Delta^\mathcal{I}$, and every $a \in N_I$ to an element in $\Delta^\mathcal{I}$. The *extension* $C^\mathcal{I}$ *of an* $\mathcal{L}$ *concept* C *in* $\mathcal{I}$ is defined as follows: $\top^\mathcal{I} = \Delta^\mathcal{I}$; $\{a\}^\mathcal{I} = \{a^\mathcal{I}\}$; $\neg C^\mathcal{I} = \Delta^\mathcal{I} \setminus C^\mathcal{I}$; $(C \sqcap D)^\mathcal{I} = C^\mathcal{I} \cap D^\mathcal{I}$; $(\exists r.C)^\mathcal{I} = \{d \in \Delta^\mathcal{I} \mid \text{there exists } e \in C^\mathcal{I} : (d,e) \in r^\mathcal{I}\}$. An interpretation $\mathcal{I}$ *satisfies an* $\mathcal{L}$ *CI* $C \sqsubseteq D$ if $C^\mathcal{I} \subseteq D^\mathcal{I}$. Moreover, $\mathcal{I}$ *satisfies* an $\mathcal{L}$ concept name assertion $C(a)$, where C is an $\mathcal{L}$ concept, if $a^\mathcal{I} \in C^\mathcal{I}$, and it satisfies an assertion of the form $r(a,b)$ iff $(a^\mathcal{I}, b^\mathcal{I}) \in r^\mathcal{I}$. We say that $\mathcal{I}$ is a *model* of an $\mathcal{L}$ KB $\mathcal{K} = (\mathcal{O}, \mathcal{D})$ if it satisfies all CIs in $\mathcal{O}$ and all assertions in $\mathcal{D}$. We say that an axiom α follows from KB $\mathcal{K}$, in symbols $\mathcal{K} \models \alpha$, if every model of $\mathcal{K}$ satisfies α.

3   LLM-Driven Process Mining in LTL$_f$

In this section, we first introduce the required technical notions and then describe the overall architecture, for LLM-driven LTL$_f$ process mining.

Given a set of propositional letters Σ, let $P \subseteq (2^\Sigma)^+$ be a set of *positive (finite) traces*, and let $N \subseteq (2^\Sigma)^+$ be a set of *negative (finite) traces*. For a *signature* $\Sigma' \subseteq \Sigma$, the LTL$_f(\Sigma')$ process mining problem for P *and* N asks to determine a *separating* LTL$_f(\Sigma')$ *formula* φ that *separates* P and N, i.e.: (i) $\sigma^+ \models \varphi$, for all $\sigma^+ \in P$; and (ii) $\sigma^- \not\models \varphi$, for all $\sigma^- \in N$. If $\Sigma' = \Sigma$, we omit Σ' and speak simply of the LTL$_f$ *process mining problem for* P *and* N. Our LLM-driven architecture addressing the separating formula generation, described in the following and illustrated in Fig. 1 (left), consists of three main steps.

First, by employing the chat interface based on one of the LLMs considered in this paper (Bard, Gemini, GPT3.5, GPT4), we prompt the chatbot with a text-based request that provides the following elements as part of the input: 1. a signature $\Sigma' \subseteq \Sigma$; 2. a set of positive traces $P \subseteq (2^\Sigma)^+$; 3. a set of negative traces $N \subseteq (2^\Sigma)^+$; (we let Σ to be implicitly determined by the positive and negative traces). The prompt asks the chatbot to provide an LTL$_f(\Sigma')$ formula that separates P and N. When $\Sigma' = \Sigma$, the prompt queries for an LTL$_f$ separating formula, omitting Σ' from the input.

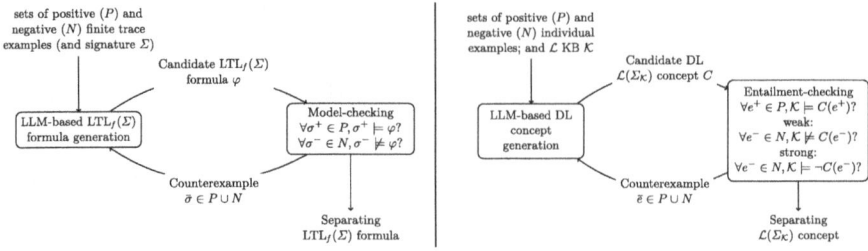

Fig. 1. LLM-driven LTL_f process mining (left) and DL concept learning (right).

Example 1. As part of the input, the prompt contains the triple (P, N, Σ'), with $P = \{\sigma_1^+, \sigma_2^+\} \subseteq (2^{\{p,q,r\}})^+$, $N = \{\sigma_1^-, \sigma_2^-\} \subseteq (2^{\{p,q,r\}})^+$, $\Sigma' = \{p, q\}$, and:

$$\sigma_1^+ = (\{r\}, \{p\}, \{q\}, \{\}, \{p, r\}, \{r\}, \{q, r\}, \{p\}, \{q, r\});$$
$$\sigma_2^+ = (\{\}, \{\}, \{p\}, \{q\}, \{\}, \{\}, \{p\}, \{\}, \{q\});$$
$$\sigma_1^- = (\{r\}, \{p, r\}, \{r\}, \{r\}, \{q, r\}, \{r\}, \{r\}, \{p, r\}, \{r\});$$
$$\sigma_2^- = (\{p, q\}, \{\}, \{\}, \{p, q\}, \{p\}, \{p\}, \{p\}, \{p\}, \{p\});$$

The prompt then requests an $LTL_f(\Sigma')$ formula that separates P and N. A correct separating $LTL_f(\Sigma')$ formula for P and N is, for instance, $G(p \rightarrow Fq)$.

As a second step, we consider the candidate formula φ obtained as part of the answer generated by the LLM, and proceed to verify it against the finite traces in P and N, to check whether the formula actually separates them, meaning that: $\sigma^+ \models \varphi$, for all $\sigma^+ \in P$; and $\sigma^- \not\models \varphi$, for all $\sigma^- \in N$. Whenever the the prompt is misunderstood, more than one candidate formula can be generated. In this case, they are all considered as valid attempts, and proceed to the evaluation phase. Moreover, we ignore minor notational variants that depart from the constructors used this paper (e.g., $\square$ and & in place of G and $\wedge$, respectively).

Example 2. As a response to the prompt illustrated in Example 1, Bard provides the formula $\neg G(pUq)$. We have $\sigma_i^+ \models \neg G(pUq)$, for $i = 1, 2$; $\sigma_i^- \models \neg G(pUq)$, for $i = 1, 2$. In contrast, a candidate formula provided by Gemini on the same prompt is $(Fp \wedge Fq)U(Gp \wedge Gq)$, for which we have that: $\sigma_i^+ \not\models (Fp \wedge Fq)U(Gp \wedge Gq)$, for $i = 1, 2$; $\sigma_i^- \not\models (Fp \wedge Fq)U(Gp \wedge Gq)$, for $i = 1, 2$.

The third and final step goes back to formula generation, when the verification module determines that a *counterexample* $\bar{\sigma}$ is found, i.e., a positive trace $\sigma \in P$ such that $\sigma \not\models \varphi$, or a negative trace $\sigma' \in N$ such that $\sigma' \models \varphi$. In this case, the procedure returns to the first step, by starting a novel conversation with the LLM interface and repeating the prompt with the original request. If, instead, the LLM has correctly identified a separating formula, the process is halted.

Clearly, the procedure sketched above does not necessarily terminate, since there is no formal guarantee that the LLM module will ever be capable of finding a correct separating formula. We ensure termination by introducing a controlled loop, taking into account when a maximum number of failed attempts is reached, after which the process is stopped. For data collection and evaluation purposes, in this paper we repeat the looping steps until the maximum number of attempts, whether or not a separating formula is correctly identified at a previous stage.

Example 3. Since σ_i^-, with $i = 1, 2$, is found to be a counterexample for the Bard candidate $\neg G(p U q)$, as shown in Example 2, we find that $\neg G(p U q)$ does not correctly separate P and N. The original request is hence prompted again to the Bard interface, in order to determine another candidate for the verification module. The list of repeatedly outputted candidates, evaluated until the limit of 5 total attempts is reached (see Sect. 5.1), is: 1. $\varphi_1 = \neg G(p U q)$, incorrect: $\sigma_{i \in \{1,2\}}^- \models \varphi_1$; 2. $\varphi_2 = AG(p \rightarrow Fq) \wedge EF(\neg p \vee Fq)$, syntactically incorrect (not in LTL_f); 3. $\varphi_3 = G(p U q) \wedge G(q U p)$, incorrect: $\sigma_{i \in \{1,2\}}^+ \not\models \varphi_3$; 3. $\varphi_4 = G(p U (q \rightarrow Fq))$, incorrect: $\sigma_{i \in \{1,2\}}^- \models \varphi_4$; 4. $\varphi_5 = G(p \rightarrow Fq)$, correct. The $\text{LTL}_f(\Sigma')$ formula $G(p \rightarrow Fq)$ thus separates P and N.

Similarly, we have seen in Example 2 that the Gemini candidate $(Fp \wedge Fq) U (Gp \wedge Gq)$ has σ_i^+, with $i = 1, 2$, as counterexamples. By repeating the attempts, until the limit of 5 is reached, we obtain: 1. $\varphi_1 = (Fp \wedge Fq) U (Gp \wedge Gq)$, incorrect: $\sigma_{i \in \{1,2\}}^+ \not\models \varphi_1$; 2. $\varphi_2 = \varphi_1$, incorrect (as above); 3. $\varphi_3 = \varphi_1$, incorrect (as above); 4. $\varphi_4 = F(p \wedge G(q U r))$, syntactically incorrect (not in the right signature Σ'); 5. $\varphi_5 = \varphi_4$, syntactically incorrect (as above). Hence, no $\text{LTL}_f(\Sigma')$ formula provided separates P and N.

4   LLM-Driven Concept Learning in DLs

In what follows, we focus on the problem of learning a separating DL concept (if it exists) by exploiting the generative capabilities of LLMs. After providing the relevant technical definitions, we illustrate the overall architecture combining a LLM-driven component with a verification step involving DL entailment checks.

Given a DL $\mathcal{L} \in \{\mathcal{ALC}, \mathcal{ALCO}\}$, let $\mathcal{K} = (\mathcal{O}, \mathcal{D})$ be a $\mathcal{L}$ KB. The *positive examples*, P, and *negative examples*, N, are subsets of $\text{ind}(\mathcal{D})$, i.e., of the set of *individual names* occurring in $\mathcal{D}$. We define the *weak $\mathcal{L}$ concept learning problem for P and N under $\mathcal{K}$* as the problem of determining a *weakly separating $\mathcal{L}(\Sigma_{\mathcal{K}})$ concept C* that *weakly separates* P and N, meaning that: (i) $\mathcal{K} \models C(e^+)$, for every $e^+ \in P$; and (ii) $\mathcal{K} \not\models C(e^-)$, for every $e^- \in N$. The *strong $\mathcal{L}$ concept learning problem for P and N under $\mathcal{K}$* requires, instead, to provide a *strongly separating $\mathcal{L}(\Sigma_{\mathcal{K}})$ concept C* that *strongly separates* P and N, i.e., such that (i) holds, as well as (ii') $\mathcal{K} \models \neg C(e^-)$, for every $e^- \in N$.

Towards LLM-assisted generation of separating concepts, we propose a three-step architecture illustrated in the following and summarised in Fig. 1 (right). As first step, similarly to the LLM-driven LTL_f process mining case, we start by prompting each of the chatbots considered in this work (Bard, Gemini, GPT3.5, GPT4) with a textual request containing the following constituents as part of the input: 1. an $\mathcal{L}$ KB $\mathcal{K} = (\mathcal{O}, \mathcal{D})$; 2. a set of positive examples $P \subseteq \text{ind}(\mathcal{D})$; 3. a set of negative examples $N \subseteq \text{ind}(\mathcal{D})$. The prompt requests the chatbot to propose an $\mathcal{L}$ concept that strongly separates P and N, or (alternatively) it queries for an $\mathcal{L}$ concept that weakly separates P and N.

Example 4. As part of the input, the prompt contains the triple $(P, N, \mathcal{K})$, defined as follows.[1] The set of positive examples is $P = \{b_1, b_2\}$, while the set of negative examples is $N = \{b_3\}$. Moreover, $\mathcal{K} = (\mathcal{O}, \mathcal{D})$ is an $\mathcal{ALCO}$ KB consisting of the ontology $\mathcal{O} = \{A_1 \sqcup A_2 \sqsubseteq A, \exists r.A \sqsubseteq B, \exists r.\{a_3\} \sqsubseteq \neg \exists r.A_1\}$ and of the dataset $\mathcal{D} = \{A_1(a_1),\ A_1(a_2),\ A_2(a_3),\ B(b_1),\ B(b_2),\ r(b_1, a_1),$ $r(b_2, a_2),\ r(b_3, a_3)\}$. The prompt then asks to provide an $\mathcal{ALCO}$ concept C that strongly separates P and N, meaning that $\mathcal{K} \models C(b_1)$ and $\mathcal{K} \models C(b_2)$, whereas $\mathcal{K} \models \neg C(b_3)$. An example of a strongly separating $\mathcal{L}(\Sigma_\mathcal{K})$ concept is, for instance, $\exists r.A_1$.

Secondly, given the proposed $\mathcal{L}$ concept C, generated by the chat interface of the LLM, we employ a suitable DL reasoner to verify whether it correctly (weakly or strongly, depending on the prompt request) separates P and N. That is, we check that: $\mathcal{K} \models C(e^+)$, for all $e^+ \in P$; and either $\mathcal{K} \not\models C(e^+)$, for all $\sigma^- \in N$, if the prompt asks for an $\mathcal{L}$ concept that weakly separates P and N; or $\mathcal{K} \models \neg C(e^-)$, for all $e^- \in N$, if the prompt asks for an $\mathcal{L}$ concept that strongly separates P and N. If the request in the prompt is misunderstood and more than one candidate concept is provided, they are all considered as valid attempts to be evaluated. We further ignore, as in the process mining case, minor notational variants of the constructors introduce above (e.g., $\cap$ or $\wedge$ in place of $\sqcap$).

Example 5. As an answer given by GPT3.5 in response to the prompt from Example 4, we have the following: $C_1 = \exists r.A \sqcap \neg \exists r.A_1$. It holds that $\mathcal{K} \models C_1(b_3)$. Moreover, $\mathcal{K} \models \neg C(b_i)$, and hence $\mathcal{K} \not\models C_1(b_i)$, for $i = 1, 2$. In contrast, GPT4 provides on the same prompt the candidate: $C_1 = \exists r.A_1$, for which we have that $\mathcal{K} \models C_1(b_i)$, for $i = 1, 2$, and $\mathcal{K} \models \neg C(b_3)$.

A *counterexample* is an example $\bar{e}$ such that: $\mathcal{K} \not\models C(\bar{e})$, with $\bar{e} \in P$; or $\mathcal{K} \models C(\bar{e})$, with $\bar{e} \in N$ (in case of weak separation)—alternatively, $\mathcal{K} \not\models \neg C(\bar{e})$, with $\bar{e} \in N$ (in case of strong separation). As our third and final step, if such a counterexample is found, we start a new conversation to repeat the initial prompt request. If, instead, a correct separating concept is found, the procedure terminates.

Similarly to what seen for the LTL$_f$ process mining case, the procedure is not guaranteed to terminate. Firstly, as seen above, it cannot be ensured that the LLM module will be able of finding a correct separating concept. In addition, such a concept does not always exist in $\mathcal{L} \in \{\mathcal{ALC}, \mathcal{ALCO}\}$ [31]. A controlled loop is again used to halt the process when a maximum number of wrong attempts is reached. However, as done in the LTL$_f$ process mining case, we ensure in this setting that the maximum number of attempts is always reached, even when a correct answer is provided at an earlier stage within this limit.

Example 6. Example 5 shows that, with respect to strong separation, each b_i, for $i = 1, 2, 3$, is a counterexample for $C_1 = \exists r.A \sqcap \neg \exists r.A_1$. Thus, C_1 does not

---

[1] The KBs used in our experiments are obtained from [4,31], and the *Concept Learning in Description Logics* Tutorial material presented at the KR 2023 Conference.

strongly separate P and N under $\mathcal{K}$. The initial prompt is then resubmitted to the GPT3.5 chatbot interface, repeatedly asking for other candidate concepts. The list of progressively outputted concepts, evaluated until the limit of 5 total attempts is reached (see Sect. 5.1), is the following: 1. $C_1 = \exists r.A \sqcap \neg \exists r.A_1$, incorrect: $\mathcal{K} \not\models C_1(b_i)$, for $i = 1, 2$, $\mathcal{K} \not\models \neg C_1(b_3)$; 2. prompt misunderstood (2 options provided): (a) $C_{2_a} = \exists r.A \sqcap \forall r.A_1$, incorrect: $\mathcal{K} \not\models C_{2_a}(b_i)$, for $i = 1, 2$, (b) $C_{2_b} = \exists r.A \sqcap \neg \exists r.A_1$, incorrect (as in Point 1); 3. $C_3 = \exists r.A \sqcap \exists r.A_1$, correct; 4. $C_4 = \exists r.A \sqcap B \sqcap r(b_1, a_1) \sqcap r(b_2, a_2) \sqcap \neg r(b_3, A_1)$, syntactically incorrect; 5. $C_5 = \exists r.(\{a_1\} \sqcap \{a_2\})$, incorrect: $\mathcal{K} \not\models C_5(b_i)$, for $i = 1, 2$. Hence, C_3 strongly separate P and N under $\mathcal{K}$.

Regarding GPT4, we have seen in Example 5 that $C_1 = \exists r.A_1$ already strongly separates P and N under $\mathcal{K}$. Moreover, for the subsequent attempts, we obtain $C_i = C_1$, with $i \in \{2, 3, 4, 5\}$, hence all the provided candidates by GPT4 are successful in strongly separating P and N under $\mathcal{K}$.

5   Proof-of-Concept Evaluation

We present the details of the proof-of-concept implementation of our architecture, as well as the results of the tests carried out using Bard, Gemini, GPT3.5, and GPT4. After describing our experimental setup, we discuss which candidates are considered as incorrect, and illustrate the implementation of our verification steps for LTL_f process mining and DL concept learning. Finally, we define the metrics involved in our data analysis, and discuss the results obtained.[2]

5.1   Experimental Setup

We carried out a set of experiments, both for LTL_f process mining and for DL concept learning, using GPT3.5, GPT4, Bard, and Gemini chatbot interfaces, based on the most recent version available at the time of our evaluation.

The starting point of the experiments is a *prompt* r, consisting of a text in natural language that includes (see Examples 1-4 above): 1. the formulation of the problem to be addressed; 2. the required inputs to that problem, including a set of positive and negative examples to be separated. This prompt is repeatedly submitted to the LLM chatbot interface, starting a separate conversation i at each step, up to a maximum of ℓ requests: in our experiments, we set $\ell = 5$ for every prompt-set. Each interaction determines a *generated answer* $a_i(r)$ by the LLM in response to the prompt r, which is collected in a dedicated text file.

From this interaction cycle, we hence obtain a *prompt set* $R = \{a_1(r), \ldots, a_\ell(r)\}$, where each generated answer $a_i(r)$ is a set of one or two *candidates*: typically, the answer consists of only one candidate, identified with it. Each prompt is also associated with an *expected response* $a_e(r)$, which is used to perform calculations with respect to certain metrics in our data analysis.

---

[2] The full set of prompts, produced outputs, calls to verification engines, and result tables, are available at https://github.com/damianoduranti/LLMknowextra.

We considered 14 prompts for LTL_f process mining and 10 prompts for DL concept learning. These prompts are grouped into 7 and 5 pairs (r_i, r_i'), respectively, according to the following criteria. 1. For LTL_f process mining, prompt r_i, with $1 \leq i \leq 7$, includes an LTL_f *process mining* problem instance (i.e., *without signature restrictions*), whereas r_i' is a variant containing the statement of an $\text{LTL}_f(\Sigma)$ *process mining* problem instance (i.e., *with signature restrictions*). 2. For learning DL concepts, the prompt r_i, with $1 \leq i \leq 5$, contains an instance of a *strong concept learning* problem, while r_i' is a variant containing a related *weak concept learning* problem instance. Hence, our data set is structured as follows. For LTL_f process mining, we have a set of 7 pairs of prompt-sets, $\{(R_1, R_1'), \ldots, (R_7, R_7')\}$, where $R_i = \{a_1(r_i), \ldots, a_5(r_i)\}$ and $R_i' = \{a_1(r_i'), \ldots, a_5(r_i')\}$, for $1 \leq i \leq 7$. For DL concept learning, we obtain a set of 5 pairs of prompt-sets: $\{(S_1, S_1'), \ldots, (S_5, S_5')\}$, where $S_i = \{a_1(s_i), \ldots, a_5(s_i)\}$ and $S_i' = \{a_1(s_i'), \ldots, a_5(s_i')\}$, for $1 \leq i \leq 5$.

We stress that no background information is provided to the LLM chatbot interface and that the prompts are submitted by creating novel conversations at each step. Indeed, aiming at progressive refinements of chained tasks, our initial attempts were based on the submission of variations or expansions of the original prompt, pinpointing when and where a counterexample was identified. However, we observed that this approach frequently led to the generation of incorrect candidates that were just minor variants of the original ones. Due to this behaviour, we opted for initiating separate conversations at each stage.

5.2   Implementation Details

We now illustrate how our candidate verification steps are implemented. First, we discuss the evaluation criteria under which we interpret and verify the answers provided by the LLMs, and in presence of which features a candidate qualifies as incorrect. Then, we describe the NUXMV encoding of the LTL_f traces provided as part of the prompts to the LLM chat interfaces. Finally, we illustrate how the DL KBs and candidate concepts are represented in OWL2 using an ontology editor and how the entailment checks are performed with a dedicated DL reasoner.

Candidate Evaluation. Formulas and concepts are considered as *incorrect candidates* if they fall into one of the following categories.

Syntactically Incorrect Candidates. Candidates that are not syntactically well-formed according to the grammars introduced in this paper. For example: $\text{AG}(p \rightarrow \text{F}q) \wedge \text{EF}(\neg p \vee \text{F}q))$, as it is not an LTL_f formula; or $B \sqcap \forall \neg r.A_1$, which is not an instance of an $\mathcal{L}$ concept, for $\mathcal{L} \in \{\mathcal{ALC}, \mathcal{ALCO}\}$. We also consider syntactically incorrect those candidates that do not respect the *signature constraint* for the $\text{LTL}_f(\Sigma)$ process mining problem, as in the case of $\text{G}(r \rightarrow \text{F}q)$, when $\Sigma = \{r\}$.

Semantically Incorrect Candidates. Syntactically correct candidates that do not correctly separate positive and negative examples under the semantics considered in this paper. This category includes (whenever the set of positive traces is non-empty) LTL_f formulas unsatisfiable under the non-strict semantics over finite traces, e.g. $\text{G}(p \rightarrow \text{X}p)$.

LTL$_f$ *Candidate Verification.* For the verification of LTL$_f$ queries we rely on an encoding into NUXMV [10]. At a glance, we first translate the LTL$_f$ formula into an equi-satisfiable LTL formula (over infinite traces) [48], we then create a symbolic transition system encoding each finite trace, and finally we model check the translated formula in the constructed transition system.

For the construction of the transition system and of the equi-satisfiable formula, we follow the approach described in [15]. The construction is as follows. For each finite trace $\sigma = (\mu_0, \ldots, \mu_{n-1})$ we create a NUXMV transition system T_σ consisting of (i) a Boolean variable v_p for each $p \in \Sigma$; (ii) a numeric variable pc having range $1, \ldots, N_\sigma$, where $N_\sigma = |\sigma| + 1$; (iii) a Boolean variable End needed to capture finite trace semantics within an infinite state semantics [15]. The behavior of such variables is as follows. Initially, pc is 1, and if pc $< N_\sigma$ in the next step it takes value pc $+ 1$, otherwise it takes value N_σ. End is initially false, and it remains false until pc takes value N_σ, and when pc $= N_\sigma$ it takes value true. Each variable v_p is such that if pc $= i$, then v_p is true if $p \in \mu_{i-1}$, otherwise it is false. The LTL$_f$ formula φ is translated into an equi-satisfiable formula f2i(φ) as follows [15]: f2i(p) $\mapsto$ p for $p \in \Sigma$; f2i($\neg\varphi$) $\mapsto$ $\neg$f2i(φ); f2i($\varphi_1 \wedge \varphi_2$) $\mapsto$ f2i($\varphi_1 \wedge$ f2i(φ_2)); f2i($X\varphi$) $\mapsto$ X(f2i(φ) $\wedge \neg$**End**); f2i($\varphi_1 U \varphi_2$) $\mapsto$ f2i(φ_1)U(f2i(φ_2) $\wedge \neg$**End**).

Then, we can verify whether a finite trace σ satisfies an LTL$_f$ formula φ by performing the model checking query to verify whether f2i(φ) holds in the symbolic transition system T_σ (written $T_\sigma \models$ f2i(φ)) [15]. Indeed, given a finite trace σ and an LTL$_f$ formula φ, it follows from the results in [15] that $\sigma \models \varphi$ iff $T_\sigma \models$ f2i(φ), where T_σ and f2i(φ) are constructed as above.

DL Candidate Verification. For the verification of the separation of DL concepts under KBs, we first encode the latter (originally in DL syntax) in the OWL2 language [26], using the Protégé editor, Version 5.5.0 [32]. The candidates provided by the LLMs are then converted into OWL2 Manchester syntax and verified using the DL QUERY plugin, Version 4.0.1, to query, for instance, checking under the given KB. We employ FACT++ 1.6.5 [53] as the reasoning engine. In the case of the strong $\mathcal{L}$ concept learning problem for P and N under $\mathcal{K}$, it is necessary to verify that every $e^+ \in P$ is listed as an instance of C under $\mathcal{K}$, and that every $e^- \in N$ is listed as an instance of $\neg C$ under $\mathcal{K}$. On the other hand, for the case of the weak $\mathcal{L}$ concept learning problem for P and N under $\mathcal{K}$, it is only necessary to check that every $e^+ \in P$ is listed as an instance of C under $\mathcal{K}$, while none of the $e^- \in N$ appears as an instance of C under $\mathcal{K}$.

5.3   Data Analysis

We now describe the metrics used in the evaluation and reported in Table 1.

The *request adequacy* measures how many times the prompt request is adequately answered by the LLM, providing a response that contains one candidate rather than two (we observed that this is the only kind of prompt misunderstanding that occurs in our experiments). It is defined by considering the complementary percentage of inadequate responses per prompt-set, and then taking their average over all the prompt-sets. The *syntactic accuracy* measures how

Table 1. Results on LTL$_f$ process mining (top) and DL concept learning (bottom).

	Request Adeq.	Synt. Acc.	Prompt-set Sem. Acc.	Total Sem. Acc.	Positive Sem. Acc.	Negative Sem. Acc.	Variab.	Complex.	Predict.
Bard	0.943	0.783	0.429	0.136	0.434	0.470	0.917	2.389	0.214
Gemini	1.000	0.786	0.286	0.129	0.386	0.429	0.457	2.750	0.250
GPT3.5	0.835	0.762	0.429	0.087	0.269	0.492	0.937	2.833	0.125
GPT4	1.000	0.986	0.929	0.643	0.786	0.800	0.605	1.277	0.631
Bard	1.000	0.880	0.700	0.400	0.460	0.620	0.860	4.348	0.071
Gemini	1.000	0.960	0.500	0.360	0.460	0.700	0.480	5.320	0.000
GPT3.5	0.980	0.743	0.500	0.137	0.257	0.390	0.903	4.500	0.000
GPT4	1.000	0.960	1.000	0.860	0.880	0.940	0.480	1.572	0.527

many times a candidate is syntactically well-formed. We define it by calculating the percentage of syntactically correct candidates in a prompt-set, and then by taking their average with respect to the number of prompt-sets.

The *prompt-set semantic accuracy* shows how many times, in a given prompt set, *at least one* candidate in a prompt-set is semantically correct. We calculate it by first associating a score of 1 when, in a prompt-set, at least one answer contains a candidate that correctly separates positive and negative examples. We then average this score over all the prompt-sets. The *total semantic accuracy* indicates how many times semantically correct answers are provided. It is defined by taking the percentage of semantically correct answers per prompt-set, and then by averaging the obtained values. We further decompose this measure to consider correct answers with respect to the positive and negative examples. The *positive* and *negative semantic accuracy* give indication of how good is the LLM to capture the positive (resp., negative) examples. They are calculated similarly to the total semantic accuracy, except that we consider the percentage of answers that are correct on the positive (resp., negative) examples.

Finally, we consider: i) *variability*, i.e., the average of the variability per prompt-set, where the latter is the ratio between the number of distinct generated candidates over the total attempts in a prompt-set; ii) *complexity*, measuring how much, on average, the size of semantically correct answers deviates from the size of the expected ones; iii) and *predictability*, representing how often, on average, the semantically correct answers coincide with the expected ones.

5.4   Results

In the following, we first present the values obtained by the LLM chatbots considered in this paper, when testing their performances on LTL$_f$ process mining and on DL concept learning problem instances, with respect to the metrics illustrated above. We then analyze the results obtained by our LLM-driven approach by comparing their prompt-set accuracy along the following dimensions: (i) in the process mining case, LTL$_f$ process mining (i.e., *without* signature restriction), on the one hand, and LTL$_f(\Sigma)$ process mining (i.e., *with* signature restriction) problem, on the other; (ii) and in the DL concept learning case, *strong* $\mathcal{L}$ concept learning, on one side, versus *weak* $\mathcal{L}$ concept learning, on the other.

Regarding the comparison of Bard, Gemini, GPT3.5 and GPT4 when performing the required tasks, we report the relevant results in Table 1 for the LTL_f process mining (top) and DL concept learning (bottom) problems. In general, we consider a higher score as an indication of better performance in all metrics, with the exception of variability and complexity (where a lower score is preferred) and predictability (where a lower score can also be interpreted as a sign of creativity in the generated answer). The results show that GPT4 consistently outperforms all Bard, Gemini, and GPT3.5 in every dimension, and that Bard, on average, performs better than GPT3.5. Surprisingly, the results indicate that Gemini has worse outcomes than Bard on multiple dimensions.

In the mining case of the LTL_f process, the request adequacy capabilities of GPT3.5 and Bard are comparable (with the latter performing slightly better than the former), while in none of our tests Gemini and GPT4 misunderstand the prompted request. Bard manages to reach the perfect performance of Gemini and GPT4 with respect to request adequacy in the case of DL concept learning. Regarding syntactic accuracy, we obtain overall positive results for all LLM for both problems, again with a better performance from GPT4, and similarly Gemini, compared to Bard, which in turn slightly surpasses GPT3.5.

Although the prompt set accuracy of Bard and GPT3.5 coincides in the LTL_f process mining case, with GPT4 quite considerably outperforming both, and with Gemini showing lower scores than any other, we find that for DL concept learning tasks, Bard behaves better than GPT3.5, which achieves the same results as Gemini. The picture is similar when we consider total semantic accuracy, with Bard and Gemini performing better than GPT3.5 in both the LTL_f and the DL cases (with a significant improvement in the latter context). Interestingly, however, Bard slightly surpasses the results obtained by Gemini. Also, here, GPT4 shows better scores by a significant factor. An analogous behavior is shown for the positive and negative semantic accuracy, with the exception that GPT3.5 appears to perform marginally better than Bard in its accuracy over negative examples in the LTL_f process mining, while Gemini is moderately more accurate than Bard on DL concept learning negative instances.

Variability and complexity are other metrics where GPT4 shows significantly better results compared to Bard and Gemini, which remain relatively close in performance. It is also worth noticing that Gemini presents a variability that is the same or lower than that of GPT4, while the complexity of its answers is comparable to that of Bard and GPT3.5. On the other hand, the lower predictability scores (calculated on semantically correct attempts) of Bard, Gemini, and GPT3.5, compared to those of GPT4, rather than implying a certain degree of valuable *impromptu* for the human user, appear to simply avoid the most obvious candidate choices. GPT4, while still providing in almost half of the successful attempts answers that deviate from the expected one, displays also a tendency to correctly identify natural options representing a solution to the problem.

Table 2. Prompt-set semantic accuracy for LTL$_f$ process mining (left) and DL concept learning (right).

	LTL$_f$ process mining	LTL$_f(\Sigma)$ process mining		Strong DL concept learning	Weak DL concept learning
GPT3.5	0.286	0.571	GPT3.5	0.400	0.600
GPT4	0.857	1.000	GPT4	1.000	1.000
Bard	0.429	0.429	Bard	0.500	0.800
Gemini	0.286	0.286	Gemini	0.500	0.600

Table 2 shows a comparison of the prompt-set accuracy in different classes of problems. In the left-hand side table, we compare the results of the LLMs on LTL$_f$ process mining problems, in the first column, and on the corresponding LTL$_f(\Sigma)$ process mining variants, in the second. We observe that all the LLMs considered perform equally well (as Bard and Gemini), or even better (as GPT3.5), in presence of signature restrictions, with GPT4 reaching a perfect score on LTL$_f(\Sigma)$ process mining. This suggests that signature restrictions can guide LLMs in the generation of an answer from the submitted prompt. In the right-hand side table, the comparison is between the prompt-set accuracy of LLMs on strong- and weak-concept learning problem instances. In this case, we notice that LLMs perform equally well (as GPT4) or better (as Bard, Gemini, and GPT3.5) on weak-concept learning problems, rather than in strong-concept learning instances. These results are in line with the observation that the stronger conditions to be met in case of strong, rather than weak, separability tend to represent an obstacle to the generative capabilities of LLMs.

In general, our analysis shows that the integration of LLMs with reasoning modules for the knowledge extraction tasks of LTL$_f$ process mining and of DL concept learning is a promising direction. Paired with verification engines, the generative capabilities of LLMs show significant potential in automated support for both reverse engineering of temporal specifications and ontology generation, as witnessed in particular by GPT4 nearly perfect results on prompt set accuracy.

6   LLM-KnowExtra Toolchain

In light of the encouraging results obtained with GPT4 in our proof-of-concept evaluation, we develop a toolchain that implements the architecture described in Sects. 3 and 4. The toolchain, named LLM-KnowExtra, is available at https:// github.com/damianoduranti/LLMknowextra. It is developed in Python and it consists of two main components, devoted to LTL$_f$ process mining and DL concept learning, respectively. Both modules are based on a shared script, in charge of setting up the interface with the GPT4 chatbot (or, by a suitable configuration of the API keys, with GPT3.5; our tool can also be modified to be compatible with Bard and Gemini, under suitable API interfaces when available).

Regarding the LTL$_f$ component of the toolchain, the finite traces to be given in input are stored in a JSON file, which can possibly include an additional signature restriction for constrained cases. After initializing communication with the LLM, the tool generates a prompt that is then submitted to GPT4. This prompt is structurally analogous to the one used in our preliminary experimental evaluation phase, except that we explicitly request that the candidate formula be in NUXMV syntax. The verification phase requires two more steps. First, the candidate formula is parsed using the PYLOGICS tool[3], and translated, by an implementation of the f2i($\cdot$) function, into a formula to be evaluated on the finite traces encoded in NUXMV. Subsequently, the obtained specification is stored in a file and verified with NUXMV against each of the traces provided in the JSON format. If the candidate specification is found to be incorrect on any of such traces, the process is stopped, and a new request is sent to the LLM (currently GPT4). The loop continues until a correct specification is found, or when a maximum number of attempts (set to 5 by default in our tests) is reached.

For the DL concept learning module of our toolchain, we provide two input files: (i) the knowledge base file, containing an ontology serialized in the OWL/XML format; (ii) the JSON file storing the positive and negative instances, as well as the type (weak/strong) of separation requested. After setting up the communication with the LLM (currently GPT4) via the API interface, the tool generates a prompt (similar to the one used in our evaluation) that contains both the full ontology in the given format, as well as the instances and the separation request. To facilitate the verification step, we require that the candidate concept be formulated in a syntax compatible with OWLREADY2[4], a Python package that allows loading, modifying, and performing reasoning tasks over OWL2 ontologies. This package, via the included HermiT reasoner, is used to perform instance checking on the candidate concept: the retrieved instances are finally compared with the positive and negative instances given as input, to evaluate whether the given concept is a successful candidate or not.

We run our toolchain on a dataset obtained by a minor extension of the one used on each LLM in our preliminary experimental evaluation and illustrated in Sect. 5. In particular, we let the toolchain perform a total of 50 prompt set cycles for each problem, over repeated sets of inputs, and where we set a bound of 5 attempts for each cycle. We executed our experiment on an Apple MacBook Pro with an M1 Pro chip and 16GB of unified memory. All computational tests were performed using Python 3.12. Table 3 reports the average computation times of our toolchain in the LTL_f process mining case (with and without signature restrictions) and in the DL concept learning case (with strong and weak separability). In particular, we show the average time required to: (i) complete one step of candidate generation and verification; (ii) obtain a response that is verified as correct; (iii) perform a full cycle of maximum five attemps, without outputting a correct candidate. As expected, the average time required for the maximum number of unsuccessful attempts is nearly fivefold the average for a single attempt, except for weak DL concept learning, where GPT4 never failed to provide at least one correct answer for each cycle. In addition, in the $LTL_f(\Sigma)$ process mining case, the average time to output a correct response is lower than the time required by a single attempt: This is explained by observing

---

[3] https://github.com/whitemech/pylogics.
[4] https://owlready2.readthedocs.io/en/latest/.

Table 3. LLM-KNOWEXTRA average computation times (in seconds).

	Single attempt	Correct response	Incorrect response
LTL$_f$ process mining	2.453	3.655	12.818
LTL$_f(\Sigma)$ process mining	2.281	1.994	12.162
Strong DL concept learning	5.331	8.106	27.287
Weak DL concept learning	4.409	6.349	0.000

that a correct response is obtained on the *first* attempt in most (85%) of the cycles, which requires even less time than the average step of a cycle.

7   Conclusion and Future Work

We propose a novel approach to LLM-driven knowledge extraction, focusing on the problems of learning process mining specifications in LTL$_f$, and concept expressions in DLs, to separate positively and negatively labeled data instances. In our framework, dedicated versions of these problems are submitted to GPT3.5, GPT4, Bard, and Gemini, exploiting their generative capabilities to output candidate specifications. The options provided are then formally verified using suitable reasoning engines. The proof-of-concept evaluation carried out in this work offers a preliminary positive assessment of our approach, supporting the adoption of LLMs in applications of LTL$_f$ process mining and DL concept learning, to complement existing symbolic techniques. Motivated by the promising performance of GPT4 in the knowledge extraction tasks, we also develop a dedicated GPT4 toolchain to automate the generation-verification cycle.

Our work paves the way for future research in multiple directions. First, we aim to extend our benchmark with the introduction of new sets of prompts as additional problem instances. We also intend to widen our experimental analysis considering the impact of prompt complexity and different data formats on the toolchain for the generation and verification of candidates. We expect differences in corpus content used for LLM training to be significant as well, and worth investigating for the improvement of dedicated knowledge extraction toolchains. In addition, we plan to determine whether Chain-of-Thought techniques, based on identification and pinpointing of incorrect solutions, can provide any benefit in guiding the reverse engineering of specifications. Finally, we intend to consider *strategy synthesis* within our framework, i.e., the task of devising sequences of actions, possibly in response to environmental conditions or other agents' choices, to reach a given goal, for automatic programming and planning applications.

Acknowledgments. This work has been partially supported by the PNRR project FAIR - Future AI Research (PE00000013), the NRRP MUR program funded by the NextGenerationEU, by the project MUR PRIN 2020 - RIPER - Resilient AI-Based Self-Programming and Strategic Reasoning - CUP E63C22000400001.

References

1. van der Aalst, W.M.P.: Process mining. Commun. ACM, pp. 76–83 (2012). https://doi.org/10.1145/2240236.2240257
2. Alur, R., Singh, R., Fisman, D., Solar-Lezama, A.: Search-based program synthesis. Commun. ACM **61**(12), 84–93 (2018). https://doi.org/10.1145/3208071
3. Angluin, D.: Queries and concept learning. Mach. Learn. **2**, 319–342 (1987). https://doi.org/10.1007/BF00116828
4. Artale, A., Jung, J.C., Mazzullo, A., Ozaki, A., Wolter, F.: Living without beth and craig: definitions and interpolants in description and modal logics with nominals and role inclusions. ACM Trans. Comput. Log. **24**(4), 34:1–34:51 (2023)
5. Baader, F., Horrocks, I., Lutz, C., Sattler, U.: An Introduction to Description Logic. Cambridge University Press, Cambridge (2017). http://www.cambridge.org/de/academic/subjects/computer-science/knowledge-management-databases-and-data-mining/introduction-description-logic?format=PB#17zVGeWD2TZUeu6s.97
6. Blum, S., Koudijs, R., Ozaki, A., Touileb, S.: Learning horn envelopes via queries from large language models. CoRR (2023). https://doi.org/10.48550/arXiv.2305.12143
7. Bshouty, N.H.: Exact learning from membership queries: some techniques, results and new directions. In: Jain, S., Munos, R., Stephan, F., Zeugmann, T. (eds.) ALT 2013. LNCS (LNAI), vol. 8139, pp. 33–52. Springer, Heidelberg (2013). https://doi.org/10.1007/978-3-642-40935-6_4
8. Bühmann, L., Lehmann, J., Westphal, P.: Dl-learner - a framework for inductive learning on the semantic web. J. Web Semant. **39**, 15–24 (2016). https://doi.org/10.1016/j.websem.2016.06.001
9. ten Cate, B., Funk, M., Jung, J.C., Lutz, C.: Sat-based PAC learning of description logic concepts. In: Proceedings of IJCAI, pp. 3347–3355. ijcai.org (2023). https://doi.org/10.24963/ijcai.2023/373
10. Cavada, R., et al.: The NUXMV symbolic model checker. In: Biere, A., Bloem, R. (eds.) CAV 2014. LNCS, vol. 8559, pp. 334–342. Springer, Cham (2014). https://doi.org/10.1007/978-3-319-08867-9_22
11. Cecconi, A., Giacomo, G.D., Ciccio, C.D., Maggi, F.M., Mendling, J.: Measuring the interestingness of temporal logic behavioral specifications in process mining. Inf. Syst. **107**, 101920 (2022)
12. Chiariello, F., Maggi, F.M., Patrizi, F.: From LTL on process traces to finite-state automata. In: Proceedings of BPM (Demos), pp. 127–131. CEUR-WS.org (2023)
13. Creswell, A., Shanahan, M., Higgins, I.: Selection-inference: exploiting large language models for interpretable logical reasoning. In: Proceedings of ICLR. OpenReview.net (2023). https://openreview.net/pdf?id=3Pf3Wg6o-A4
14. De Giacomo, G., Felli, P., Montali, M., Perelli, G.: HyperLDLF: a logic for checking properties of finite traces process logs. In: Proceedings of IJCAI, pp. 1859–1865. ijcai.org (2021)
15. De Giacomo, G., Masellis, R.D., Montali, M.: Reasoning on LTL on finite traces: insensitivity to infiniteness. In: AAAI, pp. 1027–1033. AAAI Press (2014)
16. De Giacomo, G., Vardi, M.Y.: Linear temporal logic and linear dynamic logic on finite traces. In: Proceedings of IJCAI, pp. 854–860. IJCAI/AAAI (2013)
17. Denecker, M., Kakas, A.: Abduction in logic programming. In: Kakas, A.C., Sadri, F. (eds.) Computational Logic: Logic Programming and Beyond. LNCS (LNAI), vol. 2407, pp. 402–436. Springer, Heidelberg (2002). https://doi.org/10.1007/3-540-45628-7_16

18. Fanizzi, N., Rizzo, G., d'Amato, C., Esposito, F.: DLFoil: class expression learning revisited. In: Faron Zucker, C., Ghidini, C., Napoli, A., Toussaint, Y. (eds.) EKAW 2018. LNCS (LNAI), vol. 11313, pp. 98–113. Springer, Cham (2018). https://doi.org/10.1007/978-3-030-03667-6_7

19. Fortin, M., Konev, B., Ryzhikov, V., Savateev, Y., Wolter, F., Zakharyaschev, M.: Unique characterisability and learnability of temporal instance queries. In: Proceedings of KR (2022). https://proceedings.kr.org/2022/17/

20. Fortin, M., Konev, B., Ryzhikov, V., Savateev, Y., Wolter, F., Zakharyaschev, M.: Reverse engineering of temporal queries mediated by LTL ontologies. In: Proceedings of IJCAI, pp. 3230–3238. ijcai.org (2023). https://doi.org/10.24963/ijcai.2023/360

21. Funk, M., Hosemann, S., Jung, J.C., Lutz, C.: Towards ontology construction with language models. In: Joint proceedings of the 1st workshop on Knowledge Base Construction from Pre-Trained Language Models (KBC-LM) and the 2nd challenge on Language Models for Knowledge Base Construction (LM-KBC) co-located with the 22nd International Semantic Web Conference (ISWC 2023), Athens, Greece, November 6, 2023. CEUR Workshop Proceedings, vol. 3577. CEUR-WS.org (2023)

22. Funk, M., Jung, J.C., Lutz, C.: Actively learning concepts and conjunctive queries under ELr-ontologies. In: Proceedings of IJCAI, pp. 1887–1893. ijcai.org (2021). https://doi.org/10.24963/ijcai.2021/260

23. Funk, M., Jung, J.C., Lutz, C., Pulcini, H., Wolter, F.: Learning description logic concepts: When can positive and negative examples be separated? In: Proceedings of IJCAI, pp. 1682–1688. ijcai.org (2019). https://doi.org/10.24963/ijcai.2019/233

24. Gaglione, J., Roy, R., Baharisangari, N., Neider, D., Xu, Z., Topcu, U.: Learning temporal logic properties: an overview of two recent methods. CoRR (2022). https://doi.org/10.48550/arXiv.2212.00916

25. Giorgini, P., Mazzullo, A., Robol, M., Roveri, M.: Towards large language model architectures for knowledge acquisition and strategy synthesis. In: Short Paper Proceedings of the 5th Workshop on Artificial Intelligence and Formal Verification, Logic, Automata, and Synthesis hosted by the 22nd International Conference of the Italian Association for Artificial Intelligence (AIxIA 2023), Rome, Italy, November 7, 2023. CEUR Workshop Proceedings, vol. 3629, pp. 21–29. CEUR-WS.org (2023). https://ceur-ws.org/Vol-3629/paper4.pdf

26. Grau, B.C., Horrocks, I., Motik, B., Parsia, B., Patel-Schneider, P.F., Sattler, U.: OWL 2: the next step for OWL. J. Web Semant. 6(4), 309–322 (2008)

27. Gutiérrez-Basulto, V., Jung, J.C., Sabellek, L.: Reverse engineering queries in ontology-enriched systems: the case of expressive horn description logic ontologies. In: Proceedings of IJCAI, pp. 1847–1853. ijcai.org (2018)

28. Huang, J., Chang, K.C.: Towards reasoning in large language models: a survey. In: Findings of the Association for Computational Linguistics, pp. 1049–1065. Association for Computational Linguistics (2023). https://doi.org/10.18653/v1/2023.findings-acl.67

29. Huang, W., Abbeel, P., Pathak, D., Mordatch, I.: Language models as zero-shot planners: extracting actionable knowledge for embodied agents. In: Proceedings of ICML, pp. 9118–9147. PMLR (2022), https://proceedings.mlr.press/v162/huang22a.html

30. Iannone, L., Palmisano, I., Fanizzi, N.: An algorithm based on counterfactuals for concept learning in the semantic web. Appl. Intell. 26(2), 139–159 (2007). https://doi.org/10.1007/s10489-006-0011-5

31. Jung, J.C., Lutz, C., Pulcini, H., Wolter, F.: Logical separability of labeled data examples under ontologies. Artif. Intell. **313**, 103785 (2022). https://doi.org/10.1016/j.artint.2022.103785

32. Knublauch, H., Fergerson, R.W., Noy, N.F., Musen, M.A.: The protégé OWL plugin: an open development environment for semantic web applications. In: Proceedings of the Third International Semantic Web Conference (ISWC 2004), pp. 229–243 (2004)

33. Konev, B., Lutz, C., Ozaki, A., Wolter, F.: Exact learning of lightweight description logic ontologies. J. Mach. Learn. Res. **18**(201), 201:1–201:63 (2017). http://jmlr.org/papers/v18/16-256.html

34. Liu, H., Ning, R., Teng, Z., Liu, J., Zhou, Q., Zhang, Y.: Evaluating the logical reasoning ability of ChatGPT and GPT-4. CoRR (2023). https://doi.org/10.48550/arXiv.2304.03439

35. Martins, D.M.L.: Reverse engineering database queries from examples: state-of-the-art, challenges, and research opportunities. Inf. Syst. **83**, 89–100 (2019)

36. Mateiu, P., Groza, A.: Ontology engineering with large language models. In: 25th International Symposium on Symbolic and Numeric Algorithms for Scientific Computing, SYNASC 2023, Nancy, France, September 11-14, 2023, pp. 226–229. IEEE (2023)

37. Moiseev, F., Dong, Z., Alfonseca, E., Jaggi, M.: SKILL: structured knowledge infusion for large language models. In: Proceedings of NAACL, pp. 1581–1588. Association for Computational Linguistics (2022). https://doi.org/10.18653/v1/2022.naacl-main.113

38. Monti, M., Kutz, O., Righetti, G., Troquard, N.: Improving the accuracy of blackbox language models with ontologies: a preliminary roadmap. In: Proceedings of the Joint Ontology Workshops 2024 Episode X: The Tukker Zomer of Ontology co-located with the 14th International Conference on Formal Ontology in Information Systems (FOIS 2024), Enschede, Netherlands, 15-19 July 2024. CEUR Workshop Proceedings, vol. to appear. CEUR-WS.org (2024)

39. Muggleton, S.H.: Inductive logic programming: Issues, results and the challenge of learning language in logic. Artif. Intell. **114**, 283–296 (1999). https://doi.org/10.1016/S0004-3702(99)00067-3

40. Neider, D., Gavran, I.: Learning linear temporal properties. In: Proceedings of FMCAD, pp. 1–10. IEEE (2018). https://doi.org/10.23919/FMCAD.2018.8603016

41. Olausson, T., et al.: LINC: a neurosymbolic approach for logical reasoning by combining language models with first-order logic provers. In: Proceedings of the 2023 Conference on Empirical Methods in Natural Language Processing, EMNLP 2023, Singapore, December 6-10, 2023, pp. 5153–5176. Association for Computational Linguistics (2023). https://doi.org/10.18653/v1/2023.emnlp-main.313

42. Ozaki, A.: Learning description logic ontologies: five approaches. where do they stand? Künstliche Intell. **34**(2), 317–327 (2020). https://doi.org/10.1007/s13218-020-00656-9

43. Ozaki, A., Persia, C., Mazzullo, A.: Learning query inseparable $\mathcal{ELH}$ ontologies. In: Proceedings of AAAI, pp. 2959–2966. AAAI Press (2020). https://doi.org/10.1609/aaai.v34i03.5688

44. Pallagani, V., et al.: On the prospects of incorporating large language models (LLMs) in automated planning and scheduling (APS). In: Bernardini, S., Muise, C. (eds.) Proceedings of the Thirty-Fourth International Conference on Automated Planning and Scheduling, ICAPS 2024, Banff, Alberta, Canada, June 1-6, 2024, pp. 432–444. AAAI Press (2024). https://doi.org/10.1609/icaps.v34i1.31503

45. Pan, L., Albalak, A., Wang, X., Wang, W.Y.: Logic-LM: Empowering large language models with symbolic solvers for faithful logical reasoning. CoRR (2023). https://doi.org/10.48550/arXiv.2305.12295

46. Pan, S., Luo, L., Wang, Y., Chen, C., Wang, J., Wu, X.: Unifying large language models and knowledge graphs: A roadmap. CoRR (2023). https://doi.org/10.48550/arXiv.2306.08302

47. Pešić, M., Bošnački, D., van der Aalst, W.M.P.: Enacting declarative languages using LTL: avoiding errors and improving performance. In: van de Pol, J., Weber, M. (eds.) SPIN 2010. LNCS, vol. 6349, pp. 146–161. Springer, Heidelberg (2010). https://doi.org/10.1007/978-3-642-16164-3_11

48. Pnueli, A.: The temporal logic of programs. In: 18th Annual Symposium on Foundations of Computer Science, pp. 46–57. IEEE Computer Society (1977). https://doi.org/10.1109/SFCS.1977.32

49. Settles, B.: Active Learning. Morgan & Claypool Publishers (2012). https://doi.org/10.2200/S00429ED1V01Y201207AIM018

50. Singh, I., et al.: ProgPrompt: generating situated robot task plans using large language models. In: Proceedings of ICRA, pp. 11523–11530. IEEE (2023). https://doi.org/10.1109/ICRA48891.2023.10161317

51. Song, C.H., Sadler, B.M., Wu, J., Chao, W., Washington, C., Su, Y.: LLM-planner: few-shot grounded planning for embodied agents with large language models. In: IEEE/CVF International Conference on Computer Vision, ICCV 2023, Paris, France, October 1-6, 2023, pp. 2986–2997. IEEE (2023). https://doi.org/10.1109/ICCV51070.2023.00280

52. Trajanoska, M., Stojanov, R., Trajanov, D.: Enhancing knowledge graph construction using large language models. CoRR (2023). https://doi.org/10.48550/arXiv.2305.04676

53. Tsarkov, D., Horrocks, I.: FaCT++ description logic reasoner: system description. In: Furbach, U., Shankar, N. (eds.) IJCAR 2006. LNCS (LNAI), vol. 4130, pp. 292–297. Springer, Heidelberg (2006). https://doi.org/10.1007/11814771_26

54. Valmeekam, K., Marquez, M., Sreedharan, S., Kambhampati, S.: On the planning abilities of large language models - a critical investigation. In: Advances in Neural Information Processing Systems 36: Annual Conference on Neural Information Processing Systems 2023, NeurIPS 2023, New Orleans, LA, USA, December 10 - 16, 2023 (2023). http://papers.nips.cc/paper_files/paper/2023/hash/efb2072a358cefb75886a315a6fcf880-Abstract-Conference.html

55. Yao, S., et al.: ReAct: synergizing reasoning and acting in language models. In: Proceedings of ICLR. OpenReview.net (2023). https://openreview.net/pdf?id=WEvluYUL-X

56. Ye, X., Chen, Q., Dillig, I., Durrett, G.: SatLM: satisfiability-aided language models using declarative prompting. In: Advances in Neural Information Processing Systems 36: Annual Conference on Neural Information Processing Systems 2023, NeurIPS 2023, New Orleans, LA, USA, December 10 - 16, 2023 (2023). http://papers.nips.cc/paper_files/paper/2023/hash/8e9c7d4a48bdac81a58f983a64aaf42b-Abstract-Conference.html

57. Zhang, H., Li, L.H., Meng, T., Chang, K., den Broeck, G.V.: On the paradox of learning to reason from data. In: Proceedings of IJCAI, pp. 3365–3373. ijcai.org (2023). https://doi.org/10.24963/ijcai.2023/375

FAVEL: Fact Validation Ensemble Learning

Umair Qudus$^{(\boxtimes)}$ ⓘ, Franck Lionel Tatkeu Pekarou, Ana Alexandra Morim da Silva ⓘ,
Michael Röder ⓘ, and Axel-Cyrille Ngonga Ngomo ⓘ

Data Science Group, Department of Computer Science, Paderborn University, Paderborn,
Germany
{umair.qudus,ltphen,ana.silva,michael.roeder,
axel.ngonga}@uni-paderborn.de
https://dice-research.org/

Abstract. Validating assertions before adding them to a knowledge graph is an
essential part of its creation and maintenance. Due to the sheer size of knowledge graphs, automatic fact-checking approaches have been developed. These
approaches rely on reference knowledge to decide whether a given assertion
is correct. Recent hybrid approaches achieve good results by including several
knowledge sources. However, it is often impractical to provide a sheer quantity of textual knowledge or generate embedding models to leverage these hybrid
approaches. We present FAVEL, an approach that uses algorithm selection and
ensemble learning to amalgamate several existing fact-checking approaches that
rely solely on a reference knowledge graph and, hence, use fewer resources than
current hybrid approaches. For our evaluation, we create updated versions of two
existing datasets and a new dataset dubbed FAVEL-DS. Our evaluation compares
our approach to 15 fact-checking approaches—including the state-of-the-art approach HybridFC—on 3 datasets. Our results demonstrate that FAVEL outperforms all other approaches significantly by at least 0.04 in terms of the area under
the ROC curve. Our source code, datasets, and evaluation results are open-source
and can be found at https://github.com/dice-group/favel.

Keywords: fact checking · ensemble learning · transfer learning · knowledge
management

1 Introduction

Knowledge graphs play a vital role in the web ecosystem.[1] The popularity and quantity of knowledge graphs have surged in recent years [34]. However, their usage is
bound to the assumption that each individual statement within a knowledge graph is correct. Since large knowledge graphs are generated automatically (e.g., DBpedia [2,27],
YAGO [47], and WikiData [32]), validating assertions before adding them to such a
graph is an essential part of its creation and maintenance. At the same time, the sheer
size of these graphs led to the development of automatic fact-checking approaches.
These approaches rely on reference knowledge to decide whether a given assertion is
true or false.

---

[1] http://webdatacommons.org/structureddata/2021-12/stats/stats.html.

ⓒ The Author(s), under exclusive license to Springer Nature Switzerland AG 2025
M. Alam et al. (Eds.): EKAW 2024, LNAI 15370, pp. 209–225, 2025.
https://doi.org/10.1007/978-3-031-77792-9_13

Recently, hybrid approaches achieved good results by including different knowledge sources, i.e., large reference knowledge graphs, knowledge graph embedding models, and textual corpora [39,40]. However, providing a large amount of textual knowledge for a particular field of interest is not always feasible. Likewise, the generation of embedding models for large reference knowledge graphs can incur significant costs in both runtime and computational resources [13]. At the same time, several fact-checking approaches exist, that only rely on a reference knowledge graph. While these single approaches on their own showed inferior performance when compared to hybrid approaches in recent evaluations [39,40,45], their combination was not investigated before. Our work fills this research gap.

To the best of our knowledge, our approach FAVEL is the first attempt to combine several knowledge-graph-based fact-checking approaches. Internally, it is based on an ensemble learning algorithm to combine the prediction results of the different fact-checking approaches. Ensemble learning is a powerful technique in machine learning, that holds significant promise for enhancing the overall predictive performance in comparison to single approaches. The core motivation behind adopting the ensemble method stems from the acknowledgment that diverse approaches, when combined, can collectively outperform individual models by mitigating weaknesses, such as bias-variance tradeoff, and leveraging their respective strengths. We combine this with an algorithm selection approach to automatically configure FAVEL for the given training data.

Our contributions in this paper are as follows:

- We present FAVEL, a fact-checking approach that relies on ensemble learning to combine several knowledge-graph-based fact-checking approaches. Our evaluation shows that FAVEL is able to outperform the single fact-checking approach it combines and the state of the art HybridFC.
- We propose a new dataset dubbed FAVEL-DS for the evaluation of fact-checking approaches. We created this dataset based on the DBpedia of March 2022.
- We further present BPDP 22 and FactBench Mix 22, which are updated versions of previously published datasets that we aligned to the same DBpedia version.

The remainder of this paper is structured as follows. In the following Section, we briefly explain preliminaries before we summarize the related work in Sect. 3. In Sect. 4, we explain our approach FAVEL. We evaluate our approach in Sect. 5. In Sect. 6, we present an ablation study of our approach. Finally, we conclude and discuss potential future work in Sect. 7.

2 Preliminaries

Our work focuses on knowledge graphs, in particular on knowledge graphs in the sense of the Resource Description Framework (RDF). We define them as follows:

Definition 1. (Knowledge graph) Let E be the set of all RDF resource IRIs, B the set of all blank nodes, $P \subseteq E$ the set of all RDF predicates and L the set of all literals. A knowledge graph G is a set of RDF assertions, i.e.,

$$G = \{(s, p, o) \mid s \in E \cup B, p \in P, o \in E \cup B \cup L\}, \tag{1}$$

where (s, p, o) is a triple in G comprising a subject s, a predicate p and an object o [9].

Our approach aims to fulfill the task called automatic fact checking. We formally define this task as follows:

Definition 2. (Fact checking) Given an assertion in the form of a triple, a reference knowledge graph, and/or a reference corpus, fact checking is the task of computing the likelihood that the given assertion is true or false [48].

Definition 3. (Ensemble learning) Let $\{h_1, h_2, \ldots, h_n\}$ be n individual base learners, each producing an output $h_i(\mathbf{x})$ for a given input $\mathbf{x}$. The ensemble model, denoted as $H(\mathbf{x})$, is formed by combining the outputs of these base learners. The combination can be performed through various techniques, such as averaging, voting, or weighted averaging [41]. We define the ensemble prediction $H(\mathbf{x})$ as:

$$H(\mathbf{x}) = m(h_1(\mathbf{x}), h_2(\mathbf{x}), \ldots, h_T(\mathbf{x})), \tag{2}$$

where m represents the combination or aggregation function [41].

3 Related Work

A typical approach to fact checking involves searching for evidence for a given assertion in the provided reference data. We categorized the following four fact-checking approaches based on the reference data that they use: text-based, knowledge-graph-based, embedding-based, and hybrid approaches. Table 1 gives a brief overview, which we further explain in the following.

Text-based approaches transform the given assertion into one or several search queries that they use to derive textual evidence from a reference corpus [19,48]. DeFacto [19] and its extension FactCheck [48] are two examples of this category of algorithms. In contrast to these approaches, our work relies on a knowledge graph instead of a textual corpus as reference data.

Knowledge-graph-based approaches use a knowledge graph as reference knowledge to gather evidence. This category contains a variety of algorithms, ranging from those initially designed for graph link prediction, such as Adamic Adar [1], Degree Product [42], Jaccard [29], Katz [22], and Pathent [50]. Similarly, similarity measures like SimRank [21] have been used to compare the subject and object of the given assertion. Recently, more sophisticated knowledge-graph-based fact-checking approaches emerged, e.g., path-based approaches. Path-based approaches search for paths between the subject and object of the given assertion. Identified paths receive a score to express to which extent these paths can be used as evidence for the given assertion. KL [8] takes the specificity of a path, i.e., the degree of intermediate nodes on the path, into account. Knowledge Stream [44] makes use of a line graph to determine the similarity of properties in the knowledge graph. These similarities are used in a min cost max flow algorithm to score the connection between the subject and object. REL-KL [44] is based on a combination of KL's specificity Knowledge Stream's similarity measure. PRA [26] relies on path statistics that are gathered during random walks on the reference graph. Similarly, PredPath [43] and COPAAL [49] enumerate paths that connect subject and object and use features to assign scores to them. Rule-based

Table 1. Existing fact-checking approaches.

Category	Approaches	Short Description
Text-based	DeFacto [19], FactCheck [48]	Transform a given assertion into search queries to retrieve textual evidence from a corpus.
Knowledge-graph-based	Adamic Adar [1], COPAAL [49], Degree Product [42], Jaccard [29], Knowledgestream [44], Katz [22], KL [8], PredPath [43], PRA [43], Pathent [50], REL-KL [26], SimRank [21]	Use a knowledge graph as reference knowledge and different statistical, path-, rule-, or pattern-based approaches to retrieve evidence from this graph.
	FᴀVEL (ours)	Combination of existing knowledge-graph-based approaches for improved performance.
Embedding-based	ESTHER [45], Dong et al. [11]	Utilize knowledge graph embedding models as reference data.
Hybrid	ExFakt [17], Tracy [18], Facty [28], ESTHER [45], HybridFC [40], TemporalFC [39])	Combine multiple types of reference data, such as a knowledge graph embedding model, textual evidence from a reference corpus, and paths in a reference knowledge graph.

approaches like AMIE3 [25] and RuDiK [36] mine rules from the knowledge graph. An assertion is classified as true if there is a rule that supports the existence of the assertion. KV-Eval [24] extends this general approach further by adding rules that can reject the existence of an assertion. Pattern-based approaches mine patterns similar to rule-based approaches. These patterns can be more complex than rules. Examples are GFC [30] and OGFC [31]. Our approach fits into the category of knowledge-graph-based approaches and is designed to integrate any approach of this category. Our evaluation shows that the combination of approaches can lead to better results than the usage of single approaches.

Embedding-based fact-checking approaches use knowledge graph embedding models as reference data.[2] These approaches calculate the likelihood of the existence of the given assertion based on their reference embedding model. Such approaches have been used by da Silva et al. [45] and proposed by Dong et al. [11]. In contrast to these approaches, our approach does not rely on the expensive generation of a knowledge graph embedding model.

Hybrid fact-checking approaches use more than one of the aforementioned types of reference data. ExFaKT [17] and Tracy [18] combine rules mined on a reference knowledge graph and evidence from the web. FACTY [28] relies on textual evidence and

---

[2] Some authors of knowledge graph embedding publications dubbed fact checking as a triple classification task.

paths in a reference knowledge graph. ESTHER [45] is a path-based approach, which uses a knowledge graph embedding model to determine potential paths that could serve as evidence. HybridFC [40] makes use of all three categories of reference data. TemporalFC [39] is an extension of HybridFC and focuses on volatile assertions. TemporalFC uses temporal KG embeddings in addition to the other two sources. While hybrid approaches achieve good evaluation results, their need for different types of reference data makes their usage expensive and not always applicable. In contrast, our approach only makes use of a reference knowledge graph.

Ensemble learning is used in a variety of areas to combine several existing approaches for enhancing the overall predictive performance of models [41], e.g., in the area of knowledge graphs, Speck et al. [46] proposes a combination of several entity recognition systems. To the best of our knowledge, FAVEL is the first attempt to combine knowledge-graph-based fact-checking approaches.

4   Approach

FAVEL is based on the idea of ensemble learning. Instead of using a single hybrid approach, we use multiple knowledge-graph-based fact-checking approaches in parallel and combine their results to achieve a final classification for the given assertion.

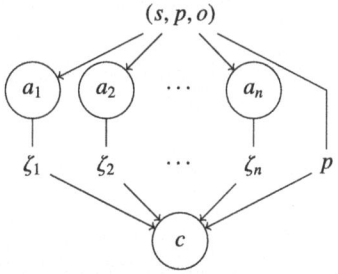

Fig. 1. Schematic overview of FAVEL.

Figure 1 gives an overview of our approach. FAVEL receives a given assertion in the form of a triple (s, p, o) as input. It forwards the assertion to the n knowledge-graph-based fact-checking approaches $\{a_1, \ldots, a_n\}$ that FAVEL is configured to use. Since these approaches are treated as black boxes, FAVEL can make use of any fact-checking approach. However, within this paper, we restrict our work to knowledge-graph-based approaches since they all use the same reference data, and no additional data is needed. The veracity scores $\{\zeta_1, \ldots, \zeta_n\}$ that the single approaches assign to the given assertion are collected. Together with the predicate p of the given assertion, these results are used as input for an ensemble classifier c. This classifier is trained to perform the final classification task. We formally define FAVEL as a fact checking function f as follows:

$$f(s, p, o) = c(a_1(s, p, o), \ldots, a_n(s, p, o), p). \tag{3}$$

There are several ensemble learning algorithms available that can be used as c within our approach [41]. We tackle this algorithm-selection problem [23] with a meta-feature-free meta-learning technique proposed by Feurer et al. [14]. This approach trains several ensemble learners and successively removes bad-performing learners until either only one learner is left or the maximum training time is reached. In the latter case, the best-performing ensemble learner is returned from the list of remaining learners. This enables FAVEL to be trained and automatically configured specifically for a given dataset.

5   Evaluation

In this section, we evaluate our approach by comparing it with 12 knowledge-graph-based, 1 text-based, and 2 hybrid contemporary state-of-the-art fact-checking approaches. To begin, we outline the datasets integral to our evaluation, followed by a detailed exposition of our experimental configuration, and then present the results and discussion.

5.1   Datasets

All our experiments are based on the DBpedia [2, 27] version March 2022.[3] We make use of three benchmarking datasets. We reuse two established datasets and update them so that their content matches our DBpedia version. In addition, we create an additional dataset—FAVEL-DS. All datasets are provided as supplementary files to this submission. The created reference graph is too large to be uploaded, but it will be accessible online along with the source code, datasets, and evaluation results after the paper is accepted.

Dataset Updates. FactBench [19] is a manually curated dataset based on DBpedia and Freebase with 10 predicates. The dataset is evenly distributed with each predicate having 150 correct statements. Gerber et al.propose the following six strategies to invalidate a correct assertion to create a false assertion [19]:

1. subject corruption with domain restriction,
2. object corruption with range restriction,
3. subject and object corruption with domain and range restrictions,
4. property corruption,
5. random subject, object and predicate corruption, and
6. temporal corruption.

For our experiments, we use the FactBench Mix dataset, which comprises a mixture of false assertions generated with strategies 1–5.

Birthplace-Deathplace (BPDP) is a dataset proposed by Syed et al. [48]. The dataset comprises 103 persons who have a birthplace and deathplace in two different countries.

---

[3] https://databus.dbpedia.org/dbpedia/mappings/mappingbased-objects/2022.03.01/.

Table 2. Post-processing statistics comprising the number of assertions of the train and test split and the number of distinct properties of BPDP 22, FactBench Mix 22 and FAVEL-DS. # True/# False assertions (Total count).

	BPDP 22	FactBench Mix 22	FAVEL-DS
Train	100/100 (200)	633/486 (1119)	380/385 (765)
Test	103/103 (206)	637/492 (1129)	163/164 (327)
Properties	2	9	11

For each of these persons, the dataset comprises 2 correct assertions and 2 wrong assertions. The latter were created by swapping the correct birth and death places.

Both datasets have been created based on older versions of the DBpedia. Hence, we update the datasets to align their content with the previously chosen DBpedia version as follows:

1. We update the entity IRIs in the dataset to match the new DBpedia version. We replace assertions of FactBench that rely on Freebase with a DBpedia-based equivalent. To this end, we derive the DBpedia IRIs for the entities and properties of these assertions. Assertions for which a replacement cannot be created are removed.
2. We verify all assertions in the datasets by checking that all true assertions occur in the chosen DBpedia version and false assertions do not occur. Assertions for which this does not hold are removed.

The updated datasets are called FactBench Mix 22 and BPDP 22, respectively. Table 2 shows the size of the updated datasets.

In addition, we create a new dataset dubbed FAVEL-DS based on the following 11 properties of the DBpedia ontology: `academicDiscipline`, `affiliation`, `award`, `birthPlace`, `chancellor`, `city`, `deathPlace`, `director`, `producer`, `productionCompany`, and `starring`. For each property, we randomly select 50 assertions from the knowledge graph as true assertions. For each of these assertions (s, p, o), we generate a false assertion (s, p, o') by replacing the object. We randomly choose the new object o' from existing triples from the knowledge graph (s, p', o') with the additional conditions that

1. $p \neq p'$,
2. o' has to fulfill the RDF-S range condition of p [5], and
3. (s, p, o') does not exist in the graph.

Our strategy to create false assertions is similar to the strategies applied by Gerber et al. [19] and Syed et al. [48] to create the FactBench Properties and the BPDP datasets, respectively. These strategies are known to create difficult negative examples since subject and object of the false assertion are connected in the knowledge graph [48]. A post processing check showed that a small number of the sampled and generated assertions had to be removed due to a repetition of assertions. The statistics of the final FAVEL-DS dataset can be found in Table 2. Figure 2 depicts the distribution of property occurrences across 3 datasets, encompassing a total of 16 properties.

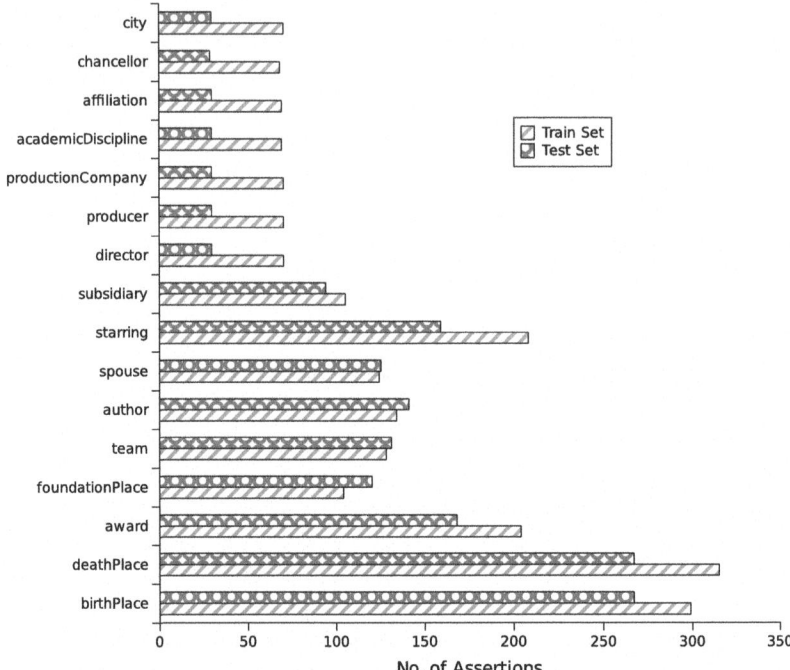

Fig. 2. Distribution of assertions with property occurrences across merged datasets.

Reference Graph. The true assertions for all three datasets origin from the Dbpedia. Hence, if we use a complete DBpedia dump or data that is derived from it the fact checking task is reduced into a simple lookup whether a given assertion exists in the reference graph. To avoid this simplification, we remove all true assertions of the three datasets from the DBpedia to create the reference graph for our evaluation.

5.2 Setup

In our evaluation, we configure FAVEL to use the following 12 knowledge-graph-based approaches: Adamic Adar [1], Degree Product [42], Jaccard [29], Katz [22], KL [8], Knowledgestream [44], Pathent [50], PRA [26], PredPath [43], REL-KL [44], Sim-Rank [21], and COPAAL [49]. We Auto-sklearn 2.0 [14] built on top of the Scikit-learn [6] as the meta-learner implementation within FAVEL. The AutoML library is configured to choose one of the following learners within a maximum runtime of 20 min:[4] AdaBoost [15], Bernoulli Naive Bayes [33], Decision Tree, Extra Trees, Gaussian Naive Bayes, Gradient Boosting [16], K-Nearest Neighbors, Linear Discriminant Analysis [20], Linear Support Vector Classification [12], Support Vector Classifica-

---

[4] We use learners that are made available by Scikit-learn [38]. For learners that do not have a bibliographic reference, we refer the interested reader to the Scikit-learn documentation at https://scikit-learn.org/stable/user_guide.html.

tion [7], Multi-Layer Perceptron, Multinomial Naive Bayes, passive-aggressive classifier, Quadratic Discriminant Analysis [20], Random Forest [4], and Stochastic Gradient Descent. We compare the performance of FAVEL to the performance of the single knowledge-graph-based approaches, the text-based approach FactCheck [48], as well as the hybrid approaches ESTHER [45] and HybridFC [40]. The latter represents the current state of the art in this area of research. Da Silva et al. [45] suggest that a system's performance can be improved if it is further combined with ESTHER. Hence, we evaluate a version of FAVEL in which it uses the input of ESTHER as 13th system. Like Qudus et al. [40], we use the English Wikipedia for approaches that rely on a reference corpus. For hybrid approaches that use a knowledge graph embedding model, we use TransE [3] to generate a model. However, the results for HybridFC with the generated TransE model were not convincing as they were very different from the results reported by Qudus et al.in their publication. Hence, we rerun HybridFC with a ConEx embedding model [10], which leads to the results reported in the next section.

Evaluation Metric. As suggested in the literature, we utilize the area under the receiver operator characteristic curve (AUC-ROC) to measure the performance of all systems [24,48,49]. We compute this score using the knowledge-base curation branch of the GERBIL framework [35,37].

5.3 Results

Table 3 shows the results of the single approaches[5], the text-based FactCheck, ESTHER, HybridFC, and the two versions of FAVEL. A Wilcoxon-signed-rank test comparing the veracity scores of FAVEL with those of HybridFC and PredPath shows that FAVEL's scores are significantly different to the others.[6] Regardless of whether ESTHER is added to the list of available fact-checking systems, FAVEL uses a Decision Tree for BPDP 22, Gradient Boosting for FactBench Mix 22, and Random Forest for FAVEL-DS.

During the evaluation, the knowledge-graph-based approaches showed a low runtime. For example, COPAAL needed 0.49 s per assertion on average.[7] In comparison, HybridFC consumed more resources. While the knowledge-graph-based module of HybridFC relies on COPAAL and, hence, has the same low runtime, the embedding-based and the text-based modules lead to a higher resource consumption. For the embedding-based module, an embedding model has to be generated. Within our evaluation, the generation of the ConEx embedding model took 33 h on a server with 2 GPUs.[8] Apart from the resource-demanding model generation, the embedding-based module of HybridFC has a very low runtime per assertion since it mainly comprises

---

[5] The results of the single approaches can also be found on GERBIL at https://gerbil-kbc.aksw. org/gerbil/experiment?id=202401120014,   https://gerbil-kbc.aksw.org/gerbil/experiment? id=202401120015, and https://gerbil-kbc.aksw.org/gerbil/experiment?id=202401120026.

[6] We use $\alpha = 0.05$.

[7] The per assertion runtimes were measured on a system with an AMD EPYC 7742 64-Core Processor 64 and 1 TB RAM.

[8] The runtime of the model generation was measured on a system with an AMD EPYC 7713 64-Core Processor, 1 TB RAM and 2 NVIDIA GeForce RTX A5000 with 24GB VRAM, each.

Table 3. AUC-ROC scores achieved by the different approaches. The best results are marked bold while the results of the best KG-based approach are underlined. T stands for text-based and H for hybrid approaches.

	Approach	BPDP 22	FactBench Mix 22	FAVEL-DS
KG-based approaches	Adamic Adar [1]	0.5000	0.6671	0.5579
	COPAAL [49]	0.5014	0.5797	0.5303
	Degree Product [42]	0.4965	0.4765	0.6096
	Jaccard [29]	0.5000	0.6729	0.5567
	Knowledgestream [44]	0.5114	0.7017	0.4859
	Katz [22]	0.4967	0.7882	0.5886
	KL [8]	0.4995	0.7097	0.4533
	PredPath [43]	0.7136	0.8719	0.7399
	PRA [26]	0.6845	0.8321	0.6942
	Pathent [50]	0.4956	0.7852	0.5872
	REL-KL [44]	0.5131	0.7294	0.4812
	SimRank [21]	0.4948	0.7213	0.5543
T	FactCheck [48]	0.4911	0.6501	0.5846
H	ESTHER [45]	0.4997	0.5855	0.5209
	HybridFC [40]	0.6811	0.8801	0.7098
Ours	FAVEL	**0.7539**	**0.9250**	**0.7718**
	FAVEL + ESTHER	**0.7539**	0.9239	0.7694

a lookup of embedding vectors. The text-based module of HybridFC needed 15 s per assertion on average. The time is mainly needed to retrieve documents and extract the textual evidence from them.

5.4   Discussion

Our evaluation results give several insights. First, on all three datasets, FAVEL shows a significantly better performance than the best-performing single knowledge-graph-based approach PredPath and the state-of-the-art approach HybridFC. Although not exactly the same, the results of HybridFC for BPDP 22 and FactBench Mix 22 are close to the results that Qudus et al. report in their evaluation on the original BPDP and FactBench Mix datasets [40].

The runtime comparison shows that the state-of-the-art approach HybridFC needs more resources concerning

1. the preparation of the system and
2. the classification of a single assertion.

HybridFC relies on a knowledge graph embedding model. The generation of this model is costly and for many models, the usage of modern GPUs is mandatory [10,13]. In practice, this resource demand could be reduced by relying on existing, pre-computed embedding models.[9] However, the average runtime of HybridFC to classify an assertion is also higher than for the knowledge-graph-based approaches. This makes FaVEL an alternative that uses fewer resources while it achieves a significantly better performance within our evaluation.

A second insight is that the configurations of FaVEL that AutoML chooses are different for all three datasets. At the same time, we didn't see a configuration of FaVEL that is very good over all three datasets. A deeper per-property analysis of the results reveals that FaVEL performs well on most properties but shows a low performance for the `city` property in the FaVEL-DS dataset. Hence, we can conclude that for different data, different ensemble strategies are better than others. This raises the question of which features of the data influence these strategies and whether there is a way to predict the best strategy for a given triple. However, we leave answering these questions to future work.

With respect to our newly created dataset FaVEL-DS, a comparison of the performance of nearly all systems on BPDP 22 and FaVEL-DS shows that both datasets seem to have a nearly equal difficulty compared to FactBench Mix 22. At the same time, FaVEL-DS is bigger than BPDP 22 and covers 11 instead of only 2 properties. Hence, we argue that FaVEL-DS is a good contribution for the future evaluation of new fact-checking approaches.

For a more comprehensive analysis of FaVEL, we utilize a concrete example drawn directly from our selected dataset and the output of all competing approaches. In the example presented in Listing 1.1, we compare the results of all approaches and FaVEL for two assertions. According to the ground truth, the left assertion is true while the right assertion is false. The example shows that some systems give higher scores to the wrong assertion (Adamic Adar, Degree Product, Katz, KL, KL-REL, Knowledgestream) while others give nearly the same scores for both (Jaccard, PredPath, Simrank, PRA). In the example, only Pathent and COPAAL give a higher score to the correct assertion. However, these two systems didn't achieve the highest scores over the complete datasets. This small example emphasizes the need of a meta learner that decides based on the given assertion which of the single approaches may give a reliable result. This is exactly the approach of FaVEL, which results in a higher score for the correct assertion in the example.

6 Ablation Study

Our previous experiments suggest that FaVEL outperforms single KG-based, text-based, or hybrid approaches. To assess the impact of performance resulting from individual KG-based approaches on FaVEL, we conduct a series of experiments wherein we systematically remove individual approaches from the ensemble setting of FaVEL and rerun the experiments. We also perform another experiment on each property of all the datasets.

---

[9] Note that this is not possible within our evaluation setup since the reference knowledge graph had to be adapted to the evaluation datasets as described in Sect. 5.1.

Listing 1.1. Example (correct and wrong assertions, from FAVEL-DS dataset.)

```
PREFIX dbr:    <http://dbpedia.org/resource/>
PREFIX dbo:    <http://dbpedia.org/ontology/>
```

```
Correct assertion:                    Wrong assertion:
dbr:Detouring_America                 dbr:George_de_Hevesy
dbo:productionCompany                 dbo:deathPlace
dbr:Warner_Bros._Cartoons             dbr:Budapest

Ground truth score: 1.0               Ground truth score: 0.0

Approach-Score-Range                  Approach-Score-Range
Adamic Adar: 0.15 [0-1]               Adamic Adar: 0.57 [0-1]
Degree Product: 2455                  Degree Product: 78451
(scale 0-100k)                        (scale 0-100k)
Jaccard: 0.002 [0-1]                  Jaccard: 0.0009 [0-1]
Katz: 0.154 [0-1]                     Katz: 0.954 [0-1]
KL: 0.0724 [0-1]                      KL: 0.107 [0-1]
KL-REL: 0.115 [0-1]                   KL-REL: 0.89 [0-1]
KS: 0.071 [0-1]                       KS: 0.14 [0-1]
Pathent: 3819                         Pathent: 3441
(scale 0-100k)                        (scale 0-100k)
PredPath: 1.0 [0/1]                   PredPath: 1.0 [0/1]
Simrank: 0.00046 [0-1]                Simrank: 0.0002 [0-1]
PRA: 0.0 [0/1]                        PRA: 0.0 [0/1]
COPAAL: 0.6 [0-1]                     COPAAL: 0.338 [0-1]
-------------------------------       -------------------------------
FaVEL Score: 0.89 [0-1]               FaVEL Score: 0.18 [0-1]
```

This experiment aims to evaluate FAVEL on a per-property basis. In this experiment, we independently merge the training and testing sets of all the datasets introduced earlier, group them by property, and separate them into multiple datasets with training and testing subsets for each property. The distribution of the datasets on the per-property basis can be found in Fig. 2. We conduct these experiments using the same setup described in Subsect. 5.2, employing knowledge-graph-based fact-checking approaches. For the first set of experiments, we set the number of iterations to 10 and compute the results' minimum, maximum, and mean. These 10 iterations take on average 200 min for each experiment on our server with the specifications described in the previous section. Table 5 and 4 show the results of our ablation study experiments.

Table 4 presents the performance of each system on a per-property basis. We can observe variations in performance across different properties among the various approaches. For example, Adamic Adar exhibits poor performance across all properties except for the `city` property, where it outperforms all other approaches with an AUC-ROC score of 0.81. PredPath, on the other hand, achieves top performance alongside FAVEL on the `deathPlace`, `author`, `spouse`, and `director` properties, with scores of 0.77, 1.0, 0.93, and 0.83, respectively. Additionally, PredPath surpasses FAVEL and all other approaches on the `foundationPlace` and `affiliation` properties, achieving scores of 0.93 and 0.90, respectively. However, it performs the worst among all other approaches on the `academicDiscipline` property. The discrepancies in performance across various properties among different approaches underscore our assumption that each approach possesses distinct advantages that FAVEL leverages. Moreover, this underscores the rationale behind our future work, where we

Table 4. Results from benchmarking FAVEL and all other approaches on a per-properties basis; the abbreviations are: AA/Adamic Adar, DP/Degree Product, KL/Knowledge Linker, and KS/KnowledgeStream. The best results are marked bold.

	FAVEL	AA	KL	KL-Rel	KS	Pathent	PRA	PredPath	Simrank	DP	Jaccard	Katz
birthPlace	**0.80**	0.63	0.56	0.55	0.53	0.68	0.77	0.75	0.57	0.48	0.65	0.67
deathPlace	**0.77**	0.68	0.63	0.64	0.64	0.72	**0.77**	**0.77**	0.65	0.59	0.63	0.75
award	**0.79**	0.52	0.52	0.56	0.52	0.59	0.69	0.78	0.68	0.52	0.52	0.63
foundationPlace	0.83	0.71	0.63	0.86	0.71	0.82	0.82	**0.93**	0.77	0.61	0.70	0.93
team	**0.92**	0.69	0.72	0.68	0.69	0.70	0.74	0.85	0.73	0.44	0.69	0.74
author	**1.00**	0.62	0.77	0.69	0.71	0.82	0.93	**1.00**	0.82	0.26	0.67	0.75
spouse	**0.93**	0.61	0.67	0.59	0.58	0.78	0.91	**0.93**	0.83	0.34	0.65	0.75
starring	**0.84**	0.52	0.76	0.75	0.75	0.81	0.70	0.82	0.67	0.51	0.53	0.79
subsidiary	**0.89**	0.78	0.82	0.87	0.82	0.87	0.86	0.83	0.81	0.62	0.82	0.83
director	**0.83**	0.46	0.50	0.54	0.48	0.76	0.76	**0.83**	0.58	0.57	0.46	0.65
producer	**0.87**	0.59	0.47	0.52	0.50	0.54	0.83	0.79	0.50	0.72	0.59	0.59
productionCompany	**0.86**	0.50	0.63	0.50	0.58	0.62	0.730	0.83	0.71	0.56	0.55	0.64
academicDiscipline	**0.79**	0.60	0.64	0.61	0.56	0.60	0.73	0.53	0.74	0.57	0.60	0.60
affiliation	0.89	0.31	0.50	0.57	0.64	0.61	0.6	**0.90**	0.45	0.55	0.34	0.57
chancellor	**0.73**	0.56	0.27	0.35	0.39	0.43	0.50	0.67	0.58	0.57	0.53	0.47
city	0.59	**0.81**	0.64	0.75	0.78	0.71	0.63	0.60	0.55	0.63	0.67	0.70

Table 5. Results of an ablation study conducted over 10 iterations of FAVEL. The 'Difference' column indicates the variance between FAVEL's overall scores in Table 3 and these average scores. w/o stands for without.

	Approach	AUC-ROC scores (10 iterations)			
		Min. score	Max. score	Avg. score	Difference
FAVEL	w/o Adamic Adar [1]	0.7705	0.7753	0.7748	0.0030
	w/o COPAAL [49]	0.7627	0.7692	0.7656	–0.0062
	w/o Degree Product [42]	0.7344	0.7344	0.7344	–0.0374
	w/o Jaccard [29]	0.7697	0.7697	0.7696	–0.0022
	w/o Knowledgestream [44]	0.7781	0.7781	0.7781	0.0063
	w/o Katz [22]	0.7622	0.7775	0.7645	–0.0073
	w/o KL [8]	0.7943	0.7943	0.7943	0.0225
	w/o PredPath [43]	0.7364	0.7364	0.7364	–0.0354
	w/o PRA [26]	0.7492	0.7525	0.7495	–0.0223
	w/o Pathent [50]	0.7474	0.7474	0.7474	–0.0244
	w/o REL-KL [44]	0.7768	0.7768	0.7768	0.0050
	w/o SimRank [21]	0.7418	0.7418	0.7418	–0.0300

aim to train the ensemble learner of FAVEL on a per-property basis and develop an algorithm selection mechanism for each property. This mechanism will enable FAVEL to determine which approach is most pertinent for a given property. By doing so, we aim to further enhance FAVEL's performance.

Table 5 demonstrates that individual systems in our ablation study have only a small impact on the overall system performance (ranging from -0.0374 on Degree Product to 0.0225 on KL). However, significant variations exist on a per-property basis. For instance, removing Adamic Adar from FAVEL would increase the overall performance by only 0.003. However, it would also forfeit FAVEL's effectiveness in handling the city property, where Adamic Adar performs exceptionally well, as indicated in Table 4.

7   Conclusion and Future Work

Within this paper, we present FAVEL—an approach that utilizes ensemble learning to combine knowledge-graph-based fact-checking approaches. Our evaluation depicts that FAVEL significantly outperforms the state-of-the-art approach HybridFC on three datasets. Notably, FAVEL requires less reference knowledge, as it does not require an additional textual corpus or a knowledge graph embedding model as supplementary reference data. Additionally, it relies on approaches with lower runtimes compared to HybridFC.

We also propose a new dataset FAVEL-DS that is bound to an explicit DBpedia version and can be used by the community in future experiments. We further updated the existing datasets BPDP and FactBench Mix creating BPDP 22 and FactBench Mix 22, two datasets which are explicitly bound to the same DBpedia version.

Our future work encompasses three main objectives. Firstly, we aim to integrate additional fact-checking approaches into our system. Secondly, we plan to focus on identifying features that can predict the optimal configuration of FAVEL for a given assertion. Finally, we intend to explore the implementation of a per-property basis ensemble learner for FAVEL, along with developing an algorithm selection mechanism to determine the best approach for each property.

Acknowledgments. This work is part of a project that has received funding from the European Union's Horizon 2020 research and innovation programme (Marie Skodowska-Curie, No. 860801), the German Federal Ministry of Education and Research (BMBF) within the project NEBULA (13N16364), the Ministry of Culture and Science of North Rhine-Westphalia (MKW NRW) within the project SAIL (NW21-059D).

References

1. Adamic, L.A., Adar, E.: Friends and neighbors on the web. Soc. Netw. **25**(3), 211–230 (2003)
2. Auer, S., Bizer, C., Kobilarov, G., Lehmann, J., Cyganiak, R., Ives, Z.: DBpedia: a nucleus for a web of open data. In: Aberer, K., et al. (eds.) ASWC/ISWC -2007. LNCS, vol. 4825, pp. 722–735. Springer, Heidelberg (2007). https://doi.org/10.1007/978-3-540-76298-0_52
3. Bordes, A., Usunier, N., Garcia-Durán, A., Weston, J., Yakhnenko, O.: Translating embeddings for modeling multi-relational data. In: Proceedings of the 26th International Conference on Neural Information Processing Systems - Volume 2, pp. 2787–2795. NIPS'13, Curran Associates Inc., Red Hook, NY, USA (2013)
4. Breiman, L.: Random Forests. Mach. Learn. **45**, 5–32 (2001). https://doi.org/10.1023/A:1010933404324
5. Brickley, D., Guha, R., McBride, B.: RDF Schema 1.1. W3C Recommendation, W3C (February 2014). http://www.w3.org/TR/2014/REC-rdf-schema-20140225/
6. Buitinck, L., et al.: API design for machine learning software: experiences from the scikit-learn project. In: ECML PKDD Workshop: Languages for Data Mining and Machine Learning, pp. 108–122 (2013)
7. Chang, C.C., Lin, C.J.: LIBSVM: a library for support vector machines. ACM Trans. Intell. Syst. Technol. **2**(3), 1–27 (2011). https://doi.org/10.1145/1961189.1961199
8. Ciampaglia, G.L., Shiralkar, P., Rocha, L.M., Bollen, J., Menczer, F., Flammini, A.: Computational fact checking from knowledge networks. PLoS ONE **10**(6), e0128193 (2015)
9. Cyganiak, R., Wood, D., Lanthaler, M.: RDF 1.1 Concepts and Abstract Syntax. W3C Recommendation, W3C (2014). http://www.w3.org/TR/2014/REC-rdf11-concepts-20140225/
10. Demir, C., Ngomo, A.-C.N.: Convolutional complex knowledge graph embeddings. In: Verborgh, R., Hose, K., Paulheim, H., Champin, P.-A., Maleshkova, M., Corcho, O., Ristoski, P., Alam, M. (eds.) ESWC 2021. LNCS, vol. 12731, pp. 409–424. Springer, Cham (2021). https://doi.org/10.1007/978-3-030-77385-4_24
11. Dong, T., Wang, Z., Li, J., Bauckhage, C., Cremers, A.B.: Triple classification using regions and fine-grained entity typing. Proc. AAAI Conf. Artif. Intell. **33**(01), 77–85 (2019). https://doi.org/10.1609/aaai.v33i01.330177, https://ojs.aaai.org/index.php/AAAI/article/view/3771
12. Fan, R.E., Chang, K.W., Hsieh, C.J., Wang, X.R., Lin, C.J.: LIBLINEAR: a library for large linear classification. J. Mach. Learn. Res. **9**, 1871–1874 (2008)
13. Ferrari, I., Frisoni, G., Italiani, P., Moro, G., Sartori, C.: Comprehensive analysis of knowledge graph embedding techniques benchmarked on link prediction. Electronics **11**(23), 3866 (2022). https://doi.org/10.3390/electronics11233866, https://www.mdpi.com/2079-9292/11/23/3866
14. Feurer, M., Eggensperger, K., Falkner, S., Lindauer, M., Hutter, F.: Auto-sklearn 2.0: hands-free autoML via meta-learning. J. Mach. Learn. Res. **23**(1), 11936–11996 (2022)

15. Freund, Y., Schapire, R.: A Decision-theoretic generalization of on-line learning and an application to boosting. J. Comput. Syst. Sci. **55**, 119–139 (1997) https://doi.org/10.1006/jcss.1997.1504

16. Friedman, J.H.: Greedy function approximation: a gradient boosting machine. The Annals of Statistics **29**, 1189–1232 (2001). https://www.jstor.org/stable/2699986

17. Gad-Elrab, M.H., Stepanova, D., Urbani, J., Weikum, G.: ExFaKT: a framework for explaining facts over knowledge graphs and text. In: WSDM, pp. 87–95. WSDM '19, ACM, New York, NY, USA (2019). https://doi.org/10.1145/3289600.3290996

18. Gad-Elrab, M.H., Stepanova, D., Urbani, J., Weikum, G.: Tracy: tracing facts over knowledge graphs and text. In: The World Wide Web Conference, pp. 3516–3520. WWW '19, Association for Computing Machinery, New York, NY, USA (2019). https://doi.org/10.1145/3308558.3314126

19. Gerber, D., et al.: DeFacto-temporal and multilingual deep fact validation. Web Semantics: Sci. Serv. Agents World Wide Web **35**(P2), 85–101 (2015). https://doi.org/10.1016/j.websem.2015.08.001

20. Hastie, T., Tibshirani, R., Friedman, J.: The Elements of Statistical Learning. SSS, Springer, New York (2009). https://doi.org/10.1007/978-0-387-84858-7

21. Jeh, G., Widom, J.: SimRank: a measure of structural-context similarity. In: Proceedings of the Eighth ACM SIGKDD International Conference on Knowledge Discovery and Data Mining (2002)

22. Katz, L.: A new status index derived from sociometric analysis. Psychometrika **18**, 39–43 (1953)

23. Kerschke, P., Hoos, H.H., Neumann, F., Trautmann, H.: Automated algorithm selection: survey and perspectives. Evol. Comput. **27**(1), 3–45 (2019). https://doi.org/10.1162/evco_a_00242

24. Kim, J., Choi, K.s.: Unsupervised fact checking by counter-weighted positive and negative evidential paths in a knowledge graph. In: Proceedings of the 28th International Conference on Computational Linguistics, pp. 1677–1686. International Committee on Computational Linguistics, Barcelona, Spain (Online) (2020). https://doi.org/10.18653/v1/2020.coling-main.147, https://www.aclweb.org/anthology/2020.coling-main.147

25. Lajus, J., Galárraga, L., Suchanek, F.: Fast and exact rule mining with AMIE 3. In: Harth, A., et al. (eds.) ESWC 2020. LNCS, vol. 12123, pp. 36–52. Springer, Cham (2020). https://doi.org/10.1007/978-3-030-49461-2_3

26. Lao, N., Cohen, W.W.: Relational retrieval using a combination of path-constrained random walks. Mach. Learn. **81**(1), 53–67 (2010)

27. Lehmann, J., et al.: DBpedia - a large-scale, multilingual knowledge base extracted from Wikipedia. Semantic Web **6**(2), 167–195 (2015). https://doi.org/10.3233/SW-140134

28. Li, F., Dong, X.L., Langen, A., Li, Y.: Knowledge verification for long-tail verticals. Proc. VLDB Endow. **10**(11), 1370–1381 (2017). https://doi.org/10.14778/3137628.3137646

29. Liben-Nowell, D., Kleinberg, J.: The link prediction problem for social networks. In: Proceedings of the Twelfth International Conference on Information and Knowledge Management (2003)

30. Lin, P., Song, Q., Shen, J., Wu, Y.: Discovering graph patterns for fact checking in knowledge graphs. In: Pei, J., Manolopoulos, Y., Sadiq, S., Li, J. (eds.) DASFAA 2018. LNCS, vol. 10827, pp. 783–801. Springer, Cham (2018). https://doi.org/10.1007/978-3-319-91452-7_50

31. Lin, P., Song, Q., Wu, Y., Pi, J.: Discovering patterns for fact checking in knowledge graphs. J. Data Inf. Q. **11**(3), 1–27 (2019). https://doi.org/10.1145/3286488

32. Malyshev, S., Krötzsch, M., González, L., Gonsior, J., Bielefeldt, A.: Getting the most out of Wikidata: semantic technology usage in Wikipedia's knowledge graph. In: Vrandečić, D., et al. (eds.) ISWC 2018. LNCS, vol. 11137, pp. 376–394. Springer, Cham (2018). https://doi.org/10.1007/978-3-030-00668-6_23

33. Manning, C., Raghavan, P., Schütze, H.: Introduction to Information Retrieval. Cambridge University Press, Cambridge (2008)
34. McCrae, J.P.: The Linked Open Data Cloud. Website (2021). https://www.lod-cloud.net/. Accessed 24 Aug 2021
35. Ngonga Ngomo, A.C., Röder, M., Syed, Z.H.: Semantic web challenge 2019. Website (2019). https://dice-group.github.io/semantic-web-challenge.github.io/. Accessed 22 May 2023
36. Ortona, S., Meduri, V.V., Papotti, P.: RuDiK: Rule discovery in knowledge bases. Proc. VLDB Endow. **11**(12), 1946-1949 (2018). https://doi.org/10.14778/3229863.3236231
37. Paulheim, H., Ngonga Ngomo, A.C., Bennett, D.: Semantic web challenge 2018. Website (2018). http://iswc2018.semanticweb.org/semantic-web-challenge-2018/index.html. Accessed 22 May 2023
38. Pedregosa, F., et al.: Scikit-learn: machine learning in Python. J. Mach. Learn. Res. **12**, 2825–2830 (2011)
39. Qudus, U., Röder, M., Kirrane, S., Ngomo, A.C.N.: TEMPORALFC: a temporal fact checking approach over knowledge graphs. In: Payne, T.R., et al. (eds.) The Semantic Web - ISWC 2023, pp. 465–483. Springer Nature Switzerland, Cham (2023). https://doi.org/10.1007/978-3-031-47240-4_25
40. Qudus, U., Röder, M., Saleem, M., Ngomo, A.C.N.: HybridFC: a hybrid fact-checking approach for knowledge graphs. In: Sattler, U., Hogan, A., Keet, M., Presutti, V., Almeida, J.P.A., Takeda, H., Monnin, P., Pirrò, G., d'Amato, C. (eds.) The Semantic Web – ISWC 2022, pp. 462–480. Springer International Publishing, Cham (2022). https://doi.org/10.1007/978-3-031-19433-7_27, https://papers.dice-research.org/2022/ISWC_HybridFC/public.pdf
41. Sagi, O., Rokach, L.: Ensemble learning: a survey. WIREs Data Min. Knowl. Discov. **8**(4), e1249 (2018). https://doi.org/10.1002/widm.1249, https://wires.onlinelibrary.wiley.com/doi/abs/10.1002/widm.1249
42. Shi, B., Weninger, T.: Fact checking in large knowledge graphs - A discriminative predicate path mining approach. CoRR **abs/1510.05911** (2015)
43. Shi, B., Weninger, T.: Discriminative predicate path mining for fact checking in knowledge graphs. Knowl.-Based Syst. **104**, 123–133 (2016)
44. Shiralkar, P., Flammini, A., Menczer, F., Ciampaglia, G.L.: Finding streams in knowledge graphs to support fact checking. In: 2017 IEEE International Conference on Data Mining (ICDM), pp. 859–864. IEEE (2017)
45. da Silva, A.A.M., Röder, M., Ngomo, A.-C.N.: Using compositional embeddings for fact checking. In: Hotho, A., et al. (eds.) ISWC 2021. LNCS, vol. 12922, pp. 270–286. Springer, Cham (2021). https://doi.org/10.1007/978-3-030-88361-4_16
46. Speck, R., Ngomo, A.C.N.: Ensemble learning of named entity recognition algorithms using multilayer perceptron for the multilingual web of data. In: Proceedings of the Knowledge Capture Conference. K-CAP 2017, Association for Computing Machinery, New York, NY, USA (2017). https://doi.org/10.1145/3148011.3154471
47. Suchanek, F.M., Kasneci, G., Weikum, G.: YAGO: a core of semantic knowledge. In: Proceedings of the 16th international conference on World Wide Web, pp. 697–706. ACM (2007)
48. Syed, Z.H., Röder, M., Ngomo, A.C.N.: FactCheck: validating rdf triples using textual evidence. In: Proceedings of the 27th ACM International Conference on Information and Knowledge Management, pp. 1599–1602. CIKM '18, Association for Computing Machinery, New York, NY, USA (2018). https://doi.org/10.1145/3269206.3269308, https://svn.aksw.org/papers/2018/CIKM_FACTCHECK/public.pdf
49. Syed, Z.H., Röder, M., Ngomo, A.-C.N.: Unsupervised discovery of corroborative paths for fact validation. In: Ghidini, C., et al. (eds.) ISWC 2019. LNCS, vol. 11778, pp. 630–646. Springer, Cham (2019). https://doi.org/10.1007/978-3-030-30793-6_36
50. Xu, Z., Pu, C., Yang, J.: Link prediction based on path entropy. Phys. A **456**, 294–301 (2016)

A Framework for Evaluating Entity Alignment Impact on Downstream Knowledge Discovery

Sarah Binta Alam Shoilee[✉], Victor de Boer, and Jacco van Ossenbruggen

Vrije Universiteit Amsterdam, Amsterdam, The Netherlands
{s.b.a.shoilee,v.de.boer,jacco.van.ossenbruggen}@vu.nl

Abstract. Entity alignment (EA) is a crucial process in integrating data from multiple sources, facilitating Knowledge Discovery (KD). Despite advances in EA techniques, selecting the appropriate algorithm for downstream KD tasks remains challenging due to several issues. These issues include domain entities alignment difficulties, the impact on KD tasks, and bias in data distribution. This paper presents a framework to address these challenges by providing a systematic approach to evaluate the impact of different EA algorithms based on three critical aspects: quality of alignment, information retrieved through alignment, and information imbalance or bias introduced through alignment. Our framework enables users to make informed decisions about algorithm selection, ensuring reliable, effective, and balanced KD. We demonstrate the application of the framework using a digital humanities case study, where the KD task involves enriching information about colonial collections. The choice of such a sensitive and historically imbalanced use-case allows us to highlight how the proposed framework helps identify suitable algorithms and to emphasis the importance of understanding the propagated information biases introduced through data alignment.

Keywords: Knowledge Discovery · Entity Alignment · Evaluation Framework · Digital Humanities

1 Introduction

Entity Alignment (EA) is a critical process in the domain of Linked Data, enabling the integration of distributed datasets by establishing identity links between instances that refer to the same real-world entity [22]. This process is fundamental to facilitating Knowledge Discovery across distributed datasets. *Knowledge Discovery* (KD) is often seen as an interdisciplinary process aimed at extracting useful and actionable knowledge from data repositories [5]. In the this paper, we define the KD task as finding (additional) useful and actionable information through EA across data sources. Selecting the appropriate EA algorithm for Knowledge Discovery presents significant challenges. The complexity arises from multiple factors, including the inherent difficulty of *domain entities*

© The Author(s), under exclusive license to Springer Nature Switzerland AG 2025
M. Alam et al. (Eds.): EKAW 2024, LNAI 15370, pp. 226–242, 2025.
https://doi.org/10.1007/978-3-031-77792-9_14

(entities that only show up in domain specific context [10]) alignment, the lack of well-defined KD task definition, and biases in data distribution.

EA algorithmic performance is often reported using benchmark datasets, which do not reflect the diversity of real-world scenarios [3]. The lack of ground truth in real-world cases further complicates the assessment of EA algorithm accuracy, leading to mistrust and hindering their application in fields like biomedicine [23] and historical research [8]. Additionally, the expectation of facilitating Knowledge Discovery through EA often lacks a clear definition of the desired discoveries, making it difficult to select algorithms that maximise gain in information for downstream tasks. Imbalanced node degree distribution [21] further exacerbates these challenges, as biases can lead to patterns that do not correspond to real phenomena [16]. EA algorithms that favour entities with dense neighborhoods (referred to as *head* entities) over entities with sparse neighborhoods (referred as *tail* entities) reinforce data imbalance and introduce bias in information retrieval [10].

To address these issues, we propose a framework designed to evaluate how links introduced through different EA techniques influence downstream Knowledge Discovery tasks, specifically in terms of gain in useful information for the user. This paper presents a method that allows a user to investigate the suitability of an EA algorithm based on three features related to the task of Knowledge Discovery. In our setting, Knowledge Discovery is implemented as a set of (SPARQL) queries that matches user competency questions. EA will establish new correspondences links between entities in distributed knowledge graphs (KGs) that may lead to additional results to such queries. The suitability of an EA algorithm is determined by a) reporting the quality of the produced links; b) counting the number of additional (correct) results for specific queries, in other words the gain in information; and c) highlighting the potential information imbalance in these results introduced by these links. By quantifying these, our framework aims to guide the selection of EA algorithms that ensure balanced and accurate Knowledge Discovery, thereby enhancing the integrity and reliability of the Knowledge Discovery process. We validate our framework through a case study in digital humanities, demonstrating its potential to support critical evaluation and informed decision-making in the selection of EA techniques.

2 Background

In this section, we further define the problem based on related work and end with stating the specific contribution of this paper.

2.1 Related Work

Domain Entity Alignment. Aligning *domain entities*, which only appear in a domain-specific context [10], presents significant alignment difficulties [8,23]. The heterogeneous nature of data sources, coupled with the absence of ground truth data, creates substantial obstacles to accurate alignment. Additionally, real-world data often exhibits sparsity [21], skewness [7], ambiguity [20], and incompleteness [18], further complicating the alignment process. Creating manually annotated ground truth data for domain entities can be counterproductive, as one of the motivations for developing automated EA systems is to address the lack of human resources for manual alignment. The manual annotation process is not only time-consuming and expensive but also prone to human error, complicating the development and validation of effective EA algorithms.

In addition, the benchmark datasets, often used to report the performance of state-of-the-art EA algorithms, typically exclude domain-specific entities [4]. This absence leads EA performance scores to inadequately reflect their effectiveness in real-world contexts. Benchmark datasets like DBpedia, Wikidata, and YAGO are constructed from structured instances from Wikipedia, where most real-world entities do not appear; even when they do, they fall under the long-tail distribution [10]. The differences in distribution between benchmark datasets and real-world datasets cause accuracy and precision scores to vary, complicating the assessment of the accuracy and quality of links produced by EA algorithms in real-world scenarios [22]. This variation can lead to potential mistrust in alignment results, hindering their application in critical fields such as biomedicine and historical research, where trust and reliability are paramount.

Lack of Task Definition. In the context of Entity Alignment, there is often an implicit expectation of facilitating Knowledge Discovery [14], yet this expectation is frequently not accompanied by a clear definition of the specific types of discoveries being sought. The absence of a well-defined information retrieval task makes it challenging to determine which EA algorithm will ultimately maximise gain in information, complicating the selection of the most appropriate algorithm for downstream Knowledge Discovery tasks. Without a formalised understanding of the desired outcomes, evaluating the effectiveness of knowledge discovery becomes infeasible. Some algorithms might excel at linking densely connected entities but perform poorly with rare or domain entities [21], leading to information bias and hindering the full potential of linked data for Knowledge Discovery. Clearly defining the goals and metrics for Knowledge Discovery by specifying the types of entities and relationships of interest and their relevant contexts is essential [2,12]. This enables researchers to better assess the strengths and weaknesses of various EA algorithms in the context of use, leading to more informed choices that enhance the overall quality and utility of the discovered knowledge.

Bias in Data and Alignment. While the benefit of sharing knowledge is universally recognised, the potential impact of unequal knowledge distribution across datasets is less frequently critiqued. This lack of balance in information, which we refer to as bias, is generally assumed a priori when producing applications that exploit connected datasets. However, such biases can significantly impact results, leading to the emergence of patterns that do not correspond to real phenomena [6,16].

An EA algorithm that predominantly favours aligning head entities may overlook rare or domain entities [3,21]. This can lead to a malformed understanding from the data, where insights and knowledge are biased towards more connected entities, masking the true diversity and complexity of the information. Selecting the right EA algorithm is therefore crucial to mitigate these biases. An algorithm that accounts for the equal distribution of information across datasets can provide a more balanced and accurate alignment, ensuring that rare entities are also appropriately linked and represented. This balance is essential for accurate Knowledge Discovery and for drawing conclusions that reflect real-world phenomena.

In certain domains, where the accuracy and completeness of data are paramount, the choice of EA algorithm can influence the outcomes of research and the validity of the insights gained. Hence, it is important to carefully evaluate and select EA algorithms that are capable of handling information biases to ensure the integrity and reliability of the Knowledge Discovery process.

2.2 Contribution of this Paper

Based on these related works, we define the problem as follows. Given a set of Entity Alignment (EA) algorithms X, the challenge for the *user* is to determine which algorithm $x \in X$ is best suited for downstream Knowledge Discovery tasks. To this end, we develop a framework that consists of three criteria to guide this selection. For any given EA algorithm, the framework enables the quantification of the following criteria:

- **Entity Alignment Efficacy**: the precision, recall, and overall performance of alignment.
- **Gain in Information**: the increase in useful and actionable information resulting from the alignment.
- **Information Bias**: the extent of information imbalance introduced into the integrated dataset due to the alignment process.

In this context, we consider two groups as potential *users* of the framework: domain researchers who might conduct linking as part of knowledge discovery for their research. The second user group consists of data managers who are responsible for managing, curating, and aligning data from one or more institutes based on domain interest and institutional strategic needs.

3   Proposed Framework

In this section, we present the proposed framework and discuss the significance of each step in quantifying Knowledge Discovery through Entity Alignment. Figure 1 outlines the complete framework.

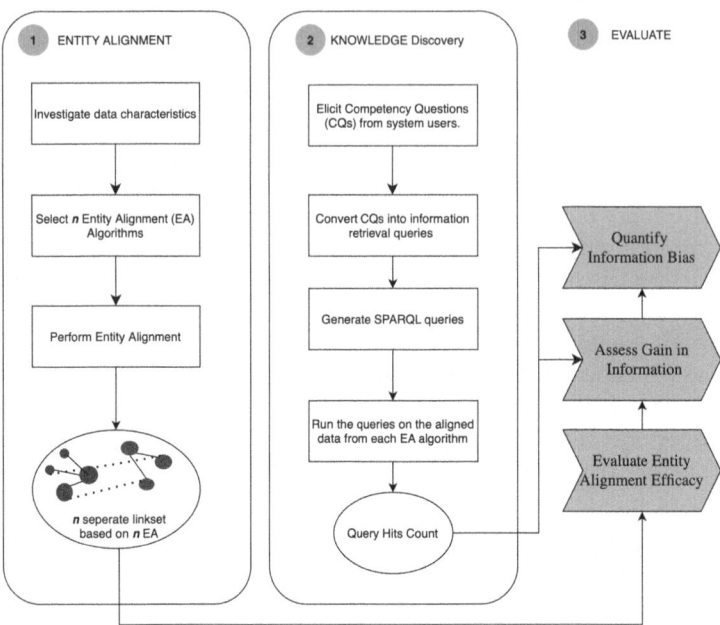

Fig. 1. Overview of the proposed Knowledge Discovery framework.

3.1 Entity Alignment

Selecting a range of algorithms based on the characteristics of the source and target KG is crucial [12]. Such KG characteristics include skewed node degree distribution of the entities, the presence of entities with missing attribute values, or entities with ambiguous labels. Assessing these characteristics enables users to determine which algorithms from the literature are possible candidates of EA for their specific problem.

In our framework, we recommend initially selecting multiple suitable algorithms rather than a single one. This allows users to compare the performance of different EA algorithms on the downstream KD task, ultimately facilitating the selection of the most appropriate algorithm.

Once the algorithms are selected, the user can proceed with alignment (consolidating links) based on each chosen EA algorithm. This will result in different versions of consolidated links corresponding to each algorithm. To understand the impact of these different versions of alignments on the downstream KD task, it is essential to define the task concretely. The process for constructing specific tasks for Knowledge Discovery is described in the next subsection.

3.2 Knowledge Discovery

To effectively define the Knowledge Discovery task, it is crucial to consider domain information needs [12]. One established method for capturing this is through Competency Questions (CQs) [19], which are natural language questions outline and define the scope of knowledge engineering. The advantage of using CQs in defining Knowledge Discovery tasks lies in their ability to be translated into information retrieval queries such as SPARQL queries, which are quantifiable.

In our framework, we propose eliciting CQs directly from system users and then converting them into information retrieval queries for Knowledge Discovery tasks. The count and approximate accuracy of the hits from these queries will serve as a proxy for "gain in information" within our framework.

3.3   Evaluation

Finally, this step involves evaluating the effectiveness of Knowledge Discovery through Entity Alignment. Our framework evaluates three aspects: (1) Entity Alignment Efficacy refers to assessment of performance of the Entity Alignment algorithms, including precision, recall, and F1-score; (2) Gain in Information reports the impact of Entity Alignment on the KD process, specifically the number of new hits generated per Competency Question after alignment and the approximate accuracy of these hits; and (3) Information bias communicates whether the knowledge is skewed towards head entities or tail entities. The metrics notation and description for these assessments is given below:

Entity Alignment Efficacy. We have integrated EA efficacy metrics into the framework to assess the performance of EA algorithms against ground truth data. Evaluation using standard classification-based metrics, i.e., precision, recall and F1-score, offers quantitative insights into algorithmic performance.

In cases where ground truth data is unavailable, precision (P_i) is estimated from manual evaluations on random samples from each algorithm's alignment.

Gain in Information. As previously mentioned, we count the number of new hits for each CQ after Entity Alignment as indicator of gain in information from that alignment algorithm, denoted as ΔH_{ij}. Here, i represents the algorithm and j represents the CQ. Related to this measure, we introduce AE_{ij}, representing the number of aligned entities that provide at least one answer to respective competency question (CQ_j) by algorithm i.

To estimate the accuracy of the gained information, we multiply ΔH_{ij} and AE_{ij} with the corresponding algorithmic precision (P_i) which is represented by estimated correct hits $(\widehat{\Delta H}_{ij}^{Correct})$ and estimated correct aligned entities $(\widehat{AE}_j^{Correct})$.

$$\widehat{AE}_{ij}^{Correct} = AE_{ij} \times P_i \tag{1}$$

$$\widehat{\Delta H}_{ij}^{Correct} = \Delta H_{ij} \times P_i \tag{2}$$

Information Bias. To understand information bias introduced through EA in KD task, we quantify the difference between the actual new hits per CQ for aligned entities and the increase in hits per CQ if aligned entities were normally distributed in the entire dataset, denoted as the *Deviation Indicator* (DI_{ij}). To calculate this, we first introduce the expected correct number of new hits for CQ_j for algorithm i, denoted by $E(\Delta H_{ij}^{Correct})$, which is calculated by the equation below.

$$E(\Delta H_{ij}^{Correct}) = \widehat{AE}_{ij}^{Correct} \times Avg(H_j) \tag{3}$$

$Avg(H_j)$ is calculated considering if all entities (of interest) from target KG were aligned with source KG, what would be the average number of result for CQ_j. We calculate DI_{ij} as the normalized difference between the estimated correct number of hits and expected correct number of hits, using the following equation.

$$DI_{ij} = \frac{\widehat{\Delta H}_{ij}^{Correct} - E(\Delta H_{ij}^{Correct})}{E(\Delta H_{ij}^{Correct})} \tag{4}$$

The Deviation Indicator is a measure for the amount to which the retrieved (correct) counts for CQ_j deviates from the expected counts, assuming that the result for CQ in the target dataset is normally distributed. This DI_{ij} value is an indicator to what extent the retrieved results are closer to average, or shifted towards the head or tail distribution. A positive value for DI_{ij} indicates that the aligned entities contributing to the result of CQ_j is from the head distribution, while a negative value suggests the opposite.

4   Case Study

In this section, we demonstrate the application of the proposed framework through a case study focused on facilitating Knowledge Discovery in museum heritage collection data for provenance research. Provenance research [17] is a multidisciplinary field that aims to trace the ownership history and historical context of cultural objects' collection, providing insights into their historical significance. Provenance research is challenging due to incomplete historical records and the substantial volume of objects requiring thorough investigation. Here, clear understanding of decision-making processes within the system is essential to mitigate the potential biases which are often inherent in the dataset.

This study considers datasets from two Dutch museums: Wereldmuseum and Museum Bronbeek. Wereldmuseum[1] focuses on ethnographic heritage objects and their collectors, whose history goes back to reflecting colonial-era state power and expansion, while Museum Bronbeek[2] primarily houses collections related to military personnel and civil servants from the Royal Netherlands East Indies Army (KNIL). By aligning these datasets based on overlapping person entities

---

[1] formerly known as National Museum of World Culture (NMVW): https://amsterdam.wereldmuseum.nl.

[2] http://museumbronbeek.nl.

in both datasets, our KD task is to identify collectors with potential colonial backgrounds and information enrichment of the objects they collected, which is significant for domain researchers, such as provenance researchers or historians.

4.1   Data Descriptions

Wereldmuseum. KG originates from heritage object collection metadata and follows the CIDOC-CRM ontology based on Linked Art recommendations. Within the context of this case study, the goal of the entity alignment is to find the same collectors who also co-exist in Bronbeek dataset. Concretely, the alignment task is to match the same individuals across KGs expressed as "actors" in various events related to objects, i.e., acquisition, transfer of custody, or production. The complete knowledge graph consists of 1,079,700,638 triples, divided into five main components: objects (976,000 entities), constituents (actors of events) (39,619 entities), historical events (70 entities), exhibitions (624 entities), and themes (438 entities). Detailed statistical distributions of these components is given in the Github repository[3].

Bronbeek Museum dataset, consisting of 15,382 constituents and 94,994 related objects was extracted from the museum's SQL database. For the purpose of this case study, it was transformed into Linked Data (LD). While we here also follow Linked Art recommendations, deviations were made to accommodate specific nuances in the dataset. The full details of the data conversion can be found in GitHub repository[4].

4.2   Entity Alignment

We first analyse the dataset characteristics in terms of density, which includes the distribution of (in-)degree per (person) entity and the distribution of attribute counts per entity. In this case study for EA task, we consider the Wereldmuseum KG (39,619 person entities) as our source KG and the Bronbeek museum(15,382 person entities) entities as our target entities. Both datasets exhibit a long-tail degree distribution, meaning a small number of actors are connected to a large number of objects, while a large number of entities are connected to only one object. Another noteworthy characteristic specific to our dataset is the low number of attribute count. Only 7,500 entities from source KG have more than two attributes, while the majority of source entities possess just one attribute value, which is the entity name or label (see Github stats for support).

Given the context of the KD task and the scope of data with limited attributes along with skewed edge-degree distribution, we opted for string-based matching approaches for alignment. For this case-study, we choose simple EA techniques to maximise recall and mitigate bias towards entities with high degree

---

[3] https://github.com/Shoilee/dh_entity_linking_v2/blob/serendipity/
knowledge_discovery/stats/stats.ipynb.

[4] GitHub repository for all codes and documentations: https://github.com/Shoilee/
dh_entity_linking_v2/tree/serendipity.

distribution or high attribute count by focusing only on entity names, in other words the string literal values of their labels.

In our approach, we consider five string matching techniques for aligning person instances across datasets based on their name's string value. 1) **Exact String Matching** is strict, focusing on precise matches, which may not always succeed due to variations in name mentions (e.g., abbreviations, full names within brackets). 2) **Initial+Surname Matching** caters to match with Wereld-museum's naming convention but may struggle with title identification (e.g., "Baron Haro van Hemert tot Dingshof", where "Baron Haro" is the title) and with inconsistency in initials(e.g., "S.J. (Sjoerd) Nauta"). 3) **Surname Matching** offers flexibility but may face challenges with correct identification of complex surnames (e.g., "Baron Haro van Hemert tot Dingshof" considered as a single-part surname "Dingshof"). 4) **Fuzzy String Matching** allows for partial matches, accommodating misspellings (e.g., using the "edit distance" metric) but may lead to significant number of false positives.

Finally, 5) **DeezyMatch** [9] deep-learning based string matching technique, initially trained on place names, which we adapted for person names using the multilingual JRC-Names dataset [11]. The positive data points for training DeezyMatch were generated when the correct alternate label exhibited a match ratio above 0.6, while negative samples were created with match ratios below 0.4, aligning with the guidance in [9]. Though we utilised default hyper-parameters and guidance provided by [9], by adjusting the training samples, the EA precision and recall can be tuned further.

The results of these alignment approaches were consolidated as owl:sameAs identity links, stored as separate linkset named graphs. We evaluate each of these strategies and these linksets in Sect. 4.4. Source codes are available in the GitHub repository.

4.3 Knowledge Discovery

As outlined in 3.2, we specify the Knowledge Discovery task that captures user needs. In our study, we captured user requirements for ethnographic heritage object provenance using Competency Questions (CQs) from previous research [15]. These CQs were elicited from interviews with museum professionals familiar with the domain. From the nine CQs provided, we selected five for analysis, refining their focus based on the dataset. Questions involving "which", "what", and rhetorical questions were transformed into counts to quantify outcomes. The resulting CQs are presented in Table 1. We then constructed SPARQL queries to answer these questions and to report gain in information. Detailed steps of this conversion is given in GitHub[5].

---

[5] https://github.com/Shoilee/dh_entity_linking_v2/blob/serendipity/README.md# knowledge-discovery.

Table 1. Competency Questions and its interpretation for Knowledge Discovery Query

CQ	Query used to report result (ΔH_{ij})
CQ-1	How many actors are involved in Wereldmuseum objects' acquisition or production events and also appear in the Bronbeek dataset?
CQ-1b	How many objects are there from the Bronbeek dataset that are collected or produced by a Wereldmuseum actor, who has a connection to at least one other object?
CQ-3	How many new pairs of actors can be formed by connecting actors through object acquisition events?
CQ-4	How many Bronbeek objects are connected to Wereldmuseum actors, where at least one (Wereldmuseum) object per actor has a known production location?
CQ-5	How many Bronbeek objects have connections with Wereldmuseum actors, where at least one (Wereldmuseum) object per actor has a connection with a historical event?
CQ-6a	How many Bronbeek objects were collected by Wereldmuseum actors and have a similar acquisition time as some of their Wereldmuseum objects?
CQ-6b	How many Bronbeek objects have unknown acquisition dates and are connected to Wereldmuseum actors, where at least one (Wereldmuseum) object per actor has a known acquisition time?
CQ-6c	How many Bronbeek objects have connections with Wereldmuseum actors, where at least one (Wereldmuseum) object per actor has a connection with a historical event and the acquisition date is similar (+/- 20 years) to other objects from the same actor?

4.4 Evaluation

Entity Alignment Efficacy. In assessing the performance of the adopted EA techniques , we aim to measure precision, recall, and F1-score.

For precision, we use estimated precision based on a random sample of 50 alignment results per strategy which are manually evaluated by the authors. The unavailability of a ground truth makes it challenging to determine recall for this task. For recall, we use the performance on a different alignment task for which we do have a ground truth. This "proxy task" aligns the same source entities (persons from Wereldmuseum) to a different set of target entities (from Wikidata instead of Bronbeek), for which a ground truth consisting of 6,178 correspondences was established previously[6]. We assume that the recall results are indicative of the recall for our task.

The efficacy for selected EA strategies on the proxy task is presented in Table 2. The results are quite expected: as the string matching strategy becomes less constrained, precision decreases while recall increases. Both Fuzzy String

---

[6] previously documented in the context of Wikidata:CopyClear initiative: https://www.wikidata.org/wiki/Wikidata:CopyClear/Museum_van_Wereldculturen/Canadian_creators_NMVW_not_in_MNBAQ.

Table 2. Recall, Precision and F1-score on Proxy Task

	instances	match	correct	Recall	Precision	F1-score
Exact Match	6178	3360	3346	0.542	0.995	0.702
Initial + Surname Match	6178	3662	3576	0.579	0.977	0.727
Surname Match	6178	8771	4991	0.808	0.569	0.668
Fuzzy String Match	6178	15028	5515	0.893	0.367	0.520
DeezyMatch	6178	3369	3359	0.544	0.997	0.704

Match and Surname Match demonstrate good recall, with later maintaining notably lower false positives compared to Fuzzy Match. Interestingly, Deezy-Match shows performance similar to Exact Match. This could be attributed to the strict criteria used for training samples (positive sample with similarity score ≥ 0.6 and negative samples having score ≤ 0.4), favoring higher accuracy over recall.

Table 3. Performance evaluation on the Wereldmuseum-Bronbeek alignment task based on 50 different random samples (selected separately from each algorithms' consolidated links)

	Total Matched	True Positive	Precision	(Proxy) Recall	F1-score
Exact Match	351	50	1	0.542	0.703
Initial+Surname Match	978	46	0.92	0.579	0.711
Surname Match	51376	9	0.18	0.808	0.294
Fuzzy String Match	3533880	0	0	0.893	0.000
DeezyMatch	318	49	0.98	0.544	0.700

In the manual evaluation of random correspondence samples, we assigned scores ranging from 0 (definitely false) to 4 (definitely correct). *Precision* was calculated by counting any score above 1 as true-match. The results are presented in Table 3. The precision scores for Exact Match, Initial+Surname Match and DeezyMatch are similar to the ones in the proxy task. However, the precision for Surname Match and Fuzzy Match are quite different due to the source-target match ratio. So, we can conclude that the precision score from the proxy task do not translate well to our actual case-study task.

As previously mentioned, to estimate recall in the current alignment task, we take the respective EA strategies' recall score from the proxy task. We also use this recall to calculate F1-score for the random sample. Due to the lower precision score in random sample the F1-score suffers for Surname Match and Fuzzy String Match, indicating the remaining three EA strategies could be a better match for current use-case. As Fuzzy Match demonstrated zero precision in random samples, we exclude this algorithm from further consideration.

Gain in Information. To quantify gain in information on the downstream KD task, we report on the count generated by each EA strategy per CQ as well as their estimated accuracy through the metrics presented in Sect. 3.3.

We utilise the questions from Table 1, and corresponding SPARQL queries to report ΔH_{ij}. We also report AE_{ij} (how many matched entities contributed to these hits). By comparing ΔH_{ij} among results yielded using different alignment strategy informing us how much information we gained.

As depicted in Table 4, the columns AE_{ij} and ΔH_{ij} demonstrate that alignment strategies based on relaxed constraints such as Surname match yield more results compared to those based on strict constraints (Exact match and Deezy-Match match). We also can see that as the complexity of the CQ increases due to the addition of more filtering criteria imposed by the query, some strict matching strategies may fail to produce any results and strategies based on relaxed constraints prove beneficial as they still generate some results. Interestingly, despite stricter matching criteria Initial+Surname match could manage to produce result for all CQ, except one.

To gain insight of the estimated accuracy of KD task, we report on $\widehat{AE}_{ij}^{Correct}$ and $\widehat{\Delta H}_{ij}^{Correct}$ as mentioned in Sect. 3.3. When comparing the estimated correct hits between Surname Match and Initial+Surname Match for certain CQs (specifically CQ-5, CQ-6a, and CQ-6c), it is notable that the correct results generated by Surname Match may be fewer than those generated by Initial+Surname Match. This occurs even though Surname Match initially aligned significantly more entities than Initial+Surname Match.

Information Bias. The Deviation Indicator (DI_{ij}), calculated using Eq. 4, assesses whether the information obtained through alignment is normally distributed for the intended entities. For example, if a CQ asks, "How many objects are there in the Bronbeek dataset under specific conditions?", it seeks to enrich information about objects' from the target dataset by aligning actors'. DI_{ij} measures the extent of second-order information bias introduced, comparing the actual hits to the average hits ($Avg(H_j)$) expected if all target entities were aligned. $Avg(H_j)$ is calculated based on entities on the right side of the '⇒' in the Query Path column in Table 4. For CQ-1b, CQ-4, CQ-5, and CQ-6b, $Avg(H_j)$ is the average number of objects per actor (12); for CQ-6a and CQ-6c, it is the average time information available per actor (18). For CQ-1 and CQ-3, average calculation is not applicable as they enrich actor information, hence no second-order information enrichment was achieved.

The last column in Table 4 lists the individual values of DI_{ij} for each CQ. This shows that indeed for some CQ-EA combinations, the actual hits are shifted toward *head* (positive score) while for other more towards *tail* (negative score). We also report on the average DI_{ij} for all CQs. We see that Exact Match and DeezyMatch exhibit the least amount of information bias, in fact they are leaning more towards the *tail*; note that, they also produce zero results for some CQ. We see that for both Initial+Surname Match and Surname Match, the average values for DI_{ij} are positive. This indicates that in general, these information

Table 4. Performance Evaluation of EA for downstream task using the framework metrics from Sect. 3.3. The Query Path column shows the (simplified) path in the KG related to CQ. Here, ⇒ represents the established link through alignment.

CQ	Query Path Source ⇒ Target	Entities AE_{ij}	$\widehat{AE}_{ij}^{Correct}$	Hits ΔH_{ij}	$\widehat{\Delta H}_{ij}^{Correct}$	Accuracy $E(\Delta H_{ij}^{Correct})$	Bias DI_{ij}
DEEZYMATCH							
CQ-1	actor ⇒ actor	87	85.26	87	85	N/A	N/A
CQ-1b	object → actor ⇒ actor ← object	36	35.28	325	319	423	−0.248
CQ-3	actor → actor ⇒ actor ← actor	2	1.96	2	2	N/A	N/A
CQ-4	place → actor ⇒ actor ← object	27	26.46	240	235	318	−0.259
CQ-5	event → actor ⇒ actor ← object	0	0	0	0	0	0.000
CQ-6a	time → actor ⇒ actor ← time	0	0	0	0	0	0.000
CQ-6b	time → actor ⇒ actor ← object	0	0	0	0	0	0.000
CQ-6c	place→actor ⇒ actor←object→ time	0	0	0	0	0	0.000
Avg.							−0.072
EXACT MATCH							
CQ-1	actor ⇒ actor	117	117	117	117	N/A	N/A
CQ-1b	object → actor ⇒ actor ← object	36	36	707	707	432	0.637
CQ-3	actor → actor ⇒ actor ← actor	0	0	0	0	N/A	N/A
CQ-4	place → actor ⇒ actor ← object	31	31	0	0	372	−1.000
CQ-5	event → actor ⇒ actor ← object	2	2	20	20	24	−0.167
CQ-6a	time → actor ⇒ actor ← time	0	0	0	0	0	0.000
CQ-6b	time → actor ⇒ actor ← object	0	0	0	0	0	0.000
CQ-6c	place → actor ⇒ actor←object→time	0	0	0	0	0	0.000
Avg.							−0.076
INITIAL+SURNAME MATCH							
CQ-1	actor ⇒ actor	277	255	277	255	N/A	N/A
CQ-1b	object → actor ⇒ actor ← object	116	107	2,153	1,981	1,281	0.547
CQ-3	actor → actor ⇒ actor ← actor	9	8	10	0	N/A	N/A
CQ-4	place → actor ⇒ actor ← object	83	76	1,841	1,694	916	0.848
CQ-5	event → actor ⇒ actor ← object	11	10	312	287	121	1.364
CQ-6a	time → actor ⇒ actor ← time	1	1	16	15	17	−0.111
CQ-6b	time → actor ⇒ actor ← object	0	0	0	0	0	0.000
CQ-6c	place→actor ⇒ actor← object→time	1	1	14	13	17	−0.222
Avg.							0.346
SURNAME MATCH							
CQ-1	actor ⇒ actor	2,680	482	2,680	482	N/A	N/A
CQ-1b	object → actor ⇒ actor ← object	1,905	343	94,298	16,974	4,115	3.125
CQ-3	actor → actor ⇒ actor ← actor	392	71	975	176	N/A	N/A
CQ-4	place → actor ⇒ actor ← object	507	91	12,776	2,300	1,095	1.100
CQ-5	event → actor ⇒ actor ← object	19	3	912	164	41	3.000
CQ-6a	time → actor ⇒ actor ← time	3	1	23	4	10	−0.574
CQ-6b	time → actor ⇒ actor ← object	7	1	163	29	15	0.940
CQ-6c	place→actor ⇒ actor ← object→time	3	1	21	4	10	−0.611
Avg.							0.997

produced through these EAs are more skewed than expected. This points at potential introduction of knowledge imbalance or bias through alignment. Note that, to generate a positive score not all the CQs generated skewed outcome (for example, CQ-6c). Surname Match exhibits more bias than the others, implying that allowing more false positives exacerbate existing data bias.

4.5 Summary

Following the framework, we quantified the quality of the produced links, their utility for the Knowledge Discovery task, and the degree of information imbalance introduced to the source dataset from the target data. In this case-study, ensuring trust and fair alignment is paramount. Therefore, for the given task, an optimal choice of alignment could be the Initial+Surname Match out of the initially selected five strategies. Despite its simplicity and potential to overlook false negatives, it manages to generate result for all CQs except one. Moreover, in sensitive cases such as provenance research-aimed at addressing past injustices-this EA with more precision may be preferable. However, this EA involves information bias for more CQ, which should be clearly communicated to the system user.

5 Discussion

Our case study demonstrates that successful alignment does not necessarily ensure an equal distribution of Knowledge Discovery opportunities across all entities. It is crucial to evaluate whether aligning different data sources genuinely results in gain in information. As evidenced by our Exact Match and DeezyMatch methods, although they achieve alignment, they often fail to answer many questions of interest to domain users. On the other hand, emphasising recall to maximise gain in information may introduce information bias through false alignments, as we have seen in the case of Surname Match and to some extent in Initial+Surname Match. Therefore, while striving to gain more knowledge, it is critical to understand whether the Knowledge Discovery opportunities are equitably distributed across the dataset or if they are merely enriching information for entities that have been historically favoured.

The primary benefit of this work is its capacity to quantify critical measures, thereby supporting transparency to the generated outcomes within a system where Knowledge Discovery is achieved through data alignment. By quantifying three dimensions-link quality, gain in information, and information bias-the proposed framework equips users with the necessary information to select the most suitable algorithm. Furthermore, in scenarios where maximising gain in information over minimising bias is favoured, data managers can explicitly communicate the extent of information imbalance introduced through this framework; this practice is coherent to tool criticism practices [1]. This explicit communication helps mitigate undesired outcomes in potential Knowledge Discovery tasks by making information imbalances in real-world knowledge graphs more visible.

While we validate the framework through a specific case study, several limitations must be acknowledged. First of all, there is a clear need to explore more advanced EA algorithms within our framework. It would be interesting to incorporate techniques such as embedding-based EA methods to analyse the interplay on alignment accuracy and biases, particularly in complex datasets with domain entities, such as this one. Nevertheless, as long as such methods produce correspondences, our framework can be applied on any sophisticated EA techniques. For the current paper, we focus on the framework and validation through simple matching strategies to demonstrate it.

The types of bias discussed in this work primarily focus on the structural biases [3] present in datasets, such as imbalanced node degree distributions. Other forms of bias, including race, gender [6], and socioeconomic disparities [13], are not explored in detail here. Though previous works already attempted these issues, these dimensions of bias yet represent significant avenues for future research, as they play a crucial role in ensuring fairness and inclusively in Knowledge Discovery systems. Investigating these biases could broaden the understanding of how alignment algorithms impact various societal groups and improve the overall fairness of data-driven systems.

Next to that, developing strategies to create or approximate ground truth data for EA evaluations will be beneficial. Methods such as crowd-sourcing annotations or leveraging domain-specific heuristics could provide valuable benchmarks for assessing algorithm performance in the absence of existing ground truths. Lastly, but most importantly, expanding the application of our framework to other domains beyond provenance research will be essential for validating its broader utility. By testing the framework in areas such as bio-medicine, historical research, and other fields that rely heavily on data quality, we can refine and adapt our approach to meet a wider range of Knowledge Discovery needs. Here, extensive user studies can give insights into the understandability of the metrics, as well as user preferences towards the weighting of the various metrics to come to a balanced selection.

6   Conclusion

In this paper, we presented a framework for evaluating Entity Alignment (EA) algorithms in the context of the Knowledge Discovery task. Our framework employs a three-tier evaluation approach that quantifies the quality of alignment, the extent of gain in information, and the degree of information imbalance introduced by the links. The ability to quantify the impact of EA on the Knowledge Discovery task provides users with the insights needed to make informed decisions. This capability is crucial for building trust in the results generated by EA systems and ensuring that Knowledge Discovery is effective and equitable. Through a case study focused on provenance research in museum heritage collections, we demonstrated the practical application of our framework.

In conclusion, our framework offers a foundation for enhancing the effectiveness of EA algorithms in multi-source data integration contexts. By continuing to refine and expand this approach, we aim to enhance the reliability, trustworthiness, and utility of Entity Alignment, ultimately contributing to the advancement of various fields that depend on integrated data.

Acknowledgement. We express our gratitude to Prof. Dr. Susan Legêne for her guidance and feedback throughout this study. This work was carried out in the context of the NWA funded project Pressing Matter (NWA.1292.19.419).

References

1. Alkemade, H., et al.: Datasheets for digital cultural heritage datasets. J. Open Humanit. Data **9**(1), 1–11 (2023)
2. Ding, P., Jun-yi, S., Mu-xin, Z.: Incorporating domain knowledge into data mining process: an ontology based framework. Wuhan Univ. J. Nat. Sci. **11**(1), 165–169 (2006)
3. Fanourakis, N., Efthymiou, V., Christophides, V., Kotzinos, D., Pitoura, E., Stefanidis, K.: Structural bias in knowledge graphs for the entity alignment task. In: Pesquita, C., et al. (eds.) The Semantic Web, vol. 13870, pp. 72–90. Springer Nature Switzerland, Cham (2023). https://doi.org/10.1007/978-3-031-33455-9_5, https://link.springer.com/10.1007/978-3-031-33455-9_5, series Title: Lecture Notes in Computer Science
4. Fanourakis, N., Efthymiou, V., Kotzinos, D., Christophides, V.: Knowledge graph embedding methods for entity alignment: An experimental review (2022). http://arxiv.org/abs/2203.09280
5. Fayyad, U., Piatetsky-Shapiro, G., Smyth, P.: From data mining to knowledge discovery in databases. AI Mag. **17**(3), 37 (1996). https://doi.org/10.1609/aimag.v17i3.1230
6. Fisher, J., Mittal, A., Palfrey, D., Christodoulopoulos, C.: Debiasing knowledge graph embeddings. In: Webber, B., Cohn, T., He, Y., Liu, Y. (eds.) Proceedings of the 2020 Conference on Empirical Methods in Natural Language Processing (EMNLP), pp. 7332–7345. Association for Computational Linguistics, Online (2020). https://doi.org/10.18653/v1/2020.emnlp-main.595, https://aclanthology.org/2020.emnlp-main.595
7. Guo, L., Sun, Z., Hu, W.: Learning to exploit long-term relational dependencies in knowledge graphs. In: Proceedings of the 36th International Conference on Machine Learning, pp. 2505–2514. PMLR (2019). https://proceedings.mlr.press/v97/guo19c.html, iSSN: 2640-3498
8. Heino, E., et al.: Named entity linking in a complex domain: case second world war history. In: Gracia, J., Bond, F., McCrae, J.P., Buitelaar, P., Chiarcos, C., Hellmann, S. (eds.) LDK 2017. LNCS (LNAI), vol. 10318, pp. 120–133. Springer, Cham (2017). https://doi.org/10.1007/978-3-319-59888-8_10
9. Hosseini, K., Nanni, F., Coll Ardanuy, M.: DeezyMatch: a flexible deep learning approach to fuzzy string matching. In: Proceedings of the 2020 Conference on Empirical Methods in Natural Language Processing: System Demonstrations, pp. 62–69. Association for Computational Linguistics (2020). https://doi.org/10.18653/v1/2020.emnlp-demos.9, https://aclanthology.org/2020.emnlp-demos.9
10. Ilievski, F., Vossen, P., Schlobach, S.: Systematic study of long tail phenomena in entity linking. In: Bender, E.M., Derczynski, L., Isabelle, P. (eds.) Proceedings of the 27th International Conference on Computational Linguistics, pp. 664–674. Association for Computational Linguistics, Santa Fe, New Mexico, USA (2018)

11. Jacquet, G., Verile, M.: JRC-names RDF: Person and organisation spelling variants as found in multilingual news articles. Dataset, European Commission, Joint Research Centre (JRC) (2015). http://data.europa.eu/89h/jrc-emm-jrc-names
12. Pinto, F.M., Santos, M.F.: Considering application domain ontologies for data mining. WSEAS Trans. Info. Sci. and App. **6**(9), 1478–1492 (2009)
13. Rahman, T., Surma, B., Backes, M., Zhang, Y.: Fairwalk: towards Fair Graph Embedding. In: Proceedings of the Twenty-Eighth International Joint Conference on Artificial Intelligence, pp. 3289–3295. International Joint Conferences on Artificial Intelligence Organization, Macao, China (2019). https://doi.org/10.24963/ijcai.2019/456, https://www.ijcai.org/proceedings/2019/456
14. Ristoski, P., Paulheim, H.: Semantic web in data mining and knowledge discovery: a comprehensive survey. J. Web Semantics **36**, 1–22 (2016). https://doi.org/10.1016/j.websem.2016.01.001, https://www.sciencedirect.com/science/article/pii/S1570826816000020
15. Shoilee, S.B.A., de Boer, V., van Oseenbruggen, J.: Polyvocal knowledge modelling for ethnographic heritage object provenance. In: Knowledge Graphs: Semantics, Machine Learning, and Languages. vol. 56, p. 127. IOS Press (2023)
16. Tiddi, I., d'Aquin, M., Motta, E.: Quantifying the bias in data links. In: Janowicz, K., Schlobach, S., Lambrix, P., Hyvönen, E. (eds.) Knowledge Engineering and Knowledge Management, pp. 531–546. Springer International Publishing, Cham (2014)
17. Tompkins, A.: Provenance Research Today. Lund Humphries (2021)
18. Van Tong, V., Huynh, T.T., Nguyen, T.T., Yin, H., Nguyen, Q.V.H., Huynh, Q.T.: Incomplete knowledge graph alignment. arXiv preprint arXiv:2112.09266 (2021)
19. Wiśniewski, D., Potoniec, J., Ławrynowicz, A., Keet, C.M.: Analysis of ontology competency questions and their formalizations in sparql-owl. Web Semant. **59**(C), 100534 (2019). https://doi.org/10.1016/j.websem.2019.100534
20. Yin, X., Huang, Y., Zhou, B., Li, A., Lan, L., Jia, Y.: Deep entity linking via eliminating semantic ambiguity with bert. IEEE Access **7**, 169434–169445 (2019). https://doi.org/10.1109/ACCESS.2019.2955498
21. Zhao, X., Zeng, W., Tang, J.: Long-tail entity alignment. In: Zhao, X., Zeng, W., Tang, J. (eds.) Entity Alignment: Concepts, Recent Advances and Novel Approaches, pp. 161–184. Springer Nature, Singapore (2023). https://doi.org/10.1007/978-981-99-4250-3_6
22. Zhao, X., Zeng, W., Tang, J., Wang, W., Suchanek, F.M.: An experimental study of state-of-the-art entity alignment approaches. IEEE Trans. Knowl. Data Eng. **34**(6), 2610–2625 (2022). https://doi.org/10.1109/TKDE.2020.3018741
23. Zheng, J.G., et al.: Entity linking for biomedical literature. BMC Med. Inform. Decis. Mak. **15**(1), S4 (2015). https://doi.org/10.1186/1472-6947-15-S1-S4

Scholarly Wikidata: Population and Exploration of Conference Data in Wikidata Using LLMs

Nandana Mihindukulasooriya[1]([⊠]) [iD], Sanju Tiwari[2] [iD], Daniil Dobriy[3] [iD],
Finn Årup Nielsen[4] [iD], Tek Raj Chhetri[5,6] [iD], and Axel Polleres[3] [iD]

[1] IBM Research, New York, USA
nandana@ibm.com
[2] Sharda University, Greater Noida, India and TIB Hanover, Hanover, Germany
tiwarisanju18@ieee.org
[3] Vienna University for Economics and Business, Vienna, Austria
daniil.dobriy@wu.ac.at
[4] DTU Compute, Technical University of Denmark, Lyngby, Denmark
axel.polleres@wu.ac.at
[5] McGovern Institute for Brain Research, Massachusetts Institute of Technology,
Cambridge, MA, USA
faan@dtu.dk
[6] Center for Artificial Intelligence (AI) Research Nepal, Sundarharaincha-09, Nepal
tekraj.chhetri@cair-nepal.org

Abstract. Several initiatives have been undertaken to conceptually model the domain of scholarly data using ontologies and to create respective Knowledge Graphs. Yet, the full potential seems unleashed, as automated means for automatic population of said ontologies are lacking, and respective initiatives from the Semantic Web community are not necessarily connected: we propose to make scholarly data more sustainably accessible by leveraging Wikidata's infrastructure and automating its population in a sustainable manner through LLMs by tapping into unstructured sources like conference Web sites and proceedings texts as well as already existing structured conference datasets. While an initial analysis shows that Semantic Web conferences are only minimally represented in Wikidata, we argue that our methodology can help to populate, evolve and maintain scholarly data as a community within Wikidata.

Our main contributions include (a) an analysis of ontologies for representing scholarly data to identify gaps and relevant entities/properties in Wikidata, (b) semi-automated extraction – requiring (minimal) manual validation – of conference metadata (e.g., acceptance rates, organizer roles, programme committee members, best paper awards, keynotes, and sponsors) from websites and proceedings texts using LLMs. Finally, we discuss (c) extensions to visualization tools in the Wikidata context for data exploration of the generated scholarly data. Our study focuses on data from 105 Semantic Web-related conferences and extends/adds more than 6000 entities in Wikidata. It is important to note that the method can be more generally applicable beyond Semantic Web-related conferences for enhancing Wikidata's utility as a comprehensive scholarly resource.

Source Repository: https://github.com/scholarly-wikidata/
    DOI: https://doi.org/10.5281/zenodo.10989709
    License: Creative Commons CC0 (Data), MIT (Code).

Keywords: Scholarly Data · Wikidata · Large Language Model

© The Author(s), under exclusive license to Springer Nature Switzerland AG 2025
M. Alam et al. (Eds.): EKAW 2024, LNAI 15370, pp. 243–259, 2025.
https://doi.org/10.1007/978-3-031-77792-9_15

1   Introduction

Scientific conferences are vital for researchers to share their research findings and advancements. It offers an opportunity to discuss research problems or limitations, a platform for networking with peers, and a platform for promoting collaboration, which is essential for learning, innovation, and problem-solving. Because of the importance of scientific conferences, we have seen tremendous growth in the number of conferences over the years [8]. For example, IEEE (Institute of Electrical and Electronics Engineers) sponsors more than 2,000[1] conferences and events annually. Similarly, ACM (Association for Computing Machinery) hosts more than 170[2] conferences annually worldwide.

Therefore, efforts have been made to capture metadata about scientific events [6–8,19] in a linked-data format as they provide valuable information. Such data can be used for (i) better understanding the progress of science overall, (ii) the evolution of particular research topics (or fields), (iii) understanding research impact (e.g. by sponsors' interest) over time, etc. The availability of scholarly metadata enables scientometrics [12], or practical tools such as recommending relevant conferences or papers to readers [6] for navigating through the fastly growing scientific output which is becoming time-consuming and almost impractical.

However, as much as the benefits these metadata about scientific events provide, there exist challenges. The primary obstacle is the collection of large-scale metadata, which is nontrivial in nature [6]. Similarly, the sustainability, which is also the focus of this paper, of the accumulated metadata constitutes the second and most significant obstacle. If the data collected is not sustainable, it may be lost over time, resulting in the loss of valuable information and efforts put into data collection. For instance, the Microsoft Academic Graph, which contained over 8 billion triples [6] with information about scientific publications and related data, was retired in December 2021[3]. While the effort was somewhat continued shortly later in OpenAlex[4], the case demonstrates sustainability issues in individual or commercial scholarly KG offerings.

We argue that collaborative, general purpose, community-driven platforms, such as Wikipedia, are generally more sustainable than such fragmented efforts: community participation is motivated by intrinsic factors, fostering a sense of belonging to the group [27]. Notably, commercial initiaves seem to recognize this, as shown by Google's declaration that it will cease operations on Freebase and transfer its content to Wikidata [22]. Wikidata, which focuses on knowledge graphs (KGs), is a sister project of Wikipedia and another example of a community-driven platform [4,25]. Wikidata currently has more 110M entities and 25K active contributors[5]. By bringing Scholarly data about scientific conferences into Wikidata, they can be seamlessly integrated with existing background knowledge through SPARQL queries. Furthermore, Wikidata benefits from a robust tooling ecosystem and widely used libraries, including entity linkers, search tools, SPARQL endpoint with high-availability, easy-to-use query editor, visualization tools, and more [3,23]. Wikidata also allows non-expert users to directly access

---

[1] https://www.ieee.org/about/at-a-glance.html.

[2] https://www.acm.org/conferences/about-conferences.

[3] https://www.microsoft.com/en-us/research/project/microsoft-academic-graph/.

[4] https://openalex.org/.

[5] https://www.wikidata.org/wiki/Wikidata:Statistics.

the KGs through search and Web UI (user interface). Therefore, the primary objective of our work is to integrate scientific conference metadata into Wikidata, a community-led platform.

After conducting an analysis of Wikidata entities related to Semantic Web conferences such as International Semantic Web Conference (ISWC), Extended/European Semantic Web Conference (ESWC), International Conference Knowledge Engineering and Knowledge Management (EKAW), International Conference on Knowledge Capture (K-CAP), SEMANTiCS, and Knowledge Graph and Semantic Web Conference (KGSWC), it was noticed that some conferences were missing and the ones that were present had only minimal information. In this project, we have extended Wikidata to include a more comprehensive set of information (e.g. see ISWC 2023[6] (Q119153957)). Within the scope of this work, we focused on the Semantic Web conferences but our method is more generally applicable and can be extended to other conference series. We note that 105 conferences we added to, updated in Wikidata is higher than the comparable related work such as Scholarly Data (35 confs)[7], ORKG (5 confs)[8] as of July, 2024.

Large language models (LLMs) have proven their language understanding capabilities with many NLP benchmarks [16]. In recent years, approaches such as in-context learning with a few-shot example have allowed them to perform many tasks such as relation or fact extraction [11,15]. Such models can be used to easily extract information from sources with natural language text, such as conference proceedings, websites, or call for papers. Nevertheless, their output can be prone to errors. In our work, LLMs are used to extract data, which is then verified by a human-in-the-loop validation to eliminate any noisy extraction and ensure accuracy.

In particular, this paper makes the following contributions.

- We analysed existing ontologies for representing scholarly data and mapped them to Wikidata to identify relevant Wikidata entities/properties as well as gaps.
- We present a method for utilizing large language models to efficiently extract conference metadata from various sources, curating them through a human-in-the-loop validation process using OpenRefine, and populating the data in Wikidata via Wikidata QuickStatements and provide an evaluation for LLM-based extractions.
- As a result of this project, we have extended over 1000 existing entities and created more than 5,000 new entities, including conferences, scientific articles, and people. These entities are now available on Wikidata and can be accessed via the Web UI or SPARQL endpoint.
- We extend visualization tools Scholia[9] and Synia[10] to better visualize the information we added to Wikidata.

---

[6] https://www.wikidata.org/wiki/Q119153957.
[7] https://bit.ly/3Vs6XNc.
[8] https://orkg.org/organizations/Event.
[9] https://scholia.toolforge.org/.
[10] https://synia.toolforge.org.

2   Related Work

Scientific events have emerged as a crucial element in scholarly communication across various scientific fields. They serve as central hubs for fostering scientific connections among various elements such as individuals (e.g., organizers and attendees), locations, activities (e.g., participant roles), and materials (e.g., conference proceedings) within the realm of scholarly discourse [8]. This section will explore the existing work in the related topic.

Different works have been done to capture and (re-)use the metadata about the scholarly events. The first work is by Fathalla et al. [8] who developed Scientific Events Ontology (OR-SEO) to capture the information of scientific events. OR-SEO is currently utilized as the framework for event pages on the OpenResearch[11] (OR) platform, which serves as a semantic wiki platform aimed at crowdsourcing metadata and supporting scholarly metadata management and utilization. Similarly, Fathalla et al. [7], in their other work, introduce a 5-star Linked Dataset containing dereferenceable IRIs (Internationalized Resource Identifier), which is an updated version of the EVENTSKG dataset. The dataset contains information about 73 different conferences in the field of computer science, such as Artificial Intelligence (AI), World Wide Web (WEB), and Software Engineering (SE). Nuzzolese et al. [21] examine the Semantic Web Dog Food (SWDF) dataset and explore its quality and sustainability challenges. The SWDF employs the Semantic Web Conference (SWC) ontology as the foundational ontology for representing data related to academic conferences. Proposed approach uses cLODg3 (conference Linked Open Data generator) [20] to regenerate the SWDF dataset based on the conference ontology, offering a sustainable solution. Scholarly data [19] initiative refactors SWDF to continue the growth of the dataset. It introduces the conference ontology, which improves SWC. The Open Research Knowledge Graph (ORKG) is one of the main knowledge graphs designed to enable the structured representation and sharing of scholarly knowledge. ORKG aims to transform scholarly publications into a structured, interconnected knowledge graph, allowing for easier data access, analysis, and reuse. Currently, ORKG has 5 conference records covering 3 conference series under the event.class[12].

In addition to scholarly ontologies, work has been done on scholarly information tooling, such as visualization and scraping. Angioni et al. [1], for example, developed an AIDA dashboard for analyzing and comparing scientific conferences. Their work uses Computer Science Ontology and the AIDA knowledge graph which enable the construction of visualization, such as top authors and organizations. Similarly, Kruger et al. [13] developed an early system named DEADLINER to extract information, such as deadline, topic, title, and program committee, from conference and workshop announcements (call for papers). The other similar work is by Fahl et al. [5] who implemented a system for scraping the CEUR Workshop Proceedings site. Fahl et al.'s [5] work also includes entity linking of event locations and proceeding editor disambiguation. Kirrane et al. [12] has presented a qualitative analysis of the main seminal papers by adopting a top-down approach and produced results with three bottom-up data-

---

[11] https://openresearch.org.
[12] https://orkg.org/organizations/Event.

driven approaches (Saffron, Rexplore, PoolParty). The analysis has been conducted on the corpus of Semantic Web papers acquired from 2006 to 2015. Several other efforts already done in earlier time and proposed Bibster, a Peer-to-Peer system [9] for transforming the bibliographic data among research community. Scholia, which our work reuses and extends, is another application by Nielsen et al. [18] that allows visualization of scholarly profiles for topics, people, organizations, species, chemicals, etc., using bibliographic and other information in Wikidata. In addition to visualization, Scholia also has functionality for scraping data from the DOI (Digital Object Identifier), NeurIPS conferences, CEUR workshop series, and Open Journal Systems. Synia is a Scholia-inspired system that creates profiles based on Wikidata, but with templates defined on a wiki. Some of Synia's profiles are displaying scholarly information [17].

Moreover, there have been several other related initiatives. Wikicite[13] is one such initiative, which focuses on a bibliographic database of source and citation metadata [24]. This work benefits from some outcomes from Wikicite such as "*is proceedings from*" (P4745) property. Efforts have also been made to maintain conference acceptance rates, such as Open Research[14] or Conference-Acceptance-Rate[15], which this work makes re-use of.

3 Overview of Scholarly Wikidata Process

The objective of our work is to extract information from structured and unstructured data sources and add it to Wikidata, specifically related to Semantic Web conferences. We leverage existing structured data and tools thereby promoting the reusability principles of the Semantic Web and further enhance them with additional information extracted from text using LLMs.

3.1 Data Sources

We considered both structured and unstructured data as input to our process. We wanted to extract information from authoritative sources, so we used the official conference proceedings and the conference website as input to our process. Table 1 illustrates the different sources we used in our process. These sources contained most of the information we wanted to extract. If the data has already been extracted and exposed in structured formats, especially in RDF (Resource Description Framework), such as DBLP and Scholarly Data, we reuse them.

3.2 Data Population with Human-in-the-Loop Validation

Figure 1 shows the process we followed to extract the data from different sources and populate data in Wikidata.

---

[13] https://meta.wikimedia.org/wiki/WikiCite.

[14] https://www.openresearch.org/wiki/ISWC.

[15] https://github.com/lixin4ever/Conference-Acceptance-Rate.

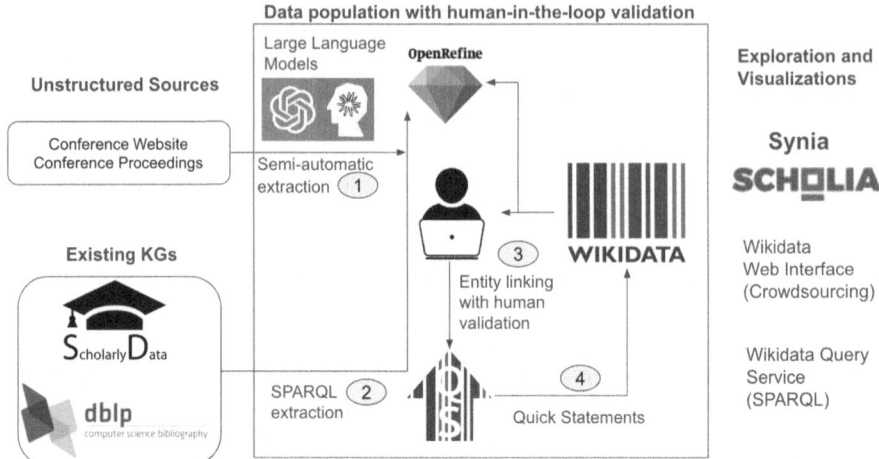

Fig. 1. Overview of the methodology. The code used for the process including the LLM prompts, SPARQL queries, OpenRefine schemas are available in the GitHub repository.

Table 1. Information extracted from different sources

Data source	Extracted Information
conference proceedings front matter	organization committee and their roles, number of papers submitted, accepted for each track of the conference, programme committees for each track with the track and their roles (PC, SPC), and prominent topics of submitted papers
conference website	important dates for each track (deadlines), number of attendee
DBLP KG	papers published in the main and poster/demo tracks along with their authors and other metadata
Scholarly Data KG	events such as tutorials, workshops, panels, and keynotes

1. Semi-automatic Extraction from Unstructured Sources. In order to extract the necessary information from unstructured sources such as conference proceedings front matter or conference websites, we have LLMs using the LangChain[16] framework. For conference front matter, the PDFs are loaded with the "PyPDFLoader" and converted into semantically meaningful chunks first using the section headings, and then using "SemanticChunker"[17]. For conference websites, given the URL, the crawler we implemented navigates to the website and traverses the contents. The main functions of the crawler are: (a) retrieving a sitemap if it exists and extending it through traversing the links, (b) extracting embedded structured data (Microdata, JSON-LD, RDFa, Open-Graph), and (c) retrieving HTML and text contents of the website pages. In both cases, relevant chunks for each extraction task are selected based on the section heading and a set of pre-defined keywords.

---

[16] https://www.langchain.com/langchain.

[17] https://python.langchain.com/docs/modules/data_connection/document_transformers/semantic-chunker/.

Finally, we formulated LLM prompts for specific extraction tasks such as extraction of the number of submitted/accepted papers for calculating the acceptance rate for each track, organization committee with their roles, programme committee with track, and roles, and conference deadlines. The Appendix[18] illustrates an example for extracting submitted/accepted papers from the proceedings front matter using in-context learning capabilities with two-shot examples. The LLM outputs are parsed to extract the output in CSV format which is used as input to OpenRefine. For further details, please refer to the implementation in GitHub repository.

2. SPARQL Query Extraction from Structured Sources. DBLP provides their data as an RDF dump with monthly releases[19], and we loaded the April 2024 snapshot into a local Virtuoso instance to create a SPARQL endpoint. Scholarly Data makes an SPARQL endpoint[20] available online. After analyzing the ontology used and the data available in each of those KGs, we formulated SPARQL queries to extract papers for each conference including both main tracks and posters and demos from DBLP and sub-events of conferences such as workshops, tutorials, panels, and keynotes from Scholarly Data. An example SPARQL query is shown in Appendix.

3. Entity Linking with Human Validation. In order to avoid creating duplicate entities and properly link to existing entities, we used the entity reconciliation functionality of OpenRefine. This step also involves disambiguation of authors having the same name. DBLP optionally provides Wikidata ID, ORCID and Google Scholar IDs when available which helps to disambiguate the authors better. Before populating Wikidata, data accuracy is ensured through human-in-the-loop validation performed using Open Refine UI.

4. Wikidata Population Using Quick Statements. We defined mappings between the CSV data and Wikidata using Open Refine schema definitions. The Appendix shows an example of such a definition. Finally, Open Refine allowed us to create Quick Statements that can be imported as a batch to Wikidata. Please refer to [14] for details.

3.3   Exploration and Visualizations

We have used Scholia and Synia as visualization tools to allow the community to explore conference-related information with visual summaries, including line charts (e.g., acceptance rates, number of participants), area charts (e.g., topics through time), maps (e.g., conference locations), graphs (e.g., co-author graphs), and timelines (e.g., important dates, deadlines). While Scholia's visualizations of Wikidata content are defined in the Scholia Web application with Jinja2 templates, Synia's visualizations are defined on wiki pages, making it easier to add new visualizations. The visualizations are generated reusing the standard graph plotting of the Wikidata Query Service and embedded on the Scholia (or Synia) webpage via Iframes. To visualize the new conference data we created new Synia template wiki pages. Table 2 illustrates some Synia

---

[18] link provided at the end of the paper.
[19] https://blog.dblp.org/2022/03/02/dblp-in-rdf/.
[20] http://www.scholarlydata.org/sparql/.

Table 2. Examples of Synia templates and examples of their corresponding pages

Description	Synia Template	Synia Page
Scientific event	https://www.wikidata.org/wiki/Wikidata:Synia:scientificevent	https://synia.toolforge.org/#scientificevent/Q119153957
Scientific events index	https://www.wikidata.org/wiki/Wikidata:Synia:scientificevent-index	https://synia.toolforge.org/#scientificevent/
Scientific Event series	https://www.wikidata.org/wiki/Wikidata:Synia:scientificeventseries	https://synia.toolforge.org/#scientificevent/Q119153957
Scientific event series index	https://www.wikidata.org/wiki/Wikidata:Synia:scientificeventseries-index	https://synia.toolforge.org/#scientificeventseries

templates and their corresponding pages. Some SPARQL queries in the Synia templates were modified from the equivalent ones in Scholia.

We also added a template with faceted views on organizational roles. For example, https://synia.toolforge.org/#scientificeventseries/Q6053150/organizationalrole/Q125207931 displays information about the Semantic Web challenge chairs through the history of ISWC. By changing the role, say for example, research track chair (Q125207937) or sponsor chair (Q125207972), similar pages can be generated for any organizer role.

Some of the template definitions of Synia were transfered to Scholia, e.g., https://scholia.toolforge.org/event/Q119153957 now displays number of participants and acceptance rate for the ISWC 2023 conference.

4   Mapping Conference Ontologies to Wikidata

Several ontologies representing conference metadata exist in literature as discussed in Sect. 2. Our objective was to use existing ontologies to understand the important aspects that should be represented in Wikidata. We utilized those conceptual models to identify the relevant Wikidata entities, properties, and qualifiers to represent scholarly data. We identified gaps in Wikidata and proceeded to add the missing elements to the platform to extend its capabilities.

Table 3. Ontologies and their coverage of conference aspects

Ontology	Conf. Metadata	Conf. Series	Topical Coverage	Roles, Committees	Sub-events	Publications	Awards	Registration	Tracks	Sponsors	Submission, Review	Attendance
Wikidata	✓	✓	✓	✓	✓	✓	✓		✓	✓	✓	✓
Scholarly Data	✓	✓	✓	✓	✓	✓	✓		✓	✓	✓	
SEO (ontology)	✓			✓	✓		✓	✓		✓	✓	
SemWeb Conference	✓			✓	✓							
SciGraph	✓	✓	✓		✓	✓			✓			
Schema.org	✓		✓	✓	✓	✓	✓			✓	✓	

4.1 Overview of Mappings to Wikidata

This section provides an overview of how existing conference ontology properties maps to Wikidata properties. Table 3 shows a comparison summary of several conference-related ontologies and Wikidata. More detailed mappings are documented in the Wiki[21] of the project.

Based on this analysis, we identified several properties that are missing in Wikidata. For example, our proposals on "number of submissions"[22] and "number of accepted contribution"[23] were accepted by Wikidata.

4.2 Extensions to Qualifier Values

Wikidata data model has a flexible way to include fine-grained information through a primary relation and a set of qualifiers. Figure 2 illustrated how the property "has programme committee member" property is used to link "Irene Celino" to "ISWC 2023" and qualifiers "applies to" and "object has role" are used to provide additional information that she was an SPC in the research track. We used Reference URL to refer to the source where the fact was extracted. Based on our analysis of the existing ontologies in literature as discussed before, we have added several qualifier values to populate richer information. Table 4 illustrates some examples of such additions. For instance, more than 60 organizer roles were added to represent the conferences we analyzed. In addition to adding new qualifiers, we also modified the property constraints appropriately considering the use case of conference metadata.

5 Data Exploration

Figure 3 shows a screenshot of Scholia displaying the topics though time for a scientific event series, the ESWC conference series. The SPARQL query behind this visualization assembles topics for each year based on the chain event series - event - proceedings - article - topic and the publication year of the article. This visualization was inspired by

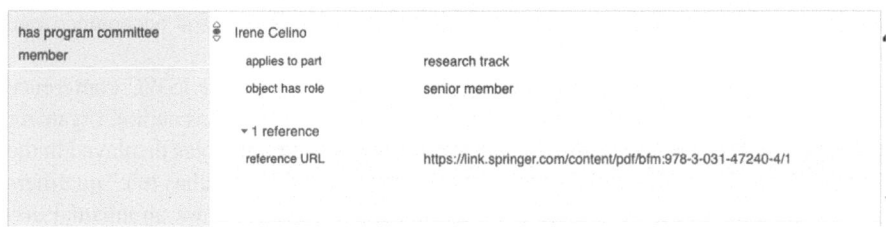

Fig. 2. An example use of qualifiers in ISWC 2023 Wikidata entity (Q119153957).

---

[21] https://github.com/scholarly-wikidata/scholarly-wikidata/wiki/Mapping-scholarly-data-ontologies-to-Wikidata.

[22] https://www.wikidata.org/wiki/Wikidata:Property_proposal/number_of_submissions.

[23] https://www.wikidata.org/wiki/Wikidata:Property_proposal/number_of_accepted_contributions.

Table 4. Qualifer values (entities) added to Wikidata based on analysis of existing scholarly data ontologies to represent the conference metadata.

property	qualifer	parent entity	entities (only labels are shown for brevity)
significant event (P793)	point in time (P585)	deadline (Q2404808)	abstract submission deadline, paper submission deadline, acceptance notification deadline, camera-ready submission deadline
organizer (P664)	object has role (P3831)	role (Q4897819)	programme chair, organization chair, workshop chair, local chair and 60 others[26]
sponsor (P859)	object has role (P3831)	sponsorship level (Q117280318)	diamond sponsor, platinum sponsor, gold plus sponsor, gold sponsor, silver plus sponsor, silver sponsor
winner (P1346)	object has role (P3831)	best paper award (Q112270830)	best research track paper award, best research track student paper award, best resource track paper award, best demo paper award, best poster paper award
has program committee member (P5804), admission rate (P5822)	applies to part (P518)	conference track (Q66087801)	research track, resources track, in-use track, posters and demos track, position paper track, evaluations and experiments track

a note in [2]: "One drawback of [Scholia] is that the topics are associated to venues as a whole and cannot be used to evaluate their temporal evolution". Any interpretation of the plots should be aware that the data behind the plot is not complete, i.e., not all conference articles are added and annotated in Wikidata. In Fig. 4 another Scholia screenshot shows the co-author graph of people associated with the ISWC conference series.

For visualizing the new conference data added, we created new Synia templates in Wikidata for scientific events and scientific event series with tables and visualizations panels. Some of Synia panels were also incorporated into Scholia. Among the panels we added were line charts for the number of participants and the acceptance rate through time for a conference series. Figure 5 illustrates the participants and acceptance rate through time for ISWC.

Figure 6 is a screenshot from a part of the Synia page about the ISWC conference series listing eight top people associated with the conference (e.g., as author, organizer or speaker) and ordered according to number of roles with unique roles displayed in the fourth column. This column shows the information from the "object has role" qualifiers for the organizer value,—if available in Wikidata. This listing shows an unequal distribution of roles where some researchers have been in several different organizational roles throughout the years while others have mostly participated in the conference as authors.

Top topics through time

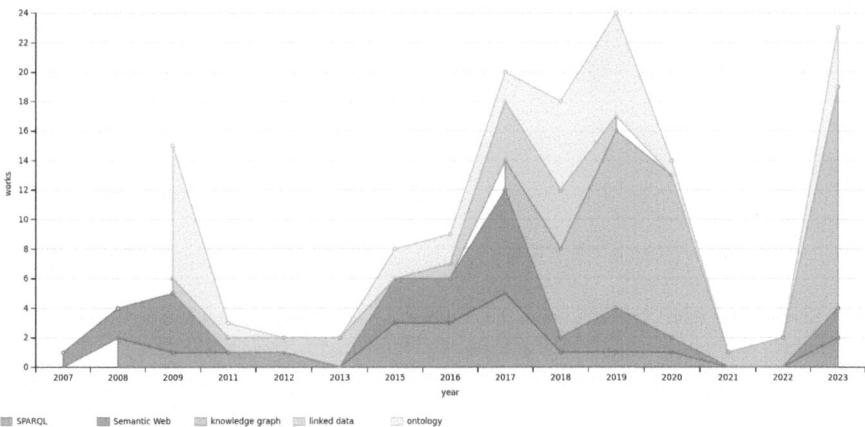

Fig. 3. Screenshot from Scholia for a panel on the page for the ESWC conference series https://scholia.toolforge.org/event-series/Q17012957#top-topic-through-time.

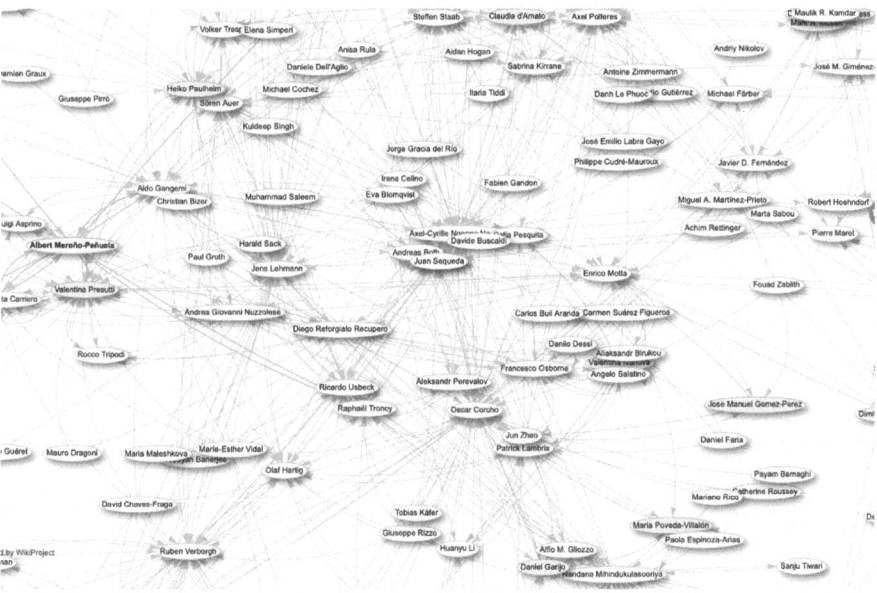

Fig. 4. Screenshot from Scholia for the co-author graph panel on the page for the ISWC 2023 conference. All co-authorships are shown, also those beyond the specific conference. It shows only part of the graph and the photographs of people have been removed (due to copyright). For the interactive complete graph, please refer to https://scholia.toolforge.org/event/Q119153957#co-authors.

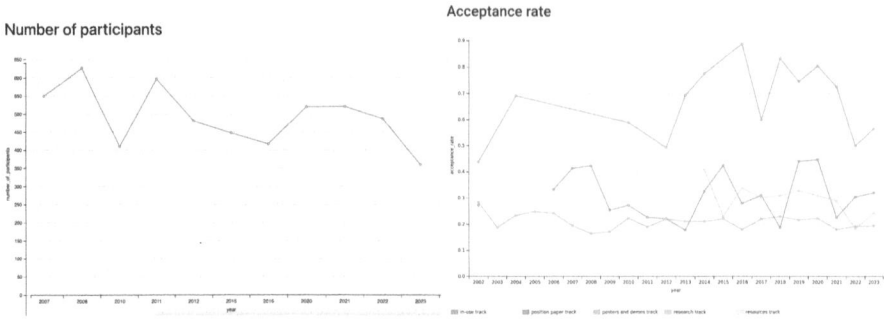

Fig. 5. Screenshots from Synia showing the number of participants and acceptance rates of ISWC through time. From https://synia.toolforge.org/#scientificeventseries/Q6053150.

Number of roles	Person	Person description	Roles	Example work
58	Jens Lehmann	German artificial intelligence researcher	author, sponsor chair	Wikidata through the Eyes of DBpedia
51	Ian R. Horrocks	Professor of Computer Science at the University of Oxford	author, editor of proceedings, programme chair, conference chair	The GRAIL concept modelling language for medical terminology
51	Axel-Cyrille Ngonga Ngomo	researcher	program committee member, author, semantic web challenge chair	An overview of the BIOASQ large-scale biomedical semantic indexing and question answering competition
50	Evgeny Kharlamov	researcher, Bosch	program committee member, author, sponsor chair	Event-Enhanced Learning for KG Completion
44	Enrico Motta	researcher, Knowledge Media InstituteThe Open University	author, editor of proceedings, general chair	Research Articles in Simplified HTML: a Web-first format for HTML-based scholarly articles
39	Ruben Verborgh	professor of Decentralized Web technology	program committee member, author, semantic web challenge chair	Interoperability and FAIRness through a novel combination of Web technologies
37	Deborah L. McGuinness	American computer scientist, professor	speaker, author, organizer	An ontology for immune epitopes: application to the design of a broad scope database of immune reactivities
34	Steffen Staab	computer scientist in Stuttgart	author, journal track chair, general chair, metadata chair, panel chair, sponsor chair, programme chair, local organization chair, organizer	Methods for Intrinsic Evaluation of Links in the Web of Data

Fig. 6. Screenshot from Synia of table with people associated with ISWC from https://synia.toolforge.org/#scientificeventseries/Q6053150.

6   Evaluation

In this section, we present evaluations for four extraction tasks using two LLMs to extract facts from preface proceedings and conference websites.

Setup. For this evaluation, we used text from 42 prefaces and websites (22 ISWC, 20 ESWC) as a test dataset. For each task, we created a human extraction ground truth by manually extracting the facts by reading the preface. Then after performing the LLM extractions, we compared system generations with the human ground truth to calculate micro precision, recall, and F1 metrics on extraction. The system is not penalized for not extracting the information that is not available in the text.

Table 5. Evaluation Results

Task	Source	Model	ISWC (2002–2023)			ESWC (2004–2023)		
			Precision	Recall	F1	Precision	Recall	F1
Submitted/accepted papers extraction	Proceedings Front Matter	gpt-4	0.92	0.87	0.89	0.94	0.95	0.95
		claude-3	1.00	0.89	0.93	0.90	0.97	0.93
Organization role extraction	Proceedings Front Matter	gpt-4	1.00	1.00	1.00	1.00	1.00	1.00
		claude-3	1.00	0.96	0.98	1.00	0.92	0.96
PC member extraction	Proceedings Front Matter	gpt-4	1.00	1.00	1.00	1.00	1.00	1.00
		claude-3	0.98	1.00	0.99	0.99	1.00	0.99
Important dates extraction	Website	gpt-4	0.8	0.97	0.88	0.81	0.94	0.87
		claude-3	0.28	0.92	0.43	0.07	0.81	0.13

Discussion. Table 5 shows that both LLMs we tested performed well in extracting submitted/accepted papers, organizational roles, and PC members from the proceeding front matter. The proceedings clearly list the organization roles and PC members and LLMs can extract them with high accuracy. On rare occasions, LLM confuses organizations (affiliations) with people. Extraction of the number of papers submitted/accepted in each track requires more natural language understanding. It was noted that sometimes the LLM extracted correct facts from its pre-trained knowledge in addition to the provided text. For example, for ISWC 2007, it generated that 7 tutorials were accepted with 1 invited tutorial even though the preface never mentioned 7 tutorials. Another example, from *"The main scientific program of ESWC 2020 contained 39 papers: 26 papers in the research track, 8 papers in the resources track, and 5 papers in the in-use track. The papers were selected out of 166 paper submissions, with a total acceptance rate of 23.5% (22% for the research track, 26% for the resources track, and 28% for the in-use track)."*, claude-3 extracted (research: 26/119, resources: 8/31, in-use: 5/18) which coincide with the acceptance ratios but with a total of 168 submissions. GPT-4 only extracted the number of accepted papers.

When extracting deadlines from websites, we noticed some hallucinations with the Claude-3 model, which extracts dates and deadlines that are not mentioned on the web page or wrongly categorizing the deadlines (e.g., identifying deadlines for poster CFPs as deadlines for full paper CFPs), resulting in a lower precision value of 28% for ISWC websites and even 7% for EWSC websites. Still, even gpt-4 could only achieve the precision of 80% and 81%, for ISWC and ESWC conferences respectively. The recall values for both models range from 81% to 97%, with claude-3 achieving the lowest performance with regard to ESWC websites (81%), mostly owing to the fact that most deadlines are identified, even though it is miscategorized. As future work, we will work on strategies to mitigate these issues with improved prompt engineering or other techniques such as chain-of-thoughts or self-verification [26].

7   Conclusions and Future Work

In this paper, we present our work on scholarly data, with a focus on Semantic Web-related conferences. The paper addresses the sustainability challenges (see Sect. 1) of metadata by bringing the metadata of scholarly events, particularly conferences, to

Wikidata. Our work uses LLMs to efficiently extract data from authoritative sources such as conference proceedings and websites and performs a human-in-the-loop validation to ensure any noisy extractions are eliminated. Our work has resulted in the creation of more than 6K entities or more than 30k statements about scientific conferences in Wikidata, providing valuable information about Semantic Web-related conferences in a machine-processable format aka KGs representation.

Impact and Reusability. By populating scholarly data in Wikidata, we make them available to a large community of both experts and non-experts using SPARQL endpoints, Web UI, and other tools such as visualizations or entity linkers. Wikidata is supported by the Wikimedia Foundation and a large community of active contributers, ensuring its sustainability. The availability of this scholarly data in Wikidata enables the community to easily access information such as acceptance rates by making simple queries. Organizers and track chairs can easily access the programme committee and other information from past years. By combining with existing data, interested parties can run ad-hoc queries, for example, the gender balance in ISWC organization committees throughout the previous years (see query[27]). The data we added to Wikidata, along with their sources, can potentially be used to generate benchmarks for tasks in scholarly domain, such as information extraction or question answering.

The data in Wikidata follows Linked Data and Findable, Accessible, Interoperable, and Reusable (FAIR) principles. In addition to the visualizations in Scholia and Synia, we also provide a Wiki page with a list of example queries[28].

Design and Technical Quality. We analyzed the existing ontologies in the scholarly data domain and used them to identify appropriate Wikidata entities, properties, and qualifiers. We followed the best practices of Wikidata by using appropirate qualifiers. For the properties we identified as gaps in Wikidata, for example, the number of submitted/accepted papers, we created property proposals following Wikidata guidelines as discussed in Sect. 4.1. We used authoritative sources such as conference proceedings or official website as our source. Furthermore, we included the reference URLs of sources where we extracted the facts to ensure their provenance is maintained and the community members can verify those. We used tools such as OpenRefine to perform entity reconciliation to avoid creation of duplicates when some entities already existed in Wikidata.

Availability. The data we generated is added to Wikidata and is already accessible via the Wikidata web user interface and SPARQL endpoint. Data in Wikidata is distributed under a Creative Commons CC0 license. Additionally, the source code for the extraction and population process is the GitHub repository under MIT license. A snapshot of the proceeding front matter links, website crawls, SPARQL query results, LLM extractions, and evaluation benchmarks are available on Zenodo, with a Creative Commons CC0 license, a persistent URI (DOI) and a canonical citation.

---

[27] https://w.wiki/9nnJ.

[28] https://github.com/scholarly-wikidata/scholarly-wikidata/wiki/Scholary-Wikidata-Query-Examples.

As future work, we are planning on creating a comprehensive benchmark for extracting triples from scholarly data communications with manually curated ground truth and making it available for the community. We believe such benchmark can help the community to evaluate new approaches and strategies for information extraction from sources using novel technologies such as LLMs.

Furthermore, based on our empirical evaluations based on the human-created ground truth, we plan to recommend a subset of extraction tasks that can be fully automated with LLMs with minimal noise so that they can be integrated to Wikidata as bots to perform large-scale extractions on all conference series.

The fact that Wikidata can be edited by anyone is quite useful for crowd-sourcing information, but also brings some challenges. Even though our experience with scholarly data on Wikidata since 2016 does not show its a common occurrence, Wikidata can be prone to vandalism and inclusion of falsified information by malicious parties [10]. Wikidata tracks the provenance of all edits with username/IP addresses, and analysis of these edits using ML models can be used to detect vandalism. Ensuring the veracity (i.e., accuracy) of crowd-sourced conference information within Wikidata poses an open challenge and we plan to formulate different strategies as part of our future work.

Appendix of the paper is available here.[29]

Acknowledgements. This research was funded in whole or in part by the Austrian Science Fund (FWF) [10.55776/COE12].

References

1. Angioni, S., Salatino, A., Osborne, F., Recupero, D.R., Motta, E.: The AIDA dashboard: a web application for assessing and comparing scientific conferences. IEEE Access 1 (2022). https://doi.org/10.1109/ACCESS.2022.3166256
2. Angioni, S., et al.: Leveraging knowledge graph technologies to assess journals and conferences at springer nature. In: Sattler, U., et al. (eds.) The Semantic Web – ISWC 2022. LNCS, vol. 13489, pp. 735–752. Springer, Cham (2022). https://doi.org/10.1007/978-3-031-19433-7_42
3. Diefenbach, D., Wilde, M.D., Alipio, S.: Wikibase as an infrastructure for knowledge graphs: the EU knowledge graph. In: Hotho, A., et al. (eds.) ISWC 2021. LNCS, vol. 12922, pp. 631–647. Springer, Cham (2021). https://doi.org/10.1007/978-3-030-88361-4_37
4. Erxleben, F., Günther, M., Krötzsch, M., Mendez, J., Vrandečić, D.: Introducing Wikidata to the linked data web. In: Mika, P., et al. (eds.) ISWC 2014. LNCS, vol. 8796, pp. 50–65. Springer, Cham (2014). https://doi.org/10.1007/978-3-319-11964-9_4
5. Fahl, W., Holzheim, T., Lange, C., Decker, S.: Semantification of CEUR-WS with Wikidata as a target Knowledge Graph. In: Joint Proceedings of TEXT2KG 2023 and BiKE 2023 (2023). https://ceur-ws.org/Vol-3447/Text2KG_Paper_13.pdf
6. Färber, M.: The Microsoft academic knowledge graph: a linked data source with 8 billion triples of scholarly data. In: Ghidini, C., et al. (eds.) ISWC 2019. LNCS, vol. 11779, pp. 113–129. Springer, Cham (2019). https://doi.org/10.1007/978-3-030-30796-7_8

---

[29] https://github.com/scholarly-wikidata/scholarly-wikidata/blob/fa6bbdc78f69df81ae12b45d8537e1977eee8aa6/docs/EKAW_2024_Paper_Appendix.pdf.

7. Fathalla, S., Lange, C., Auer, S.: EVENTSKG: a 5-star dataset of top-ranked events in eight computer science communities. In: Hitzler, P., et al. (eds.) ESWC 2019. LNCS, vol. 11503, pp. 427–442. Springer, Cham (2019). https://doi.org/10.1007/978-3-030-21348-0_28

8. Fathalla, S., Vahdati, S., Lange, C., Auer, S.: SEO: a scientific events data model. In: Ghidini, C., et al. (eds.) ISWC 2019. LNCS, vol. 11779, pp. 79–95. Springer, Cham (2019). https://doi.org/10.1007/978-3-030-30796-7_6

9. Haase, P., et al.: Bibster – a semantics-based bibliographic peer-to-peer system. In: McIlraith, S.A., Plexousakis, D., van Harmelen, F. (eds.) ISWC 2004. LNCS, vol. 3298, pp. 122–136. Springer, Heidelberg (2004). https://doi.org/10.1007/978-3-540-30475-3_10

10. Heindorf, S., Potthast, M., Stein, B., Engels, G.: Vandalism detection in Wikidata. In: Proceedings of the 25th ACM International Conference on Information and Knowledge Management, CIKM 2016, Indianapolis, IN, USA, October 24-28, 2016, pp. 327–336. ACM (2016)

11. Khorashadizadeh, H., Mihindukulasooriya, N., Tiwari, S., Groppe, J., Groppe, S.: Exploring in-context learning capabilities of foundation models for generating knowledge graphs from text. In: Joint Proceedings of the Second International Workshop on Knowledge Graph Generation From Text and the First International BiKE Challenge co-located with 20th Extended Semantic Conference (ESWC 2023), Hersonissos, Greece, May 29th, 2023. CEUR Workshop Proceedings, vol. 3447, pp. 132–153. CEUR-WS.org (2023)

12. Kirrane, S., et al.: A decade of semantic web research through the lenses of a mixed methods approach. Semantic Web 11(6), 979–1005 (2020). https://doi.org/10.3233/SW-200371

13. Kruger, A., et al.: Deadliner: building a new niche search engine. In: Proceedings of the Ninth International Conference on Information and Knowledge Management, pp. 272–281 (2000). https://doi.org/10.1145/354756.354829

14. Mihindukulasooriya, N.: Dblp to wikidata: Populating scholarly articles in Wikidata. In: Proceedings of the ISWC 2024 Posters, Demos and Industry Tracks co-located with the 23rd International Semantic Web Conference (ISWC2024) (2024)

15. Mihindukulasooriya, N., Tiwari, S., Enguix, C.F., Lata, K.: Text2kgbench: a benchmark for ontology-driven knowledge graph generation from text. In: Payne, T.R., et al. (eds.) The Semantic Web - ISWC 2023, Part II. LNCS, vol. 14266, pp. 247–265. Springer, Cham (2023). https://doi.org/10.1007/978-3-031-47243-5_14

16. Min, B., et al.: Recent advances in natural language processing via large pre-trained language models: a survey. ACM Comput. Surv. 56(2), 1–40 (2023)

17. Nielsen, F.Å.: Synia: aisplaying data from Wikibases. In: Wiki Workshop 2023 proceedings (2023). https://doi.org/10.48550/ARXIV.2303.15133

18. Nielsen, F.Å., Mietchen, D., Willighagen, E.: Scholia and scientometrics with wikidata. In: Scientometrics 2017, pp. 237–259 (2017)

19. Nuzzolese, A.G., Gentile, A.L., Presutti, V., Gangemi, A.: Conference linked data: the scholarlydata project. In: Groth, P., et al. (eds.) ISWC 2016, Part II. LNCS, vol. 9982, pp. 150–158. Springer, Cham (2016). https://doi.org/10.1007/978-3-319-46547-0_16

20. Nuzzolese, A.G., Gentile, A.L., Presutti, V., Gangemi, A.: Generating conference linked open data in one click. In: ISWC (Posters & Demos) (2016)

21. Nuzzolese, A.G., Gentile, A.L., Presutti, V., Gangemi, A.: Semantic web conference ontology - a refactoring solution. In: Sack, H., Rizzo, G., Steinmetz, N., Mladenić, D., Auer, S., Lange, C. (eds.) ESWC 2016. LNCS, vol. 9989, pp. 84–87. Springer, Cham (2016). https://doi.org/10.1007/978-3-319-47602-5_18

22. Pellissier Tanon, T., Vrandečić, D., Schaffert, S., Steiner, T., Pintscher, L.: From freebase to Wikidata: the great migration. In: Proceedings of the 25th International Conference on World Wide Web, pp. 1419–1428. WWW '16, International World Wide Web Conferences Steering Committee, Republic and Canton of Geneva, CHE (2016). https://doi.org/10.1145/2872427.2874809

23. Rossenova, L., Duchesne, P., Blümel, I.: Wikidata and wikibase as complementary research data management services for cultural heritage data. In: Wikidata 2022: Wikidata Workshop 2022, Proceedings of the 3rd Wikidata Workshop 2022 co-located with the 21st International Semantic Web Conference (ISWC2022) (2022)

24. Taraborelli, D., Dugan, J.M., Pintscher, L., Mietchen, D., Neylon, C.: Wikicite 2016 Report. Technical report, Wikimedia Foundation (2016). https://doi.org/10.6084/M9.FIGSHARE. 4042530

25. Vrandečić, D., Krötzsch, M.: Wikidata: a free collaborative knowledgebase. Commun. ACM **57**(10), 78-85 (2014). https://doi.org/10.1145/2629489

26. Weng, Y., et al.: Large language models are better reasoners with self-verification. In: Findings of the Association for Computational Linguistics: EMNLP 2023, Singapore, December 6-10, 2023, pp. 2550–2575. Association for Computational Linguistics (2023). https://doi.org/10.18653/V1/2023.FINDINGS-EMNLP.167

27. Xu, B., Li, D.: An empirical study of the motivations for content contribution and community participation in Wikipedia. Inf. Manag.t **52**(3), 275–286 (2015). https://doi.org/10.1016/j.im.2014.12.003

Understanding Inflicted Injuries in Young Children: Toward an Ontology Based Approach

Fatima Maikore[1]($\boxtimes$) (ID), Suvodeep Mazumdar[2]($\boxtimes$) (ID), Amaka Offiah[3] (ID),
Anthony Hughes[1] (ID), Sneha Roychowdhury[1] (ID), Katie Hocking[4],
and Vitaveska Lanfranchi[1] (ID)

[1] Department of Computer Science, University of Sheffield, Sheffield, UK
{f.maikore,ajhughes3,s.roychowdhury,v.lanfranchi}@sheffield.ac.uk
[2] Information School, University of Sheffield, Sheffield, UK
s.mazumdar@sheffield.ac.uk
[3] School of Medicine and Population Health, Sheffield, UK
a.offiah@sheffield.ac.uk
[4] Sheffield Teaching Hospitals NHS Foundation Trust, Sheffield, UK
katie.hocking3@nhs.net

Abstract. Investigations of children experiencing inflicted injuries is often initiated once admitted into the emergency department for injuries, and involves understanding the complex interactions between a range of different factors such as health conditions, fracture biomechanics, family history, carer accounts and so on. In this position paper, we propose the use of ontologies for capturing case details of such radiology investigations in order to create an initial knowledge base of retrospective cases, to be further expanded in the future. We discuss how we developed our ELECTRICA (ELEctronic knowledge base for Clinicians, Trainers and Researchers in Child Abuse) ontology and the different components of the ontology. We are currently using this ontology to create a knowledge base, to be used as a vision demonstrator to access larger datasets. In the longer term, we would like to use the knowledge base to support clinicians and radiologists in making decisions on current cases by offering a mechanism for searching historical cases, identifying similar cases and offering insights on risk factors that may have been missed during investigations.

Keywords: inflicted injury · non-accidental injury · child abuse · ontology development

1 Introduction and Context

Protecting children from all forms of violence is their fundamental right, yet inflicted injuries among children are a significant global challenge, recognised as a specific target in the 2030 agenda for Sustainable Development. As such, the

M. Alam et al. (Eds.): EKAW 2024, LNAI 15370, pp. 260–270, 2025.
https://doi.org/10.1007/978-3-031-77792-9_16

UN SDG 16.2 on 'end abuse, exploitation, trafficking and all forms of violence against and torture of children' incorporates indicators such as 16.2.1[1] to ensure the physical or psychological safety of children. Child maltreatment, or the abuse and neglect that occurs to children under 18 years of age is a major public health issue that causes severe socio-economic burden [24] and has enormous impact on individual health, as well as long term persisting mental health issues well into adulthood [12]. Nearly 3 in 4 children, or over 300 million children aged 2–4 years regularly suffer physical punishment and/or psychological violence at the hands of parents and carers[2].

Despite protections such as the United Nations Human Rights' Convention on the Rights of Child, over a billion children experience some form of emotional, physical or sexual violence every year, and one child dies as a result of violence every five minutes[3]. The scale and nature of this challenge is truly global, yet it is incredibly difficult to form a global assessment of the problem, particularly due to sparse and incomplete data (e.g. most recent (from 2019) data for indicator 16.2.1 is available from only 18 of 191 member states). It is in this context that we investigate one of the issues captured within this indicator: inflicted injury among young children. Child maltreatment is also difficult to identify as is often not the primary reason for a child to visit a doctor [12]. Vulnerable children who experience maltreatment are often brought to health services, particularly emergency care and therefore investigating the use of the emergency department offers insights into physical abuse and child maltreatment [34]. In emergency departments, inflicted trauma is often falsely reported, sexual abuse not mentioned and/or emotional abuse is not displayed or witnessed [23,35]. While some screening tools for recognising potential child maltreatment exist (such as SPUTOVAMO [32], ESCAPE [8]), there is a general lack of consistency in policies across departments [26], varying adherence to guidelines [17,28] and suboptimal diagnostic accuracy of tools [18,31].

Factors such as inconsistent history, injuries incompatible with history and/or developmental level of the child, interactions between child and parents, reason for visit, delay in seeking medical help, findings of head-to-toe examinations can offer signals that raise doubts about the safety of the child [12]. Other factors such as age, repeated attendance and injury type may not have as strong predictive capabilities in identifying potential physical abuse [35]. Within radiology, identifying abusive fractures is further challenged due to limited understanding of fracture biomechanics, scarcity of clinicians specialising in child protection and a lack of existing knowledge base. Often, domain expertise and experience in clinical practice is essential to determine potential cases of inflicted injuries, drawing inferences and reasoning based on evidence and arguments.

---

[1] Indicator 16.2.1 refers to 'Proportion of children aged 1–17 years who experienced any physical punishment and/or psychological aggression by caregivers in the past month'. More details at https://sdgs.un.org/goals/goal16#targets_and_indicators.

[2] World Health Organization, Child Maltreatment, https://www.who.int/news-room/fact-sheets/detail/child-maltreatment.

[3] Violence against children, https://sdgs.un.org/topics/violence-against-children.

For example, inconsistent narratives in the accounts of (guilty?) carers, biomechanics of fractures not consistent with carer history, previous history of injuries and so on. In order to support clinical staff in the process of identification of inflicted injuries, it is therefore important to have some methods to highlight potential inconsistencies, or areas of concern. At the same time, due to the varied practice, and the lack of existing knowledge bases, it has been difficult to establish a standardised approach to the problem.

Ontologies can be used to represent knowledge about a particular domain, as concepts and relationships, and allows this knowledge to be captured in a machinereadable format along with explicit semantics between concepts within that ontology [19]. This, in turn, provides a shareable, common understanding of the domain knowledge. In the biomedical domain, ontologies have been well developed, studied and implemented [29], however they are typically broad and span many medical topics. Therefore, ontologies in highly specific and interdisciplinary domain areas, such as detection of inflicted injuries in young children, have been created to capture more nuanced medical knowledge [21,25]. However, while ontologies like the International Classification of External Causes of Injuries (ICECI) [21] and the Adverse Childhood Experiences Ontology (ACESO) [25] capture external causes and adverse experiences, they do not directly represent the specific medical details of inflicted injuries, such as fractures. These ontologies focus more on classifying incidents or high-level experiences that lead to injuries but lack a detailed representation of the physiological outcomes of these injuries, particularly in young children. Therefore, more specialised ontologies or extensions are needed to capture the medical specifics of injuries like fractures in this context. To address this gap, we propose the ELECTRICA ontology, which represents the detailed medical and physiological aspects of injuries, with a focus on fractures in children under 2 years of age.

In this paper, we present our approach toward developing a knowledge base of retrospective cases of suspected inflicted fractures in children, who attended the Sheffield Childrens' Hospital using an ontology. The knowledge base aims to capture a detailed description of trauma in children, including (but not limited to) injuries sustained, mechanisms of injury and sociodemographic status providing an evidence base against which children presenting with suspected abusive injuries can be compared. Herein, we present how we developed the ELECTRICA ontology and introduce our ontology - we expect this initial work to support access to a large bank of previous cases, and eventually the development of a comprehensive clinical decision support tool for radiologists.

2   Related Work

We divide previous research into two sections: existing work related to biomedical ontologies that capture specific knowledge about children, adolescents, adverse events, and injury, and discussing the available technologies for the detection of inflicted injuries. We extended our search into the adverse childhood events literature to gather a broader view of the available methodologies.

2.1 Existing Ontologies

To the best of our knowledge, the earliest known injury classification ontology is the ICECI [21]. This is a taxonomical vocabulary for the classification of injuries. The vocabulary was developed to systematically describe injuries and the circumstances of injury occurrences, primarily to aid in injury prevention efforts. Although this ontology captures a wide range of information regarding people, locations, and injury types involved with an injury, it fails to capture practical information about a victim's hospital visit. For example, the clinical findings, observations, and image studies taken by a clinician can all provide pertinent knowledge about potential inflicted injuries and patterns of abuse. Works, such as ACEO [25], have looked at how public services may capture knowledge with regards to adverse childhood experiences. This work is focused on early intervention and mental health and may be insufficient to capture more nuanced information regarding a clinician's view of an observed child in their care, and how that child has presented in a clinical setting. Other works from public services have focused on capturing knowledge with regards to the environmental factors that may affect children's lives [15]. The Children's Health Exposure Analysis Resource (CHEAR) was designed to help health researchers access data regarding potential environmental and genetic information and how this may influence particular conditions and experiences. This valuable resource offers great value in capturing environmental factors that may contribute to a child's abuse, however more work is required to capture data with respect to the child's presentation in a clinical setting.

2.2 Inflicted Injury Detection

There is a small but growing area of research regarding the automation of inflicted injury detection. In a systematic review [14], the authors discovered that there is a small body of literature researching child abuse differentiation through the use of natural language processing techniques. Currently, seven related studies have been synthesized in this review, none of which make use of state of the art techniques, i.e., transformer-based large language models (LLM). The authors also raised concerns regarding several elements of these papers; risk of bias, overfitting, and dataset size. This leaves interesting opportunities for creating new baselines and benchmarks in this particular area. It is noticeable that the systematic review does not focus on the use of ontologies or knowledge graphs to enhance the cited systems. Recent works have focused on identifying Adverse Childhood Experiences (ACEs) in a wide range of modalities, including electronic health records [3,4,10,33,34], social media [37] and knowledge graphs [1,2,6]. Attempts to establish automated methods of tracking can be seen through the use of manual ontology mappings [36]. Authors utilise their ACE ontology, and then map this to a wider vocabulary, such as Unified Medical Language System (UMLS). Once a dataset is then tagged with UMLS concepts, the authors can then infer ACE concepts and relationships from the mappings. Other works on detecting ACEs focus on natural language processing

(NLP) methodologies such as pretrained embedding techniques [3,16] LLMs to mine and extract information from large quantities of electronic health records (EHR), using lexicons and vector distance metrics to identify similar terms [4]. This allows the capture of more dynamic relationships between salient terminology related to ACEs. State of the art techniques now necessitate the use of LLMs, as seen in works that extract social determinates of health factors from EHR [10]. Researchers demonstrate the efficacy of LLM based extraction methodologies and how further finetuning of embeddings yields improved results compared to knowledge crafted features like bag of words [3].

3   Methodology

We aim to create the ELECTRICA (ELEctronic knowledge base for Clinicians, Trainers and Researchers in Child Abuse) ontology as a standardized framework for organizing, integrating, and analyzing relevant information from diverse sources regarding child abuse. The development of the ontology followed the NeON (Networked Ontologies) methodology, which emphasizes collaboration and interoperability among multiple ontologies to represent complex domains. This approach encourages the reuse of existing knowledge resources, thereby enhancing interoperability and reducing redundancy [30]. More specifically, owing to the specific nature of the domain, we combined scenarios 1 (co-developing the ontology) and 8 (reusing existing ontologies) of the NeON methodology to develop our ontology. Below we describe the different stages of the NeON methodology that we used to develop our ontology.

The **specification** phase of the ELECTRICA ontology involved working closely with pediatric radiologists who are core members of the project team. Their expertise was crucial in identifying the clinical need and defining the ontology's boundaries and focus areas. Together, we formulated essential competency questions to guide the development, ensuring the ontology's relevance and utility in clinical settings. For the **scheduling** phase, we developed a Gantt chart to effectively track the project timeline and milestones, ensuring that all tasks were clearly defined and assigned to the appropriate team groups. In the **conceptualisation** phase, we meticulously analysed reports from investigations of suspected child abuse cases to identify the primary classes and relationships for the ontology. Using this understanding, we created an online form as a pilot to support domain experts in (manually) capturing relevant data from these documents, iteratively developing the form based on their feedback. The key classes and properties were revised multiple times to fine-tune the concepts, ensuring that class names were precise and that the characteristics modelled accurately reflected the unique anatomy of children under 2 years old, who are the primary subjects in the data represented by the ontology. In the **Formalisation** stage, we selected OWL 2 as the language for implementing the ELECTRICA ontology, for it's high expressiveness and reasoning capabilities. To ensure comprehensive coverage of key concepts, we identified additional ontologies to reuse by searching through ontology repositories such as BioPortal and the Ontology Lookup Service (OLS), identifying the Ontology for General Medical Science (OGMS) [22],

Basic Formal Ontology (BFO) [7] and Systematized Nomenclature of Medicine - Clinical Terms (SNOMED-CT) [9] as the most relevant ontologies for our use case. We used Protégé to implement and refine the ontology (**Implementation** phase). The **Evaluation** phase involved multiple methods to ensure its accuracy, consistency, and completeness such as getting feedback from domain experts, annotating datasets to ontology classes and using SPARQL queries to answer competency questions, using the OntOlogy Pitfall Scanner! (OOPS!) [20] to identify potential modelling pitfalls, and using Pellet [27] and HermiT [11] reasoners to check for logical inconsistencies.

4   ELECTRICA Ontology

The ELECTRICA ontology[4] is a comprehensive framework designed to systematically represent concepts from reports of child abuse or suspected child abuse cases. The scope of the ontology is anticipated to cover concepts related to fractures identified from radiological investigations, the characteristics of those fractures, and the family and medical history of the patients. At the time of writing this paper, the ontology includes 1,141 classes and 63 properties.

Figure 1 shows a high-level view of the ELECTRICA ontological model focusing on the fracture class, its hierarchy and the associated classes that show the characteristics. The fracture classes are represented by blue boxes; the classes showing characteristics of the fracture and how they are identified are represented in yellow; while classes for describing the case details, patient and family history are represented in green. Reused classes from external ontologies are represented in red boxes.

The ELECTRICA ontology meticulously organises fracture data, emphasising anatomical details relevant to children under two years old. At the core of the ontology's structure is the main class 'Fracture', which is further subdivided into specific subclasses based on fracture location and type. For example, 'Fracture of the Skull' is further divided into classes such as 'Fracture of the left frontal bone', 'Fracture of the left occipital bone', and similarly for the right side of the skull. The 'Fracture of the Facial Bone' class is subdivided into specific classes for different parts of the facial structure. This includes classes like 'Fracture of the left angle of the mandible', 'Fracture of the right condylar process of the mandible', and 'Fracture of the nasal bone', among others. The ontology includes classes for fractures of the neck or trunk, such as 'Fracture of the cervical spine' which is further broken down into detailed classes for each vertebra (e.g., 'Fracture of the C1 Vertebra', with subclasses for specific parts of C1 like 'Fracture of the anterior arch of C1'). This pattern continues for other spinal regions like the thoracic and lumbar spine, as well as the sacral spine or coccyx. For instance, the 'Fracture of the C2 Vertebra' class includes specific subclasses such as 'Fracture of the dens of C2,' which is particularly relevant in young children due to the higher incidence of atlantoaxial instability and the unique developmental characteristics of their cervical spine.

---

[4] The ontology is accessible on BioPortal.

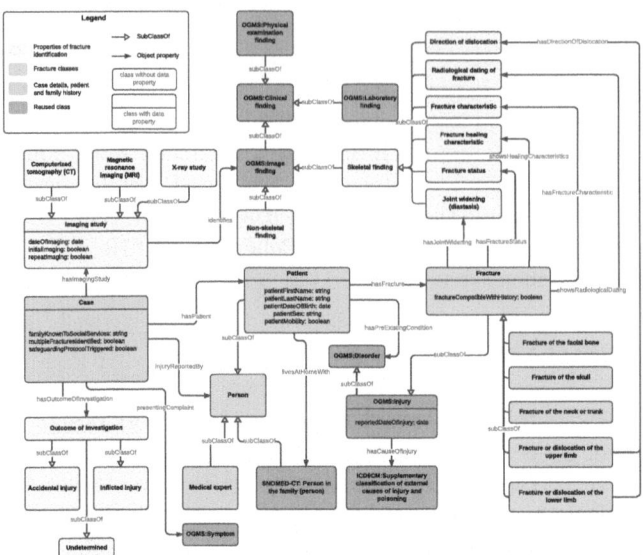

Fig. 1. ELECTRICA Ontology - main classes of the ontology. A higher resolution image is available at: https://github.com/fatibaba/electrica/blob/main/ELECTRICAdatamodel.pdf

The ontology also covers fractures of the upper and lower limbs, detailing fractures of bones such as the humerus, radius, ulna, femur, tibia, and fibula, down to the specific phalanges of the fingers and toes. Each class is hierarchically organised to detail the exact location and type of fracture, crucial for accurately documenting and analysing fractures in young children. This is particularly important in young children, where growth plate fractures can significantly impact future bone growth and overall development.

Fracture characteristics within ELECTRICA are detailed through the 'fracture characteristics' class, which includes subclasses such as 'buckle', 'greenstick', 'transverse', 'Oblique', and the various Salter-Harris types (Salter-Harris type 1–5) that are critical for understanding fractures that involve the growth plates. Some features, such as 'metaphyseal corner' fractures, are particularly indicative of potential inflicted injuries, as they are often associated with inflicted trauma in young children [13]. The ontology also includes classifications such as 'avulsion', 'comminuted depressed fracture' and others, each representing different patterns and severity of fractures.

We also captured additional features of fractures depending on the specific bone involved. For instance, the 'sutural diastasis' class is included for detailed documentation of skull fractures, where separation along the sutures can occur. Similarly, the 'direction of dislocation' class is utilised for injuries involving dislocations in the upper or lower limbs, allowing for precise characterisation of the dislocation's direction, such as anterior, posterior, superior, or inferior.

In addition to the detailed classification of fractures, we introduced classes for other key radiological findings that enhance fracture descriptions. The 'radiological dating of fracture' class categorises fractures by age, as determined through radiographic analysis, with subclasses like '0–2 weeks', '2–4 weeks', and beyond, aiding in understanding the injury's timeline and healing process. Another vital class, 'fracture healing characteristic', includes subclasses such as 'lamellar bridging bone of hard callus' and 'subperiosteal new bone formation', which detail specific healing features and provide insights into the repair stage and quality.

To enhance the specificity of the ELECTRICA ontology, we introduced the 'fracture status' class to distinguish between possible and definite fractures, aiding in the differentiation of suspected and confirmed fractures. We also capture detailed patient and family history to provide comprehensive case context, utilising data properties such as 'familyKnownToSocialServices' and 'safeguardingProtocolTriggered'. Other properties like 'totalPreviousHospitalVisits' track prior hospital visits, while 'multipleFracturesIdentified' documents cases with multiple fractures. To address pre-existing conditions affecting fracture susceptibility, the 'hasPreExistingCondition' object property links to the 'disorder' class, facilitating detailed medical condition documentation that aids in assessing the injury.

We leveraged existing ontologies to create a robust and interoperable ELECTRICA framework. BFO served as the upper-level ontology, structuring ELECTRICA and facilitating integration with OGMS, which also uses BFO. Following NHS England's guidance, we aligned ELECTRICA with SNOMED-CT to enhance clinical data capture. Additionally, we incorporated the 'causes of injury' class and its subclasses from the International Classification of Diseases, Version 9 - Clinical Modification [5] to enrich injury causation details. This reuse and extension of established ontologies ensure ELECTRICA's comprehensiveness and compatibility with healthcare systems, promoting interoperability and standard adherence.

We conducted a comprehensive evaluation of the ELECTRICA ontology, confirming its accuracy, consistency, and completeness through multiple validation methods. We executed SPARQL queries on sample annotated dataset, which effectively demonstrated the ontology's capability to retrieve and represent the required information, aligning with the identified competency questions. Additionally, using the OntOlogy Pitfall Scanner! (OOPS!) and reasoners like Pellet and HermiT, we identified and resolved several issues including missing labels, minor hierarchical inconsistencies, and redundant classes, thereby ensuring the ontology's robustness, user-friendliness, and logical consistency. We have documented the results of this validation in detail on the GitHub repository[5]. The evaluation process confirmed that ELECTRICA is robust and fit for purpose.

---

[5] The ELECTRICA GitHub repository.

5 Future Work

With our initial ontology, we are currently using a web-based interface that allows clinicians, paediatricians, and emergency care staff to input historical cases and populate our ontology, thereby developing a comprehensive knowledge base. For example, a paediatrician treating a child with unexplained fractures could enter details such as the type of injury, patient age, and clinical history. This information would then be mapped onto relevant concepts in the ontology, such as specific injury types, mechanisms of trauma, and patient demographics, enriching the knowledge base with every case added. Although this process is currently manual and requires significant effort-thus far, we have a sample of 63 retrospective cases-we are working to extend our network of international paediatric radiologists who can contribute to this effort.

In the future, we plan to develop NLP models to automate case entry and scale up to larger datasets, improving the efficiency of ontology population. For instance, these NLP models could extract relevant information from clinical notes or radiology reports and automatically classify cases into our ontology. Once the ontology is populated with a larger volume of data, we will apply data science techniques to analyse the knowledge base and develop predictive models. These models could help clinicians identify patterns, predict the likelihood of certain injuries based on clinical history, or guide decision-making in paediatric emergency care.

Acknowledgments. The work described in this paper was funded by The Children's Hospital Charity funded ELECTRICA project (CA19008)

Disclosure of Interests. The authors have no competing interests to declare that are relevant to the content of this article.

References

1. Ammar, N., Shaban-Nejad, A.: Explainable artificial intelligence recommendation system by leveraging the semantics of adverse childhood experiences: proof-of-concept prototype development. JMIR Med. Inform. **8**(11), e18752 (2020). https://doi.org/10.2196/18752, https://medinform.jmir.org/2020/11/e18752/
2. Ammar, N., et al.: SPACES: explainable multimodal AI for active surveillance, diagnosis, and management of adverse childhood experiences (ACEs). In: 2021 IEEE International Conference on Big Data (Big Data), pp. 5843–5847. IEEE, Orlando, FL, USA (2021). https://doi.org/10.1109/BigData52589.2021.9671303, https://ieeexplore.ieee.org/document/9671303/
3. Annapragada, A.V., Donaruma-Kwoh, M.M., Annapragada, A.V., Starosolski, Z.A.: A natural language processing and deep learning approach to identify child abuse from pediatric electronic medical records. PLoS ONE **16**(2), e0247404 (2021). https://doi.org/10.1371/journal.pone.0247404

4. Bejan, C.A., et al.: Mining 100 million notes to find homelessness and adverse childhood experiences: 2 case studies of rare and severe social determinants of health in electronic health records. J. Am. Med. Inform. Assoc. **25**(1), 61–71 (2018). https://doi.org/10.1093/jamia/ocx059, https://academic.oup.com/jamia/article/25/1/61/3940211

5. BioPortal, N.: International classification of diseases, version 9-clinical modification, **2021** (2021)

6. Brenas, J.H., Shaban-Nejad, A.: Proving the correctness of knowledge graph update: a scenario from surveillance of adverse childhood experiences. Front. Big Data **4**, 660101 (2021). https://doi.org/10.3389/fdata.2021.660101, https://www.frontiersin.org/articles/10.3389/fdata.2021.660101/full

7. Ceusters, W., Smith, B.: Aboutness: towards foundations for the information artifact ontology (2015)

8. Dinpanah, H., Pasha, A.A.: Potential child abuse screening in emergency department; a diagnostic accuracy study. Emergency **5**(1) (2017)

9. El-Sappagh, S., Franda, F., Ali, F., Kwak, K.S.: Snomed CT standard ontology based on the ontology for general medical science. BMC Med. Inform. Decis. Mak. **18**, 1–19 (2018)

10. Fu, Y., et al.: Extracting social determinants of health from pediatric patient notes using large language models: novel corpus and methods. In: Proceedings of the 2024 Joint International Conference on Computational Linguistics, Language Resources and Evaluation (LREC-COLING 2024), pp. 7045–7056. ELRA and ICCL, Torino, Italia (2024). https://aclanthology.org/2024.lrec-main.618/

11. Glimm, B., Horrocks, I., Motik, B., Stoilos, G., Wang, Z.: Hermit: an owl 2 reasoner. J. Autom. Reason. **53**(3), 245–269 (2014). https://doi.org/10.1007/s10817-014-9303-3

12. Hoedeman, F., et al.: Recognition of child maltreatment in emergency departments in Europe: should we do better? PLoS ONE **16**(2), e0246361 (2021)

13. Kleinman, P.K.: Diagnostic Imaging of Child Abuse. Cambridge University Press, Cambridge (2015)

14. Lupariello, F., Sussetto, L., Di Trani, S., Di Vella, G.: Artificial intelligence and child abuse and neglect: a systematic review. Children **10**(10), 1659 (2023). https://doi.org/10.3390/children10101659, https://www.mdpi.com/2227-9067/10/10/1659

15. McCusker, J., McGuinness, D., Pinheiro, P.: Children's Health Exposure Analysis Resource (2019). https://bioportal.bioontology.org/ontologies/CHEAR

16. Mikolov, T., Chen, K., Corrado, G., Dean, J.: Efficient Estimation of Word Representations in Vector Space (2013). http://arxiv.org/abs/1301.3781, arXiv:1301.3781 [cs]

17. Offiah, A., Hall, C.: Observational study of skeletal surveys in suspected non-accidental injury. Clin. Radiol. **58**(9), 702–705 (2003)

18. Offiah, A., Moon, L., Hall, C., Todd-Pokropek, A.: Diagnostic accuracy of fracture detection in suspected non-accidental injury: the effect of edge enhancement and digital display on observer performance. Clin. Radiol. **61**(2), 163–173 (2006)

19. Patel, A., Debnath, N.C.: A comprehensive overview of ontology: fundamental and researchdirections. Curr. Mater. Sci. **17**(1), 2–20 (2024). https://doi.org/10.2174/2666145415666220914114301, https://www.eurekaselect.com/208820/article

20. Poveda-Villalón, M., Gómez-Pérez, A., Suárez-Figueroa, M.C.: Oops!(ontology pitfall scanner!): an on-line tool for ontology evaluation. Int. J. Semant. Web Inf. Syst. (IJSWIS) **10**(2), 7–34 (2014)

21. Samson, T.: International Classification of External Causes of Injuries (2010). https://bioportal.bioontology.org/ontologies/ICECI
22. Scheuermann, R.H., Ceusters, W., Smith, B.: Toward an ontological treatment of disease and diagnosis. Summit Transl. Bioinform. **2009**, 116 (2009)
23. Schouten, M.C., et al.: The value of a checklist for child abuse in out-of-hours primary care: to screen or not to screen. PLoS ONE **12**(1), e0165641 (2017)
24. Sethi, D., Bellis, M., Hughes, K., Gilbert, R., Mitis, F., Galea, G.: European report on preventing child maltreatment. World Health Organization, Regional Office for Europe (2013)
25. Shaban-Nejad, A., Brenas, J.: Adverse Childhood Experiences Ontology (2019). https://bioportal.bioontology.org/ontologies/ACESO
26. Sidebotham, P., Biu, T., Goldsworthy, L.: Child protection procedures in emergency departments. Emerg. Med. J. **24**(12), 831–835 (2007)
27. Sirin, E., Parsia, B., Grau, B.C., Kalyanpur, A., Katz, Y.: Pellet: a practical owl-dl reasoner. J. Web Semant. **5**(2), 51–53 (2007). https://doi.org/10.1016/j.websem.2007.03.004, https://www.sciencedirect.com/science/article/pii/S1570826807000169, Software Engineering and the Semantic Web
28. Sittig, J.S., Uiterwaal, C.S., Moons, K.G., Nieuwenhuis, E.E., van de Putte, E.M.: Child abuse inventory at emergency rooms: Chain-ER rationale and design. BMC Pediatr. **11**, 1–7 (2011)
29. Smith, B.: Biomedical ontologies. In: Elkin, P.L. (ed.) Terminology, Ontology and their Implementations. Health Informatics, pp. 125–169. Springer, Cham (2023). https://doi.org/10.1007/978-3-031-11039-9_5, https://link.springer.com/10.1007/978-3-031-11039-9_5
30. Suárez-Figueroa, M.C., Gómez-Pérez, A., Fernandez-Lopez, M.: The neon methodology framework: a scenario-based methodology for ontology development. Appl. Ontol. **10**(2), 107–145 (2015)
31. Sugar, N.F.: Diagnosing child abuse (2008)
32. Teeuw, A.H., Kraan, R.B., van Rijn, R.R., Bossuyt, P.M., Heymans, H.S.: Screening for child abuse using a checklist and physical examinations in the emergency department led to the detection of more cases. Acta Paediatr. **108**(2), 300–313 (2019)
33. Tiyyagura, G., et al.: Development and validation of a natural language processing tool to identify injuries in infants associated with abuse. Acad. Pediatr. **22**(6), 981–988 (2022). https://doi.org/10.1016/j.acap.2021.11.004, https://linkinghub.elsevier.com/retrieve/pii/S1876285921005404
34. Wildeman, C., Emanuel, N., Leventhal, J.M., Putnam-Hornstein, E., Waldfogel, J., Lee, H.: The prevalence of confirmed maltreatment among us children, 2004 to 2011. JAMA Pediatr. **168**(8), 706–713 (2014)
35. Woodman, J., Lecky, F., Hodes, D., Pitt, M., Taylor, B., Gilbert, R.: Screening injured children for physical abuse or neglect in emergency departments: a systematic review. Child: Care Health Dev. **36**(2), 153–164 (2010)
36. Wu, J., Smith, R., Wu, H.: Adverse childhood experiences identification from clinical notes with ontologies and NLP (2022). http://arxiv.org/abs/2208.11466, arXiv:2208.11466 [cs]
37. Wu, J., Smith, R., Wu, H.: Ontology-driven self-supervision for adverse childhood experiences identification using social media datasets. In: Proceedings of the 1st Workshop on Scarce Data in Artificial Intelligence for Healthcare, pp. 5–10. SCITEPRESS - Science and Technology Publications, Vienna, Austria (2022). https://doi.org/10.5220/0011531100003523, https://www.scitepress.org/DigitalLibrary/Link.aspx?doi=10.5220/0011531100003523

A Review and Comparison of Competency Question Engineering Approaches

Reham Alharbi[1,2](✉) ⓘD, Valentina Tamma[1](✉) ⓘD, Floriana Grasso[1](✉) ⓘD,
and Terry R. Payne[1](✉) ⓘD

[1] University of Liverpool, Liverpool, UK
{R.Alharbi,V.Tamma,Floriana,T.R.Payne}@liverpool.ac.uk
[2] Taibah University, Madinah, Kingdom of Saudi Arabia
rfalharbi@taibahu.edu.sa

Abstract. Competency Questions (CQs) are essential in ontology engineering; they express an ontology's functional requirements through natural language questions, offer crucial insights into an ontology's scope, and are pivotal for various tasks, such as ontology reuse, testing, requirement specification, and pattern definition. CQ engineering approaches are gaining prominence, transitioning from manual to automatic methods, each with its own purpose, techniques, outcomes, and limitations. In this paper, we provide a review aimed at positioning these approaches within a formal categorisation, highlighting the main challenges among them and facilitating the inclusion of upcoming initiatives, and propose a benchmark to support comparative studies.

Keywords: CQs authoring · CQs generating · Ontology development phases

1 Introduction

Competency Questions (CQs) [30] are natural language questions that characterise the scope of the knowledge represented by an ontology. They model the functional requirements that an ontology-based information system should satisfy to achieve its intended purpose. Within the early stages of ontology development, they are used to suggest possible concepts and relationships the ontology should model [42–44,49,51], can be used in subsequent phases to verify and validate the knowledge encapsulated in the ontology [13,31], recommend ontologies for reuse [2,4,11], and facilitate the consumption of ontology content, such as generating APIs [22]. Their significance has been further highlighted in a small number of recent surveys that have targeted ontology developers to identify their practical usage [3,40].

In addressing problems related to CQs formulation in the ontology development cycle, previous studies have focused on artefacts and processes that can enhance the quality of CQs after they have been manually authored. This

© The Author(s), under exclusive license to Springer Nature Switzerland AG 2025
M. Alam et al. (Eds.): EKAW 2024, LNAI 15370, pp. 271–290, 2025.
https://doi.org/10.1007/978-3-031-77792-9_17

approach has taken several forms, such as the use of *Controlled Natural Language (CNL)*, which has been used for authoring ontologies [20,41], tests from CQs [15,21], and the CQs themselves [8,13,32,48,54]. However, ontology developers must first manually develop the CQs and then verify their compliance with respect to the CNL templates/archetypes, therefore they have little effect on the initial authoring effort required. Indeed, the manual process of authoring and verifying CQs is both tedious and time-consuming. It demands a significant level of intellectual effort and attention to detail, which can be a substantial burden on developers. The intricate nature of this task not only extends the development timeline but also increases the potential for errors, which might necessitate further revisions and checks. The complexity and the challenges associated with this process highlight the need for more streamlined approaches or tools that can either automate this process or assist developers in creating and validating these CQs more efficiently.

In this paper, we conduct a scoping review [10,39] over a small, curated set of studies to comprehend how CQs are formulated or constructed in the ontology development process. Scoping reviews are emerging as a useful tool for identifying and analysing knowledge gaps in a body of work [10]; therefore by conducting this review we map the salient issues that have been identified in the curated set of papers, which can then be used to conduct a comprehensive systematic review [34]. Thus, we address the following research questions:

RQ1: What are the main dimension(s) that characterise approaches for CQ formulation?

RQ2: What is the level of automation employed by different methodologies in constructing CQs?

RQ3: What are the methods and materials used for validating approaches for constructing (semi)-automatically CQs?

Therefore, our work primarily focuses on categorising CQ formulation approaches, identifying the resources used for each approach, and the validation criteria used to assess the relevance of the CQs with respect to the ontology whose construction they support. The contribution of this paper is twofold:(i) a categorisation of the current state of CQ formulation / construction approaches, together with their purpose, validation criteria, outcomes, and limitations, (ii) a proposal for a CQ benchmark along with its task specifications and evaluation criteria, that emerges as a key requirement from the scoping review as a gap in the literature. In this paper, we provide an overview of the methods underlying CQ construction approaches in Sect. 2, including recent ones where CQs are generated exploiting Generative AI, followed by the methodology used to conduct the scoping review (Sect. 3). We then discuss the research questions and how they fit the literature within the review (Sect. 4). A proposal for benchmarking is then presented based on the initial findings of the review in Sect. 5, prior to giving our conclusions in Sect. 6.

2 CQ Generation Guideline and LLM Based Approaches

CQs are integral to the specification of requirements in many ontology development methodologies [42–44,49,51] and to the validation of ontological artifacts, as they facilitate the evaluation of the ontological commitments that have been made and are generally accepted as a standard verification technique for ontologies [14,21,31,33]. However, the guidelines for creating these CQs are only vaguely specified.

For example, Gruninger and Fox [30], state that CQs should specify the requirements for an ontology and thus serve as a mechanism for characterising the ontology scope. These CQs act as constraints on what the ontology can be, rather than determining a particular design with its corresponding ontological commitments. They can also be used to assess whether the ontology meets the specified requirements based on the ontological commitments identified. The Ontology 101 [42] development process proposes a set of questions to aid ontology developers in creating their CQs. These include: *"Does the ontology contain enough information to answer these types of questions? Does the answer require a particular level of detail or representation of a particular area?"*. These CQs are intended as a sketch for further development and do not need to be exhaustive.

Rao et al. [45] propose knowledge elicitation techniques to guide ontology developers in creating CQs. For example, the 20-questions technique involves domain experts generating questions they consider important in their domain, which the ontology should be able to answer. Additionally, through card sorting, ontology engineers can identify criteria important to domain experts for grouping similar cases, thus forming the basis of the CQs. These questions are expected to ensure that the ontology can respond to queries about these concepts.

With the advent of Large Language Models (LLMs) and Generative AI, a new shift emerged in 2023 with the possible automation of knowledge engineering activities, in particular the formulation of CQs [6,9,17,46]. A number of approaches have emerged that exploit LLMs to (partially) automate the formulation process, and that differ with respect to the nature of the knowledge resources used in the prompts. In particular, we classify approaches into:

1. **Reverse engineering of CQs:** this is a reversal of the traditional workflow, where a knowledge source that was previously built from a set of ontologies, themselves engineered from CQs [17] is used to generate new CQs for a different purpose.
2. **Retrofitting CQs:** in this context, an ontology exists, but no associated CQs are published as part of its documentation. In this case the aim is to identify possible CQs that were used in the development of the ontology, thus enabling its reuse for future uses [6].
3. **Generating CQs:** these are studies that generate CQs, either from a set of class and property names [46], or generate CQs from a corpus of text describing a domain [9].

3   Research Method

This review's protocol is informed by the methodological framework provided in [10], and is illustrated in Fig. 1. Having identified the initial *Research Questions* in Sect. 1, the next stage is that of *Identifying Relevant Studies*. This involved conducting a manual search process using the following archives: IEEE[1], ACM Digital Library,[2] ScienceDirect,[3] Springer,[4] and Elsevier.[5] Additional research papers were included in the search by examining the 'related work' and 'reference list' sections of each of the papers indentified. We also included general and academic search engines such as Google Search and Google Scholar to identify other relevant papers. Furthermore, we considered the citations to certain papers by using the 'cited by' option in Google Scholar to include papers not identified by the previous methods.

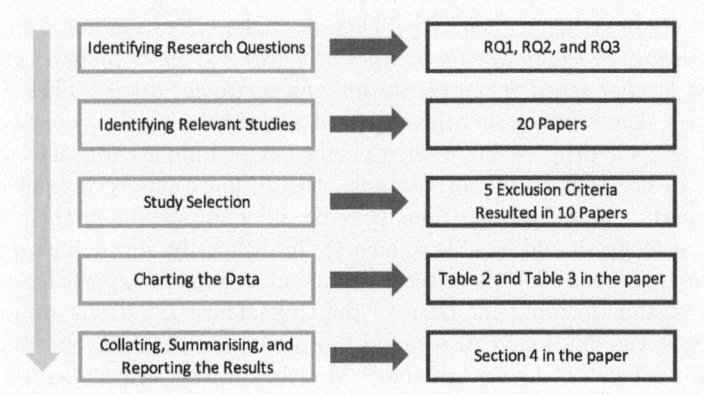

Fig. 1. Methodological framework stages of the *Scoping Study*, summarised from [10].

The field of CQ engineering is broad and interdisciplinary, encompassing areas such as computer science, educational assessment, and human resources. Therefore, for the purposes of the scoping review, the search was focused on the use of *CQs in ontology engineering*. As a result, studies such as [50] were excluded as they focus on CQs in the educational field. Therefore, different combinations of search terms were formulated using keywords deemed pertinent to the study, resulting in the identification of 20 papers:

CQs (generation OR retrofitting OR revering OR authoring) AND (Knowledge graph OR Ontology), CQs (templates OR archetypes OR patterns).

---

[1] https://ieeexplore.ieee.org/Xplore/home.jsp.
[2] https://dl.acm.org/.
[3] https://www.sciencedirect.com/.
[4] https://www.springer.com/.
[5] https://www.elsevier.com/.

For the third stage (Fig. 1), *Study Selection,* a screening process was employed to identify and eliminate studies that did not address our research questions. This in turn comprised three stages: 1) title and abstract screening, 2) full-text screening, and 3) filtering based on inclusion and exclusion criteria. As a result, 10 papers were included in the final analysis (Table 1). Papers were excluded if they violated one or more of the following criteria:

1. The paper is not written in English.
2. The paper is irrelevant to the Semantic Web domain.
3. The paper primarily focuses on building an ontology or a part of an ontology from CQs, emphasising the evaluation of development methods rather than the CQs themselves.
4. The paper primarily focuses on authoring test-form CQs, where the main interest lies in evaluating the testing method rather than the CQs themselves.
5. The paper primarily focuses on the formalisation of CQs into SPARQL or description logics queries, emphasising the formalisation methods rather than the CQs themselves.

Table 1. Studies included in the Scoping Review.

[24]	2011	Using Goal Modelling to Capture Competency Questions in Ontology-based Systems
[48]	2014	Towards Competency Question-Driven Ontology Authoring
[15]	2014	CQChecker: A tool to check ontologies in OWL-DL using competency questions written in controlled natural language
[54]	2019	Analysis of Ontology Competency Questions and their formalization in SPARQL-OWL
[32]	2019	CLaRO: A Controlled Language for Authoring Competency Questions
[8]	2021	Assessing and Enhancing Bottom-up CNL Design for Competency Questions for Ontologies
[9]	2023	Automating the Generation of Competency Questions for Ontologies with AgOCQs
[6]	2024	An Experiment in Retrofitting Competency Questions for Existing Ontologies [8]This paper was originally released as a pre-print in 2023 [5].
[17]	2024	RevOnt: Reverse engineering of competency questions from knowledge graphs via language models
[46]	2024	Can LLMs Generate Competency Questions

A specific form was designed for the fourth stage of the reviewing framework, *Charting the Data,* for the data extraction process, given the research questions targeted in this scoping review. The form includes the following characteristics:(i) title and year of publication, (ii) automation level, (iii) knowledge resource, (iv) evaluation measure, (v) ground truth, (vi) outcomes. This data is presented across two tables, with Table 2 appearing in the discussion on RQ2, and Table 3 appearing in the discussion on RQ3. The final stage, *Collating, Summarizing, and Reporting the Results* is presented in Sect. 4.

4   Results and Discussion

4.1   RQ1: Categorising CQs Formulation/Construction Approaches

In this section, we address the first research question: *What are the main dimension(s) that characterise approaches for CQ formulation?* This question is necessary to set the scene for the other two research questions that delve in automation support and validation used within the categories identified here.

There are several dimensions or characteristics that on first inspection appear orthogonal (such as automation and validation), but that on closer inspection are inter-dependent (Table 2). One of the most significant characteristics is regarding the level of automation involved in the approaches. It is possible to consider approaches as purely *manual* (i.e. with no structured support), beyond that of following best practice when authoring CQs [30,42], or *semi-automated*, involving structured or established resources such as patterns [48,54], templates [8,15,32], and archetypes [15]. These can be also categorised as *filler-based questions*, as they aim to support developers in manually deriving CQs from specific ontologies [15], or support the manual formulation of CQs for ontology development [8,32,48,54]. This contrasts with the automatic construction (or generation) of CQs using a defined resource such as an ontology or Knowledge Graph (KG), and text corpora [6,9,17,46] as input to a generative LLM system. The ultimate goal of such approaches is to produce CQs that cover a specific domain and can be integrated into various ontology development phases, depending on the knowledge source exploited in each method.

It is noteworthy that research in CQ engineering initiated with authoring approaches, which spanned from 2011 to 2021-the most recent study available at the time of this review. Conversely, generating approaches emerged in 2023 and have swiftly evolved, encompassing four distinct methods as of June 2024.

4.2   RQ2: Constructing CQs

In this subsection, we address the question: *What is the level of automation employed by different methodologies in constructing CQs?* The traditional approach whereby CQs are authored manually, following established guidance [30,42] is naturally a barrier to effective participation in ontology development, as it heavily relies on the collaboration of the domain experts, and it gives rise to inconsistencies and *ad hoc* solutions. In response to this challenge, a variety of resources and approaches emerged that were designed to facilitate the semi-automated authoring of CQs, based on an associated *Controlled Natural Language (CNL)* [36], which typically works as a set of *'filler-based questions'*. CNLs can be used to either evaluate the CQs that were manually derived from specific ontologies [15] or support the manual authoring of CQs [8,32,48,54]. These approaches however lack a formal evaluation with respect to their accuracy in the context of CQ authoring and use, evaluation which has been attempted in other contexts [28,29].

A complete automation of the generation of CQs for a prospective ontology, in contrast, is a non-trivial task. Approaches are beginning to emerge that exploit the use of Large Language Models (LLMs) and Generative AI models to propose viable CQs based on the careful crafting of prompts that include resources such as selected triples [6, 46] or Knowledge Graph fragments [17]. Each of the studies is discussed below, grouped by the level of automation involved, based on the characterisation in Table 2.

Manual Approaches. In 2011, Fernandes et al. [24] proposed applying the Tropos methodology [16], a development approach for Multi-Agent systems, to enable the definition of CQs through goal modelling. Their methodology begins with an early requirements activity to analyse organisational goals, followed by a late requirements phase where CQs are captured and linked to these goals. This approach aims to provide a consistent process for ontology engineers to develop ontologies from scratch, addressing the gap between the definition of CQs and the ontology modelling process.

Semi-automatic Approaches. Ren et al. [48] analysed the structure of CQs and defined a set of 19 CQ *archetypes*, syntactic patterns of CQs that would be instantiated by the ontology developer; e.g., "Which [CE1] [OPE] [CE2]?", where CE1 and CE2 are class expressions (or individuals, in certain cases), and OPE corresponds to an object property expression. However, out of these 19 patterns, 14 were merged with types of ontology elements, specifically OWL classes and object properties, resulting in a 1:1 mapping attribute [32]. This restricts their use to only OWL ontologies with certain limited formalisation patterns. For example, a simple subclass request such as *"What are the types of diagnosis?"* from DemCare_CQ_8 [18] has no applicable pattern [32]. Although these patterns assist ontology engineers in formulating machine-processable CQs for ontology testing, their adequacy or coverage has still to be investigated.

Table 2. Levels of Automation and resources required for CQ construction.

Automation	Study	Approach	Resource
Manual	Fernandes et al. [24]	Methodology to define CQs	N/A
Semi- Automatic	Ren et al. [48]	Pattern of CQs	Pizza ontology CQs
	Bezerra et al. [15]	Template of CQs	Pizza ontology CQs & Software ontology CQs
	Wiśniewski et al. [54]	Pattern of CQs	234 CQs for 5 ontologies
	Keet et al. [32]	Template of CQs	Wiśniewski et al. [54] patterns
	Antia et al. [8]	Template of CQs	CLaRO templates [32] & new dataset of 92 CQs
Automatic	Antia and Keet [9]	Corpus-based method for generating CQs	Corpus of Covid-19 research articles
	Alharbi et al. [6]	LLM + Ontology	CORAL [25] & CQs dataset [54]
	Ciroku et al. [17]	LLM + KG	(WDV dataset)
	Rebboud et al. [46]	LLM + Ontology	5 Ontologies & SBERT

The use of patterns was also used by Wiśniewski et al. [54], who proposed a total of 106 patterns, identified by analysing 234 CQs for five ontologies using various natural language processing tasks. This was primarily achieved through a pattern detection process to distinguish between entity and predicate chunks. An *Entity Chunk (EC)* refers to a noun or noun phrase that describes an object (entity) represented in the ontology as either a class or an individual, whereas a *Predicate Chunk (PC)* refers to a verb or phrasal verb that represents the relationship between entities in the ontology. These patterns vary in sentence structure to accommodate different question formulation preferences (e.g., "Who" and "Where" question types, omitted in [13]), and thus provide better coverage for CQs than previous studies [15,48]. Furthermore, the authors deemed that these patterns could be used to specify requirements for an ontology. These patterns were analysed and utilised to design a template-based CNL. Given the limited number of CQ patterns, a template-based approach was adopted for the CNL at this stage, instead of specifying a grammar.

Bezerra et al. [15] proposed 14 patterns to function as a CNL through templates for CQs. For example, "Does <class> + <property> <class>?", which could be filled with vocabulary taken from the ontology. The authors considered these patterns as support for the *Ontology Requirements Specification* phase; i.e., for creating and processing CQs written in natural language. However, the patterns emerging from this study were considered to have limited coverage as they are based specifically on CQ sets taken from the Pizza ontology [19]. Thus, there was the risk that this introduced both *domain bias* and *CQ author bias* (as the Pizza CQs were created *after* the ontology had been developed). It also risked exhibiting *prejudiced* patterns [54]; an example of these is were the CQ "Which pizza has 'hot' as spiciness?" is created as it fits with the Pizza ontology's "hasSpiciness" data property. However, a better CQ would have been to use the more linguistically natural phrase "Which pizzas are hot?" that is fully agnostic with respect to how it is represented in the ontology (whether or not it is with a data property, object property, or a class) [54].

CLaRO [32] is a CNL that is based on templates for use in authoring CQs. Keet et al. transformed the 106 patterns identified by Wiśniewski [54] into 93 main templates (plus 41 variants) using CNL. The additional CNL features specifically addressed issues such as: (i) singular/plural forms; (ii) the use of personal pronouns in patterns; (iii) removal of redundant words in text chunks; and (iv) synonym usage. CLaRO's 134 templates were evaluated, demonstrating coverage of about 90% of the test sets in [54]. It also has the potential to fulfil the objectives outlined in Wiśniewski et al. [54] to streamline the formalisation of ontology content requirements.

CLaRO was subsequently expanded by Antia et al. [8], incorporating an additional 120 main templates (with an additional 12 variants). This new dataset of 92 CQs generated 27 new templates and 7 more variants, significantly increasing the domain coverage and enhancing the effectiveness of the CNL. The resulting CLaRO v2 has since evolved and now includes a total of 147 templates and 59 variants, achieving 94.1% coverage. However, to effectively use CQs, ontology developers must initially create them manually and subsequently verify their compliance using these templates. The consistent dependence on a manual process highlights a prevalent challenge in the solutions developed so far [9].

Automatic Approaches. One of the first automated approaches originated in 2023, with the proposal of `AgOCQs`, a corpus-based method for generating CQs, that uses a domain text corpus as a knowledge resource to extract text, which is then pre-processed using NLP techniques and fed into a pre-trained language model to generate questions [9]. These questions then undergo filtering to remove duplicates and meaningless questions through semantic grouping. Although the inclusion of a corpus as a source of knowledge had been prevalent in other domains such as expert systems and data mining [37,53], what made this novel was its application in CQ generation.

The possibility of exploiting generative AI has resulted in new ways to automate the construction of CQs, by formulating different prompts that exploit the source domain content as a knowledge resource. One of the first examples of this originally appeared in 2023 [5] with the proposal of `RETROFIT-CQs`, which *retrofits* CQs from existing ontologies by utilising those ontologies as a knowledge resource to extract triples in the form of (`'subject'`, `'predicate'`, `'object'`) [6]. These triples then provide a contextual boundary, which is fed into a variety of different LLMs via specifically designed prompts to generate CQs for each triple. The generated questions are filtered to remove duplicates and semantically paraphrased questions. An advantage of this approach was that it specifically addressed scenarios were there was a lack of CQs for existing ontologies, that could hamper their use during different ontology development phases such as for *Ontology Reuse* [2,4,11] and *Ontology Exploitation* [22].

The use of LLMs was also explored by Rebboud et al. [46] and Ciroku et al. [17]. Both studies proposed methods for generating CQs but with different objectives. Although the approach proposed by Rebboud et al. shared some similarities with that of Alharbi et al. [6] in that ontologies were used as a knowledge resource for LLM prompts, their processing mechanisms were notably different. Specifically, Rebboud et al. [46] parsed an ontology to extract classes, properties, and schema, represented as triples: (`'Classes'`, `'Properties'`, `'Classes'`). These elements were then divided into batches of 20 classes, with each iteration processing all of these classes and the related properties that connect them to other classes corresponding to their domain or range. However, it is unclear if the splitting mechanism used is manual or automatic, how the process ensures that all related classes appear in the same iteration, or how repetitions are handled. Although this approach addressed the problem resulting from a lack of CQs for existing ontologies [46], the ultimate goal was to assess the capabilities of LLMs in several tasks, such as using CQs as inputs to generate parts of an ontology. This is similar to the contributions of studies such as NeOn-GPT [23], and others [26,27], which are outside the scope of this review.

RevOnt [17] was proposed by Ciroku et al. as a method for extracting CQs from knowledge graphs (KGs) using pre-trained language models. The primary aim of this approach was to assist ontology developers in CQs elicitation, therefore, targeting the *Requirement Specification* phase in ontology development methodologies, in a similar way to other approaches such as AgoCQs [9]. The process began with the verbalisation of data from the KG, specifically using the WDV dataset[6] for Wikidata entries, followed by the abstraction of these verbalisations into triples; i.e. (`'subject'`, `'predicate'`, `'object'`). For each triple, this abstraction is then fed into a pre-trained language model as context, accompanied by three facts: the class of the subject, the property, and the class of the object. The resulting CQs are grammatically vetted and subjected to a filtration process for similarity and paraphrase detection.

Although various efforts can facilitate the construction of CQs, investigating different methods for validating the outputs remains a significant challenge in comparing methods and choosing which to elaborate in practice.

4.3 RQ3: Validating CQ Construction Approaches

Having previously explored the approaches taken in generating CQs, and in particular, the level of automation involved in such approaches, we now address the question: *What are the methods and materials used for validating approaches for constructing (semi)-automatically CQs.* The way in which validation is conducted very much depends on the automation approach taken; validation is not typically conducted when using most of the manual and semi-automated methods beyond that which is human-based (i.e. primarily verifying that the CQs appear valid with knowledge of the domain, but without any specified resource used to determine the *ground truth*). Automated techniques, on the other hand, may exploit evaluation measures with a dataset used to provide a ground truth against which accuracy or precision/recall etc. can be determined. These factors contribute to the difficulty in generating systematic comparisons across different approaches. In particular, the automated approaches employ various knowledge resources and are evaluated against diverse benchmarks. Moreover, the evaluation measures themselves inherently differ among these methods. Furthermore, these methods target different phases of ontology development, where the generated CQs can be further elaborated. These methods are discussed in much more depth below, together with the other characterisation listed in Table 3.

Manual and Semi-automatic Authoring Methods. Few of different ways of analysing CQ and validating manual and semi-automatic authoring methods have been considered. Approaches that are based on patterns of CQs [48,54] analyse the CQs to propose the patterns but perform no specific validation. Template based approaches [15,32] result in a reduction of patterns (as in some cases, more than one pattern can emerge in one template, and sometimes there may be one pattern that can generate more than one template), but again, no specific validation is typically used. However, CLaRO [32] was validated against subset of the CQs used to derive the patterns, which were then transformed into the templates that were used to build it. A similar approach was adopted when

---

[6] https://github.com/gabrielmaia7/WDV.

Table 3. Evaluation Measures and Resources required for Validation.

	Study	Evaluation Measure	Ground Truth	Outcomes
Manual & Semi-Automated	Fernandes et al. [24]	Case Study	N/A	Method
	Ren et al. [48]	Human-based evaluation	N/A	Patterns
	Bezerra et al. [15]	Human-based evaluation	N/A	CNL-templates
	Wiśniewski et al. [54]	N/A	N/A	Patterns
	Keet et al. [32]	Human-based evaluation	N/A	CNL (CLaRO) templates
	Antia et al. [8]	Human-based evaluation	N/A	CNL (CLaRO v2) templates
Automated	Antia and Keet [9]	Human-based evaluation	CLaRO templates	CQs & their templates
	Alharbi et al. [6]	SBERT & Human-based evaluation	Existing CQ	CQs
	Ciroku et al. [17]	BLEU & Human-based evaluation	Annotated benchmark	CQs CQs
	Rebboud et al. [46]	SBERT	Existing CQ	CQs

CLaRO was extended to CLaRO2 [8] through the addition of more CQs, which resulted in more patterns and in turn more templates.

The extent to which there has been an uptake of these *filler-based questions* (i.e. patterns, templates, and archetypes, etc.) for authoring CQs across the Ontology Engineering community is not clear. An investigation into this could identify the practical barriers that hinder the applicability of these efforts to different contexts; however, a challenge remains, as to how these methods could be compared within a common framework.

Automatic Authoring Methods. The automatic CQ generation methods differ from the manual or semi-automatic ones in that they utilise additional resources in the generation method itself, and such resources can also form the basis of *ground truth* and thus be used for validation. For example, Antia and Keet [9] validated their CQs by matching them with the CLaRO templates [32]. Specifically, if the abstract form of a question matched a template, it was classified as a CQ. AgOCQs was applied to Covid-19 research articles, and the generated CQs underwent a human-based evaluation through a survey targeting specific user groups and individuals. In this evaluation, 73% of the user group and 69% of the ontology experts judged all the CQs to provide clear domain coverage. According to the authors in [9], AgOCQs were designed to assist ontology developers in scoping the ontology and identifying the domain, which suggests that they target the *Requirement Specification* phase of ontology development.

The use of LLMs poses its own challenges, as well as opportunities. The validity of the CQs generated by RETROFIT-CQs [7] was evaluated through two

approaches: (i) a human-based evaluation, and(ii) using a comparative evaluation with ground truth resources. For the human-based evaluation, an ontology
developer assessed the correctness and quality of the CQs for their ontology,
with RETROFIT-CQs achieving a precision over 0.75 based on developer evaluation. Furthermore, the developer noted that the generated CQs not only reflected
the model's ontology representation but also expose unintended modelling outcomes, which could have been included in the initial requirements elicitation
phase. The comparative evaluation compared the generated CQs against the
ground truth resources, which in this case constituted CQs for three ontologies
selected from existing CQ datasets [25,54]. SBERT [47] was used to measure
semantic similarity, with RETROFIT-CQs achieving a recall of 0.99. Further investigations explored the affect of changing the LLM parameters (such as varying
the creativity parameter to assess the impact on the resulting CQs).

The notion of a comparative assessment was also used in other approaches
[17,46], where ontologies were used as part of the ground truth. In one approach
[46], semantic similarity between the generated CQs and the ontologies themselves was measured using SBERT [47]. Although in these studies, low precision
scores were reported, it was noted that this should not be interpreted as the
result of irrelevant questions being generated, but rather that new CQs were
discovered that could make a valuable addition to the ground truth dataset.
For the second approach [17], the ground truth consisted of manually curated
CQs for the WDV dataset[7] (for a detailed discussion on verbalisation evaluation, interested readers can refer to [17]). The results were variable, depending
on the type of CQ generated (for example, higher quality scores were attributed
to questions where the answer was the object of the triple).

5   CQsBEN - A Benchmark for CQs Formulation Approaches

Validating manual (or semi-automatic) approaches has always been challenging
due to the subjective nature of ontology engineering, and the fact that CQs
typically need to address a requirement that is difficult to quantify. However,
as more automatic, generative methods emerge, there is a growing desire to
perform comparative evaluations between them. To date, few common resources
have been used by different studies (one exception being Dem@Care [18], which
has been used by several studies [6,46]), and there is little consistency on the
use of evaluation measures to assess the CQs. Furthermore, different studies have
addressed different phases in the ontology development lifecycle, and thus cannot
be directly compared. For example, some approaches target the *Requirement
specification* phase [9,17], whereas others address the context where ontologies
have missing or non-existent CQs [6,46].

Developing a multi-purpose benchmark is fundamentally different from creating a specific benchmark that could be used for a single purpose (for example, LOVBench [35], which is used ontology term ranking). It should serve all

---

[7] https://zenodo.org/records/10370725.

types of formulation/construction approach used (including those that are manual, semi-automated, and fully automated), and thus include criteria that assess them, starting from specifying the tasks to providing evaluation criteria. In the discussion below, we have identified three main tasks for such a benchmark, which we refer to as CQsBEN: (i) **Poor/Incorrect Requirements:** This category addresses common errors in question formulation that hinder effective query processing and data retrieval; (ii) **Scoping CQs:** Such questions may help to define the domain, but are not used for querying. These require specialised handling to aid the definition of a domain; (iii) **Verified CQs:** These CQs can be directly queried and can serve as benchmarks for system capabilities.

Each of these main tasks includes subtasks, where Table 4 indicates their relevance to either one or both of the authoring or generating CQs approaches. The subtasks for each main task, along with their description, are as follows:

1 **Poor/Incorrect Requirements:**
   (a) Linguistic Perspectives:
       i. *Identify Ambiguous Questions:* Create a repository of CQ examples that exhibit ambiguity in wording or context.
       ii. *Develop Clarity Guidelines:* Formulate standards/templates to help rephrase ambiguous questions for improved clarity and specificity.
   (b) Question Types:
       i. *Classify Question Types:* Systematically categorise CQs into types such as narrative, factual, or descriptive, and assess their suitability in different contexts.
       ii. *Evaluate Contextual Appropriateness:* Develop criteria to measure the effectiveness of question types within their intended contexts.
   (c) Domain Knowledge:
       i. *Align Questions with Domain Relevance:* Establish a review process to ensure questions are pertinent wrt the relevant domain knowledge.
       ii. *Refine Focus Through Filtering:* Implement a mechanism to exclude questions that, while correct, are irrelevant to the task at hand.
   (d) Incorrectness:
       i. *Fact-Check Information:* Set up a robust protocol for verifying the factual accuracy of CQs.
       ii. *Correct Erroneous Inputs:* Introduce a correction mechanism for adjusting factually incorrect CQs.
2 **Scoping CQs:**
   (a) *Catalogue Scoping CQs:* Document all CQs that contribute to defining the scope of the information domain.
   (b) *Analyse for Domain Contribution:* Analyse how these CQs help in shaping the understanding of the domain.
   (c) *Integrate into Information Architecture:* Develop strategies to utilise scoping CQs for enhancing the structure of information repositories.
3 **Verified CQs:**
   (a) *Database of Verified CQs:* Maintain an updated list of CQs that can be directly transformed into SPARQL queries.

(b) *Transformation into Queries:* Convert verified CQs into effective SPARQL queries.

(c) *Testing and Validation:* Conduct rigorous testing to ensure the queries retrieve accurate and relevant data.

(d) *Documentation and Examples:* Create detailed documentation and examples of successful CQ transformations for training and reference.

CQsBEN aims to refine the process of handling CQs, ensuring that they adhere to practices for developing high-quality CQs, which are clear, relevant, and effectively transformable into queries. By addressing each category with specific subtasks, we can significantly improve the accuracy and efficiency of CQs engineering approaches.

Each task and subtask has its own evaluation measure to assess an approach's performance related to these tasks. These evaluation measures include: (i) **Subjective Evaluation:** This will be related to tasks that identify poor CQs, evaluating their relevance and accuracy; (ii) **Similarity Matching:** Techniques such as SBERT will be employed for calculating performance metrics-precision, recall, and F1-score-for tasks that involve identifying scoping CQs;(iii) **Testing for Verified CQs:** Involves similarity matching for the CQs and unit/acceptance

Table 4. Summary of Tasks, Subtasks, CQs Engineering Approaches, and Evaluation Measures.

Task	Subtask	Automation Approach	Evaluation Measure
Poor/Incorrect Requirements	Linguistic Perspectives	*Manual/Semi-Automatic*	Subjective
	Question Types	*Automatic*	Subjective
	Domain Knowledge	*Automatic*	Subjective
	Incorrectness	*Automatic*	Subjective
Scoping CQs	Catalogue	*Automatic*	Similarity
	Scoping CQs		Matching
	Analyse for Domain	*Automatic*	Subjective
	Contribution		
	Integrate into	*Automatic*	Subjective
	Information		
	Architecture		
Verified CQs	Database of	*Manual/Semi-Automatic*	Subjective
	Verified CQs	*Automatic*	Similarity
			Matching
	Transformation	*Manual/Semi-Automatic*	Subjective
	into Queries	*Automatic*	Verified SPARQL
	Testing and Validation	*Manual/Semi-Automatic*	Verified SPARQL
		Automatic	Verified SPARQL
	Documentation	*Manual/Semi-Automatic*	Subjective
	and Examples	*Automatic*	Subjective

testing for the corresponding SPARQL queries. Table 4 specifies the evaluation measures for each subtask identified earlier.

It is not trivial to collect a dataset for CQsBEN as open-sourced repository data often lack essential components, especially the design documents and testing programs. We identify two main implementation steps to organise the process; (i) **Gathering all Published Requirements:** Collecting and documenting all existing requirements related to tasks such as CORAL [25], the CQs dataset [54], along with individual ontologies that have published their CQs; (ii) **Categorisation According to Tasks:** organising the requirements based on their respective tasks to streamline the benchmark design process. Table 5 displays the initial set of proposed datasets, individual ontologies, along with the number of CQs. However, this is not the final set, as we continue to communicate with developers to encourage them to share their CQs and expand this list.

Table 5. Datasets/ ontologies and their Corresponding Number of CQs.

Datasets/Ontologies	Num. CQs
CORAL [25]	834
CQs dataset [54]	234
WDV-CQ [17]	1786
DOREMUS [1]	218
NORIA-O [52]	55
Odeuropa [38]	13
Polifonia [12]	247

6   Conclusion

In this paper, we have conducted a *Scoping Study* on CQ engineering approaches, exploring the main dimension(s) that characterise approaches for CQ formulation, as well as considering differing levels of automation assumed by the approaches studied. Furthermore, we have examined the resources used by approaches that formulate CQs, and considered the validation criteria used to assess the relevance of the CQs with respect to their corresponding ontology. By doing so, this allows researchers to align their work with established approaches and have assisted ontology developers in selecting and adapting existing approaches to meet their specific needs. Our findings offer a structured overview that not only encapsulates the diversity of existing approaches but also clarifies their application contexts and limitations.

Furthermore, the study reports on the current state of the art in CQ engineering approaches, with a focus on the level of automation assumed, as well as the types of approach for which different types of validation criteria are meaningful. This level of reporting is crucial for positioning new methods relative to the state-of-the-art, thereby facilitating innovation and refinement in the field. Through this analysis, it has become evident that while fully automated approaches have emerged as a recent trend, designed to overcome the limitations of earlier methods. However, the field is still in its infancy, and as yet no single approach has emerged as consistently superior. This underscores the necessity for a standardised framework to assess these emerging methodologies.

Addressing this need, we have proposed a benchmark, CQsBEN, for CQs that includes detailed task specifications and evaluation criteria. It is designed to serve as a robust standard for assessing and comparing the effectiveness of different CQ engineering approaches, and should enable not only the comparison across the existing methods, but also provide a foundation for evaluating upcoming initiatives in CQ engineering.

References

1. Achichi, M., Lisena, P., Todorov, K., Troncy, R., Delahousse, J.: DOREMUS: a graph of linked musical works. In: Vrandečić, D., et al. (eds.) ISWC 2018. LNCS, vol. 11137, pp. 3–19. Springer, Cham (2018). https://doi.org/10.1007/978-3-030-00668-6_1

2. Alharbi, R.: Assessing candidate ontologies for reuse. In: Proceedings of the Doctoral Consortium at ISWC 2021 (ISWC-DC), pp. 65–72 (2021). https://api.semanticscholar.org/CorpusID:244895203

3. Alharbi, R., Tamma, V., Grasso, F.: Characterising the gap between theory and practice of ontology reuse. In: Proceedings of the 11th on Knowledge Capture Conference. K-CAP '21, pp. 217–224. Association for Computing Machinery (2021)

4. Alharbi, R., Tamma, V., Grasso, F.: Requirement-based methodological steps to identify ontologies for reuse. In: Islam, S., Sturm, A. (eds.) CAiSE 2024. LNCS, vol. 520, pp. 64–72. Springer, Cham (2024). https://doi.org/10.1007/978-3-031-61000-4_8

5. Alharbi, R., Tamma, V., Grasso, F., Payne, T.: An experiment in retrofitting competency questions for existing ontologies (2023). https://arxiv.org/abs/2311.05662

6. Alharbi, R., Tamma, V., Grasso, F., Payne, T.: An experiment in retrofitting competency questions for existing ontologies. In: Proceedings of the 39th ACM/SIGAPP Symposium on Applied Computing. SAC '24, pp. 1650–1658. Association for Computing Machinery (2024)

7. Alharbi, R., Tamma, V., Grasso, F., Payne, T.: The role of Generative AI in competency question retrofitting. In: Extended Semantic Web Conference, ESWC2024. Hersonissos, Greece (2024)

8. Antia, M., Keet, C.M.: Assessing and enhancing bottom-up CNL design for competency questions for ontologies. In: Proc. of the Seventh International Workshop on Controlled Natural Language (CNL 2020/21), pp. 1–11. Association for Computational Linguistics (ACL) (2021)

9. Antia, M., Keet, C.M.: Automating the generation of competency questions for ontologies with AgOCQs. In: Ortiz-Rodriguez, F., Villazón-Terrazas, B., Tiwari, S., Bobed, C. (eds.) KGSWC 2023. LNCS, vol. 14382, pp. 213–227. Springer, Cham (2023). https://doi.org/10.1007/978-3-031-47745-4_16

10. Arksey, H., O'Malley, L.: Scoping studies: towards a methodological framework. Int. J. Soc. Res. Methodol. 8(1), 19–32 (2005)

11. Azzi, S., Assi, A., Gagnon, S.: Scoring ontologies for reuse: an approach for fitting semantic requirements. In: Garoufallou, E., Vlachidis, A. (eds.) MTSR 2022. LNCS, vol. 1789, pp. 203–208. Springer, Cham (2023). https://doi.org/10.1007/978-3-031-39141-5_17

12. de Berardinis, J., et al.: The Polifonia ontology network: building a semantic backbone for musical heritage. In: Payne, T.R., et al. (eds.) ISWC 2023. LNCS, vol. 14266, pp. 302–322. Springer, Cham (2023). https://doi.org/10.1007/978-3-031-47243-5_17

13. Bezerra, C., Freitas, F.: Verifying description logic ontologies based on competency questions and unit testing. In: Proceedings of the IX Seminar on Ontology Research and I Doctoral and Masters Consortium on Ontologies, vol. 1908, pp. 159–164 (2017)

14. Bezerra, C., Freitas, F.: Verifying description logic ontologies based on competency questions and unit testing. In: ONTOBRAS, pp. 159–164 (2017)

15. Bezerra, C., Santana, F., Freitas, F.: CQChecker: a tool to check ontologies in OWL-DL using competency questions written in controlled natural language. Learn. Nonlinear Models 12(2), 115–129 (2014)

16. Bresciani, P., Perini, A., Giorgini, P., Giunchiglia, F., Mylopoulos, J.: Tropos: an agent-oriented software development methodology. Auton. Agent. Multi-agent Syst. 8(3), 203–236 (2004)

17. Ciroku, F., de Berardinis, J., Kim, J., Meroño-Peñuela, A., Presutti, V., Simperl, E.: Revont: reverse engineering of competency questions from knowledge graphs via language models. J. Web Semant. 82, 100822 (2024)

18. Dasiopoulou, S., Meditskos, G., Efstathiou, V.: Semantic knowledge structures and representation. Technical report D5.1, FP7-288199 Dem@Care: Dementia Ambient Care: Multi-sensing Monitoring for Intelligence Remote Management and Decision Support (2012). http://www.demcare.eu/downloads/D5.1SemanticKnowledgeStructures_andRepresentation.pdf

19. Debellis, M.: A practical guide to building owl ontologies using protégé 5.5 and plugins (2021). https://www.researchgate.net/publication/351037551_A_Practical_Guide_to_Building_OWL_Ontologies_Using_Protege_55_and_Plugins

20. Denaux, R., Dimitrova, V., Cohn, A.G., Dolbear, C., Hart, G.: Rabbit to OWL: ontology authoring with a CNL-based tool. In: Fuchs, N.E. (ed.) CNL 2009. LNCS (LNAI), vol. 5972, pp. 246–264. Springer, Heidelberg (2010). https://doi.org/10.1007/978-3-642-14418-9_15

21. Dennis, M., van Deemter, K., Dell'Aglio, D., Pan, J.Z.: Computing authoring tests from competency questions: experimental validation. In: d'Amato, C., et al. (eds.) ISWC 2017. LNCS, vol. 10587, pp. 243–259. Springer, Cham (2017). https://doi.org/10.1007/978-3-319-68288-4_15

22. Espinoza-Arias, P., Garijo, D., Corcho, O.: Extending ontology engineering practices to facilitate application development. In: Corcho, O., Hollink, L., Kutz, O., Troquard, N., Ekaputra, F.J. (eds.) EKAW 2022. LNCS, vol. 13514, pp. 19–35. Springer, Cham (2022). https://doi.org/10.1007/978-3-031-17105-5_2

23. Fathallah, N., Das, A., De Giorgis, S., Poltronieri, A., Haase, P., Kovriguina, L.: Neon-GPT: a large language model-powered pipeline for ontology learning. In: Extended Semantic Web Conference, ESWC2024. Hersonissos, Greece (2024)

24. Fernandes, P.C.B., Guizzardi, R.S., Guizzardi, G.: Using goal modelling to capture competency questions in ontology-based systems. J. Inf. Data Manag. 2(3), 527 (2011)

25. Fernández-Izquierdo, A., Poveda-Villalón, M., García-Castro, R.: CORAL: a corpus of ontological requirements annotated with lexico-syntactic patterns. In: Hitzler, P., et al. (eds.) ESWC 2019. LNCS, vol. 11503, pp. 443–458. Springer, Cham (2019). https://doi.org/10.1007/978-3-030-21348-0_29

26. Funk, M., Hosemann, S., Jung, J.C., Lutz, C.: Towards ontology construction with language models. In: Proceedings of the KBC-LM'23: Knowledge Base Construction from Pre-trained Language Models Workshop at ISWC. CEUR Workshop Proceedings (2023)

27. Gangemi, A., Lippolis, A.S., Lodi, G., Nuzzolese, A.G.: Automatically drafting ontologies from competency questions with frodo. Stud. Semant. Web 55, 107–121 (2022)

28. Gao, T., Fodor, P., Kifer, M.: High accuracy question answering via hybrid controlled natural language. In: 2018 IEEE/WIC/ACM International Conference on Web Intelligence (WI), pp. 17–24 (2018). https://doi.org/10.1109/WI.2018.0-112

29. Gao, T., Fodor, P., Kifer, M.: Knowledge authoring for rule-based reasoning. In: Panetto, H., Debruyne, C., Proper, H.A., Ardagna, C.A., Roman, D., Meersman, R. (eds.) OTM 2018. LNCS, vol. 11230, pp. 461–480. Springer, Cham (2018). https://doi.org/10.1007/978-3-030-02671-4_28

30. Grüninger, M., Fox, M.S.: The role of competency questions in enterprise engineering. In: Rolstadås, A. (ed.) Benchmarking — Theory and Practice. IAICT, pp. 22–31. Springer, Boston, MA (1995). https://doi.org/10.1007/978-0-387-34847-6_3

31. Keet, C.M., Ławrynowicz, A.: Test-driven development of ontologies. In: Sack, H., Blomqvist, E., d'Aquin, M., Ghidini, C., Ponzetto, S.P., Lange, C. (eds.) ESWC 2016. LNCS, vol. 9678, pp. 642–657. Springer, Cham (2016). https://doi.org/10.1007/978-3-319-34129-3_39

32. Keet, C.M., Mahlaza, Z., Antia, M.-J.: CLaRO: a controlled language for authoring competency questions. In: Garoufallou, E., Fallucchi, F., William De Luca, E. (eds.) MTSR 2019. CCIS, vol. 1057, pp. 3–15. Springer, Cham (2019). https://doi.org/10.1007/978-3-030-36599-8_1

33. Kim, H.M., Fox, M.S., Sengupta, A.: How to build enterprise data models to achieve compliance to standards or regulatory requirements (and share data). J. Assoc. Inf. Syst. 8, 105–128 (2007)

34. Kitchenham, B.A., Charters, S.: Guidelines for performing systematic literature reviews in software engineering. Technical report. EBSE-2007-001, Keele University and Durham University (2007)

35. Kolbe, N., Vandenbussche, P.Y., Kubler, S., Le Traon, Y.: Lovbench: ontology ranking benchmark. In: Proceedings of the Web Conference 2020. WWW '20, pp. 1750–1760. Association for Computing Machinery (2020)

36. Kuhn, T.: A survey and classification of controlled natural languages. Comput. Linguist. 40(1), 121–170 (2014). https://doi.org/10.1162/COLI_a_00168

37. Li, Q., Li, S., Zhang, S., Hu, J., Hu, J.: A review of text corpus-based tourism big data mining. Appl. Sci. 9 (2019). https://doi.org/10.3390/app9163300

38. Lisena, P., et al.: Capturing the "semantics of smell": the Odeuropa data model for olfactory heritage information. In: Groth, P., et al. (eds.) ESWC 2022. LNCS, vol. 13261, pp. 387–405. Springer, Cham (2022). https://doi.org/10.1007/978-3-031-06981-9_23

39. Mays, N., Roberts, E., Popay, J.: Synthesising research evidence. In: Fulop, N., Allen, P., Clarke, A., Black, N. (eds.) Studying the Organisation and Delivery of Health Services: Research Methods. Routledge, London (2001)

40. Monfardini, G.K.Q., Salamon, J.S., Barcellos, M.P.: Use of competency questions in ontology engineering: a survey. In: Almeida, J.P.A., Borbinha, J., Guizzardi, G., Link, S., Zdravkovic, J. (eds.) ER 2023. LNCS, vol. 14320, pp. 45–64. Springer, Cham (2023). https://doi.org/10.1007/978-3-031-47262-6_3

41. Namgoong, H., Kim, H.: Ontology-based controlled natural language editor using CFG with lexical dependency. In: Aberer, K., et al. (eds.) ISWC ASWC 2007. LNCS, vol. 4825, pp. 353–366. Springer, Heidelberg (2007). https://doi.org/10.1007/978-3-540-76298-0_26

42. Noy, N.F., McGuinness, D.L.: Ontology development 101: a guide to creating your first ontology. Technical report, Stanford Knowledge Systems Laboratory Technical report KSL-01-05 (2001)

43. Poveda-Villalón, M., Fernández-Izquierdo, A., Fernández-López, M., García-Castro, R.: LOT: an industrial oriented ontology engineering framework. Eng. Appl. Artif. Intell. **111**, 104755 (2022)

44. Presutti, V., Daga, E., Gangemi, A., Blomqvist, E.: Extreme design with content ontology design patterns. In: Proceedings of the 2009 International Conference on Ontology Patterns, vol. 516, p. 83–97 (2009)

45. Rao, L., Reichgelt, H., Osei-Bryson, K.: Knowledge elicitation techniques for deriving competency questions for ontologies. In: Proceedings of the Tenth International Conference on Enterprise Information Systems (ICEIS 2008), vol. ISAS-2, pp. 105–110. Barcelona, Spain (2008)

46. Rebboud, Y., Tailhardat, L., Lisena, P., Troncy, R.: Can LLMs generate competency questions? In: Extended Semantic Web Conference, ESWC2024. Hersonissos, Greece (2024)

47. Reimers, N., Gurevych, I.: Sentence-BERT: sentence embeddings using Siamese BERT-networks. In: Proc. of the 2019 Conference on Empirical Methods in Natural Language Proceedings and the 9th International Joint Conference on Natural Language Proceedings (EMNLP-IJCNLP), pp. 3982–3992. Association for Computational Linguistics (2019)

48. Ren, Y., Parvizi, A., Mellish, C., Pan, J.Z., van Deemter, K., Stevens, R.: Towards competency question-driven ontology authoring. In: Presutti, V., d'Amato, C., Gandon, F., d'Aquin, M., Staab, S., Tordai, A. (eds.) ESWC 2014. LNCS, vol. 8465, pp. 752–767. Springer, Cham (2014). https://doi.org/10.1007/978-3-319-07443-6_50

49. Sequeda, J.F., Briggs, W.J., Miranker, D.P., Heideman, W.P.: A pay-as-you-go methodology to design and build enterprise knowledge graphs from relational databases. In: Ghidini, C., et al. (eds.) ISWC 2019. LNCS, vol. 11779, pp. 526–545. Springer, Cham (2019). https://doi.org/10.1007/978-3-030-30796-7_32

50. Sitthisak, O., Gilbert, L., Davis, H.C.: Transforming a competency model to parameterised questions in assessment. In: Cordeiro, J., Hammoudi, S., Filipe, J. (eds.) WEBIST 2008. LNBIP, vol. 18, pp. 390–403. Springer, Heidelberg (2009). https://doi.org/10.1007/978-3-642-01344-7_29

51. Suárez-Figueroa, M.C., Gómez-Pérez, A., Fernández-López, M.: The neon methodology framework: a scenario-based methodology for ontology development. Appl. Ontol. **10**(2), 107–145 (2015)

52. Tailhardat, L., Chabot, Y., Troncy, R.: NORIA-O: an ontology for anomaly detection and incident management in ICT systems. In: Meroño Peñuela, A., et al. (eds.) ESWC 2024. LNCS, vol. 14665, pp. 21–39. Springer, Cham (2024). https://doi.org/10.1007/978-3-031-60635-9_2

53. Tseng, Y.H., Ho, Z.P., Yang, K.S., Chen, C.C.: Mining term networks from text collections for crime investigation. Expert Syst. Appl. **39**(11), 10082–10090 (2012)
54. Wiśniewski, D., Potoniec, J., Ławrynowicz, A., Keet, C.M.: Analysis of ontology competency questions and their formalizations in SPARQL-OWL. J. Web Semant. **59**, 100534 (2019)

A Generic Framework to Better Understand and Compare FAIRness Measures

Philippe Lamarre[1](✉), Jennie Andersen[1], Alban Gaignard[2],
and Sylvie Cazalens[1]

[1] INSA Lyon, CNRS, Ecole Centrale de Lyon, Universite Claude Bernard Lyon 1,
Université Lumière Lyon 2, LIRIS, UMR5205, 69621 Villeurbanne, France
{philippe.lamarre,jennie.andersen,sylvie.cazalens}@insa-lyon.fr
[2] Nantes Université, CNRS, INSERM, l'institut du thorax, 44000 Nantes, France
alban.gaignard@univ-nantes.fr

Abstract. In recent years, the adoption of the FAIR principles has achieved notable success. This progress has led to the development of numerous assessment tools originating from diverse fields of application, thus addressing diverse object types, interpretations and implementations. Given the plethora of proposals available, it is crucial for users to precisely understand these measures, compare them effectively, make informed choices, and accurately interpret the obtained measurements. To meet these needs, we propose a model to formally represent and analyze measures. Besides the benefit of homogenization, it allows for the formal definition of three characteristic quantities: coverage, granularity and impact. Our experiments show how these quantities (i) contribute to explain different scores obtained by digital artifacts using two different state-of-the-art assessment engines, (ii) enable a comparative study of different FAIRness measures, independently of any digital artifact.

Keywords: FAIR data · FAIR assessments · Trustworthiness

1 Introduction

These recent years, the FAIR principles [40] have been increasingly adopted to assess the Findability, Accessibility, Interoperability and Reusability of their digital resources. Due to this widespread adoption, they have been specialized or even extended to meet the needs of very different scientific communities. For instance, the RDA FAIR4RS working group derived these principles to specifically address research software [5], typically targeting their usability and reusability within other software. These principles have also been adapted in the context of AI [24], ontology development and semantic artifacts [9,32], data analysis workflows [38].

To support people in these assessment tasks, numerous tools have been developed, originating from diverse fields of application. They may address different types of objects, stem from different interpretations or implementations of the principles. Consequently, besides the varied terminology, one can notice that sometimes sub-principles are skipped, and how indicators are expressed or implemented changes from tool to tool.

M. Alam et al. (Eds.): EKAW 2024, LNAI 15370, pp. 291–308, 2025.
https://doi.org/10.1007/978-3-031-77792-9_18

In addition, when scores are provided, they may not range in the same interval; the functions used to aggregate the scores, as well as the weights assigned to the various indicators, may also differ. In some way, this diversity is understandable as the FAIR principles are not restrictive guidelines.

However, with numerous and diverse available tools, a user may be faced to different questions such as: "Why does my resource get such a score with this tool?", "Why does it receive a higher score with tool A compared to tool B?", "What are the differences between tools A and B?", "Which one fits my needs better?". To answer such questions, some studies already have proposed comparisons of tools, based on metrics used, on characteristics of the tools themselves, on the measurements obtained with numerous datasets [7,19,27,29,34,35,42]. These studies are clearly useful, but it is still a challenge to interpret scores, understand and compare FAIRness measures, and make informed choices.

This article aims to tackle this challenge based on two main points. First we observe that the FAIR principles alone do not offer a sufficient framework to take into account the multiplicity of variations found from a measure to another. A generic model enabling some homogenization in representing the measures could help. Second, to our knowledge, there are no formally defined characteristic quantities to reflect the salient features of a given measure, that could help both its understanding and its comparison with other ones.

The remainder of the paper is organized as follows. In Sect. 3, we propose a formal model based on the tree structure of the FAIR principles, coping with the variability of FAIRness measures. Section 4 outlines the methodology for representing a given measure within this model, using FAIR-checker and F-UJI as examples. Section 5 introduces the coverage rate, granularity and impact quantities facilitating the comparison of FAIRness measures. We show with experimental results in Sect. 6 how our framework can be used to provide better insights on diverging FAIR assessments. We propose concluding remarks in Sect. 7.

2   Related Work

The FAIR principles were published in 2016 [40] as general guidelines for the publication of digital resources to make them Findable, Accessible, Interoperable, and Reusable. Since then, numerous methods and tools have been developed to assess FAIRness, each with its own interpretation of the principles. This variety of interpretations and of types of assessment methods (automated tools, checklists, self-assessment questionnaires, etc.) makes them difficult to compare. To address this problem, the FAIR data maturity model [4] proposes a set of indicators that express measurable aspects of the FAIR principles and on which future evaluation tools can be based. Although this initiative proposes consensual definitions adopted by large multi-disciplinary communities, it is still challenging to compare FAIRness measures in their whole, from implementations to evaluation results.

Several comparisons have already been conducted. Slamkov *et al.* [34] compare five questionnaires and checklists: ARDC's tool [3], CSIRO's tool [10], SATIFYD [11], EUDAT checklist [25], and the SHARC grid [12], according to their main characteristics (type, documentation, dependency to a specific repository, automated score computation) and the results obtained on seven datasets.

Another comparison [35] focuses on three automated tools: F-UJI [13], FAIR Evaluation Service[1] [41] and FAIR-Checker [17]. They are all compared based on distinguishing aspects (documentation, availability of the code, format and log of the results...). Then the author focus on F-UJI and the FAIR Evaluation Service to compare their metrics/indicators in detail, first by focusing on their expression in natural language and then on the experimental results obtained on three datasets. Since then, both F-UJI and FAIR-Checker have changed. Wilkinson *et al.* [42] highlight that some of the differences between the results of F-UJI and the FAIR Evaluation Service are not due to the metrics themselves, but to their different ways of collecting the metadata to be evaluated. This applies to all automated tools and contributes to the difficulty of comparison.

Krans *et al.* [27] provide a detailed qualitative comparison of ten tools, both questionnaires and automated tools. They mostly focus on their prerequisites, the ease and effort to use them, the type and quality of the outputs. They also test them on two datasets and observe a large variability in the FAIRness scores obtained, thus showing the difficulty in interpreting the questions in questionnaires and the differences in the implementations of the principles for automated tools. Candela *et al.* [7] provide an overview of twenty FAIR assessment tools, analyzing distinguishing features such as target, adaptability, methodology... They particularly document the divergence between declared intents of metrics and what is actually assessed. Some tools have also been compared in the context of domain-specific FAIRness assessment [19]. In particular, they compare the overall FAIRness score obtained on some datasets by the FAIR Evaluation Service, F-UJI, FAIRshake [8], and a self-assessment based on the FAIR data maturity model. They observe that the tools obtain scores close enough to consider that they give similar levels of FAIRness, especially if of the same kind (questionnaire or automated tool).

Recently, Moser *et al.* [29] propose a brief comparison of the FAIRness measures rather than the tools based on them. They focus on the FAIR Maturity Indicators [41], the FAIR Data Maturity Model [4], FAIRsFAIR metrics used in F-UJI [13] and FAIR metrics for EOSC [20]. They compare their numbers of indicators, and their structures: some metrics define indicators for intermediate principles (A1 and R1) while others do not. They also highlight that some of them give different importance to their indicators.

The main objective of our proposal is closer to this latter work, while we aim to push further the comparison of the scores and of the importance of each element in the measures. We propose an innovative approach with a generic model and formal definitions of characteristic quantities.

---

[1] With the metric collection: "All Maturity Indicator Tests as of May 8, 2019".

3   A Generic Model to Represent FAIRness Measures

Our aim is to define a simple unifying framework expressive enough to represent and compare as many measures as possible. We focus on the importance they give to the FAIR principles and sub-principles. In this view our analysis of FAIRness measures has identified three notions to describe their tree-like organization: principles, indicators and implementations. As the ways scores are computed vary a lot, we propose a generic representation. It ensures that if a score is computed for a given principle or sub-principle by a measure, the score computed through its representation in the model is the same. Hence, we propose to represent a measure of FAIRness as a tuple

$$\mathcal{M} = (V, E, \text{FAIR}, \Diamond, w, v_{max}, D) \tag{1}$$

where elements (V, E, FAIR) refer to the structure and $(\Diamond, w, v_{max}, D)$ to the score computation. They are detailed in the following.

3.1   Modeling of the Structure

Obviously, the different measures of FAIRness rely on the hierarchy of the FAIR principles [40]. Here, the term *principle* is used in a broad sense, i.e. it refers to both principles and sub-principles. We represent them as a tree, illustrated with the tree of ellipses in Fig. 1. Its sets of nodes is denoted $P = \{\text{FAIR}, \text{F}, \text{F1}, \text{F2}, \text{F3}, \text{F4}, \text{A}, \text{A1}, \text{A1.1} \ldots\}$ and its edges, $E_P = \{(\text{FAIR}, \text{F}), (\text{F}, \text{F1}), (\text{F}, \text{F2}) \ldots\}$.

Then, these principles are refined into several measurable criteria, expressed in natural language, which we call indicators. An indicator is named a "metric" in FAIR-Checker [17] and F-UJI [13], a "FAIRness assessment question" in O'FAIRe [1], a "maturity indicator test" in the FAIR Evaluation Service [41], or a "check" in FOOPS! [18]. Given $\mathcal{M}$, a measure of FAIRness, we denote $I(\mathcal{M})$ its set of indicators. Finally, in the case of an automated tool, indicators are associated to implementations, belonging to set $Imp(\mathcal{M})$, allowing a resource to be evaluated on them.

Hence, the structure of a measure $\mathcal{M}$ is represented by a directed rooted tree (V, E, FAIR), simply adding indicators and implementations to the FAIR principle tree, where:

- FAIR is the root of $\mathcal{M}$ and of the FAIR principles tree;
- V are the nodes of the tree, such that $V = P \cup I(\mathcal{M}) \cup Imp(\mathcal{M})$;
- E are the edges, where $E \subseteq E_P \cup (P \times I(\mathcal{M})) \cup (I(\mathcal{M}) \times Imp(\mathcal{M}))$;
- there can only be one implementation per indicator.

In this model, an implementation can be linked only to an indicator, which in turn can be linked only to a principle, any principle, not just to the leaves of the FAIR principles tree. This is intended to simplify the representation and to ease understanding and comparison of the measures. Methods for representing existing measures that do not comply with these constraints (e.g. with hierarchies of indicators, or additional sub-principles.) are discussed in Sect. 4.

To manipulate a measure such defined, we introduce the usual notions of children and of descendants: Let $n \in V$ be a node of the tree, then $\text{children}_{\mathcal{M}}(n)$ is the set of children of n in $\mathcal{M}$, and $\text{desc}_{\mathcal{M}}(n)$ is the set of descendants of n in $\mathcal{M}$.

3.2   Computing the Scores

Intuitively, we consider that, given some resource d to evaluate, the score at a node is obtained by a weighted aggregation of the scores obtained by its children. The score of an indicator comes directly from executing its implementation imp for d, so we assume a family of evaluation functions, eval_d such that $\text{eval}_d(imp)$ denotes the obtained score. In this view, we detail the elements $(\Diamond, w, v_{max}, D)$ of the representation of a measure.

- $\Diamond : \mathcal{P}(\mathbb{R}^+ \times \mathbb{R}^+) \rightarrow \mathbb{R}^+$ is an aggregation function, producing a new (aggregated) score from some pairs $(score, weight)$. It can be either a weighted sum (noted SUM) or a weighted average (noted AVG). It returns 0 in case of an empty set.
- $w : V \rightarrow \mathbb{R}^+$ is a weighting function. Given a node n, $w(n)$ represents the importance of n with respect to its siblings in the hierarchy.
- $v_{max} : Imp(\mathcal{M}) \cup I \rightarrow \mathbb{R}^+$ is a function where, with i being an implementation or and indicator, $v_{max}(i)$ is the maximum value that can be obtained for i whatever the resource. For example, in F-UJI, $v_{max}(\text{R1-01MD}) = 4$.
- D is a function such that, given any implementation imp, $D(imp)$ is a set of expressions of the form: $\forall d, \text{eval}_d(imp) \geqslant v \Rightarrow \text{eval}_d(imp') \geqslant v'$. Intuitively, for any resource, if executing imp results in a value above v, one is warranted that implementation imp' results in a value above some minimal value v'. This is what we call a dependency between imp and imp'.

Keeping these notations, the computation of the score obtained by some resource d can then be expressed as follows.

$$\text{score}(\mathcal{M}, d) = \text{score}(\mathcal{M}, \text{eval}_d, \text{FAIR}) \tag{2}$$

Function score, for a given node $n \in V$, is expressed for any function eval by:

$$\text{score}(\mathcal{M}, \text{eval}, n) = \begin{cases} \text{eval}(n) & \text{if } n \in Imp(\mathcal{M}) \\ \underset{n' \in \text{children}_{\mathcal{M}}(n)}{\Diamond} (w(n'), \text{score}(\mathcal{M}, \text{eval}, n')) & \text{otherwise} \end{cases} \tag{3}$$

4   Strategies for Representing Measures in Our Approach

Some measures fit the proposed model directly. For others, several strategies may be applied. We explain and illustrate them with several FAIRness measures or tools, five automated tools and two questionnaires. Two tools can be used to assess any digital resources: FAIR-Checker [15,17], and FAIR Evaluation Service [2,41] (the indicators from the collection "All Maturity Indicator Tests as of May 8, 2019"). One tool focuses on datasets: F-UJI [13,14]. Two tools are designed specifically for ontologies: FOOPS! [16,18] and O'FAIRe [1,30]. We also consider two online questionnaires with ARDC's questionnaire [3] and SATIFYD [11], allowing to make a self-assessment of a digital resource.

All these tools can be used online, and, except for O'FAIRe, they allow to assess any digital resource. We rely preferably on the online tools of the measures since they are more up-to-date and since it is sometimes easier to understand how the global score is computed. We first discuss the representation of the structure, then the computation of the scores.

4.1   Representing the Structure of Measures

The backbone of the structure chosen to represent measures is the full FAIR principles tree. In addition, several indicators can be attached to a same principle, an indicator can only be attached to a single principle, and an implementation to a single indicator.

For measures that fully or partially use the FAIR principles tree without additional principles, the indicators remain linked to principles just as they are. Notice that the entire FAIR principles tree is kept in the representation of the measure. This applies for example to particular questionnaires which are not detailed according to each sub-principle, but only according to F, A, I and R. Some measures define and use new sub-principles with associated indicators. These sub-principles are not represented but their indicators are considered and linked to their closest parent in the FAIR principles tree.

Some measures have several level of refinement for their indicators. For instance, Wilkinson *et al.* [41] propose a level of "maturity indicator", and then of "maturity indicator test", F-UJI [13] has a level of "metrics" and then of "tests".

To map these measures to our model, we reduce them to a single level, focusing on the one highlighted by their measuring tools, the one for which a score is clearly given. Hence, in F-UJI, we keep their "metrics" as indicators, and in [41] the "maturity indicator test" since it is the only one to appear clearly in their online tool.

Notice that, even though some of the particularities of some measures are not reproduced in their representation within the generic model, the strategies maintain the characteristic elements necessary for understanding the importance they give to the FAIR principles. This is also the case for their score calculation which we explain in the following.

4.2   Representing the Scores Computations

In our model, the scoring method is a weighted aggregation of the evaluations and scores obtained all along the tree structure, as explained in Sect. 3.2. We seek to represent the scoring of existing measures in this way, ensuring that if a score is computed for a given (sub-)principle by the measure, the score computed through its representation in the generic model is the same.

Measures that compute a score already have maximum values $v_{max}(i)$ for implementations and indicators, an aggregation function $\Diamond$, and possibly a set D of dependencies between implementations. These elements are all kept as is in the representation of the measure. Only the weights remain to be defined.

We first explain how to determine the weight of each implementation and indicator, and then how to deduce the weights on the other nodes.

First, all the implementations get a weight equal to 1. Then, the weights of the indicators may be explicit or implicit in the initial expression of a measure. When explicit, such as in O'FAIRe, where the weights correspond to their "credits", they are kept as is in the representation. When implicit, such as in FOOPS!, they are set to 1. Some approaches, for example the FAIR Evaluation Service, do not compute any score. This means that each user is supposed to choose how to use the results obtained for implementations and indicators. We assume they all have a weight equal to 1 and that the score is computed with a weighted average.

It is now possible to weight the other nodes of the representation. If the measure uses a sum to compute the overall score, the weight of other nodes is set to 1. If the measure computes the score with a (weighted) average, then the weight assigned to a node is the sum of the weights of its children. According to this, a principle to which no indicator is attached gets a weight equal to 0 (respectively 1) in case of a weighted average (respectively a sum).

For questionnaire approaches, where the score is usually computed first on F, A, I, and R, and then globally as an average of these four scores, the weights of the main principles F, A, I, and R are 1, and the weights of the other principles are 0, since the indicators are not detailed according to them.

4.3   Examples of Representation in the Generic Model

FAIR-Checker computes the score of a resource by averaging the values obtained for the indicators. It implicitly uses weights, so we set them to 1 for all the indicators. Then, the weight of a node is assigned the sum of the weights of its children. This is illustrated in Fig. 1a: an example of dependency in FAIR-checker is that the implementation of the indicator *I-I1* associated to principle I1 delegates evaluation to the one implementing *F2A* associated to F2. Hence, in Fig. 1a, we represent these full delegations by equalities.

F-UJI computes the score using an unweighted sum. Its representation is illustrated in Fig. 1b[2]. Notice that in F-UJI, value v_{max} varies from 1 to 4, thus assigning different importance to indicators. Dependencies between implementations, if any, are not taken into account.

---

[2] According to https://www.f-uji.net/index.php?action=methods, assessment details of *FsF_A2_01M* are excluded from the F-UJI implementation.

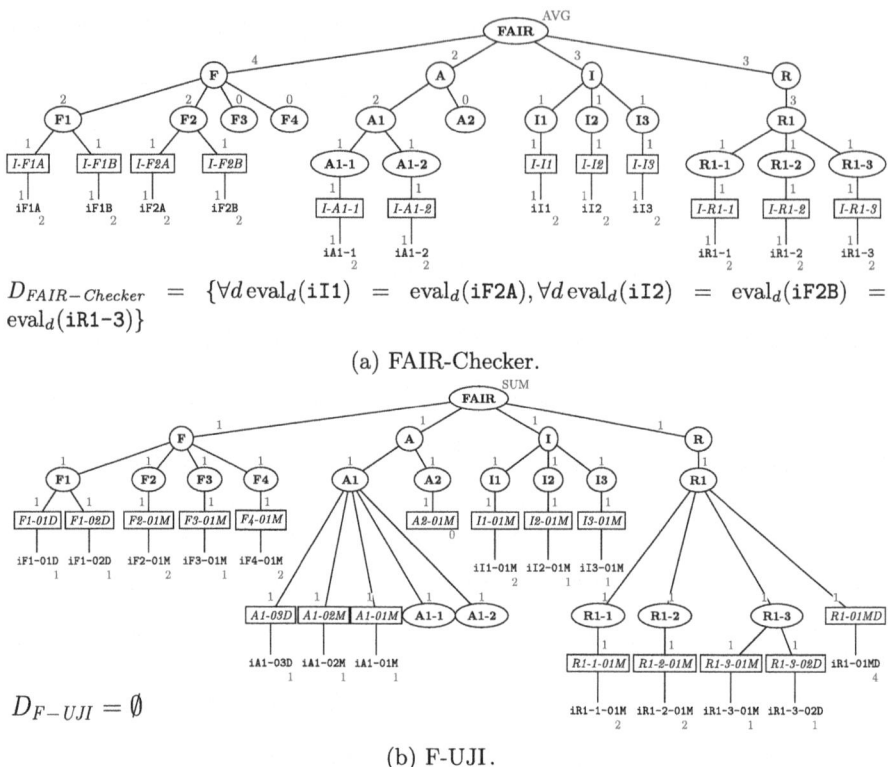

$D_{FAIR-Checker}$ = {$\forall d\, \mathrm{eval}_d(\mathtt{iI1})$ = $\mathrm{eval}_d(\mathtt{iF2A}), \forall d\, \mathrm{eval}_d(\mathtt{iI2})$ = $\mathrm{eval}_d(\mathtt{iF2B})$ = $\mathrm{eval}_d(\mathtt{iR1-3})$}

(a) FAIR-Checker.

$D_{F-UJI} = \emptyset$

(b) F-UJI.

(p): principle – $\boxed{id}$: indicator – im: implementation

w: weight – v: v_{max} (under the leaves)

Fig. 1. Representation of measures.

5 Characteristic Quantities for Measure Analysis

To highlight the salient traits of a measure we formally define three characteristic quantities: coverage rate, granularity and impact.

Coverage Rate. The role of the *coverage rate* is to measure to what extent a FAIRness measure covers the FAIR principles. Several definitions may apply. The simplest would be the proportion of leaves of the FAIR principles tree that have an indicator. However, this does not take into account the specificity of all FAIRness measures, since some of them have an indicator on A1 for instance but neither on A1.1 nor A1.2. Such measures should have a higher coverage rate than others not covering any of these three principles. Therefore, one could consider all principles and calculate the proportion of them having an indicator in their descendants. This would correct the previous problem but it may lead to significantly high scores. For instance, with only one indicator on the principle A1.1, the coverage rate is of 20%.

Hence, we propose a coverage rate that takes into account the hierarchical aspect of the FAIR principles, where each sibling in the tree counts the same in the calculation of the coverage rate, in particular, each of the four main principles counts as 25%. Let $\mathcal{M}$ be a FAIRness measure. This *coverage rate* computed following the hierarchy is defined recursively:

$$\text{cover}(\mathcal{M}) = \text{cover}(\mathcal{M}, \text{FAIR}) \tag{4}$$

where, for a given node $n \in P$:

$$\text{cover}(\mathcal{M}, n) = \begin{cases} 0 & \text{if } \text{children}_{\mathcal{M}}(n) = \emptyset \\ \dfrac{\left(\displaystyle\sum_{n' \in \text{children}_{\mathcal{M}}(n) \cap P} \text{cover}(\mathcal{M}, n')\right) + \text{local}(\mathcal{M}, n)}{|\text{children}_{\mathcal{M}}(n) \cap P| + \text{local}(\mathcal{M}, n)} & \text{else} \end{cases} \tag{5}$$

with $\text{local}(\mathcal{M}, n) = 1$ if $\text{children}_{\mathcal{M}}(n) \cap I(\mathcal{M}) \neq \emptyset$ and $\text{local}(\mathcal{M}, n) = 0$ otherwise.

This definition is based on the indicators specified by the measure. Adapting it to account for implementations instead, is not particularly challenging and is not presented here. However, notice that this adaptation would produce different results in only one scenario: if at least one indicator is not implemented. Such occurrences exist and are quite natural. For instance, an indicator may prove to be unimplementable.

Granularity. *Granularity* complements the previous definition. Intuitively, granularity evaluates the extent to which the indicators provide a detailed description of each principle. Higher granularity indicates a more thorough exploration of the principles and a finer-grained analysis of the FAIRness of the resources. Practically, we quantify granularity as the average number of indicators per principle that has at least one indicator.

$$\text{gran}(\mathcal{M}) = \text{gran}(\mathcal{M}, \text{FAIR}) \tag{6}$$

where, for $n \in P$:

$$\text{gran}(\mathcal{M}, n) = \frac{|\text{desc}_{\mathcal{M}}(n) \cap I(\mathcal{M})|}{|\{p \in (\{n\} \cup \text{desc}_{\mathcal{M}}(n)) \cap P, \text{children}_{\mathcal{M}}(p) \cap I(\mathcal{M}) \neq \emptyset\}|} \tag{7}$$

Similarly to the coverage rate, granularity is based on the indicators specified by the measure and can be easily adapted to account for implementations. The results provided by these two definitions diverge in the same scenarios as those presented for coverage rate.

Impact. Last but not least, the *impact* of a principle intuitively quantifies the percentage of score that a digital element obtains when the executions of all the implementations used by the measure to evaluate this principle are successful. More precisely, the evaluations of all the implementations under the principle are assumed to be at their maximum value (i.e. $v_{max}(imp)$ for a given imp) disregarding all others, which are typically set to 0, except in presence of dependencies. Thus we first introduce function $best_n$:

$$best_n(imp) = \begin{cases} v_{max}(imp) & \text{if } imp \in \text{desc}(n) \\ 0 & \text{otherwise} \end{cases} \tag{8}$$

Then, the impact of a node n, noted impact(n) is defined as the ratio of the node's best score to the maximum achievable score according to the measure:

$$\text{impact}(\mathcal{M}, n) = \frac{score(\mathcal{M}, best_n, \text{FAIR})}{score(\mathcal{M}, best_{FAIR}, \text{FAIR})} \tag{9}$$

6   Experimental Results

Through these experiments, we show that digital resources may obtain very different FAIR assessment scores when using different FAIR evaluation engines. In addition, we show that our proposed framework can precisely document the coverage and the relative importance, for each tool, of both fine-grained FAIR indicators as well as global principles, thus providing insights for users and tool developers on possible evaluation biases. Additional material is available online[3].

6.1   Do FAIR Assessment Engines Reach Consensus on FAIRness ?

Table 1 reports FAIR assessments for a collection of 10 scientific digital resources with F-UJI and FAIR-Checker. In this selection we aim at covering diverse domains with different types of resources such as datasets, ontologies, software, or training material. These resources are exposed on the web through diverse modalities such as institutional open data platforms, community specific registries (bioinformatics tools, machine learning models), e-learning platforms, legacy websites or raw metadata.

The F-UJI scores, expressed in percentages, have been manually collected from the tool's web interface [14]. The FAIR-Checker scores have been collected through the tools API [15]. Since FAIR-Checker only provides fine-grained scores as reported in Sect. 4, we computed a global score as a percentage based on the maximum achievable score.

Table 1 shows a relatively good agreement between the two engines for the first 5 entries (50% of our collection) with a standard deviation lower than 10. The best agreement appears for very high or very low scores. However,

---

[3] https://github.com/ICG4FAIR/ICG4FAIR.

Table 1. Multiple FAIR assessments, ranked by standard deviation.

Resource	F-UJI (%)	FAIR-Checker (%)	Std dev
Dataset (PANGAEA) [31]	91	91.70	0.49
Gene Ontology (OLS) [21]	18	16.70	0.92
Dataset (Harvard Dataverse) [23]	75	79.20	2.97
Dataset (Kaggle) [26]	60	70.80	7.64
Online course (Moodle) [28]	4	16.70	8.98
Dataset (Governmental platform) [22]	52	70.80	13.29
Dataset (WHO) [39]	27	50.00	16.26
Training material (TeSS) [36]	39	70.80	22.49
Bioinformatics tool (bio.tools) [6]	18	54.20	25.60
Dataset (RDF metadata) [33]	43	87.50	31.47

Table 2. Evaluating a bio.tools catalogue record.

	FAIR Score (%)	F (%)	A (%)	I (%)	R (%)
F-UJI	18.8	**35.7**	33	**0**	10
FAIR-Checker	54.2	**75**	50	**66.7**	16.7

the evaluations provided by F-UJI appear to be more fine-grained. The online course (Moodle) and Gene Ontology both obtain the FAIR-Checker minimal score (16.7%) but obtain different scores with F-UJI, 4% and 18% respectively, suggesting that F-UJI evaluation is more detailed. In addition, for the second half of our resource collection, the FAIR assessment scores begin to diverge with a standard deviation ranging from 13.29 to 31.47. Globally, we observe that FAIR-Checker provides higher scores compared to F-UJI.

For the last two entries the scores are very different with a standard deviation greater than 25. It is completely reasonable to wonder why the evaluation results are so different. Is it due to the way FAIR assessment engine retrieve metadata, as described in [37]? Is it due to the engine inner implementations? In the next paragraphs, we feed our model and compare FAIR-Checker and F-UJI, when evaluating a record of a bioinformatics tools catalogue.

Now we focus on how impact, granularity and coverage quantities can help in understanding divergent FAIR assessments. Table 2 reports very different results for the global FAIR assessment of the bio.tools record [6]. If we explore in more details each individual principle, we highlight that findability and interoperability scores greatly differ.

Table 3 shows how F-UJI and FAIR-Checker differ in terms of impact, granularity and coverage. We can see that reusability has a highest impact (41.67%) on the global assessment score compared to FAIR-Checker (25%). This could contribute to the explanation of the very low global FAIR assessment of the bio.tools record in F-UJI (Table 2) compared to FAIR-Checker. The findability of the bio.tools record is better scored in FAIR-Checker (75%) compared to F-UJI

Table 3. Comparing F-UJI and FAIR-Checker (FC)

	Impact		Granularity		Coverage	
	F-UJI	FC	F-UJI	FC	F-UJI	FC
F	29.17	33.33	1.25	**2**	100	**50**
A	12.5	16.67	2	1	66.67	50
I	**16.67**	**25**	1	1	100	100
R	**41.67**	**25**	1.25	1	100	100

(35.7%). However, this result should be interpreted with caution due to a poor coverage of F principles in FAIR-Checker (50%), compared to F-UJI (100%). In addition, despite a low coverage, we show that this findability principle has still the higher impact (33.33%) in the global assessment, which is questionable. Regarding the interoperability, we observe 0% for F-UJI and 66.7% for FAIR-Checker. Both engines have the same granularity (1) for an 100% coverage, meaning that the two engines, by design, do not agree on the indicators for interoperability, or that their implementation greatly vary.

6.2 Comparison of Measures Based on the Characteristic Quantities

We illustrate the use of the characteristic quantities of the measures introduced in Sect. 5 to objectively highlight their salient features and some of their differences. All the seven measures considered in Sect. 4 have been represented in the generic model following the methodology explained in the same section.

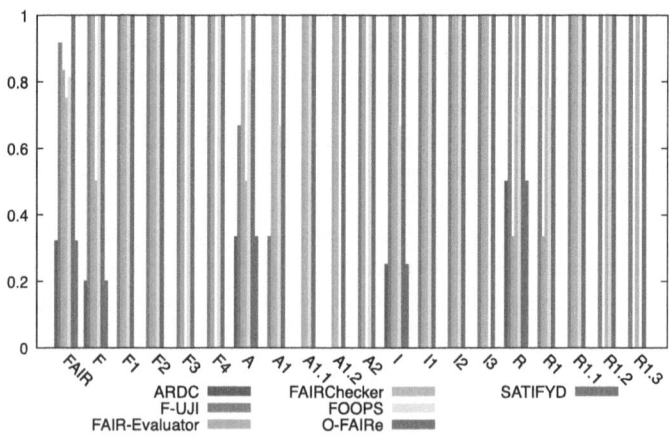

Fig. 2. Coverage rates.

Coverage Rates and Granularities. The coverage rate and granularity of all the principles, for all the seven measures are shown in Fig. 2 and Fig. 3 respectively. Notice that concerning O'FAIRe, the coverage rate and the granularity do not consider that some of its indicators are not (yet) implemented. This does not change the coverage value since all principles have at least one indicator implemented. However, the granularity is slightly overestimated.

We highlight some elements of the analysis that can be made. First, the coverage rate of root FAIR is rarely equal to 1. This means that the majority of the measures we are studying do not cover one or several principles. Only O'FAIRe covers all the principles. Questionnaires ARDC and SATIFYD have a low coverage rate because they only consider principles F, A, I and R. For FOOPS!, the coverage rate of A is 0.83. Its value is 0.66 for A1 whereas one would expect less because the value is 1 for A1.1 while A1.2 is not covered at all. This is because of the presence of an indicator directly related to A1. We also observe that some sub-principles are not covered by each measure. By design, R1.2 and R1.3 are not covered in the FAIR Evaluation Service as well as F3 and F4 for FAIR-Checker. In addition some principles such as A1.2 and R1.3 are covered by only a few number of tools, which questions on the technical feasibility of their implementation.

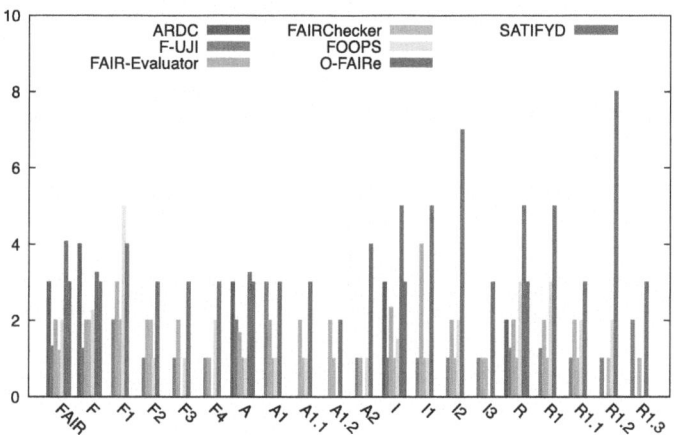

Fig. 3. Granularities

O'FAIRe gets the highest granularity at root FAIR, with high scores for I2 and R1.2 in particular. This is because it defines many indicators may be to address the great variety of vocabularies and meta-data present within the semantic artifacts it usually assess. As for FOOPS!, the high value of granularity of F1 shows an important care in providing indicators for this principle. Notice too that the granularity of R1 is higher than the granularities of R1.1, R1.2, R1.3. This is because several indicators have been directly attached to R1.

In fact, granularity and coverage rate are complementary and should sometimes be considered together before drawing a conclusion. For example, the promising granularity of F for ARDC could mean that F has been paid a lot of attention. However, the coverage rate of F for ARDC is quite low, meaning that there is no analysis according the sub-principles of F. Contrary to what one could expect, the analysis is not that precise.

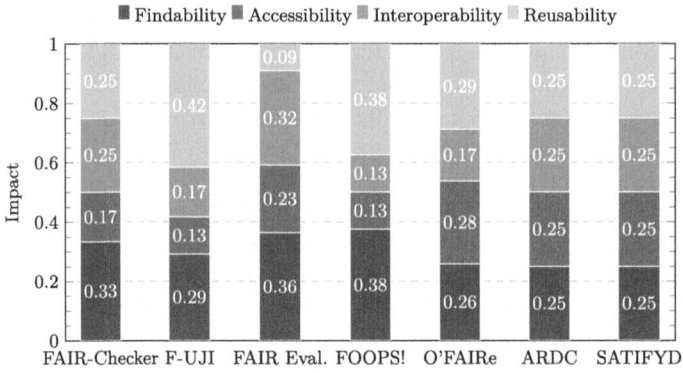

Fig. 4. Impact of the main FAIR Principles for different measures

Impacts. We then compare the different measures according to the impact, i.e. the importance they give to each principle. Figure 4 illustrates the impact of the four main principles, but we could compare the measures in detail according to each sub-principles. Findability and reusability are the most important principles in general. Reusability even count as 42% of the maximum score for F-UJI. But unlike the other measures, it is of really low importance for FAIR Evaluation Service, which does not consider R1.2 and R1.3. Overall, these measures are not balanced, with a principle that is almost three times more important than another one for three of the five considered measure.

Apart from O'FAIRe, none of the measures really think about the importance to give to each indicator. This is understandable from the point of view of certain measures. For instance, Wilkinson *et al.* [41] insist on the fact that FAIRness measures should (only) act as an incentive to improve the FAIRness of digital resource, that "there is no intrinsic value in an evaluation score" and that we should not declare a resource FAIR or non-FAIR. However, even if the measures are only intended to push for improvement in the FAIRness of resources, it is a unfortunate that there is no order of priority. The SHARing Rewards and Credit (SHARC) Interest Group [12] and the Data Maturity Model Working Group [4], both from the Research Data Alliance, developed this idea that some criteria are more important than others by categorizing them as "useful", "important" or "essential".

Table 4. Impacts for FAIR-Checker, with and without dependencies.

	F	F1	F2	F3	F4	A	I	I1	I2	I3	R	R1	R1.1	R1.2	R1.3
noDep	0.33	0.16	0.16	0	0	0.16	0.25	0.08	0.08	0.08	0.25	0.25	0.08	0.08	0.08
Dep	**0.58**	0.16	**0.41**	0	0	0.16	**0.50**	**0.16**	**0.25**	0.08	**0.41**	**0.41**	0.08	0.08	**0.25**

Dependency Aware Analysis. Our last experiment highlights the importance of taking into account dependencies. Those may stem from different codes checking a same property, or from full code delegations such as those expressed by FAIR-Checker and presented in Fig. 1a. The first line in Table 4 shows impacts without dependencies. For the second one, to compute the impact of a node n, instead of systematically set to 0 the implementations that do not belong to desc(n), their values are set according to the dependencies. For example, when computing impact(F), implementation i$I1$ is set to v_{max}(iF2A), which further increases the impact of F. Hence, the success of all the indicators belonging to desc(F) ensures to obtain not 33% but 58% of the maximum possible score. From a user point of view, such analysis is quite important, since it reveals that F is much more central to this measure than one might think at first glance.

7   Concluding Remarks

In this paper, we introduce a generic model and three computable and objective quantities aimed at more precisely interpreting FAIRness measures, comparing tools and possibly revealing evaluation biases. By adapting the hierarchy of principles, our framework could be repurposed for different evaluations, including IT security or energy footprint. Our experiments show that our framework i) contributes to explain different scores obtained by the same digital artifacts using different assessment engines and ii) facilitates the setup of comparative studies of various FAIRness metrics. Our approach is intended to be generic and to cover a large spectrum of FAIR assessment use cases. However, some of our choices induce some inaccuracies concerning granularity. Indeed, F-UJI divides its indicators into a new level of tests that would increase its granularity if considered. We provide experimental results on a limited set of resources. We are convinced that our generic model would benefit from larger scale experiments, with more FAIR assessment tools. As future works, we intend to more deeply analyze links between indicators or implementations, and to conduct larger scale experiments. This would require time, expertise, and would clearly be facilitated by involving tool development teams. We thus aim at contributing to collective initiatives tackling the challenges of harmonizing FAIR assessment frameworks.

Acknowledgments. This work is supported by the ANR DeKaloG (Decentralized Knowledge Graphs) project, ANR-19-CE23-0014, CE23 - Intelligence artificielle.

Disclosure of Interests. The authors have no competing interests to declare that are relevant to the content of this article.

References

1. Amdouni, E., Bouazzouni, S., Jonquet, C.: O'faire makes you an offer: metadata-based automatic fairness assessment for ontologies and semantic resources. Int. J. Metadata Semant. Ontol. **16**(1), 16–46 (2022)
2. Fair evaluation service. https://w3id.org/AmIFAIR. Accessed 18 Apr 2024
3. Australian Research Data Commons (ARDC): FAIR self assessment tool: Ardc online questionnaire (2022). https://ardc.edu.au/resource/fair-data-self-assessment-tool/. Accessed 22 Apr 2024
4. Bahim, C., et al.: The fair data maturity model: an approach to harmonise fair assessments. Data Sci. J. **19**, 41–41 (2020)
5. Barker, M., et al.: Introducing the fair principles for research software. Scientific Data **9** (2022). https://api.semanticscholar.org/CorpusID:252878844
6. Sample bioinformatics tool. https://bio.tools/bwa. Accessed 22 Apr 2024
7. Candela, L., Mangione, D., Pavone, G.: The fair assessment conundrum: Reflections on tools and metrics. Data Sci. J. **23**, 33 (2024). https://api.semanticscholar.org/CorpusID:270073165
8. Clarke, D.J., et al.: Fairshake: toolkit to evaluate the fairness of research digital resources. Cell Syst. **9**(5), 417–421 (2019)
9. Corcho, Ó., et al.: A maturity model for catalogues of semantic artefacts. Scientific Data **11** (2023). https://api.semanticscholar.org/CorpusID:258615711
10. Csiro misc questionnaire. https://web.archive.org/web/20210813120307/, http://5stardata.csiro.au/. Accessed 11 July 2024
11. Data Archiving and Networked Services (DANS): Satifyd online questionnaire (2019). https://satifyd.dans.knaw.nl/. Accessed 22 Apr 2024
12. David, R., et al.: Fairness literacy: the Achilles' heel of applying fair principles. CODATA Data Sci. J. **19**(32), 1–11 (2020)
13. Devaraju, A., Huber, R.: An automated solution for measuring the progress toward fair research data. Patterns **2**(11), 100370 (2021). https://doi.org/10.1016/j.patter.2021.100370, https://www.sciencedirect.com/science/article/pii/S2666389921002324
14. F-uji misc tool. https://www.f-uji.net/index.php?action=test. Accessed 16 Apr 2024
15. Fair-checker tool. https://fair-checker.france-bioinformatique.fr. Accessed 18 Apr 2024
16. Foops! misc tool. https://foops.linkeddata.es/FAIR_validator.htm. Accessed 18 Apr 2024
17. Gaignard, A., Rosnet, T., De Lamotte, F., Lefort, V., Devignes, M.D.: Fair-checker: supporting digital resource findability and reuse with knowledge graphs and semantic web standards. J. Biomed. Semant. **14**(1), 1–14 (2023). https://doi.org/10.1186/s13326-023-00289-5
18. Garijo, D., Corcho, O., Poveda-Villalón, M.: FOOPS!: an ontology pitfall scanner for the fair principles. In: International Semantic Web Conference (ISWC) 2021: Posters, Demos, and Industry Tracks. CEUR Workshop Proceedings, vol. 2980. CEUR-WS.org (2021). http://ceur-ws.org/Vol-2980/paper321.pdf
19. Peters-von Gehlen, K., Höck, H., Fast, A., Heydebreck, D., Lammert, A., Thiemann, H.: Recommendations for discipline-specific fairness evaluation derived from applying an ensemble of evaluation tools. Data Sci. J. **21**, 7–7 (2022)
20. Genova, F., et al.: Recommendations on FAIR metrics for EOSC. Publications Office of the European Union (2021)

21. Gene ontology. https://www.ebi.ac.uk/ols4/ontologies/go. Accessed 22 Apr 2024
22. Sample gouvernemental dataset. https://www.data.gouv.fr/en/datasets/donnees-relatives-a-lepidemie-de-covid-19-en-france-vue-densemble/. Accessed 22 Apr 2024
23. Sample harvard dataset. https://dataverse.harvard.edu/dataset.xhtml?persistentId=https://doi.org/10.7910/DVN/JGO6VI. Accessed 22 Apr 2024
24. Huerta, E.A., et al.: Fair for AI: an interdisciplinary and international community building perspective. Sci. Data **10** (2022). https://api.semanticscholar.org/CorpusID:260201856
25. Jones, S., Grootveld, M.: How fair are your data? (2021). https://doi.org/10.5281/zenodo.5111307
26. Sample kaggle dataset. https://www.kaggle.com/datasets/imdevskp/corona-virus-report. Accessed 22 Apr 2024
27. Krans, N., Ammar, A., Nymark, P., Willighagen, E., Bakker, M., Quik, J.: Fair assessment tools: evaluating use and performance. NanoImpact **27**, 100402 (2022)
28. Sample moodle course. https://moodle.polytechnique.fr/course/index.php?categoryid=1018. Accessed 22 Apr 2024
29. Moser, M., Werheid, J., Hamann, T., Abdelrazeq, A., Schmitt, R.H.: Which fair are you? a detailed comparison of existing fair metrics in the context of research data management. In: Proceedings of the Conference on Research Data Infrastructure, vol. 1 (2023)
30. O'faire misc tool. https://agroportal.lirmm.fr/landscape#fairness_assessment. Accessed 18 Apr 2024
31. Sample pangaea dataset. http://doi.org/10.1594/PANGAEA.908011. Accessed 22 Apr 2024
32. Poveda-Villalón, M., Espinoza-Arias, P., Garijo, D., Corcho, Ó.: Coming to terms with fair ontologies. In: International Conference Knowledge Engineering and Knowledge Management (2020). https://api.semanticscholar.org/CorpusID:225078634
33. Sample RDF metadata. https://data.rivm.nl/meta/srv/eng/rdf.metadata.get?uuid=1c0fcd57-1102-4620-9cfa-441e93ea5604&approved=true. Accessed 22 Apr 2024
34. Slamkov, D., Stojanov, V., Koteska, B., Mishev, A.: A comparison of data fairness evaluation tools. In: Budimac, Z. (ed.) Proceedings of the Ninth Workshop on Software Quality Analysis, Monitoring, Improvement, and Applications, Novi Sad, Serbia, September 11-14, 2022. CEUR Workshop Proceedings, vol. 3237. CEUR-WS.org (2022). https://ceur-ws.org/Vol-3237/paper-sla.pdf
35. Sun, C., Emonet, V., Dumontier, M.: A comprehensive comparison of automated fairness evaluation tools. In: 13th International Conference on Semantic Web Applications and Tools for Health Care and Life Sciences, pp. 44–53 (2022)
36. Sample training material. https://tess.elixir-europe.org/materials/make-your-research-fairer-with-quarto-github-and-zenodo. Accessed 22 Apr 2024
37. Van De Sompel, H., Soiland-Reyes, S.: Fair signposting: exposing the topology of digital objects on the web. In: International FAIR Digital Objects Implementation Summit 2024. TIB Open Publishing (2024)
38. de Visser, C., et al.: Ten quick tips for building fair workflows. PLOS Comput. Biol. **19** (2023). https://api.semanticscholar.org/CorpusID:263224298
39. Sample who dataset. https://data.who.int/dashboards/covid19/data. Accessed 22 Apr 2024
40. Wilkinson, M.D., et al.: The FAIR guiding principles for scientific data management and stewardship. Sci. Data **3**(1), 1–9 (2016)

41. Wilkinson, M.D., et al.: Evaluating fair maturity through a scalable, automated, community-governed framework. Sci. Data **6**(1), 174 (2019)
42. Wilkinson, M.D., Sansone, S.A., Marjan, G., Nordling, J., Dennis, R., Hecker, D.: FAIR Assessment Tools: Towards an "Apples to Apples" Comparisons (2023). https://doi.org/10.5281/zenodo.7463421

ORKA: An Ontology for Robotic Knowledge Acquisition

Mark Adamik[1]([✉])[ID], Romana Pernisch[1,2][ID], Ilaria Tiddi[1][ID],
and Stefan Schlobach[1][ID]

[1] Vrije Universiteit Amsterdam, Amsterdam, The Netherlands
m.adamik@vu.nl
[2] Discovery Lab, Elsevier, Amsterdam, The Netherlands

Abstract. Most autonomous agents operating in the real world use perception capabilities and reasoning mechanisms to acquire new knowledge of the environment, where perception capabilities include both the physical sensor devices and the software-based perception pipelines involved in the process. For autonomous agents to be able to adjust and reason over their own perception, knowledge of the sensors and the corresponding perception algorithms is required. We present the Ontology for Robotic Knowledge Acquisition (ORKA), that models the perception pipeline of a robotic agent by representing the sensory, algorithmic and measurement aspects of the perception process, thereby unifying the agent's sensing with the characteristics of the environment and facilitating the grounding process. The ontology is based on the alignment between SSN and OBOE, linked to external databases as additional knowledge sources for robotic agents, populated with instances from two different robotic use-cases, and evaluated using competency questions and comparisons to related ontologies. A proof of concept use-case is presented to highlight the potential of the ontology.

Keywords: Robotic Knowledge Acquisition · Robotic Perception · Ontologies

1 Introduction

Robots are increasingly used across various sectors due to advancements in AI and robotics. They range from service robots in restaurants and hospitals through home-use robots like vacuum cleaners to agricultural robot. Yet, their utility and effectiveness is hindered by a lack of common-sense knowledge and understanding of the world [12]. Presently, the common-sense knowledge robots use is implicitly embedded within specialised control programs designed for various robots and applications.

Fundamentally, the autonomy and behaviour of robotic systems are shaped by their ability to perceive their surroundings, as many of the decisions these agents make are based on the interpretation of data acquired through sensors. However, the data acquired through the sensors is essentially meaningless without the background knowledge that allows for the interpretation of this information. In order to efficiently organise the data and transform it into knowledge

M. Alam et al. (Eds.): EKAW 2024, LNAI 15370, pp. 309–327, 2025.
https://doi.org/10.1007/978-3-031-77792-9_19

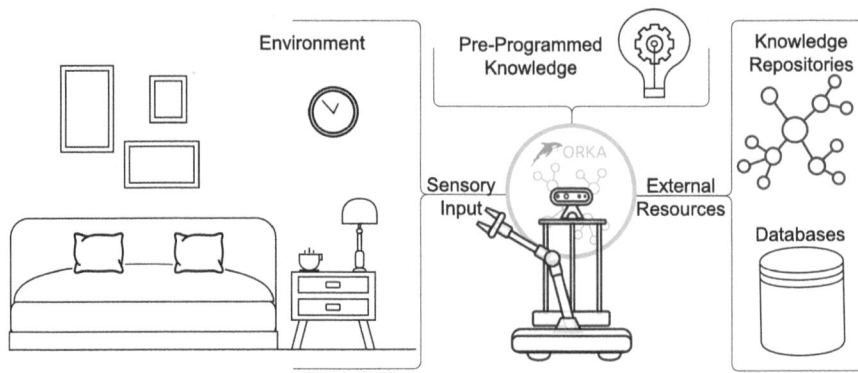

Fig. 1. The three main sources of knowledge for a robotic agent is the pre-programmed knowledge that the robotic system engineers equipped the robot with; the sensory information the robot acquires through the sensory data, and external sources the robot can connect to. The purpose of ORKA is to unify these knowledge sources.

that could be acted on, the agent needs to have some prior information about what type of perception capabilities it is equipped with, which sensors expose them, and what aspects of the environment are observed by them.

Let us consider a robot equipped with two sensors: a 2D planar LiDAR for depth sensing and an RGB camera for vision. For a robot to interpret these signals, knowledge of the robotic engineer is needed to consider that the floating point values returned by the LiDAR represent distances represented (usually) in units of millimeters, and that the integer values of the camera represent pixel values corresponding to colour intensities on three channels. This knowledge, often implicit and only considered by system designers, is usually not directly available to the robot to reason over, thereby limiting its potential utility and its adaptability to unforeseen circumstances.

Similarly, considering the perception pipelines designed for these sensors, engineers are typically knowledgeable about the semantics of an algorithm's output – e.g. bounding boxes of generic, deep learning-based object recognition algorithms such as Yolo [27] indicate the location of the object, where the object class label represent the type of the object and the values represent the certainty of such classification. Outputs of a specifically-purposed gaze detector algorithm [3] instead indicate the detected face, as well as the yaw and pitch angles of the detected direction. To enable agents to autonomously process this information and independently select the optimal algorithms aligned with both the sensory abilities and the task at hand, awareness of their own perception capabilities is required.

Knowledge-enabled robotics [13] aims at supporting robotic agents to such type of reasoning using knowledge representation formalisms, to ultimately allow for a shared understanding between robotics agents and their environments. We distinguish three fundamental sources of knowledge for robotic agents as depicted in Fig. 1: (i) *pre-programmed knowledge* that the robot is initially equipped with, and is designed by the engineers of the robot, also encompassing the knowledge representation and reasoning abilities such as ontologies or internal simulations [11,33] that robots use to infer new knowledge; (ii) *sensorial knowledge* gathered by the sensors and perception algorithms that create observations of the environment[1]; and (iii) the *external knowledge* that robots can access from databases [36] and knowledge graphs [14] to further improve their reasoning abilities. One of the most notable example of the latter source of knowledge is the Semantic Web, as one of the primary vision of it is to *"make the web more accessible to computers"* [6]. The problem of linking of robotic perception to the resources of the Semantic Web has been identified and formalised in our previous work [4].

While some studies have proposed models that represent parts of such knowledge in the form of ontologies (e.g. robot capabilities and sensory devices [19,24]), none of these attempts represent all the above described sources of knowledge, and unify the knowledge acquisition process of a robotic agent from sensors to understanding the environment. Furthermore, the represented characteristics of objects are somewhat limited in these ontologies [5]. With this in mind, we use the Ontology101 methodology [25] to define the Ontology for Robotic Knowledge Acquisition (ORKA). ORKA ties together the domains of sensors, perception processes, measurement properties of percepts and the properties of the perceived objects to jointly represent the domain of robotic perception. The contributions are two-fold. First, we present ORKA, a domain ontology for robotic knowledge acquisition, as well as the most important design choices made for it in the form of competency questions. Second, we model the robotic knowledge acquisition process from sensory data to real world entities, showing with two practical examples how such ontology-driven representation can benefit different use-cases. To the best of our knowledge, this is the first attempt to create an ontology that unifies and ties together the different sources of knowledge and their interrelations in artificial agents. The ontology and the related resources are available online[2].

2   Related Work

One of the most comprehensive knowledge-based robotics framework is KnowRob [33], which includes the Socio-physical Model of Activities (SOMA) [10], the Semantic Robot Description Language (SRDL) [19] and RoboSherlock [8,9]. Designed to aid robots performing manipulation activities in home environments, SOMA presents a very fine-grained model for objects and their social (e.g. cleaningness) and physical (e.g. colour) qualities. The SRDL module includes a detailed taxonomy of the different sensory capabilities, as well as some software categories. RoboSherlock is a cognitive vision system that provides an image processing-based perception pipeline to accommodate different perceptual processes. While KnowRob and its extensions provide an

---

[1] We consider a "perception algorithm" any algorithmic process that results in new observations about characteristics of objects.

[2] https://github.com/Dorteel/orka.

extensive and more detailed knowledge representing perception, they still lack a fine-grained taxonomy of characteristics entities can possess, the links of these characteristics to the robot sensory capabilities, and knowledge about the measurement standards of the sensory devices as well as the algorithms used to acquire these information.

Alternative knowledge-driven approaches include the Perception and Manipulation Knowledge (PMK) framework [24], which represents both object properties and algorithms; the OpenRobots Ontology (ORO) [20], a knowledge management platform allowing cognitive robots to perform reasoning on previously acquired knowledge, and the Ontology-based Unified Robot Knowledge (OUR-K) [21], exhibiting integrating low-level sensory data with perceptual features (e.g. colour, texture, SIFT features), which are further associated with concepts and perception algorithms. The project seems however discontinued and no further details could be explored. There are some ontologies attempting to link sensory data to higher-level robot capabilities in specific domains, e.g. the culturally-aware assistive robots in the CARESSES project[3], including objects and qualities such as colour and size, but restricting the perception modelling to the audio (e.g. speed and pitch of the speech); the OROSU ontology [15] representing medical devices and their sensing measurements in the robotic surgery domain; and the ROSETTA ontology [32], where sensors and sensory characteristics are designed for industrial robotics tasks.

Other ontologies for sensing outside the robotics domain include the Semantic Sensor Network (SSN) [16] and its extension Sensor, Observation, Sample and Actuator (SOSA) ontology [17], both describing sensors and their observations, as well as sensing processes according to Semantic Web standards. While both ontologies offer a comprehensive overview of the sensory observation processes, they do not provide details about the specific sensors and corresponding algorithms employed by robots, as well regarding the characteristics of the environment the measurements are obtained about. The RDF Data Cube Vocabulary [2] allows to describe statistical data and their measurements, but places a greater emphasis on statistical data, and lacks the capabilities to describe sensory data accurately. Similarly, the Ontology of units of Measure and related concepts (OM) [29] provides a vocabulary to tie together different measurement units, with a focus on science and engineering [28]. The Extensible Observation Ontology (OBOE) [22] describes ecological observations of entities using measurement standards and characteristics. Its structure of the observations is easily transferable to the robotics domain, and the characteristics described in the ontology are quite comprehensive.

As shown, various ontologies incorporate aspects relevant to our problem (e.g. sensors, processes, characteristics), but none fully includes a suitable representation connecting the characteristics of an environment with the sensory devices used by a robot. Additionally, perception algorithms are hardly represented, and the algorithms' capabilities and the semantics of their outputs is missing. These ontologies also do not make use of external knowledge sources.

---

[3] http://caressesrobot.org/ontology/.

To overcome these limitations, we present the Ontology for Robotic Knowledge Acquisition (ORKA) in the next section.

3 Ontology for Robotic Knowledge Acquisition

Following the Ontology101 [25] Ontology Engineering methodology, we start by defining the domain, purpose and intended use of the ontology articulated through a list of competency questions. Subsequently, we evaluate some of the ontologies of Sect. 2 for possible re-use. Finally, we present the structure of ORKA, as well as the Semantic Web Rule Language (SWRL) rules that augment its reasoning capabilities.

3.1 Domain and Scope

The ontology serves as a shared vocabulary for researchers and engineers working in the domain of robotic perception and cognitive robotics, and defines the interrelations between the concepts used to describe sensors, subjects of sensors and the algorithms that operate on the sensory data. ORKA also includes an integration of additional knowledge sources, that could be used to improve the perception of the autonomous agent with common-sense information (as shown later in Sect. 5). The domain represented by ORKA is the complete process of knowledge acquisition for autonomous robotic agents, with a special focus on the link between robotic perception and environment.

We show the primary components the ontology should encompass using a few practical examples. An ideal ontology would allow autonomous robotic agents to consider that the sensors provide information about specific aspects of the environment, and the entities contained within. For example, some sensors such as IMU-s or encoders provide information about the state of the robot, and not of the objects it perceives; cameras provide information about objects that are detected, but not their measurements; LiDARs provide information about distances, but not colours. At the level of algorithms, not all sensory data and processes need to be considered at all times. For example, gaze-detection algorithms would only be required in contexts where a human is present, and sensors used for mapping are usually not required during manipulation tasks. Defining contexts for the different sensory applications could therefore help the agent decide which of the sensory data is relevant for the given task. Additionally, algorithms come in different model sizes, and ideally an agent should have the knowledge to deploy the most suitable model depending on the task at hand. Finally, in order to give a meaning to the sensor readings in the form of entities of the real-world, it is important that the measurement systems are also specified within the ontology. These entities can be also aligned with external, common-sense knowledge sources to acquire further knowledge: e.g. an object recognition algorithm returning contradicting or incorrect information could use a knowledge source such as WikiData [35] and DBpedia [7] to augment the object detection algorithm with the necessary information.

Restricting scope, we limit our investigation to the perception of physical objects, and refer the inclusion of other entities, such as events and actions, to future work.

Competency Questions. Using the above defined use-cases and scope limitations, we derive a set of competency questions to be answered by ORKA.

CQ1: Given a robotic agent, what sensors and perception algorithms is it equipped with?

CQ2: Which characteristics of the environment do the sensors and their associated algorithms from CQ1 observe?

CQ3: What units of measurement is the data from the sensors and perception algorithms in CQ1 provided in?

CQ4: What observable and observed characteristics do given entities possess?

CQ5: What characteristics do the sensors from CQ1 possess?

CQ6: What characteristics do the algorithms from CQ1 possess?

CQ7: Which algorithm is the most suitable for a given context?

CQ8: What external knowledge sources are available to the robot and what do they describe?

CQ9: What characteristics of given entities are described in the external knowledge sources?

CQ1-3 define the foundational concepts the ontology aims to connect, namely the sensors, algorithms, the characteristics, and the measurement standards. **CQ4** focuses on the characteristics which are implied by the existence of entities. **CQ5** and **CQ6** aim to describe the characteristics of sensors and algorithms respectively. Derived from the use-cases described earlier, these competency questions focus on the characteristics that are relevant to the knowledge acquisition process. This is reflected by **CQ7**, which requires the characteristics of **CQ5** and **CQ6** to determine the utility of the sensors and algorithms in a given context. As the comprehensive representation of a context is outside of the current scope of ORKA, the simplified representation of context includes (i) a task at hand, (ii) the perceived objects, and (iii) the objects required for completing the task. We disregard the particular task and instead use the required objects as a substitute for it. **CQ8** and **CQ9** address the third source of knowledge described in Sect. 1, i.e. the inclusion of external sources of knowledge in the knowledge-acquisition process. **CQ8** aims to describe the knowledge sources available to the robot, whereas **CQ9** focuses on the potential knowledge that these sources could offer.

3.2   Re-using Existing Ontologies

With the domain and scope defined, we consider the re-use of existing ontologies. As mentioned in Sect. 2, existing ontologies allow the modelling of sensors [16,17,19], observations [16,17,22] or tasks and activities [33], but none is comprehensive and flexible enough represent our robotic knowledge acquisition process sufficiently.

The sources that describe most of the required components are the SSN [16] ontology when considered together with the system capabilities module of SSN, which aims at extending SSN to capture the capabilities of sensors and measurement. The choice for reusing SSN was further reinforced with the fact that an alignment module is provided for OBOE [22], which in turn describes the characteristics the observations concern, and the associated units of measurements. Although some components are not included in the alignment of SSN and OBOE (i.e. the specific sensors, the coupling between the sensors and the measured characteristics, the specific algorithms that comprise the perception pipelines, the external knowledge sources, etc.), it is deemed as the best candidate to serve as a backbone of ORKA.

3.3 Core Structure

We followed a top-down approach, starting with the definitions of the most general concepts. These considerations resulted in a core ontology (Fig. 2), specified below.

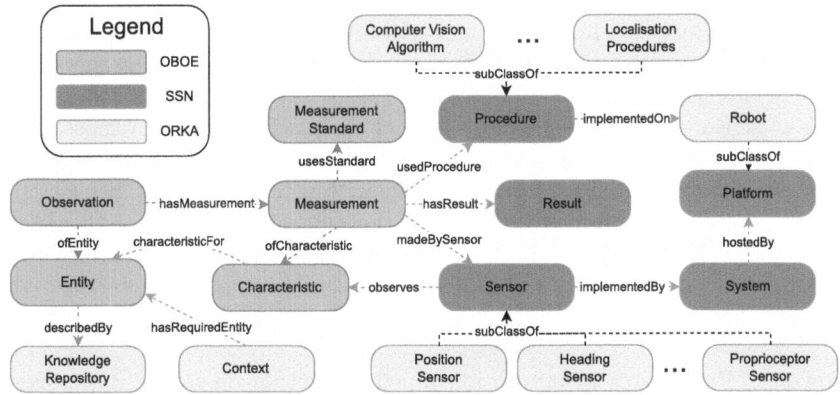

Fig. 2. A simplified view of the core classes of the ORKA ontology as well as the OBOE and SSN alignment.

Robots and Sensors. In accordance with SSN, in ORKA robots are considered a special class of `sosa:Platform`, that host `sosa:Sensor` devices and have specific `sosa:Procedures` implemented on. To address **CQ1**, sensors are also defined as a subclass of the `oboe:Entity` class, and sensor instances are linked to an instance of the `orka:Robot` class using the `sosa:hosts` predicate. SSN is extended with a detailed classification of sensors following [30]. A sensor can be either a `ProprioceptorSensor`, i.e. a device that can measure different characteristics related to the body of the agent that equips the sensors, or an `ExteroceptorSensor`, which measures the external environment. The former includes encoders that measure the position of the motor devices, or sensors that measure the orientation or acceleration of the robot, such as tilt sensors or IMU devices. Using this classification the robot can infer that the sensory information concerns robots:

$$\text{ProprioceptorSensor} \sqsubseteq \forall\text{observesCharacteristic.}(\forall\text{characteristicFor.Robot})$$

An additional classification is provided based on the characteristic of the environment or the objects that they measure (e.g. position, heading, etc.). As an example, the sensory activation of a bumper sensor always implies the `Location` characteristic of an entity.

$$\text{Bumper} \sqsubseteq \forall\text{observesCharacteristic.Location}$$

In determining the proprioception abilities of sensors, some adjustments are made. For example, some sensors indicated as exteroceptors in [30], such as compass and encoders, are modelled as proprioceptors in ORKA, as they indicate the orientation or position of the robot (or components of the robot).

To be able to address **CQ5**, the most important characteristics that influence the knowledge acquisition capabilities of the sensors are inserted as specific data properties. Some of these properties, such as `range, resolution, accuracy, precision` are included in the system capabilities module of SSN, while others, such as `sampling rate` of the sensors needed to be added. These properties are appended directly to the bottom levels of the `Sensor` class, where the specific models of sensors are represented, as general class axioms. Simpler sensors such as a thermometer or a bumper provide the characteristics of the entity without the necessity of a specialised perception algorithm to be utilised. Sensors that provide more complex information about the environment, such as a camera, require a specialised (often computer vision) algorithm, here collectively called `Procedures`, that allows for the observation of additional characteristics that the sensor does not provide on its own.

Procedures. Procedures in perception pipelines, acting on sensor data, significantly affect autonomous agents' capabilities. To capture this impact, we incorporate both algorithms and deep learning models together with their properties into the ontology. In order to address **CQ6**, certain characteristics (i.e. model size, inference speed, memory usage) and model versions of existing algorithms are included. For `ObjectDetection` algorithms, we currently use the official values of YoloV5[4] as implemented on our LocoBot mobile robot, and represent these as data property assertions on the class level, in the form of general class axioms. These characteristics also play a role in answering **CQ7**. Depending on the context, models with higher detection speed could be preferred in some scenarios, whereas higher precision could be required in others. In the current version of the ontology, a greater emphasis is placed on the `ComputerVisionAlgorithm` class, which follows an application-based taxonomy [18]. Future work will be aimed at expanding it to a wider variety of algorithms.

---

[4] https://pytorch.org/hub/ultralytics_yolov5/.

Entity. The entities correspond to the phenomena being observed. In the case of robotic agents situated in a real-world environment, the properties of these entities could help distinguish between individuals, and address cognitive robotics problems such as object permanence and occlusion. Following OBOE, in ORKA entities are the main subjects of the observations, holding observable and inherent characteristics that describe the objects (**CQ4**) through the `hasCharacteristic` predicate. In ORKA, instances of `Sensor` and `Robot` are also considered as entities.

Observations and Measurements. Observations are made of entities belonging to the `Entity` class. In case of a sensor, an `Measurement` is produced that serves as an input to a `Procedure`. In the case of an object detection algorithm, the observation concerns a single entity, which has multiple corresponding `Measurements` involving the label and bounding-box that the procedure provides. Measurements concern a single `Characteristic`. A precision (where applicable) can also be assigned, as it can be inferred from the *precision* characteristics of the sensor device performing the measurement. The measurements also correspond to a measurement standard.

Measurement Standards. The sensory knowledge acquisition process starts with the recording of the observations of the physical world. In order to address **CQ3**, the measurement units contained within the ontology are adopted from OBOE. However, as OBOE does not include some units that are relevant for robotic perception (e.g. pixel values or binary events such as a switch or bumper), ORKA also incorporates these as measurement standards. Finally, instead of organising units into *base units* and *derived units* as in [22,29], ORKA defines a hierarchy of unit classes based on the characteristics they define.

The `oboe:usesStandard` data property is used to link measurement standards to their corresponding measurements. In order to derive the measurement standards associated with the different sensors and procedures, the current implementation of ORKA uses SWRL rules [1]. The reason behind this modelling choice is that sensors are difficult to describe in general terms regarding their outputs, as different implementations of the same sensor type might produce outputs corresponding to different measurements. For example, a specific model of ultrasonic sensor might be implemented so that its output is the raw data in terms of time taken for the sound to reflect, while another model might also implement the conversion from the time taken to estimate the distance. Furthermore, some procedures, such as object recognition algorithms, produce multiple measurements corresponding to different units of measure, and therefore a single unit cannot be associated with a single procedure.

Characteristics. Given **CQ2**, a core aspect of ORKA is to represent characteristics of an object, that could be considered common-sense (i.e. where an object was observed, when it was observed or what it looks like). Given the lack of a comprehensive catalogue of the specific qualities (such as size, colour, shape, weight, etc.) that a particular endurant might have [23], we use three main sources of characteristics to include in ORKA. Firstly, we adopt the Characteristics defined in OBOE. Secondly, we examine the set of current perception algorithms, and sensors as well as their measurements to identify some of the main object characteristic to include in ORKA. This includes `ObjectType` (corresponding to the label), and `Location` as the output of the YOLO algorithm, or `Size` and `Colour` as the output of the point cloud processing algorithm that processes the images produced by the depth camera. This category also includes the measurement units discussed above, where we consider the SI quantity dimensions described in [34] (e.g. length, mass, time), and their derived characteristics (e.g. height, width, depth, weight, age, etc.). Lastly, we include some of the properties that external knowledge sources utilise to characterise objects, and which could be acquired from an external knowledge base. Specifically, WikiData [35] was used to examine additional characteristics, such as material, density, hardness and names of different shades of colours.

Observable characteristics are divided into two main categories: those related to objects and those related to the environment, influencing object perception (e.g. brightness impacting colour). The categories include the `SpatialCharacteristic` class, which outlines an object's location and orientation within a specific reference frame, and the `VisualCharacteristic` class, which defines an object's colour and pattern. This division suggests that robots without visual sensors should not be assigned tasks requiring the observation of visual characteristics. For example, a TurtleBot with only LiDAR is not suited for tasks involving visual characteristics, and its tasks should be adjusted or performed by robots equipped with the necessary sensors.

External Knowledge Sources. To be able to answer **CQ8** and **CQ9**, external knowledge sources, and links to these sources should be described. Currently, we limit our investigation to common-sense databases part of the Semantic Web that could be useful for robotic agents to operate in an environment, and use WikiData and DBpedia to serve as a proof-of-concept external knowledge graphs. As these sources contain information about general characteristics of the class of the entities being observed (e.g. colour, material, weight, etc.), but not about specific instances of entities that exist around the robotic agent, links are established between the characteristics included in ORKA and their counterpart in the external knowledge source. We link the resources using the `hasDBpediaURI` and `hasWikiDataURI` data properties. Furthermore, the `sparqlEndpoint` data property is used to provide some information on how to query these external knowledge sources. In the current version of ORKA, the corresponding links have been established manually. Automatic entity linking procedures could be considered in future work.

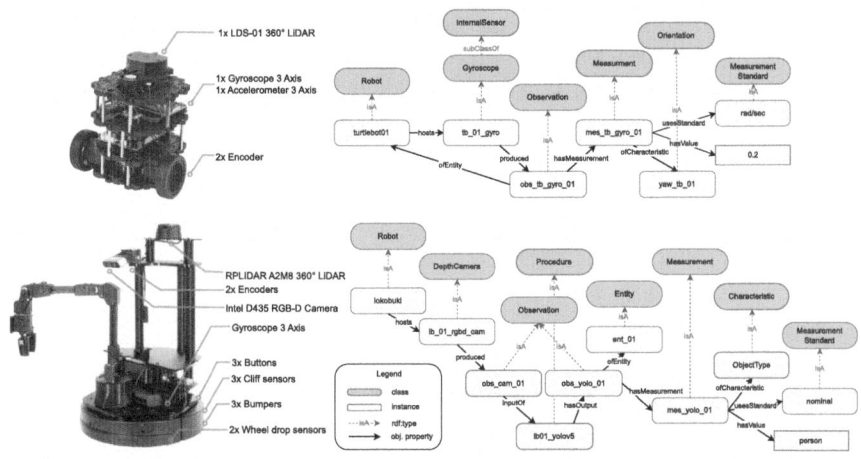

Fig. 3. A depiction of a TurtleBot 3 Burger (https://emanual.robotis.com/docs/en/platform/turtlebot3/features/) (top) and a LocoBot WX250s (https://www.trossenrobotics.com/locobot-wx250.aspx) (bottom) and their sensors, together with instances of sensory readings represented in the ontology. As the gyroscope of the TurtleBot is a proprioceptor, a SWRL rule automatically derives that the observations correspond to the entity that hosts the sensor. For the purposes of this paper, the manipulator of LocoBot is excluded from the consideration.

Individuals. The last step is the population of the ontology with individuals. We consider two mobile robots and their perception abilities to be instantiated: a LocoBot WX250s, and a TurtleBot3 Burger. The two robots are equipped with a total of 21 sensors and four procedures, a YoloV5 object detector, a gaze detector based on L2CS-Net [3], a point cloud segmentation algorithm and a Simultaneous Localisation and Mapping (SLAM) algorithm. A depiction of the robots as well as the corresponding sensors and an example of the sensory process can be seen in Fig. 3. For every sensor and algorithm within the ontology, instances of test observations have been added to demonstrate and validate the competency questions.

3.4 SWRL Rules

To address **CQ4** and **CQ7**, ORKA needs to infer certain relationships, such as which algorithms can detect the `ObjectType` and other characteristics of the given entities (**CQ4**) or which entities required for a context (**CQ7**). As OWL-DL does not allow for the desired expressivity, we use SWRL rules [1] to infer the required information. The rules can be divided into rules concerning sensory data, and rules concerning procedures. As an example for the former, if the sensor is a `ContactSensor` and produces an `Observation` with a `Measurement` of`Characteristic` `Location`, then it should also be inferred that, since the measurement range of any `ContactSensor` is 0, the value of the location should be equated with the given sensor's `Location`. An example for the latter is the rule concerning **CQ7**, which aids in the deduction of the `canDetect` object property between a `Procedure` and an `Entity`. This involves establishing a connection between an entity required for a task, instantiated as an instance, and potential algorithms capable of completing the task by detecting the specified entity. The complete list of implemented SWRL rules with comments is available in the online repository[5].

4   Evaluation

The current version of ORKA contains 443 classes, 74 object properties and 39 data properties. The ontology is populated with 54 instances, representing the two robots, their capabilities, observations and measurements, as well as some of the observed entities. The competency questions that articulate the requirements of the ontology are evaluated using several SPARQL queries and the Pellet reasoner [31]. The queries were evaluated using the individual instances of the two robotic agents presented in Sect. 3. A complete list of the queries as well as the results can be found in the repository.

Listing 1 illustrates the query for **CQ1** which returns the sensors a robotic agent is equipped with, and the implemented perception algorithms. The query is used to give a complete list of the sensory knowledge acquisition capabilities of any robotic agent.

```
SELECT ?s ?r ?a
WHERE { ?r sosa:hosts ?s .
        ?a orka:implementedOn ?r}
```

Listing 1: Query corresponding to CQ1.

In practice, the list could be used by other robotic agents in a multi-agent setting, or by a user to assess which robot might be appropriate in which situation. The queries corresponding to **CQ3** and **CQ8** follow a similar pattern, with entities being returned that have a `hasMeasurementStandard` and `describedBy` properties defined respectively. Listing 2 presents a more complex query corresponding to **CQ2**, and returns the individuals of the class `Characteristic` corresponding to sensors and procedures answered to **CQ1**.

---

[5] https://github.com/Dorteel/orka.

```
SELECT ?item ?charType
WHERE {    {?a orka:implementedOn ?r .
            ?a orka:observesCharacteristic ?c .
            ?c a ?charType .
            BIND(?a AS ?item)}
     UNION {?r sosa:hosts ?s .
            ?s orka:observesCharacteristic ?c .
            ?c a ?charType .
            BIND(?s AS ?item) }}
```

Listing 2: Query for the characteristics measured by sensors and algorithms (CQ2).

When evaluating competency questions **CQ4-6**, the sub-properties of `hasObservableCharacteristic`, `hasSensorCharacteristic` and `hasAlgorithmCharacteristic` are queried respectively. In a practical scenario, specific characteristics such as `VisualCharacteristics` of entities, `maxRange` of a sensor or `DetectionSpeed` of a procedure would be the subject of the query.

To determine the context in which certain perception algorithms are better than others, as per **CQ7**, two simple contexts are presented. A *HRI-Dialogue* context has a human face as a required entity, whereas a *Fetch-Object* context has the object to be fetched as the required entity. The query of Listing 3 formulates this question and returns the entities that match the description of the context, except where none of the algorithms can detect the required entity. Context requirements can be modified to select among algorithm characteristics, such as opting for an object detector version that prioritises higher detection speed over accuracy.

```
SELECT ?c ?e (COALESCE(?a, "None") AS ?alg)
WHERE { ?c orka:hasRequiredEntity ?e .
        OPTIONAL {
            ?a orka:implementedOn ?r .
            ?a orka:canDetect ?e .}}
```

Listing 3: Query used to evaluate CQ7.

CQ9 necessitates listing the characteristics of entities from an external knowledge source, useful for correcting mislabelled entities by comparing the detected entity's characteristics with those described in the external source. The ontology is not able to evaluate this competency question by relying only on SPARQL queries, as it would require a dynamic construction of URI's within the query. However, using the `describedBy` property as in **CQ8** allows for the access of these external resources using an additional Python script that incorporates SPARQL queries.

Table 1. Ontology comparison with respect to CQ1 – CQ9. As KnowRob represents an infrastructure consisting of multiple ontologies, the aggregated capabilities are considered. Checkmarks in parentheses indicate partial answering of CQ-s.

Ontology	CQ1	CQ2	CQ3	CQ4	CQ5	CQ6	CQ7	CQ8	CQ9
KnowRob [33]	✓	✓		✓	✓	✓	✓		
PMK [24]	✓			✓	✓				
CARESSES				✓					
ROSETTA [32]	✓	✓	✓	✓	✓	✓			
ORO [20]				✓					
OROSU [26]	✓								
SSN [16]	✓	✓			✓	✓			
ORKA	✓	✓	✓	✓	✓	✓	✓	✓	(✓)

Comparison. A comparison was performed to evaluate ORKA with respect to the other publicly available ontologies. Table 1 shows the coverage of the CQs not just of ORKA (last row), but also of some of the related ontologies described in Sect. 2. ORKA answers all but the last CQ completely, making it the most fitting ontology for the envisioned domain and scope. KnowRob [33] and ROSETTA [32] is able to answer six CQs, making them the next best-performing ontologies. PMK [24] and SSN/SOSA [17] can each answer four CQs, and ORO [20], OROSU [15] and CARESSES are only able to address one CQ. With the low coverage of CQs by these domain-specific ontologies, creating alignments with ORKA to extend the knowledge acquisition capabilities can be envisioned, which could ultimately serve as a unifying vocabulary for robot capabilities.

Finally, we compare ORKA and the relevant domain ontologies in terms of size. Table 2 includes the number of classes and instances for the sensors, algorithms, characteristics, measurement units, as well as sensor and algorithm characteristics that were considered within each ontology. These values are acquired using the DL Query plugin of the Protégé ontology development software.

Recall that ORKA is in its preliminary version, and the number of its individuals is expected to increase. We consider the ones we have enough to prove the ontology generalisability to a further extent. Further robot scenarios outside of the ones described in this paper are left for future work.

Threats to Validity. The competency questions that guided the development process were driven by the use-cases, and do not yet cover the entirety of the robotic knowledge acquisition domain. Currently, ORKA is limited to the individuals representing the two robots and their associated sensory devices. For the ontology to be more complete, more links between characteristics and the sensor devices outside of the individuals in the ontology should be included. Moreover, the comparison only considers the freely available versions of the considered ontologies. More up-to-date versions could appear that would render the com-

Table 2. Number of classes and instances (in parentheses) of ORKA and comparison ontologies.

	KnowRob [33]	PMK [24]	CARESSES	ROSETTA [32]	ORO [20]	OROSU [15]	ORKA (ours)
Sensors	21	4 (4)	-	22 (69)	-	4 (4)	37 (18)
Algorithms	63	14 (7)	-	45 (61)	-	13 (5)	22 (4)
Characteristics	27	47 (40)	1	78 (245)	3 (13)	-	120 (11)
Measurement Units	-	-	-	2 (2)	-	-	17
Sensor Characteristics	6	2 (2)	-	13 (86)	-	-	9
Algorithm Characteristics	1	-	-	-	-	-	8
External Sources	-	-	-	-	-	-	2

parison out-of-date. ORKA's interoperability could be improved by introducing alignment modules with ontologies within and outside the domain. Lastly, our design methodology does not include evaluation with external validators, which also poses a threat to the validation of ORKA. Yet, we show how ORKA can support the well-known issue of robotics knowledge-driven perception, enabling further studies and first prototypes in the future.

5   Example Use-Case

This section provides a proof-of-concept with ORKA, where a robot detects different objects, and using an external knowledge graph (i.e. WikiData) verifies and corrects the label provided by the object detection algorithm, performing the perceived-entity linking task [4].

The example concerns the task of fetching an orange for the user from a basket. With the use of a camera, an object detection algorithm and a colour detection procedure, an observation graph is produced containing the entities and their recognised properties. As the fetching task contains an object that is recognisable with several of the available object detection algorithms, the one with the fastest inference speed is chosen. However, as the object detection algorithm incorrectly recognises two of the fruits (a grapefruit and a lemon) as an orange, a colour detection procedure is initiated to further distinguish the objects. The colour detection algorithm utilises a modified MASK-RCNN segmentation algorithm to calculate the mean pixel value for each segment, and consults ORKA to retrieve the corresponding closest available colour.

The robot associates each entity with a colour, and uses the corresponding colour property (wdt:P462) of WikiData to refer to the colour of an orange (accessed through the hasWikiDataURI data property) and therefore disambiguate which entity has the colour of an orange as described in WikiData. An overview of the process as well as the outputs are shown in Fig. 4. This example serves as a proof-of-concept, showcasing the utility and potential of ORKA to represent the knowledge and reasoning capabilities of the agents.

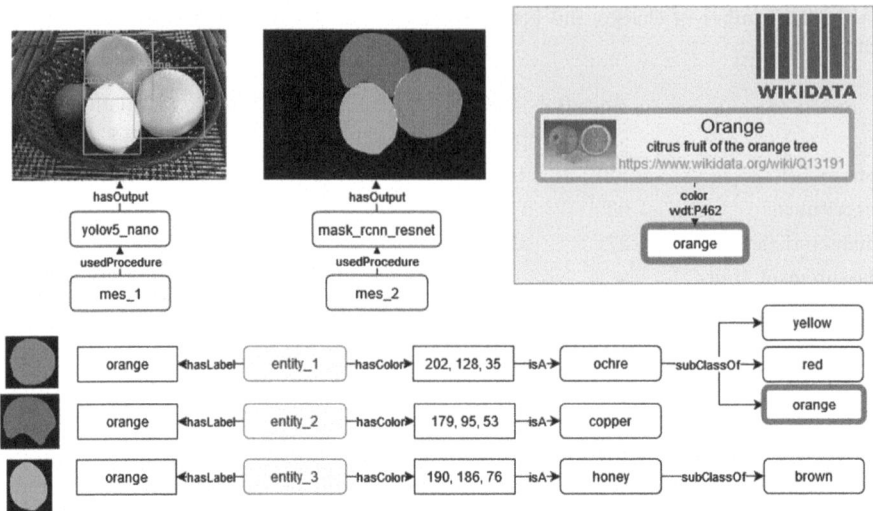

Fig. 4. A simplified depiction of the robot view, the created observation graphs concerning the entities, and the matched property. As only one entity has the colour matching the description found in the external knowledge source, the entity is chosen.

6   Conclusion and Future Work

We have presented the Ontology for Robotic Knowledge Acquisition (ORKA), a model that represents the sensory and algorithmic capabilities of robotic agents with regard to the perception of the environment. We have shown that the model is capable of capturing several characteristics of objects, and allows for the linking of object classes and characteristics to external knowledge sources. The reasoning of the model has been reinforced with SWRL rules to allow for the automatic inference of characteristics captured by the algorithms. The model is evaluated using the competency questions formulated as SPARQL queries, and a proof of concept is presented that showcased the potential of ORKA. In the future, we intend to extend the model to address problems such as the representation of time, using e.g. multiple descriptions of the same location. In order to promote re-usability and compatibility, alignment modules will be provided to allow other knowledge-driven robotic systems to use and integrate ORKA. A modularisation of ORKA will be introduced, which allows users and developers to focus on the aspects relevant to their respective applications. Furthermore, we plan to use ORKA in combination with physical robots, and test the ontology reasoning capabilities as more observations and measurements are made. We encourage experts and users in the field of knowledge representation and robotics to use and revise our model so that ORKA can become a community effort in the future.

References

1. SWRL: A Semantic Web Rule Language Combining OWL and RuleML. https://www.w3.org/submissions/SWRL/
2. The RDF Data Cube Vocabulary (2014). https://www.w3.org/TR/vocab-data-cube/
3. Abdelrahman, A.A., Hempel, T., Khalifa, A., Al-Hamadi, A., Dinges, L.: L2CS-Net : fine-grained gaze estimation in unconstrained environments. In: 2023 8th International Conference on Frontiers of Signal Processing (ICFSP), pp. 98–102 (2023). https://doi.org/10.1109/ICFSP59764.2023.10372944
4. Adamik, M., Pernisch, R., Tiddi, I., Schlobach, S.: Advancing robotic perception with perceived-entity linking. In: The Semantic Web - ISWC 2024 - 23rd International Semantic Web Conference, Baltimore, The United States of America, 11–15 November 2024, Proceedings. Lecture Notes in Computer Science, Springer (2024)
5. Adamik, M., Schlobach, S.: Towards a definition and conceptualisation of the perceived-entity linking problem. In: RobOntics 2023 Ontologies for Autonomous Robotics 2023. CEUR Workshop Proceedings, CEUR-WS (2023)
6. Antoniou, G., Groth, P., van Harmelen, F., Hoekstra, R.: A Semantic Web Primer, 3rd edn. MIT Press, Cambridge (2012)
7. Auer, S., Bizer, C., Kobilarov, G., Lehmann, J., Cyganiak, R., Ives, Z.: DBpedia: a nucleus for a web of open data. In: Aberer, K., et al. (eds.) ASWC/ISWC -2007. LNCS, vol. 4825, pp. 722–735. Springer, Heidelberg (2007). https://doi.org/10.1007/978-3-540-76298-0_52
8. B'alint-Bencz'edi, F., et al.: RoboSherlock: cognition-enabled robot perception for everyday manipulation tasks. arXiv (2019)
9. Beetz, M., Balint-Benczedi, F., Blodow, N., Nyga, D., Wiedemeyer, T., Marton, Z.C.: RoboSherlock: unstructured information processing for robot perception. In: 2015 IEEE International Conference on Robotics and Automation (ICRA), pp. 1549–1556. IEEE, Seattle, WA, USA (2015). https://doi.org/10.1109/ICRA.2015.7139395
10. Beßler, D., et al.: Foundations of the socio-physical model of activities (SOMA) for autonomous robotic agents (2020). https://doi.org/10.48550/arXiv.2011.11972
11. Blum, C., Winfield, A.F.T., Hafner, V.V.: Simulation-based internal models for safer robots. Front. Robot. AI **4**, 74 (2017). https://doi.org/10.3389/FROBT.2017.00074
12. Davis, E., Marcus, G.: Commonsense reasoning and commonsense knowledge in artificial intelligence. Commun. ACM **58**(9), 92–103 (2015). https://doi.org/10.1145/2701413
13. Engel, U. (ed.): Robots in Care and Everyday Life: Future, Ethics, Social Acceptance. SpringerBriefs in Sociology, Springer, Cham (2023). https://doi.org/10.1007/978-3-031-11447-2
14. Fischer, L., et al.: Which tool to use? Grounded reasoning in everyday environments with assistant robots. In: Proceedings of the 11th International Workshop on Cognitive Robotics (2018)
15. Gonçalves, P.J., Torres, P.M.: Knowledge representation applied to robotic orthopedic surgery. Robot. Comput.-Integr. Manuf. **33**, 90–99 (2015). https://doi.org/10.1016/j.rcim.2014.08.014, citation Key: orosu
16. Haller, A., et al.: The modular SSN ontology: a joint W3C and OGC standard specifying the semantics of sensors, observations, sampling, and actuation. Semant. Web **10**(1), 9–32 (2018). https://doi.org/10.3233/SW-180320

17. Janowicz, K., Haller, A., Cox, S.J.D., Phuoc, D.L., Lefrancois, M.: SOSA: a lightweight ontology for sensors, observations, samples, and actuators. J. Web Semant. **56**, 1–10 (2019). https://doi.org/10.1016/j.websem.2018.06.003

18. Khanday, N.Y., Sofi, S.A.: Taxonomy, state-of-the-art, challenges and applications of visual understanding: a review. Comput. Scie. Rev. **40**, 100374 (2021). https://doi.org/10.1016/j.cosrev.2021.100374

19. Kunze, L., Roehm, T., Beetz, M.: Towards semantic robot description languages. In: 2011 IEEE International Conference on Robotics and Automation, pp. 5589–5595 (2011). https://doi.org/10.1109/ICRA.2011.5980170

20. Lemaignan, S., Ros, R., Mösenlechner, L., Alami, R., Beetz, M.: Oro, a knowledge management platform for cognitive architectures in robotics. In: 2010 IEEE/RSJ International Conference on Intelligent Robots and Systems, pp. 3548–3553 (2010). https://doi.org/10.1109/IROS.2010.5649547

21. Lim, G.H., Suh, I.H., Suh, H.: Ontology-based unified robot knowledge for service robots in indoor environments. IEEE Trans. Syst. Man Cybern. Part A: Syst. Hum. **41**, 492–509 (2011). https://doi.org/10.1109/TSMCA.2010.2076404

22. Madin, J., Bowers, S., Schildhauer, M., Krivov, S., Pennington, D., Villa, F.: An ontology for describing and synthesizing ecological observation data. Eco. Inform. **2**(3), 279–296 (2007). https://doi.org/10.1016/j.ecoinf.2007.05.004

23. Masolo, C., Borgo, S.: Qualities in formal ontology (2010)

24. Diab, M., Akbari, A., Din, M.U., Rosell, J.: PMK—a knowledge processing framework for autonomous robotics perception and manipulation (2019). https://www.mdpi.com/1424-8220/19/5/1166

25. Noy, N.F., McGuinness, D.L.: Ontology Development 101: A Guide to Creating Your First Ontology (2001)

26. Prestes, E., Fiorini, S., Carbonera, J.: Core Ontology for Robotics and Automation (2014)

27. Redmon, J., Divvala, S.K., Girshick, R.B., Farhadi, A.: You only look once: unified, real-time object detection. In: 2016 IEEE Conference on Computer Vision and Pattern Recognition, CVPR 2016, Las Vegas, NV, USA, 27–30 June 2016, pp. 779–788. IEEE Computer Society (2016). https://doi.org/10.1109/CVPR.2016.91

28. Rijgersberg, H., Wigham, M., Top, J.L.: How semantics can improve engineering processes: a case of units of measure and quantities. Adv. Eng. Inform. **25**(2), 276–287 (2011). https://doi.org/10.1016/j.aei.2010.07.008

29. Rijgersberg, H., Van Assem, M., Top, J.: Ontology of units of measure and related concepts. Semant. Web **4**(1), 3–13 (2013). https://doi.org/10.3233/SW-2012-0069

30. Siciliano, B., Khatib, O. (eds.): Springer Handbook of Robotics. Springer, Heidelberg (2008). http://dx.doi.org/10.1007/978-3-540-30301-5

31. Sirin, E., Parsia, B., Grau, B.C., Kalyanpur, A., Katz, Y.: Pellet: a practical OWL-DL reasoner. J. Web Semant. **5**(2), 51–53 (2007). https://doi.org/10.1016/j.websem.2007.03.004

32. Stenmark, M., Malec, J.: Knowledge-based instruction of manipulation tasks for industrial robotics. Robot. Comput.-Integr. Manuf. **33**, 56–67 (2015). https://doi.org/10.1016/j.rcim.2014.07.004, citation Key: rosetta

33. Tenorth, M., Beetz, M.: KnowRob: a knowledge processing infrastructure for cognition-enabled robots. Int. J. Robot. Res. **32**(5), 566–590 (2013). https://doi.org/10.1177/0278364913481635

34. Thompson, A.: Guide for the Use of the International System of Units (SI)

35. Vrandečić, D., Krötzsch, M.: Wikidata: a free collaborative knowledgebase. Commun. ACM **57**(10), 78–85 (2014). https://doi.org/10.1145/2629489
36. Waibel, M., et al.: RoboEarth. IEEE Robot. Autom. Mag. **18**(2), 69–82 (2011). https://doi.org/10.1109/MRA.2011.941632

Transformers in the Service of Description Logic-Based Contexts

Angelos Poulis[1]([⊠]) (iD), Eleni Tsalapati[2] (iD), and Manolis Koubarakis[3,4] (iD)

[1] Department of Computer Science, Boston University, Boston, USA
apoulis@bu.edu
[2] Athens Technology Center, Chalandri, Greece
etsalapati@atc.gr
[3] AI Team, Department of Informatics and Telecommunications,
National and Kapodistrian University of Athens, Athens, Greece
koubarak@di.uoa.gr
[4] Archimedes/Athena RC, Marousi, Greece

Abstract. Recent advancements in transformer-based models have initiated research interests in investigating their ability to learn to perform reasoning tasks. However, most of the contexts used for this purpose are in practice very simple: generated from short (fragments of) first-order logic sentences with only a few logical operators and quantifiers. In this work, we construct the natural language dataset, DELTA_D, using the description logic language $\mathcal{ALCQ}$. DELTA_D contains 384K examples, and increases in two dimensions: i) reasoning depth, and ii) linguistic complexity. In this way, we systematically investigate the reasoning ability of a supervised fine-tuned DeBERTa-based model and of two large language models (GPT-3.5, GPT-4) with few-shot prompting. Our results demonstrate that the DeBERTa-based model can master the reasoning task and that the performance of GPTs can improve significantly even when a small number of samples is provided (9 shots). We open-source our code and datasets.

Keywords: Reasoning · Knowledge Bases · Natural Language Processing

1 Introduction

Description Logic (DL) languages [1] are fragments of first-order logic (FOL) that have evolved into one of the main formalisms for the representation of conceptual knowledge in a precise and well-defined manner. An expressive and decidable DL language that, besides the standard Boolean operators, supports existential, universal, and numerical constraints is $\mathcal{ALCQ}$. For instance, in $\mathcal{ALCQ}$ one can formally express sentences like the ones appearing in Fig. 1.

A. Poulis and E. Tsalapati—This work was performed while the author was with the AI Team, Dept. of Informatics and Telecommunications, National and Kapodistrian University of Athens.

M. Alam et al. (Eds.): EKAW 2024, LNAI 15370, pp. 328–345, 2025.
https://doi.org/10.1007/978-3-031-77792-9_20

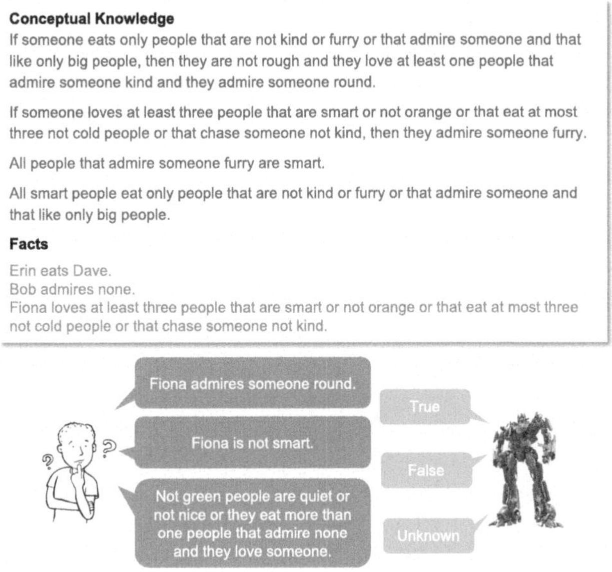

Conceptual Knowledge

If someone eats only people that are not kind or furry or that admire someone and that like only big people, then they are not rough and they love at least one people that admire someone kind and they admire someone round.

If someone loves at least three people that are smart or not orange or that eat at most three not cold people or that chase someone not kind, then they admire someone furry.

All people that admire someone furry are smart.

All smart people eat only people that are not kind or furry or that admire someone and that like only big people.

Facts

Erin eats Dave.
Bob admires none.
Fiona loves at least three people that are smart or not orange or that eat at most three not cold people or that chase someone not kind.

Fiona admires someone round.   True

Fiona is not smart.   False

Not green people are quiet or not nice or they eat more than one people that admire none and they love someone.   Unknown

Fig. 1. An example from DELTA$_D$, where the context contains three sentences of high linguistic complexity level and the true and false sentences are of reasoning depth 3.

The formal apparatus of DLs allows us to perform deductive reasoning tasks, such as *entailment checking*, i.e., deciding whether a sentence or a set of sentences, logically implies another. With the latest advancements in transformer-based language models (TLMs) a new strand of research has emerged that investigates if TLMs can learn to carry out such tasks over *contexts* expressed in natural language [3,5,7,21]. However, in most cases, the contexts used were either composed of rather short sentences, simple in structure (i.e., their formal representations contain only a few logical operators and quantifiers) [3,18,22], or they were of limited size [5,22].

The fundamental question that this work seeks to answer is: *"How well can TLMs perform inference over contexts produced from an expressive DL language, like $\mathcal{ALCQ}$?"*. A natural subsequent research question, in line with the literature [3,20] but here is for a higher expressivity, is: *"Is the performance of TLMs affected by the reasoning depth required to perform the inference process?"*. A third research question that arises is whether the fragment of the formal language is enough to measure the reasoning ability of a model. For instance, all sentences appearing in Fig. 1 can be expressed formally within $\mathcal{ALCQ}$, but some are linguistically more complex than others, and one would expect that contexts containing mainly that complex sentences would be hard to process. Hence, the third research question of this paper is: *"Is the performance of TLMs affected by the linguistic complexity of the context?"*.

As discussed by Madusanka et al. [13], the most suited reasoning problem to evaluate the impact of language constructs (like quantifiers and negation) is *textual entailment checking*, i.e., entailment checking with natural language, in a purely logical sense, eliminating the influence of any background or commonsense knowledge. Hence, to answer the above research questions, we have constructed the synthetic dataset $DELTA_D$ (DEscription Logics with TrAnsformers) of 384K examples (context-question-answer-depth-linguistic complexity) based on $\mathcal{ALCQ}$, where the question is the statement that we check whether it is logically deduced from the context, under the open world assumption. Besides the isolation of the commonsense/background knowledge, the synthetic nature of the dataset allows us to perform a systematic study of the performance of the TLMs, as $DELTA_D$ gradually increases both in reasoning and in linguistic complexity. Additionally, it allows us to eliminate obvious statistical features (e.g., the correlation between the answer "False" and the word "not" in the sentence in question).

We systematically tested the textual entailment checking ability of supervised fine-tuned DeBERTa and few-shot prompting on large language models (GPT-3.5, GPT-4) over $DELTA_D$. Our results show that the performance of the DeBERTa-based model, $DELTA_M$, is marginally affected when the reasoning depth increases or decreases (differently from [22]) with average accuracy 97.5%. Also, we show that the length of the sentences does not affect the performance of the model (accuracy 99.7% in max. reasoning depth and max. length of sentence). To ensure that $DELTA_M$ does not overfit on $DELTA_D$, and inspired by [27], we changed the probability distributions used for the dataset generation and the accuracy of the model remained equally good. Additionally, testings to similar datasets [20] returned good results.

To check the impact of semantics on $DELTA_M$, we followed the approach of Tang et al. [21] and we generated a symbolic "translation" of $DELTA_D$ by translating the pool of words used for the generation of the synthetic dataset to symbols. Zero-shot testings of $DELTA_M$ show that, in contrast to Tang et al. [21], it performs equally well, while GPTs' accuracy is slightly decreased. This indicates the performance of $DELTA_M$ is not affected by the semantics of the dataset (an expected result as the dataset is nonsensical). However, further translation of the dataset to resemble the language used to describe DL sentences led to a significant reduction in the accuracy of the models. Finally, good testing results of $DELTA_M$ on a real-world scenario (fuel cell system diagnostics) demonstrate the potential of TLMs to be utilized in rule-based system diagnostics.

Overall, we make the following contributions:

(C_1) We provide the first large, description logic benchmark of 384K examples. This is a significant contribution because building large benchmarks over expressive logic languages, like $\mathcal{ALCQ}$, is a challenging task as it requires performing query answering with logic reasoners, a process that can be very time-consuming (~ 1 min. for KBs with long subsumption axioms/facts of our dataset). Both the dataset and the code for its generation are openly available[1].

---

[1] https://github.com/angelosps/DELTA.

(C_2) We show that TLMs can perform entailment checking with very high accuracy over synthetic natural language contexts generated from $\mathcal{ALCQ}$ sentences. This demonstrates the potential of TLMs to be utilized for scalable reasoning tasks over vast KBs, thus bypassing formal representations required by traditional knowledge-based systems.

(C_3) We show that the performance of TLMs is not affected by the length of the sentences.

(C_4) We show that DeBERTa-based models are not affected by the vocabulary of the dataset.

($C5$) We show how these contributions can be leveraged in a real-world use-case scenario.

2 Background on Description Logics

We can use $\mathcal{ALCQ}$ [1] to represent knowledge about a domain by defining three types of entities: individuals (e.g., *John*), concepts (e.g., *Postdoc*, i.e., the concept describing the entities that are postdocs) and roles (e.g., *teaches*). A *concept expression* C can be formed using these entities, Boolean constructors ($\sqcap$, $\sqcup$, $\neg$), quantifiers ($\forall$, $\exists$), and number restrictions ($\leq, \geq$) recursively as follows: $C, D := A \mid \top \mid \bot \mid \neg C \mid C \sqcap D \mid C \sqcup D \mid \forall R.C \mid \exists R.C \mid \geq nR.C \mid \leq nR.C$, where A is an atomic concept, R an atomic role, $\top$ the top concept, which has every individual as an instance, and $\bot$ the dual of $\top$. In this way, one can represent formally complex concept expressions, such as all entities that *"have a Ph.D., teach at most two postgraduate courses and are not academics"* ($\exists hasDegree.PhD \sqcap \leq 2teaches.PostgrCourse \sqcap \neg Academic$). *Subsumption axioms* in $\mathcal{ALCQ}$ have the form $C \sqsubseteq D$ and describe relationships between concept expressions. For example, one can describe formally that all postdocs are described by the aforementioned concept as *Postdoc* $\sqsubseteq \exists owns.PhD \sqcap \leq 2teaches.PostgrCourse \sqcap \neg Academic$. We denote with *LHS* (left-hand side) the concept expression that appears on the left of the subsumption symbol ($\sqsubseteq$) in a subsumption axiom and with *RHS* (right-hand side) the concept expression that appears on the right. *Assertional axioms* or, simply, *facts* describe knowledge about named individuals, i.e., that are *instances* of some concept (expression) and have the form $C(a)$ or $R(a, b)$, where a, b individuals. Using complex expressions one can construct very complex facts. An $\mathcal{ALCQ}$ *knowledge base* (KB) is a set of subsumption axioms and a set of facts.

Delta-closure$(\mathcal{K}, t)$ of a KB $\mathcal{K}$ is the set of subsumption axioms and facts that are inferred from $\mathcal{K}$ within time t seconds using the InferredOntologyGenerator[2], and no more axioms or facts are inferred after t[3]. Given a KB $\mathcal{K}$ and a subsumption axiom or a fact a, we say that $\mathcal{K}$ *entails* a (subsumption axiom or fact) if every model of $\mathcal{K}$ (i.e., if every interpretation

---

[2] https://www.cs.ox.ac.uk/isg/tools/HermiT//download/0.9.2/owlapi/javadoc/org/semanticweb/owl/util/InferredOntologyGenerator.html.

[3] Hence, Delta-closure can be calculated only for KBs that InferredOntologyGenerator can calculate all axioms and facts within t.

that satisfies all subsumption axioms and facts of $\mathcal{K}$) is also a model of a. *Entailment checking* can be considered as the prototypical reasoning task for querying knowledge: we check whether some statement is necessarily true, presuming the statements of the knowledge base. Following the semantics of DLs, we make the *open-world assumption*, i.e., missing information is treated as unknown. Supposing that $\mathcal{K}$ is transformed in negated normal form, we consider *inferrence depth*, or simply *depth* of a with respect to $\mathcal{K}$, $depth(a, \mathcal{K})$, as the size of the *justification* [9] for a, i.e., the *minimum* number of subsumption axioms and facts in $\mathcal{K}$ that can be used to logically deduce that a is true or false. If none of the two can be deduced, the answer is "unknown" and a is not characterized by any depth.

3   Dataset Generation

We investigate the ability of transformers to perform textual entailment checking over $\mathcal{ALCQ}$ KBs expressed in natural language with respect to two dimensions: i) the depth $\mathcal{D}$ of the sentences (i.e., subsumption axioms/facts in question), henceforth mentioned as *queries*, with respect to the corresponding KB, ii) the linguistic complexity level $\mathcal{L}$ (defined in Sect. 3.1) of the knowledge required to answer the queries. To achieve this, each *example* in the dataset DELTA$_D$ is a 5-tuple $\langle \mathcal{T}, \mathcal{Q}, \mathcal{A}, \mathcal{D}, \mathcal{L} \rangle$, where $\mathcal{T}$ is the *context* containing $\mathcal{ALCQ}$ axioms (subsumption axioms/facts) expressed in natural language, $\mathcal{Q}$ the query expressed in natural language, henceforth mentioned as *question*, $\mathcal{A}$ is the *answer* which can be either *true, false,* or *unknown*, and $\mathcal{D}$ the depth of $\mathcal{Q}$, if $\mathcal{A}$ is *true* or *false*, otherwise it is denoted as **na**. $\mathcal{L}$ is the linguistic complexity of the KB[4].

The pipeline for the generation of the dataset is presented in Fig. 2. For the generation of an example (described in detail in Sect. 3.1) of linguistic complexity level n ($\mathcal{L} \leq n$) and depth m ($\mathcal{D} \leq m$), we first generate a KB $\mathcal{K}$ using a specially crafted probabilistic context-free grammar (denoted in Fig. 2 with $\mathcal{ALCQ}$-n PCFG) for producing subsumption axioms and facts of maximum linguistic complexity n. Then, *Delta-closure*$(\mathcal{K}, t)$ is calculated for $t = 5$ s. KBs that required more than 5 s to calculate the Delta-closure, were discarded.

From the Delta-closure we calculate, as described in Sect. 3.2, *true* (answer=true), *false* (answer=false) and *unknown* (answer=unknown) queries, which eventually will formulate the sentences in question. A KB is kept only if it can produce queries with all three types of answers at all depths up to m, otherwise a new one is generated. Once this process is completed, the generated queries (subsumption axioms/facts) along with the original $\mathcal{K}$, are translated into natural language statements $\mathcal{Q}$ and into the context $\mathcal{T}$, respectively, by utilizing a set of natural language templates, as described in Sect. 3.3.

3.1   KB Generation

To create diverse contexts and to avoid overfitting to a specific vocabulary, we have defined two different pools of terms, *Pool A* and *Pool B*. Pool A contains

---

[4] DELTA$_D$ also contains the justification for each answer to be used for future work or by the research community for other downstream tasks, such as proof generation.

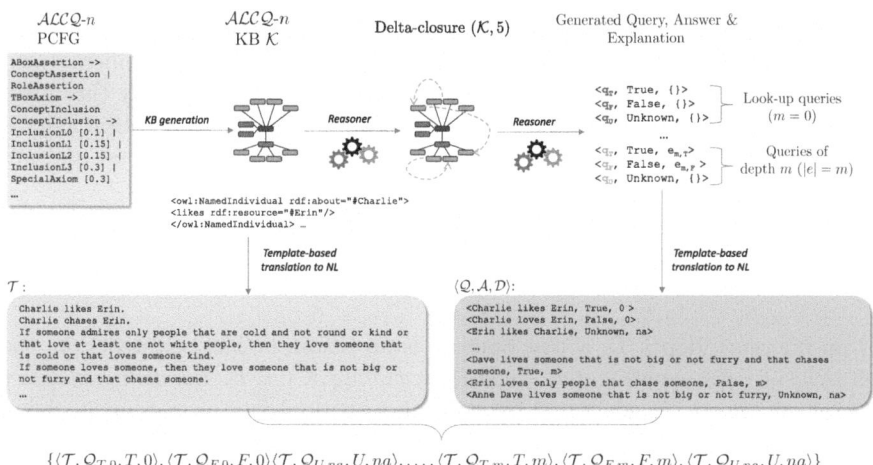

Fig. 2. Data generation pipeline for examples with n-level context and answers of minimum inference depth $\leq m$

14 atomic concepts, 5 roles, and 8 individuals, mostly taken from RuleTaker dataset [3] (in RuleTaker the subsumption axioms are simple conjunctive implications, where the concept names are named "attributes", the roles "relations" and the individual names "entities"). Pool B contains 8 atomic concepts, 8 roles, and 8 individuals. Both pools can be found in Appendix 1[5].

From each pool, we generate 20 datasets (40 in total) of 1000 KBs each, of various inference depths and axiom lengths.

To obtain KBs of different linguistic complexity levels, we have manually crafted four types ($\mathcal{L} = 0, 1, 2, 3$) of PCFGs, based on the number of constructors and quantifiers appearing in their axioms. In general, a concept of linguistic complexity $\mathcal{L}$ contains $\mathcal{L}$ Boolean constructors and at most $\mathcal{L} + 1$ quantifiers.

An $\mathcal{L}$-type PCFG produces KBs of linguistic complexity level $\mathcal{L}$ with axioms that their one side (e.g., LHS) is of linguistic complexity $\mathcal{L}$ and their other side (e.g., RHS) of at most $\mathcal{L}-1$, but also contains simpler axioms, of smaller linguistic complexity levels. For instance, KBs of level $\mathcal{L} = 0$ contain only very simple facts or subsumption axioms that do not contain any Boolean constructors but can contain one quantifier, such as *Enthusiastic* $\sqsubseteq \exists supports.Enthusiastic$ (translated in natural language as "Enthusiastic people support someone enthusiastic"), but KBs of level $\mathcal{L} = 3$ can contain subsumption axioms as complex as the first one appearing in Fig. 1.

It is important to discern the notion of linguistic complexity of a sentence from its length. We do not focus here only on sentences that contain, for instance, multiple conjunctions but rather on sentences with a more complex structure (with quantifiers as well), leading to increased linguistic complexity.

---

[5] The complete Appendix can be accessed at https://github.com/angelosps/DELTA/blob/master/Appendix.pdf.

To keep KBs processable by reasoners, the subsumption axioms can contain up to seven atomic concepts and up to two nested quantifiers (e.g., $\exists likes.(\exists loves.Cat)$), which describes the entities that like some entity that loves some cat). All KBs are rather small (with a minimum of 3 subsumption axioms and 1 fact, and a maximum of 14 subsumption axioms and 12 facts) and are checked for satisfiability and consistency with HermiT.

3.2   Query Generation

For an inference depth $\mathcal{D}$, a *true query* q is an axiom or fact selected from the Delta-closure of a consistent $\mathcal{K}$, such that $depth(q, \mathcal{K}) = \mathcal{D}$. An *unknown query* (answer=unknown) is generated by creating a random fact or statement (using the corresponding PCFG) such that it does not belong to the Delta-closure of $\mathcal{K}$ and is consistent with $\mathcal{K}$. A *false query* (answer=false) can be generated in three ways:

- From an inconsistent $\mathcal{K}$: for every $a \in \mathcal{K}$ if $\mathcal{K} \setminus \{a\}$ is consistent, then a is a false query over the KB $\mathcal{K} \setminus \{a\}$.
- From a consistent $\mathcal{K}$: i) By negating a true query q with $depth(q, \mathcal{K}) = \mathcal{D}$ (and applying De Morgan's laws). ii) By automatically generating an appropriate axiom or fact a such that $\mathcal{K} \cup \{a\}$ is inconsistent and $depth(a, \mathcal{K}) = \mathcal{D}$. For instance, suppose that a KB $\mathcal{K}_1$ contains the axioms $(\forall admires.\bot)(Anne)$ and $\forall admires.\bot \sqsubseteq \forall likes.Quiet$ which in natural language are translated into: "Anne admires none", "All people that admire none like only quiet people". Then, the fact $(\exists likes.\neg Quiet)(Anne)$ stating that "Anne likes someone who is not quiet" forms a false query for $\mathcal{K}$.

The disadvantage of the first approach is that it requires calling the reasoner multiple times, a time-consuming process, especially in KBs with long axioms (e.g., $\mathcal{L} = 3$ KBs). Hence, we used the two latter approaches.

We set the reasoning depth limit to five (i.e., $\mathcal{D} = 0, 1, 2, 3, 5$) following the literature [3]. Extending this further would require longer times for the dataset generation.

3.3   Data Translation to Natural Language

The KBs and queries were translated to natural language using templates. The templates were created based on the user-friendly Manchester syntax for $\mathcal{ALCQ}$ [8]. Following this syntax, the intersection ($\sqcap$) and union ($\sqcup$) operators, are translated as "and" and "or", respectively, the existential ($\exists$) quantifier is translated as "someone" or "something" (depending on whether the pool is about people or things), the universal ($\forall$) as "only", and the number restrictions $\leq, \geq$ as "at most" and "at least". Also, we use the word "that" for intersections and nested quantifiers. For instance, the fact $(\exists likes.(\forall likes.Kind))(Bob)$ is translated as "Bob likes someone that likes only kind people".

Following the template-based approach suggested by Tafjord et al. [20], the axioms of the form $C \sqsubseteq D$ are, roughly, translated into natural language in four different ways: i) "If C then D"; ii) "People/Things that are C are D", iii) "All people/things that are C are D"; iv) If $C = \top$ and $D = \forall R.C'$ this is translated as "Someone/something can R only people/things that are C'". A fact $C(a)$ is translated as "a is C". To ensure grammatical correctness of the resulting sentences, we used the grammar checker LanguageTool[6].

3.4 The Dataset DELTA$_D$

At the end, the examples of the same depth and level from both pools are merged. This results in 20 datasets of 2000 KBs each, with each resulting dataset containing sentences from both vocabularies. From each KB we generated three queries (true, false, unknown) for each depth ($\mathcal{D} = \{0, 1, 2, 3, 5\}$), i.e., from each KB we generated $3 \times (d + 1)$, $d \in \mathcal{D}$, queries. So, in total, the dataset contains $\Sigma_{d \in \mathcal{D}} 3 \times (d + 1) \times 2000 \times (\mathcal{L}_{max} + 1) = 384K$ examples, as we generate KBs for each linguistic complexity level ranging from zero up to $\mathcal{L}_{max} = 3$.

3.5 Statistical Features

As it is thoroughly discussed by Zhang et al. [27], it is impossible to eliminate all statistical features that exist in data, besides, some of them inherently exist in logical reasoning problems.

However, DELTA$_D$ is balanced with respect to some of the most obvious features: i) *KB size*: From the same KB we extract all three types of questions (true, false, unknown); ii) *Inference depth*: We keep a KB only if it can provide all three types of questions with the same inference depth; iii) *Formulation of the question*: The translation to natural language is implemented in such a way that the word "not" appears almost equal number of times in true questions (52.39%), false questions (50.71%) and unknown questions (46.60%); iv) *Average length in words*: True questions 10.85, false questions 8.97, unknown questions 10.34.

4 Experiments

We systematically tested the entailment checking ability of supervised fine-tuned DeBERTaV3-large, due to its recent advancements in NLU tasks [6]. We also tested in zero-shot and few-shot prompting the models GPT-3.5 (gpt-3.5-turbo-1106) and GPT-4 (gpt-4) from OpenAI, as they have demonstrated strong performance across various reasoning benchmarks [15]. Our limited resources did not allow us to test the performance of other models, such as the Llama family; we plan to do this in future work.

---

[6] https://pypi.org/project/language-tool-python/.

Table 1. Accuracy of DELTA models on Test (own), on $D_{5,3}$ dataset, and slices of $D_{5,3}$ per depth.

	DELTA$_{0,3}$	DELTA$_{1,3}$	DELTA$_{2,3}$	DELTA$_{3,3}$	DELTA$_{5,3}$
Test (own)	100.0	99.8	99.8	99.6	99.7
$D_{5,3}$	61.2	90.5	95.2	99.3	99.8
$\mathcal{D} = $ N/A	100.0	99.4	99.5	99.2	99.7
$\mathcal{D} = 0$	100.0	100.0	100.0	100.0	100.0
$\mathcal{D} = 1$	43.4	100.0	100.0	100.0	100.0
$\mathcal{D} = 2$	24.5	73.1	99.5	100.0	100.0
$\mathcal{D} = 3$	34.1	71.3	99.5	100.0	100.0
$\mathcal{D} = 4$	29.2	77.6	84.5	99.5	100.0
$\mathcal{D} = 5$	19.3	76.5	83.5	98.5	99.5

Table 2. Accuracy of DELTA models on Test (own) across all levels.

	$\mathcal{D} = 0$	$\mathcal{D} \leq 1$	$\mathcal{D} \leq 2$	$\mathcal{D} \leq 3$	$\mathcal{D} \leq 5$
$\mathcal{L} = 0$	100.0	99.7	99.4	98.9	98.9
$\mathcal{L} \leq 1$	100.0	99.7	99.6	99.7	99.5
$\mathcal{L} \leq 2$	99.9	99.5	99.7	99.7	99.6
$\mathcal{L} \leq 3$	100.0	99.8	99.8	99.6	99.7

4.1 DeBERTa-Based Models

Evaluation Setup. We fine-tuned the DeBERTaV3-large to predict true/false/unknown (i.e., multi-class sentence classification) for each example. A context-question pair was supplied to the model as [CLS] context [SEP] question [SEP]. We used accuracy as the evaluation metric. The test data has an equal balance of true/false/unknown answers, hence the baseline of random guessing is 33.3%. The specifics of the chosen hyper-parameters, which we maintained consistently throughout our experiments, can be found in Appendix 3.

For each combination of depth and level, we trained different models in subsets of DELTA$_D$. A model DELTA$_{i,j}$ is trained on examples of reasoning depth up to i and of linguistic complexity level up to j. For instance, the model DELTA$_{3,2}$ has been trained to depths up to 3 and linguistic complexity levels up to 2. The final model DELTA$_M$ has been trained in all depths and all linguistic complexity levels, i.e., DELTA$_M$ = DELTA$_{5,3}$. We split each dataset into 70% training, 10% validation, and 20% test sets.

Evaluation Results. Table 1 illustrates the performance of DELTA models when trained on up to $\mathcal{L} \leq 3$ linguistic complexity over the various inference depths (the results for smaller levels are presented in Appendix 4). For instance, the column DELTA$_{0,3}$ shows the performance of the model trained on all levels of depth 0. "Test (own)" represents the (held out) test set of the dataset that the model has been trained. The $D_{5,3}$ dataset has questions from all inference depths ($\mathcal{D} \leq 5$) of all levels ($\mathcal{L} \leq 3$). "Depth N/A" refers to the unknown questions, as these are not provable. "$\mathcal{D} = 0$" to "$\mathcal{D} = 5$" lines represent subsets of D$_{5,3}$ of 0-reasoning depth to 5-reasoning depth, respectively.

It is observed that models trained even on $\mathcal{D} \leq 2$ datasets generalize quite well in larger depths (83.5% for questions of $\mathcal{D} = 5$), while when trained on $\mathcal{D} \leq 3$ datasets they show impressive generalization ability (98.5% for questions of $\mathcal{D} = 5$). Finally, we observe that the model trained on $\mathcal{D} \leq 5$ datasets almost masters (99.5–100%) the reasoning task for all reasoning depths and linguistic complexity levels.

Table 2 demonstrates the performance of each model DELTA$_{i,j}$ when tested on "Test (own)". For instance, the cell that corresponds to $\mathcal{D} = 0$, $\mathcal{L} = 0$ shows the accuracy of the model DELTA$_{0,0}$. We observe that for all depths $\mathcal{D} = 0$ to $\mathcal{D} \leq 5$ the models are robust across levels, hence increasing lengths do not affect their performance.

We, also, partitioned the dataset to the various depths, i.e., we extracted from DELTA$_D$ five datasets which contain only data of depth $\mathcal{D} = i$ (of all levels) and *not* up to i. Additionally, we trained a model on a set of $3,200$ examples specifically at depth 3 for all lengths ($\mathcal{L} \leq 3$). The accuracy when tested in questions of depth 3 was 97.5%, it slightly dropped when tested in questions of smaller depths ($\mathcal{D} = 1, 2$) to $\sim 94.5\%$, except for the look-up questions ($\mathcal{D} = 0$), where the accuracy reached 99.0%. The model showed even better performance ($\sim 97.8\%$) in larger depths ($\mathcal{D} = 4, 5$). Differently from the findings of Tian et al. [22], these results demonstrate the model's capacity for generalization across both lesser and greater reasoning depths than those encountered during training.

Zero-Shot Performance of DELTA$_M$ on Other Distributions

Results for Tweaked Dataset. We generated the new dataset DELTA$_T$ by changing the probability distributions of the PCFG for $\mathcal{L} = 3$ as follows: We increased the probability of the universal quantifier ($\forall$) from 0.33 to 0.70 and the probability of the disjunction ($\sqcup$) from 0.50 to 0.80. DELTA$_T$ contains $1,200$ examples of up to reasoning depth $\mathcal{D} \leq 5$. This tweaking has resulted in sentences with 0.8/sentence disjunctions and 0.62/sentence universals. The accuracy of DELTA$_M$ on DELTA$_T$ was 100.0% for both true and false questions, 98.9% for unknown questions, and, 99.6% overall. As it is evident, the model is robust according to this tweaked distribution.

Results for ProofWriter Dataset [20]. The reason for choosing this dataset is that it was generated in a similar way (using PCFGs) as DELTA$_D$, it is under the open-world assumption and, partly, we have used the same pool of terms (Pool A). DELTA$_M$ demonstrated very high accuracy in true questions (95.2%) and false questions (94.0%) but low accuracy (50.8%) in unknown questions. On average the accuracy was 75.7%. The very high accuracy for true/false questions is a surprising result as although DELTA$_D$ and ProofWriter have many common types of subsumption axioms/facts, ProofWriter also contains subsumption axioms that involve individual names (e.g. "If Bob is blue then Erin is red") and negated role assertions (e.g., "Bob does not like Erin"), which are not supported by $\mathcal{ALCQ}$ and therefore are not contained in the training set of DELTA$_M$. The low performance of DELTA$_M$ in unknown questions can be attributed to the different generation processes among the two datasets. According to the generation process described in Tafjord et al. [20] (the source code is not openly available), the unknown questions in ProofWriter contain terms that appear in the context, whereas, as described in Sect. 3.2, unknown questions in DELTA$_D$ are formulated by choosing random terms from the corresponding pool, thus they can be completely irrelevant to the context. Hence, we can assume this is a statistical feature that DELTA$_M$ may have learned.

Results for Use Case Scenario. We utilized the ontology subsumption axioms and facts (generated from lab experiments) from [23] and generated $1,500$ examples for fuel cell system diagnostics. The context contained subsumption axioms of the form "If a system is in a state that is described by a low voltage value that is a result of an observation made by some voltage sensor that is a reliable sensor then the system is under some flooding" and facts of the form "v1 is a high voltage value". Again, DELTA$_M$ performed particularly well (94.0% accuracy). Detailed results of this section along with a sample of this dataset are presented in Appendix 6. Also, the full dataset is openly available.

Handcrafted Quality Tests of DELTA$_D$. To test the quality of DELTA$_D$ on which DELTA$_M$ is trained, we created simple test examples based on some of the most important knowledge base equivalences according to Rudolph [17]. The examples are demonstrated in detail in Appendix 7.

For instance, for the conjunction subsumption axiom, we provided the context: "Anne is red and green" and the two (true) questions: "Anne is red" and "Anne is green". The model performs well overall, answering correctly 24/29 questions, however, it seems that in contexts involving number restrictions, it returned the answer "unknown", which in two out of the six cases was wrong (notice though that the set of questions with number restrictions in DELTA$_D$ was balanced with respect to their answers).

Additionally, it failed to learn the property $A \sqcap B \sqsubseteq A \sqcup B$. This became evident through the test: "Context: Anne is red and green. Question: Anne is red or green. Answer: True. Prediction: Unknown". To find where it fails in the reasoning chain, we asked the model the intermediate sentences "Anne is green" and "Anne is red", to which it returned (correctly) the answer "true", but it returned "unknown" to the question "If someone is red and green then they are red or green". Whereas, in the test "Context: Anne is red. Anne is green. Question: Anne is red or green." the predicted answer was, again, falsely, "unknown".

Table 3. Average accuracy per inference depth of 0-shot, 9-shot GPT-3.5, GPT-4 on 100 examples from DELTA of linguistic complexity level $\mathcal{L} \leq 1$.

	0-shot GPT-3.5	0-shot GPT-4	9-shot GPT-3.5	9-shot GPT-4
$\mathcal{D} = \text{N/A}$	60	98	75	92
$\mathcal{D} = 0$	74	80	84	89
$\mathcal{D} = 1$	68	57	82	86
$\mathcal{D} = 2$	48	42	60	77
$\mathcal{D} = 3$	43	32	67	80
$\mathcal{D} = 4$	44	30	57	83
$\mathcal{D} = 5$	46	40	57	83

4.2 GPT Models

Evaluation Setup. We tested GPT-3.5-turbo and GPT-4 models from the chat completion API provided by OpenAI.

Our examples were limited to linguistic complexity $\mathcal{L} \leq 1$, due to the models' context width limit: contexts of $\mathcal{L} \geq 2$ could not fit in the window for the few-shot setting. For the same reason, the maximum number of training shots was limited to 9. To enforce deterministic behavior to the models we set `temperature=0`. To make the responses less verbose we set `max_tokens=3`.

As transformer-based language models undergo pre-training through a certain form of language modeling objective, the most common approach to evaluate these models in the zero/few-shot setting is by employing prompt engineering techniques. To formulate our prompt, we used the guidelines[7] from OpenAI and our approach was based on the deductive reasoning prompts presented in [21]. The prompt that we concluded in was the following: {"role": "system", "content": "You are an assistant capable of logical reasoning. Answer to the given question with 'True', 'False', or 'Unknown' based on the context."}.

Evaluation Results. In Table 3 we present the average accuracy (over 100 examples) per inference depth of 0-shot and 9-shot prompting for GPT-3.5 and GPT-4. It is noted that GPT-4 has good performance (max 92% and min 77%) with just 9 shots. Also, it is evident that the models consistently struggle in increased inference depths, demonstrating that our dataset is challenging even for $\mathcal{L} \leq 1$.

4.3 Tests on Symbolic Data

To test the effect of the semantics of the words on the performance of DELTA$_M$ we created the dataset SoftSymbolic, generated by replacing consistently in the test set the words appearing in pools A and B with symbols. Specifically, all individuals (e.g., **Anna**) were replaced with an a_i symbol (e.g., a_3), all classes (e.g., **smart**) with an C_j symbol (e.g., C_2), and all roles (e.g., **likes**) with an R_k symbol (e.g., R_5), where i, j, k is some ID number. The average performance of

---

[7] https://platform.openai.com/docs/guides/prompt-engineering/strategy-write-clear-instructions.

Table 4. Average accuracy per inference depth of DELTA$_M$ on SoftSymbolic and HardSymbolic datasets.

	Soft.	Hard.
$\mathcal{D} = $ N/A	84.1	71.8
$\mathcal{D} = 0$	100.0	58.3
$\mathcal{D} = 1$	99.3	39.4
$\mathcal{D} = 2$	96.3	45.5
$\mathcal{D} = 3$	94.9	68.2
$\mathcal{D} = 4$	97.0	66.6
$\mathcal{D} = 5$	95.6	60.7

Table 5. Average accuracy per inference depth of 0-shot, 9-shot GPT-3.5, GPT-4 on 100 examples (per depth) of the HardSymbolic dataset.

	0-shot GPT-3.5	0-shot GPT-4	9-shot GPT-3.5	9-shot GPT-4
$\mathcal{D} = $ N/A	26	100	70	86
$\mathcal{D} = 0$	72	32	58	78
$\mathcal{D} = 1$	70	2	38	64
$\mathcal{D} = 2$	60	2	64	66
$\mathcal{D} = 3$	56	4	64	54
$\mathcal{D} = 4$	54	0	64	58
$\mathcal{D} = 5$	60	2	70	46

DELTA$_M$ over the SoftSymbolic dataset was 95.3%, hence in contrast to Tang et al. [21], we conclude that DELTA$_M$ is not affected by the lack of semantics.

To check if the models can perform over purely logical examples, we generated the HardSymbolic dataset, which resulted from the SoftSymbolic by also utilizing the DL terminology: the word "some" (corresponding to the existential quantifier) was translated as "exists", the "if ... then ..." (corresponding to subsumption) as "is subsumed by", etc. Samples of each dataset are presented in Appendix 5.

The performance of DELTA$_M$ on both Soft and Hard Symbolic datasets is presented in Table 4. We observe that the average performance of DELTA$_M$ over the HardSymbolic dataset dropped to 58.5%. This is an expected result, as the structure of the tested sentences was very different from the sentences in which DELTA$_M$ had been trained. We tested also the GPT models on (100 examples of) the HardSymbolic dataset where they showed similar performance to DELTA$_M$. The results are shown in Table 5. The average accuracy of GPT-3.5 0-shot was 57%, 9-shot 61%; and GPT-4 20%, 65%, respectively. Hence, TLMs seem to struggle with purely logical datasets.

5   Related Work

Multiple surveys [10, 25, 26] in the literature describe the most recent research developments on the use of transformers for reasoning tasks. One of the first datasets generated for this purpose was from Clark et al. [3] with RuleTaker,

Table 6. The state-of-the-art benchmarks for deductive reasoning with transformers.

	#Questions	Avg size of sentences	Max size of sentences	Formal Language	Generating method
bAbI Task15 [24]	2K	4.6	5	Role assertions	Synthetic
Ruletaker [3]	520K	6.13	27	Conj. Sub. Axioms	Synthetic
ProofWriter [20]	500K	6.07	27	Conj. Sub. Axioms	Synthetic
BirdsElectricity [20]	5K	12.85	22	Conj. Sub. Axioms	Synthetic
ParaRules$_{C/O}$ [20]	40K	11.85/11.78	37/37	Conj. Sub. Axioms	Synthetic & Manual
FOLIO [5]	1.4K	10.58	59	FOL\ Num. Restr.	Synthetic & Manual
PrOntoQA [18]	5.8K	6.2	20	FOL\ Num. Restr.	Synthetic
LogicNLI [22]	30K	8.72	25	FOL\ Num. Restr.	Synthetic
DELTA $_D$ ($\mathcal{L}_0$/ $\mathcal{L}_1$/ $\mathcal{L}_2$/ $\mathcal{L}_3$)	96K each, 384K (in total)	6.28/9.71/ **13.08/13.08**	26/38/ 50/**62**	$\mathcal{ALCQ}$	Synthetic

demonstrating the potential of transformers to perform logical question answering under CWA by training TLMs on synthetic datasets. However, their approach was limited to short expressions of simple conjunctive subsumption axioms. Tafjord et al. [20], generated the ProofWriter datasets (under CWA and OWA) and with a T5 [16]-based model fined-tuned on ProofWriter showed that TLMs can generate proofs with high accuracy (94.8% for depth 5). We generated DELTA$_D$ based on the approach for the generation of the datasets RuleTaker and ProofWriter, i.e., using PCFGs. However, DELTA$_D$ is different from these datasets as i) $\mathcal{ALCQ}$ is a much more expressive logic language hence we produced new PCFGs; ii) we have defined different PCFGs for each linguistic complexity level (which has not been done for any other dataset in the literature); iii) it is balanced regarding the aspects discussed in Sect. 3.5.

In more expressive contexts, Ontañón et al. [14] showed that TLMs perform well (up to 90.5%) over contexts generated by propositional logic and a small subset of FOL. Han et al. [5], with the FOLIO dataset (1.4K), generated from FOL sentences -but without number restrictions-, tested the ability of various TLMs for the same reasoning task and concluded that RoBERTa [12] performed best among all tested models (including GPT-3 and Codex) but still, the performance was low. Tian et al. [22] introduced the much richer synthetic dataset LogicNLI (30K), under OWA for diagnosing TLMs' ability in FOL reasoning, showing that even their best-performing model does not learn to perform reasoning tasks and cannot generalize to different scenarios. Schlegel et al. [19] generated a very simple dataset (containing a single conjunction) for satisfiability checking and showed that models that perform well on hard problems do not perform equally well on easier ones, concluding that transformers cannot learn the underlying reasoning rules rather than they tend to overfit to patterns in the generated data. Also, Zhang et al. [27], and Tian et al. [22] achieved similar results. Bang et al. [2] studied ChatGPT's [11] deductive reasoning ability on bAbi task 16 [24] and EntailmentBank [4], performing merely well. In addition, differently from our results (where the performance decrease was small), Tang et al. [21] showed that TLMs perform significantly better when using natural language instead of symbolic representations of logical facts and subsumption axioms.

Most of the aforementioned benchmarks are composed of short sentences; the ones with longer sentences (avg. 13 words/sentence) are small ($\leq$ 40K), while none of them has examples with numerical restrictions. This is better demonstrated with Table 6, where we present the metrics of the datasets that are most relevant to DELTA$_D$ (Entailment Bank is omitted as it does not conform to a specific formal language). A work that is close to our research is that of He et al. [7], who tested the ability of TLMs, and specifically of RoBERTa, to perform natural language inference tasks over existing OWL2 ontologies (e.g., FoodOn, Shema.org). However, the task studied is different: in He et al. [7], given two concept expressions C and D the TLM is asked to infer if one entails/contradicts the other, while in this work TLMs decide if a sentence can be inferred from *a set of* subsumption axioms and facts, i.e., a KB.

Relevant to our research is also the work of Madusanka et al. [13], who investigated the effects of the various types of quantifiers on the performance of TLMs. As the generated dataset is not openly available, it is hard to assess its complexity and hence its relevance to DELTA$_D$. However, it is worth noting that they do not investigate systematically the aspects that we have focused on in this work (inference depth, linguistic complexity).

6   Conclusions and Future Work

We generated the only large dataset (384K) in the literature that targets expressive DLs (namely, $\mathcal{ALCQ}$), enjoys both high expressivity and high linguistic complexity, and is publicly available for further understanding the functionality of TLMs. We showed that DELTA$_M$, can carry out entailment checking over expressive synthetic datasets with very high accuracy, regardless of the linguistic complexity of the context. Differently from recent results in the literature, our model *has* learned to generalize on unseen reasoning depths, smaller or greater. Zero-shot tests showed that DELTA$_M$ is mostly robust to other distributions. Tests with the GPT family showed that GPT-4 can have significant performance with only a few shots. The high accuracy of zero-shot tests in a real-world scenario demonstrates the potential of TLMs for performing reasoning tasks, bypassing the need for formal representations by domain experts.

Our qualitative tests revealed the need for systematic evaluation techniques for synthetically generated datasets. This will be our next step in future work. Furthermore, we plan to explore the upper limit of the expressivity of the logic language so that a TLM can perform reasoning tasks with high accuracy. Finally, we will expand our evaluation section with other state-of-the-art generative models.

Acknowledgments. This work has been partially supported by project MIS 5154714 of the National Recovery and Resilience Plan Greece 2.0 funded by the European Union under the NextGenerationEU Program.

References

1. Baader, F., Calvanese, D., McGuinness, D.L., Nardi, D., Patel-Schneider, P.F. (eds.): The Description Logic Handbook: Theory, Implementation, and Applications. Cambridge University Press, Cambridge (2003)
2. Bang, Y., et al.: A multitask, multilingual, multimodal evaluation of ChatGPT on reasoning, hallucination, and interactivity. CoRR abs/2302.04023 (2023). https://doi.org/10.48550/ARXIV.2302.04023
3. Clark, P., Tafjord, O., Richardson, K.: Transformers as soft reasoners over language. In: Bessiere, C. (ed.) Proceedings of the Twenty-Ninth International Joint Conference on Artificial Intelligence, IJCAI 2020. pp. 3882–3890. ijcai.org (2020). https://doi.org/10.24963/ijcai.2020/537
4. Dalvi, B., et al.: Explaining answers with entailment trees. In: Conference on Empirical Methods in Natural Language Processing (2021). https://api.semanticscholar.org/CorpusID:233297051
5. Han, S., et al.: FOLIO: natural language reasoning with first-order logic. CoRR abs/2209.00840 (2022). https://doi.org/10.48550/ARXIV.2209.00840
6. He, P., Gao, J., Chen, W.: DeBERTaV3: improving DeBERTa using ELECTRA-style pre-training with gradient-disentangled embedding sharing. In: The Eleventh International Conference on Learning Representations, ICLR 2023, Kigali, Rwanda, 1–5 May 2023. OpenReview.net (2023). https://openreview.net/pdf?id=sE7-XhLxHA
7. He, Y., Chen, J., Jimenez-Ruiz, E., Dong, H., Horrocks, I.: Language model analysis for ontology subsumption inference. In: Rogers, A., Boyd-Graber, J., Okazaki, N. (eds.) Findings of the Association for Computational Linguistics: ACL 2023. pp. 3439–3453. Association for Computational Linguistics, Toronto, Canada (2023). https://doi.org/10.18653/v1/2023.findings-acl.213, https://aclanthology.org/2023.findings-acl.213
8. Horridge, M., Drummond, N., Goodwin, J., Rector, A.L., Stevens, R., Wang, H.: The Manchester OWL syntax. In: Grau, B.C., Hitzler, P., Shankey, C., Wallace, E. (eds.) Proceedings of the OWLED*06 Workshop on OWL: Experiences and Directions, Athens, Georgia, USA, 10–11 November 2006. CEUR Workshop Proceedings, vol. 216. CEUR-WS.org (2006). https://ceur-ws.org/Vol-216/submission_9.pdf
9. Horridge, M., Parsia, B., Sattler, U.: Laconic and precise justifications in OWL. In: Sheth, A., et al. (eds.) ISWC 2008. LNCS, vol. 5318, pp. 323–338. Springer, Heidelberg (2008). https://doi.org/10.1007/978-3-540-88564-1_21
10. Huang, J., Chang, K.C.: Towards reasoning in large language models: a survey. In: Rogers, A., Boyd-Graber, J.L., Okazaki, N. (eds.) Findings of the Association for Computational Linguistics: ACL 2023, Toronto, Canada, 9–14 July 2023, pp. 1049–1065. Association for Computational Linguistics (2023). https://doi.org/10.18653/V1/2023.FINDINGS-ACL.67

11. Liu, Y., et al.: Summary of ChatGPT-related research and perspective towards the future of large language models. Meta-Radiology **1**(2), 100017 (2023). https://doi.org/10.1016/j.metrad.2023.100017

12. Liu, Y., et al.: RoBERTa: a robustly optimized BERT pretraining approach. CoRR abs/1907.11692 (2019). http://arxiv.org/abs/1907.11692

13. Madusanka, T., Zahid, I., Li, H., Pratt-Hartmann, I., Batista-Navarro, R.: Not all quantifiers are equal: probing transformer-based language models' understanding of generalised quantifiers. In: Bouamor, H., Pino, J., Bali, K. (eds.) Proceedings of the 2023 Conference on Empirical Methods in Natural Language Processing, pp. 8680–8692. Association for Computational Linguistics, Singapore (2023). https://aclanthology.org/2023.emnlp-main.536

14. Ontañón, S., Ainslie, J., Cvicek, V., Fisher, Z.: LogicInference: a new dataset for teaching logical inference to seq2seq models. CoRR abs/2203.15099 (2022). https://doi.org/10.48550/arXiv.2203.15099

15. OpenAI: GPT-4 technical report. CoRR abs/2303.08774 (2023). https://doi.org/10.48550/ARXIV.2303.08774

16. Raffel, C., et al.: Exploring the limits of transfer learning with a unified text-to-text transformer. J. Mach. Learn. Res. **21**, 140:1–140:67 (2020). http://jmlr.org/papers/v21/20-074.html

17. Rudolph, S.: Foundations of Description Logics, pp. 76–136. Springer, Heidelberg (2011). https://doi.org/10.1007/978-3-642-23032-5_2

18. Saparov, A., He, H.: Language models are greedy reasoners: a systematic formal analysis of chain-of-thought. In: The Eleventh International Conference on Learning Representations, ICLR 2023, Kigali, Rwanda, 1–5 May 2023. OpenReview.net (2023). https://openreview.net/pdf?id=qFVVBzXxR2V

19. Schlegel, V., Pavlov, K.V., Pratt-Hartmann, I.: Can transformers reason in fragments of natural language? In: Goldberg, Y., Kozareva, Z., Zhang, Y. (eds.) Proceedings of the 2022 Conference on Empirical Methods in Natural Language Processing, EMNLP 2022, Abu Dhabi, United Arab Emirates, 7–11 December 2022, pp. 11184–11199. Association for Computational Linguistics (2022). https://aclanthology.org/2022.emnlp-main.768

20. Tafjord, O., Dalvi, B., Clark, P.: ProofWriter: generating implications, proofs, and abductive statements over natural language. In: Zong, C., Xia, F., Li, W., Navigli, R. (eds.) Findings of the Association for Computational Linguistics: ACL/IJCNLP 2021, Online Event, 1–6 August 2021. Findings of ACL, vol. ACL/IJCNLP 2021, pp. 3621–3634. Association for Computational Linguistics (2021). https://doi.org/10.18653/v1/2021.findings-acl.317

21. Tang, X., et al.: Large language models are in-context semantic reasoners rather than symbolic reasoners. CoRR abs/2305.14825 (2023). https://doi.org/10.48550/ARXIV.2305.14825

22. Tian, J., Li, Y., Chen, W., Xiao, L., He, H., Jin, Y.: Diagnosing the first-order logical reasoning ability through LogicNLI. In: Moens, M., Huang, X., Specia, L., Yih, S.W. (eds.) Proceedings of the 2021 Conference on Empirical Methods in Natural Language Processing, EMNLP 2021, Virtual Event/Punta Cana, Dominican Republic, 7–11 November 2021, pp. 3738–3747. Association for Computational Linguistics (2021). https://doi.org/10.18653/v1/2021.emnlp-main.303

23. Tsalapati, E., et al.: Enhancing polymer electrolyte membrane fuel cell system diagnostics through semantic modelling. Expert Syst. Appl. **163**, 113550 (2021). https://doi.org/10.1016/J.ESWA.2020.113550

24. Weston, J., Bordes, A., Chopra, S., Mikolov, T.: Towards AI-complete question answering: a set of prerequisite toy tasks. In: Bengio, Y., LeCun, Y. (eds.) 4th International Conference on Learning Representations, ICLR 2016, San Juan, Puerto Rico, 2–4 May 2016, Conference Track Proceedings (2016). http://arxiv.org/abs/1502.05698
25. Yang, Z., Du, X., Mao, R., Ni, J., Cambria, E.: Logical reasoning over natural language as knowledge representation: a survey. CoRR abs/2303.12023 (2023). https://doi.org/10.48550/arXiv.2303.12023
26. Yu, F., Zhang, H., Wang, B.: Nature language reasoning, a survey. CoRR abs/2303.14725 (2023). https://doi.org/10.48550/ARXIV.2303.14725
27. Zhang, H., Li, L.H., Meng, T., Chang, K., den Broeck, G.V.: On the paradox of learning to reason from data. CoRR abs/2205.11502 (2022). https://doi.org/10.48550/arXiv.2205.11502

Additive Counterfactuals for Explaining Link Predictions on Knowledge Graphs

Roberto Barile[1]([✉]) [iD], Claudia d'Amato[1,2] [iD], and Nicola Fanizzi[1,2] [iD]

[1] Dipartimento di Informatica, University of Bari Aldo Moro, Bari, Italy
{roberto.barile,claudia.damato,nicola.fanizzi}@uniba.it
[2] CILA – University of Bari Aldo Moro, Bari , Italy

Abstract. The goal of *Link Prediction* is to predict missing facts in *Knowledge Graphs* that are inherently incomplete. *Embedding Models* are generally adopted for this purpose since they are effective and scalable. However, they lack both interpretability and, more importantly, explainability, which is crucial in many tasks and domains. To fill this gap, post-hoc explanations are often computed. A post-hoc explanation for an embedding-based link prediction typically consists of discovering facts that made the prediction possible. Methods that can yield such explanations tend to provide *subtractive counterfactual explanations*, i.e., identify facts whose removal has the greatest impact on the predictions. However, since KGs are incomplete, there may be facts that are missing in the KG but useful for the explanations. Therefore, we formalize a new complementary approach based on the generation of plausible and meaningful *additional facts* to be used for providing explanations. Specifically, we propose IMAGINE, that can generate *additive counterfactual explanations*, by identifying the *additional facts* that most affect predictions. It builds on a *post-training* technique, which is used to assess the impact of adversarial modifications of the knowledge graph, and on *Graph Summarization*, that is used for identifying plausible additions. We present a comparative experimental study that proves the effectiveness of the proposed solution through quantitative and qualitative evaluations.

1 Introduction

Knowledge Graphs (KGs) [18] have emerged as tools for representing, navigating, and querying the growing flood of data by making knowledge in diverse domains accessible to both humans and machines. A KG is a multi-relational graph composed of entities and relations, represented as nodes and edges, respectively. KGs are often integrated with ontologies that formally define classes and relations, allowing for advanced inference capabilities. There are several examples of large KGs, including enterprise products [14,33] and open collections [2,7,28].

Despite their proven utility, the inherent incompleteness of KGs often complicates their use, also as a consequence of the *Open World Assumption* (OWA) typically made with KGs, as they result from complex, incremental, and distributed building processes [18]. This underpins the importance of *Link*

M. Alam et al. (Eds.): EKAW 2024, LNAI 15370, pp. 346–363, 2025.
https://doi.org/10.1007/978-3-031-77792-9_21

Prediction (LP) and *Triple Classification* tasks, which aim at completing KGs by inferring missing facts and deciding on the truth of triples, respectively.

In this paper, we focus on LP that is generally tackled by means of *Machine Learning* (ML) methods grounded in *Knowledge Graph Embeddings* (KGEs) [30], which gained prominence because of their superior scalability. KGE models are based on vectors in low-dimensional spaces learned to represent elements in the KGs: they allow complex tasks to be framed as linear algebra operations. However, such models tend to operate as "opaque boxes" whose predictions are difficult to explain, potentially undermining their credibility and the confidence in their predictions. This opacity can be particularly problematic in fields where understanding the predictions is critical, such as finance, healthcare, and pharmacology. For example, relations among certain drugs and side effects may be predicted, hence it becomes crucial having clues in order to possibly identify potential associations [27] and/or understanding if such predictions may influence decisions on investment in a specific drug research.

In this regard, *Explainable Artificial Intelligence* (XAI) [17] methods come into focus to enhance the transparency and comprehensibility of ML models. They can be divided into 1) *post-hoc* methods, useful for computing explanations given the predictions, and 2) *clear box* methods, that can produce predictions along with their explanations [17]. We will focus on the former approach because, unlike the latter, these methods can be independent of the LP method adopted. In the *post-hoc* setting, one of the most valuable types of explanation is the *Counterfactual Explanation* (CE), which indicates potential changes in the knowledge source that could alter the outcome [16]. CEs can be divided into *Subtractive CEs* (SCEs), for identifying salient removal changes in the knowledge source, and *Additive CEs* (ACEs), for identifying significant additional information [9]. A *post-hoc* explanation for a prediction typically consists of a specific set of facts that allow such an inference.

Several promising *post-hoc* explanation methods have emerged, e.g., [36], KELPIE [31] and KELPIE++ [4], but all of them are grounded on SCEs identifying facts that most significantly affect the predictions if removed. Hence, only existing facts in the KG are considered for generating explanations for the computed predictions. However, as KGs are to be considered inherently incomplete, there may be facts that are missing from the KG that are actually useful for building the explanations. In this perspective, ACEs could be computed for identifying *additional facts* that would most affect the predictions if inserted, thus providing complementary explanations to SCEs. This view is also coherent with the OWA characterizing the KGs in logic contexts, for which there could be plausible facts missing in the KG. Even more so, cognitive studies proved that, when considering explanations, people tend to prefer ACEs to SCEs because the former foster creative thinking [9].

Motivated by these observations, we formalize a new *post-hoc* explanation method that is grounded in the generation of plausible and meaningful *additional facts* to be employed for providing explanations. Specifically, we propose

IMAGINE a framework for providing *Additive Counterfactual Explanations* for LP, which is able to generate them, by identifying the *additional facts* that most affect the predictions. It builds on a *post-training* technique introduced in [31] and that is used for assessing the impact of adversarial modifications of the KG, and on *Graph Summarization* [10], that is adopted for identifying plausible *additional facts*.

It is worth noting that the explanations computed by IMAGINE can be complementary to SCEs, since both SCEs and ACEs can help to understand the rationale behind the predictions by analyzing how the predictions are affected, that is, either improve or worsen. Ideally, ACEs are those showing to improve the predictions. For this purpose, we perform a comparative experimental analysis in order to verify this hypothesis. It is also worthwhile to be mentioned that valuable ACEs that are computed by IMAGINE could be exploited for understanding how the KG could be refined by optimizing the LP performances. Moreover, even if the generated *additional facts* may not be valid, they still provide valuable insights, as they help to understand how the model behaves under different conditions.

The contributions of the paper are the following:

- we formalize IMAGINE, to the best of our knowledge, the first solution for computing *Additive Counterfactual Explanations* to LP in KGs, to be meant as complementary to solutions for generating SCEs
- IMAGINE offers insights for understanding how the KG could be refined by optimizing the performance of LP
- we experimentally prove, via quantitative and qualitative evaluation, that IMAGINE provides valuable explanations, by adopting state-of-the-art evaluation protocols
- we experimentally prove that ACEs and SCEs approaches are complementary and argue on the need for the formalization of a new unified framework.

The rest of the paper is organized as follows. Section 2 reviews the most recent and effective approaches for explaining LPs. Section 3 presents fundamentals for the paper understanding. Section 4 details our new solution, IMAGINE, that computes explanations to LP results by adopting ACEs. Section 5 illustrates the comparative experimental study, proving the effectiveness of IMAGINE via quantitative and qualitative evaluation. Section 6 summarizes the achievements and challenges and outlines future research directions.

2   Related Works

In this section, we survey *post-hoc* methods for explaining LP on KGs. The firstly emerged methods [29,36] explain a prediction by providing a single fact, that is a statement: subject, predicate, object. Specifically, in [36] the fact whose insertion or removal most significantly poisons the prediction is returned, while CRIAGE [29], employs approximated *Influence Functions* [23], and focuses on facts featuring as object either the subject or the object of the prediction. Hence, we regard these methods as providing SCEs.

With the goal of targeting more complex explanations, methods explaining a prediction by providing a path from the subject to the object of the prediction have been proposed. CROSSE [38] and APPROXSEMANTICCROSSE [11] find paths exploiting the KG topology and (semantic) similarity of entities and/or relationships with respect to the predicted fact, while, LINKLOGIC [22] adopts a perturbation based approach.

More recent methods explain a prediction by returning a specific set of relevant facts, rather than a single fact or a path. KELPIE [31] employs a novel *post-training* process for adversarial re-training of the KGE adopted for the prediction, but that can be tailored to different KGE models. KELPIE++ [4] integrates *Graph Summarization* into KELPIE making it more efficient, while enhancing the effectiveness of the explanations and allowing the presentation/inspection of the explanations at different level of granularity. Both KELPIE and KELPIE++ provide two synergistic types of explanations: *necessary* and *sufficient* explanations. A *necessary* explanation for a prediction is a set of facts without which the model would not have made the same inference. Such *necessary* explanations fall into the SCEs category. In contrast, a *sufficient* explanation is a set of facts that makes the model able to replicate the given prediction for other entities. The process for finding *sufficient* explanations involves generating *additional facts* by starting with those involving the subject of the prediction and then corrupting these facts with other entities. Similarly, KGEx [3] utilizes subgraph sampling and *Knowledge Distillation*, while [5] provides explanation by performing *abduction* on a logical theory learned through an ML approach.

Differently, KE-X [39] frames KGE score functions in a *Message Passing* framework to identify the facts that maximize the *Information Gain* with respect to the prediction. Notably, GENI [1] also includes schema axioms in the explanations by evaluating whether predicate embeddings satisfy certain numerical properties. However, it is restricted to translational and bilinear KGE models and as such it is not fully model agnostic. KGEXPLAINER [26] uses a greedy search algorithm based on perturbations. All the aforementioned methods build explanations from facts in the KG, whereas our goal is to employ missing but reasonable facts. In *sufficient* explanations, facts missing in the KG are involved. Yet the original facts from which *additional facts* are derived as perturbations ultimately contribute to the explanations, but not the *additional facts*. Instead, we aim to provide explanations consisting of actual *additional facts*.

Further methods providing explanations not consisting in a specific set of relevant facts or a path have been also proposed. In [21] *Horn Clauses* are mined to explain predictions, while FEABI [19] can build interpretable vectors from entity embeddings based on propositional features extracted from the KG via *Feature Selection*. In this category, also [6], XTRANSE [37], and GNNEXPLAINER [35] tailored for *Graph Neural Networks* are worthwhile to be mentioned, but they adopt a *clear box* approach rather than the *post-hoc* approach that is our target.

Finally, other *post-hoc* methods that are independent of the task requiring the explanation have been proposed. They can be categorized along their yielding: a) *saliency explanations*, to identify relevant input features as explanations; or b) *prototypes*, to identify training samples as explanations. Methods grounded on *saliency explanation*, such as SHAP [25], are popular, but struggle with predictions based on KGEs whose numerical features lack interpretability. Solutions based on *prototypes* are more suitable for LPs, but often *Influence Functions* are employed to the purpose [20] which pose scalability issues as in CRIAGE.

In contrast with the surveyed methods, IMAGINE focuses on CEs rather than *clear-box* approaches and *post-hoc* methods alternative to CEs, and as such it is applicable to any KGE model. Moreover, being focused on ACEs, IMAGINE is able to exploit the inherent incompleteness of KGs for explanations purposes.

3 Fundamentals

In this section, we present the fundamentals that are functional for the full understanding of proposed explanation solution IMAGINE. In Sect. 3.1 we introduce KGs more formally and the basics of KGE methods. In Sect. 3.2 we recall the notion of quotient graph that is employed by IMAGINE for performing graph summarization, which is exploited for identifying plausible additions.

3.1 Knowledge Graphs and Embedding-Based Link Prediction

A KG is a graph-based data structure $\mathcal{G}(\mathcal{V}, \mathcal{R})$ with $\mathcal{V}$ representing a set of nodes, also known as entities, and $\mathcal{R}$ representing a set of predicates. In the adopted RDF model, a KG is a collection of triples in the format $\langle s, p, o \rangle$, i.e., *subject*, *predicate*, and *object* where $s, o \in \mathcal{V}$ and $p \in \mathcal{R}$.

Various models have been proposed for representing KGs in low-dimensional vector spaces, by learning for each entity and predicate in the KG a unique numerical vector (or *embedding*) in a given space. Several embedding spaces can be employed, e.g., real, point-wise, complex, discrete, Gaussian, manifold. Without loss of generality, in the following we focus on real embeddings (vectors will be denoted in **bold**).

All these models represent each entity $e \in \mathcal{V}$, and each predicate $p \in \mathcal{R}$ by means of an embedding vector $\mathbf{e} \in \mathbb{R}^k$, and $\mathbf{p} \in \mathbb{R}^i$, respectively, where $k, i \in \mathbb{N}$ are hyperparameters. Moreover, each model is associated to a *scoring function* $f : \mathbb{R}^k \times \mathbb{R}^i \times \mathbb{R}^k \to \mathbb{R}$: for each triple $\langle s, p, o \rangle$, the score $f(\langle s, p, o \rangle)$ measures the likelihood of such statement. In the following, we report formulations where higher values convey more plausibility; symmetric formulations can be derived for models where lower scores convey higher likelihood. The embeddings and parameters are learned from the KG by minimizing a loss function grounded on f. To this purpose, the set of triples encoded by $\mathcal{G}$ is divided into a training set $\mathcal{G}_{train}$, a validation set $\mathcal{G}_{val}$ and a test set $\mathcal{G}_{test}$. Beyond entity and predicate embeddings, models can also learn shared parameters that are not explicitly connected to any KG element, and similar to the weights of neural layers.

Given an incomplete triple $\langle s, p, ? \rangle$, LP is performed by computing

$$o = \arg\max_{e \in \mathcal{V}} f(\langle s, p, e \rangle)$$

Moreover, the *rank* of a triple $\langle s, p, o \rangle$ in $\mathcal{G}_{test}$ can be computed as:

$$\text{rank}(\langle s, p, o \rangle) = |\{e \in \mathcal{V} \mid f(\langle s, p, e \rangle) >= f(\langle s, p, o \rangle)\}|$$

Computing the rank of each triple in $\mathcal{G}_{test}$ is required for evaluating the LP performance.

3.2 Quotient Graphs and Simulations

In the context of KGs, quotient graphs aim at summarizing the data graph with a higher-level topology [10]. These are based on the notion of quotient set. Given a set X and an equivalence relation $\sim$, X is partitioned into disjoint subsets of equivalent elements. The *quotient set* $X/_\sim$ contains all such subsets, the *equivalence classes*. Then, considering the KG $\mathcal{G}(\mathcal{V}, \mathcal{R})$, its *quotient graph* $\mathcal{Q}(\mathcal{V}/_\sim, \mathcal{R})$ w.r.t. a relation $\sim$ over $\mathcal{V}$ will have $\mathcal{V}/_\sim$, quotient set of $\mathcal{V}$ w.r.t. $\sim$, as its node set and the same predicate set $\mathcal{R}$. Note that the triples in $\mathcal{Q}$ can be defined in different ways, depending on the desired level of preservation of the structure in $\mathcal{G}$.

In our approach, we will focus on a simulation of the original $\mathcal{G}$, requiring a proper level of abstraction, with no need to preserve its complete structure. A *simulation* [18] is a binary relation from a graph $\mathcal{G}$ to a graph $\mathcal{G}'$ that ensures that any path between connected nodes in $\mathcal{G}$ also exists in $\mathcal{G}'$.

Formally: a *simulation* from a graph $\mathcal{G}(\mathcal{V}, \mathcal{R})$ to a graph $\mathcal{G}'(\mathcal{V}', \mathcal{R}')$ is a relation $S \subseteq \mathcal{V} \times \mathcal{V}'$ such that for each triple $\langle x, y, z \rangle$ in $\mathcal{G}$, if $x \, S \, x'$ exists for some $x' \in \mathcal{V}'$, then there exists $z' \in \mathcal{V}'$ such that $z \, S \, z'$ and $\langle x', y, z' \rangle$ is a triple in $\mathcal{G}'$. $\mathcal{G}'$ *simulates* $\mathcal{G}$ when a simulation exists from $\mathcal{G}$ to $\mathcal{G}'$. Adopting simulation, a triple $\langle X, y, Z \rangle$ exists in $\mathcal{Q}$ if and only if, for some $x \in X$ and $z \in Z$, the triple $\langle x, y, z \rangle$ is in the original graph $\mathcal{G}$.

An alternative way for defining the triples in $\mathcal{Q}$ is bisimulation [4]. It is a stricter condition than simulation, indeed, adopting bisimulation, a triple $\langle X, y, Z \rangle$ exists in $\mathcal{Q}$ if and only if, for each $x \in X$ and $z \in Z$, the triple $\langle x, y, z \rangle$ is in the original graph $\mathcal{G}$.

4   The Proposed Approach

We introduce IMAGINE, a search-based method for generating the ACEs for link predictions on KGs. ACEs consist of *additional triples*, i.e., those that are neither explicitly stated in, nor entailed by, the KG, yet are not assumed to be false under OWA.

Specifically, given a true triple $t = \langle s, p, o \rangle$, IMAGINE returns as explanation a set of *additional triples* featuring s that is relevant for t, i.e., leading the LP method to modify its rank. Moreover, we consider an *additional triple* as useful if it is likely part of an explanation.

IMAGINE builds on KELPIE [31] and KELPIE++ [4], that generate effective CEs involving multiple triples, apply to arbitrary KGE models and offer resources for the reproducibility of experiments. As for them, IMAGINE is a *post-hoc* explanation method applicable to any KGE model, but differently from KELPIE and KELPIE++, that are grounded on computing SCEs, to the best of our knowledge, IMAGINE is the first solution providing ACEs.

IMAGINE is organized around the following four main components:

- Triple Builder that generates a set of useful *additional triples* featuring s;
- Pre-Filter that selects the most useful *additional triples*;
- Explanation Builder that combines the pre-filtered triples into *candidate explanations* and identifies sufficiently relevant ones;
- Relevance Engine that computes the relevance of a *candidate explanation*.

IMAGINE overall architecture is similar to the one of the aforementioned systems, but a novel solution for the Triple Builder component is formalized, since it is the component in charge of generating *additional triples* by computing ACEs. Moreover, a modified version of the Relevance Engine is proposed by formalizing a new alternative way for computing the relevance score of candidate explanations. The Pre-Filter and Explanation Builder are adaptations of their counterparts in KELPIE and KELPIE++. Each of these components is described below.

4.1   Triple Builder

For its purpose, the Triple Builder relies on *Graph Summarization*, leveraging the notion of quotient graph (see Sect. 3.2). Essentially, this component abstracts the KG and reconstructs the original one. The reconstruction contains *additional triples*, as the abstraction does not perfectly preserve the original structure. Specifically, each entity inherits triples from the other entities in its equivalence class. The rationale behind this component is that *additional triples* formed with entities in the same equivalence class of the subject of the prediction are more useful than those originating from entities of different types.

In Algorithm 1 we formalize the logic of the Triple Builder. It requires as input the training KG, its quotient and the prediction to be explained. Moreover, we define $\mathcal{G}_{train}^{s}$ as the subgraph of $\mathcal{G}_{train}$ containing the triples featuring s as subject or as object. The quotient graph $\mathcal{Q}_{train}$ of $\mathcal{G}_{train}$ is computed initially and independently of the specific prediction. IMAGINE computes it by exploiting a function $\tilde{Cl}$ to determine the equivalence relation. Such function Cl is defined in [11] as returning the classes to which an entity can be proven to belong to; while, its approximated version $\tilde{Cl}$ simplifies the required reasoning service (*realization*).

As an example, we report in Fig. 1a a very simple KG regarding cities/nations and its quotient graph in Fig. 1b. Moreover, we consider the prediction $\langle Milan, located\ in, Italy \rangle$. Initially, the algorithm identifies the quotient node S

Algorithm 1: Generate additional triples featuring s

Input: triple $\langle s, p, o \rangle$; training graph $\mathcal{G}_{train}$; quotient graph $\mathcal{Q}_{train}$
Output: set of additional triples $\mathcal{A}^s$

1 $S \leftarrow$ get_quotient_node$(s, \mathcal{Q}_{train})$;
2 $triples \leftarrow \{\}$;
3 **foreach** $\langle S, r, O \rangle \in Q_s^{train}$ **do**
4 $\quad$ **if** $s \in S$ **then**
5 $\quad\quad \lfloor \ \mathcal{A}^s \leftarrow \mathcal{A}^s \cup \{\langle s, r, o \rangle \mid o \in O\}$;
6 $\quad$ **if** $s \in O$ **then**
7 $\quad\quad \lfloor \ \mathcal{A}^s \leftarrow \mathcal{A}^s \cup \{\langle o, r, s \rangle \mid o \in S\}$;
8 $\mathcal{A} \leftarrow \mathcal{A}^s \setminus \{\langle s, p, o \rangle\}$;
9 $\mathcal{A} \leftarrow \mathcal{A}^s \setminus \mathcal{G}_{train}^s$;

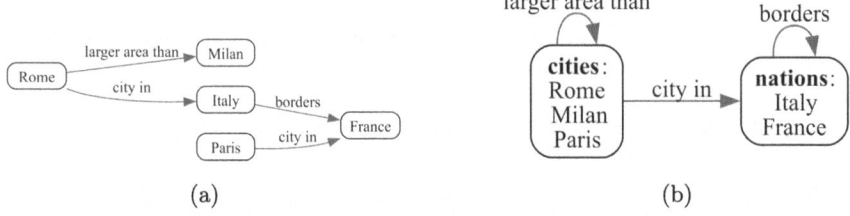

(a)   (b)

Fig. 1. A very simple KG regarding cities/nations (a) and its quotient graph (b)

in $\mathcal{Q}_{train}$ containing the prediction's subject s (Line 1). In our example, s is *Milan*. Then, the output set $\mathcal{A}^s$ is initialized as an empty set (Line 2). The process then continues by analyzing each quotient triple $\langle S, p, O \rangle$ in $\mathcal{Q}_{train}^s$ (Line 3). Specifically, if s is in S, then Triple Builder extends $\mathcal{A}^s$ with the triples $\langle s, r, o \rangle$ for all o in O (Lines 4–5). Conversely, if s is in O, the triples $\langle o, r, s \rangle$ for all o in O are added to $\mathcal{A}^s$ (Lines 6–7). For instance, for the quotient triple $\langle \{Milan, Rome, Paris\}, city\ in, \{Italy, France\} \rangle$, the additional triples $\{\langle Milan, city\ in, Italy \rangle, \langle Milan, city\ in, France \rangle\}$ arise. Finally, the algorithm removes any redundant triples from $\mathcal{A}^s$. Specifically, it removes the input triple t itself (Line 8) and, additionally, it removes the triples already occurring in $\mathcal{G}_{train}^s$ (Line 9).

4.2 Pre-filter and Explanation Builder

The Pre-Filter aims at reducing the complexity for the following stages. Therefore, this component filters $\mathcal{A}^s$ to obtain $\mathcal{F}^s$ by selecting k triples based on a topological measure of utility. Specifically, the utility of a triple $\langle s, p, e \rangle$ is the length of the shortest-path from e to the o. The Explanation Builder, on the other hand, searches for the set of triples in $\mathcal{F}^s$ that make up an acceptable explanation X^* in terms of a given relevance threshold. Specifically, it enumerates all possible candidate explanations (each denoted as X) as sets/combinations of the

---

Algorithm 2: Assess the *relevance* of a *candidate explanation*, I-mode

---

Input: prediction $\langle s, p, o \rangle$, candidate explanation X;
                 training graph $\mathcal{G}_{train}(\mathcal{V}, \mathcal{R})$, KGE model m;
Output: relevance rel

1 $\mathcal{V} \leftarrow \mathcal{V} \cup \{new\}$;
2 $\mathcal{G}_{train}^{new} \leftarrow$ get_triples($\mathcal{G}_{train}^{s}, new$);
3 **new** $\leftarrow$ rand_init();
4 **new** $\leftarrow$ post_train($\mathcal{G}_{train}^{new} \cup X$);
5 $rank_{new} \leftarrow m.\text{rank}(\langle new, p, o \rangle)$;
6 **new** $\leftarrow$ rand_init();
7 **new** $\leftarrow$ post_train($\mathcal{G}_{train}^{new}$);
8 $rank'_{new} \leftarrow m.\text{rank}(\langle new, p, o \rangle)$;
9 $rel \leftarrow (rank'_{new} - rank_{new})/(rank'_{new} - 1)$;

---

triples $\mathcal{F}^s$, then it determines whether any X can be accepted as X^* after invoking the Relevance Engine. Moreover, it features heuristic conditions to prune the space of candidate explanations.

In addition, IMAGINE features an extended version of the Explanation Builder, as suggested in KELPIE++. It includes a step that summarizes $\mathcal{F}^s$ to speed up the search and possibly lead to more effective explanations. The summarization can be performed using two different modes: *simulation* and *bisimulation*.

4.3   Relevance Engine

This component assesses the relevance of a candidate explanation X by estimating the rank variation of the triple $\langle s, p, o \rangle$ after adding the triples in X to G_{train} and re-training the KGE model. Ideally, the relevance of X should be assessed by re-training all the KGEs from scratch on the entire KG. However, this approach is impractical. Hence, we employ *post-training*, which is a scalable approximation of re-training introduced in KELPIE. It consists in adding a new entity to G_{train} and learning the corresponding embedding using a set of samples obtained by adapting the triples in G_{train}^s to the new entity, while keeping all the other embeddings and parameters fixed. The process of *post-training* a new entity is more efficient than the ideal complete re-training, as it solely optimizes one embedding on a limited set of triples.

We formalize the process of IMAGINE in Algorithm 2. Firstly, the Relevance Engine adds a new entity called *new* to the training graph (Line 1). Note that, *new* is an actual entity that the algorithm adds to the set $\mathcal{V}$, rather than a variable representing an element of $\mathcal{V}$. Then, it computes the set $\mathcal{G}_{train}^{new}$ of triples for this entity, starting from $\mathcal{G}_{train}^s$ (Line 2). Specifically, for each triple $\langle s, r, q \rangle$ ($\langle q, r, s \rangle$) in $\mathcal{G}_{train}^s$, it adds to $\mathcal{G}_{train}^{new}$ the triple $\langle new, r, q \rangle$ ($\langle q, r, new \rangle$). Next, it initializes a random embedding formalized as **new** for such entity (Line 3). Then it optimizes this embedding by performing *post-training* on the set $\mathcal{G}_{train}^{new} \cup X$ (Line 4). Then, it computes the $rank_{new}$ of the triple $\langle new, p, o \rangle$, which is obtained by injecting the new entity in the prediction to explain (Line 5).

Algorithm 3: Assess the *relevance* of a *candidate explanation*, W-mode

---

Input: prediction $\langle s, p, o \rangle$, candidate explanation X;
   training graph $\mathcal{G}_{train}(\mathcal{V}, \mathcal{R})$, KGE model m;
Output: relevance rel

1 $\mathcal{V} \leftarrow \mathcal{V} \cup \{new\}$;
2 $\mathcal{G}_{train}^{new} \leftarrow \text{get\_triples}(\mathcal{G}_{train}^{s}, new)$;
3 $\mathbf{new} \leftarrow \text{rand\_init}()$;
4 $\mathbf{new} \leftarrow \text{post\_train}(\mathcal{G}_{train}^{new} \cup X)$;
5 $rank_{new} \leftarrow m.\text{rank}(\langle new, p, o \rangle)$;
6 $\mathbf{new} \leftarrow \text{rand\_init}()$;
7 $\mathbf{new} \leftarrow \text{post\_train}(\mathcal{G}_{train}^{new})$;
8 $rank'_{new} \leftarrow m.\text{rank}(\langle new, p, o \rangle)$;
9 $rel \leftarrow rank_{new} - rank'_{new}$

Next, the Relevance Engine can compute relevance by comparing $rank_X$ with the rank of the prediction. However, an additional step is made (as introduced in KELPIE) for making the process more robust to random fluctuations. Specifically, it re-initializes randomly the embedding **new** (Line 6). Then, it optimizes such embedding, but performing *post-training* on the set $\mathcal{G}^{new}$ rather than $\mathcal{G}_{train}^{new} \cup X$ (Line 7). Next, it computes again $rank_{new}$ (Line 8). Finally, it computes the relevance rel as the rank improvement divided by the ideal rank improvement, which is $rank'_{new} - 1$ (Line 9). To clarify, for a triple with rank k, achieving a rank improvement of $k - 1$ leads to rank it as first, which is the ideal rank.

We also introduce a variant of IMAGINE that we dub W-IMAGINE (to signify *worsen*).

W-IMAGINE differs from IMAGINE solely in the Relevance Engine component. We can formalize the process related to W-IMAGINE in Algorithm 3. The Relevance Engine involves a *post-training* process that is analogous to Algorithm 2, but it computes relevance as a measure of rank worsening, specifically: $rank_{new} - rank'_{new}$. W-IMAGINE allows for assessing the impact of *additional* triples on top ranked predictions, while IMAGINE is tailored to those not ranked as first.

5 Experimental Evaluation

We illustrate the experimental evaluation that was carried out, specifying the experimental setup and presenting quantitative and qualitative results.

5.1 Experimental Setting

We performed the experiments on three datasets: DB50K [32], DB100K [13] and YAGO4-20 [4]. We report statistics on the datasets in Table 1. The common aspect of these datasets is that, along with the RDF triples, they contain OWL (specifically OWL2-DL) statements, including class assertions and other schema

Table 1. Statistics of the datasets

	Entities	Relations	Train triples	Valid triples	Test triples
DB50K	24620	351	32194	123	2095
DB100K	98776	464	587688	49172	49114
YAGO4-20	96910	70	555182	69398	69398

Table 2. Performance of the LP models

	DB100K		DB50K		YAGO4-20	
	H@1	*MRR*	*H@1*	*MRR*	*H@1*	*MRR*
COMPLEX	0.346	0.408	0.410	0.457	0.197	0.245
CONVE	0.301	0.363	0.112	0.144	0.151	0.197
TRANSE	0.061	0.143	0.282	0.367	0.048	0.095

axioms involving classes and relationships. The integration of these datasets with the corresponding ontologies is described in [4]. We employed the reasoner HERMIT [15] offline to materialize the implicit class assertions in the KGs. Next, the function $\tilde{Cl}$ was implemented as a simple lookup over explicit class assertions.

IMAGINE supports any KGE model; therefore, we performed our experiments on TRANSE [8], CONVE [12] and COMPLEX [34] as these represent the three major families of models: *Geometric*, *Deep Learning*, and *Tensor Decomposition* KGE models. Table 2 reports the LP performance of each model that we measured on each dataset in terms of typical measures, namely *Mean Reciprocal Rank (MRR)* and *Hits-at-1 (H@1)* in their filtered variant.

We adopt a methodology building on the one introduced for the evaluation of CRIAGE (also utilized for KELPIE and KELPIE++). It measures effectiveness as the impact on the LP performance achieved through the CEs. Both SCEs and ACEs can either improve or worsen such results, so we conduct experiments that measure effectiveness in terms of both improvement and worsening of LP performance. For evaluating the improvement, for each KGE model, we randomly sample a set P of 100 test triples that are not ranked first by the LP method. After extracting explanations for all predictions in P, we add their triples and re-train the model. As the original model did not yield those predictions, their initial $H@1$ is 0.0, while their MRR is less than 1.0. However, if the extracted explanations are indeed able to increase the rank, the re-trained model should lead to improve both metrics. Therefore, the effectiveness is assessed as the increase in the $H@1$ and MRR over P. Conversely, for evaluating the worsening of LP results we select a set P of 100 test set triples that are ranked first by the LP method, randomly, for each model. After extracting explanations for all predictions in P, we add their triples and re-train the KGE model. As the original model correctly led to those predictions, their initial values for both $H@1$ and MRR are 1.0. However, if the extracted explanations are indeed able to decrease the rank, the re-trained model should lead to a worsening of both metrics. Therefore, the effectiveness is assessed as the decrease in $H@1$ and MRR over P. In both scenarios, the variations of the metrics are denoted as $\Delta H@1$ and ΔMRR, respectively.

Ideally, ACEs are those leading to an improvement. To demonstrate that IMAGINE can generate valuable ACEs, we compare it to baseline methods that, similarly to IMAGINE, generates CEs aimed at improving LP results, but is based on existing triples and thus returns SCEs. This baseline methods, which we call I-KELPIE and I-KELPIE++ (to emphasize its focus on *improving* LP performance), are built on KELPIE and KELPIE++. Whilst KELPIE and KELPIE++ identify the sets of triples whose removal most decreases the ranks of the triples in P, I-KELPIE and I-KELPIE++ extract the sets of triples whose removal most increases the ranks of the triples in P. Intuitively, it identifies noisy triples, or those that encode information that is meaningful but unrelated to the predictions. The goal is to show that the *additional triples* generated by IMAGINE are indeed useful, as they contribute more significantly to improving LP performance than noise removal. We include experiments for each summarization approach adopted in the Explanation Builder: no summarization, *bisimulation*, and *simulation*. Essentially, we compare IMAGINE, I-KELPIE, and I-KELPIE++.

Even more so, for demonstrating that the ACEs generated by IMAGINE are valuable, we set up experiments measuring the worsening of LP performance. We compare W-IMAGINE with KELPIE and KELPIE++. We aim at proving that SCEs worsen LPs more than ACEs, thus implying that *additional triples* are not a source of noise to the extent that adding triples worsen LPs more than removing existing triples, even if in the Relevance Engine we focus on the rank decrease. We experiment with the different summarization solutions in the Explanation Builder. Essentially, we compare KELPIE, KELPIE++, W-IMAGINE, and W-IMAGINE++.

The code, the datasets, the trained models utilized in our study are openly accessible on GitHub[1] along with a detailed account of the hyperparameters adopted in all the phases of the experiments and a description of the execution environment.

5.2   Outcomes of the Evaluation

In Table 3 we report the effectiveness metrics for the experiments measuring the improvement of LP performance. IMAGINE consistently outperforms I-KELPIE and I-KELPIE++ in any setup of the Explanation Builder. The only exception is observed on explaining predictions on DB50K made with CONVE. Moreover, IMAGINE, when equipped with simulation or bisimulation, always performs on par or better than its version with no summarization in the Explanation Builder.

In Table 4 we report the effectiveness metrics for the experiments measuring the worsening of LP performance. In this setting, KELPIE and KELPIE++ consistently achieve a larger worsening than W-IMAGINE. We remark that the explanations computed by IMAGINE can be complementary to those computed

---

[1] https://github.com/rbarile17/imagine.

Table 3. Outcomes of the experiments: *improvement* setting

KGE model	Method	Summarization	DB100K		DB50K		YAGO4-20	
			$\Delta H@1$	ΔMRR	$\Delta H@1$	ΔMRR	$\Delta H@1$	ΔMRR
COMPLEX	IMAGINE	–	0.510	0.458	**0.580**	**0.536**	0.520	0.501
	IMAGINE	*bisimulation*	0.510	0.458	**0.580**	**0.536**	0.520	0.501
	IMAGINE	*simulation*	**0.540**	**0.490**	0.450	0.433	**0.560**	**0.537**
	I-KELPIE	–	0.130	0.100	0.030	−0.029	0.080	0.037
	I-KELPIE++	*bisimulation*	0.130	0.100	0.030	−0.029	0.080	0.037
	I-KELPIE++	*simulation*	0.090	0.062	0.030	−0.037	0.100	0.061
CONVE	IMAGINE	–	0.420	0.347	0.040	0.012	0.280	0.285
	IMAGINE	*bisimulation*	0.420	0.347	0.040	0.012	0.280	0.285
	IMAGINE	*simulation*	**0.430**	**0.378**	0.030	0.008	**0.510**	**0.502**
	I-KELPIE	–	0.100	0.036	0.040	0.014	0.050	0.036
	I-KELPIE++	*bisimulation*	0.100	0.036	0.040	0.014	0.050	0.036
	I-KELPIE++	*simulation*	0.140	0.081	**0.050**	**0.025**	0.040	0.030
TRANSE	IMAGINE	–	**0.140**	0.165	0.360	0.292	0.330	0.332
	IMAGINE	*bisimulation*	**0.140**	0.165	0.360	0.292	0.330	0.332
	IMAGINE	*simulation*	**0.140**	**0.184**	0.410	0.333	**0.400**	**0.409**
	I-KELPIE	–	0.080	0.073	0.090	0.036	0.030	0.002
	I-KELPIE++	*bisimulation*	0.080	0.073	0.090	0.036	0.030	0.002
	I-KELPIE++	*simulation*	0.040	0.023	0.100	0.039	0.020	0.001

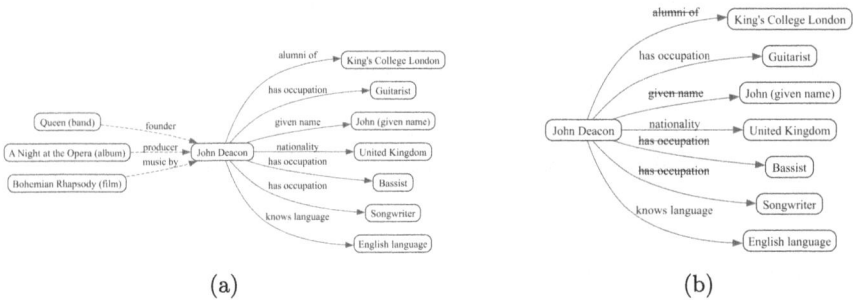

(a)   (b)

Fig. 2. Subgraphs with an ACE (a) and an SCE (b) for the prediction ⟨*John Deacon, member of, Queen (band)*⟩. ACE triples are identified by dashed edges, SCE triples by edges with strike-through labels.

as SCEs. Indeed, IMAGINE outperforms the methods providing SCEs on non top-ranked test triples. In contrast, W-IMAGINE as an adaption of IMAGINE to be employed for generating ACEs that impact top-ranked test triples underperforms the methods providing SCEs. The latter observation highlights that removing triples has a greater impact on correct predictions than *additional triples*, but also indicates that *additional triples* do not introduce such significant noise that adding them degrades performance more than removing existing triples. Finally,

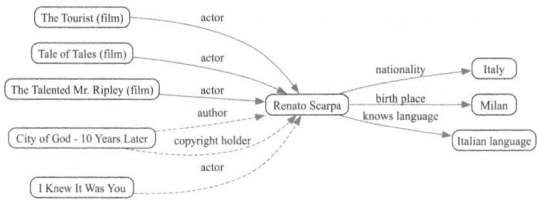

Fig. 3. Subgraph featuring *Renato Scarpa* for the prediction ⟨*Renato Scarpa, has occupation, actor*⟩. ACE triples are indicated by dashed edges.

Table 4. Outcomes of the experiments: *worsening* setting

KGE model	Method	Summarization	DB100K		DB50K		YAGO4-20	
			$\Delta H@1$	ΔMRR	$\Delta H@1$	ΔMRR	$\Delta H@1$	ΔMRR
COMPLEX	W-IMAGINE	–	−0.280	−0.203	−0.210	−0.156	−0.390	−0.268
	W-IMAGINE	*bisimulation*	−0.280	−0.203	−0.210	−0.156	−0.390	−0.268
	W-IMAGINE	*simulation*	−0.310	−0.247	−0.280	−0.224	−0.450	−0.340
	KELPIE	–	−0.710	−0.614	−0.980	**−0.960**	−0.610	−0.469
	KELPIE++	*bisimulation*	−0.710	−0.614	−0.980	**−0.960**	−0.610	−0.469
	KELPIE++	*simulation*	**−0.780**	**−0.673**	**−0.990**	−0.952	**−0.650**	**−0.540**
CONVE	W-IMAGINE	–	−0.180	−0.112	−0.060	−0.040	−0.240	−0.170
	W-IMAGINE	*bisimulation*	−0.180	−0.112	−0.060	−0.040	−0.240	−0.170
	W-IMAGINE	*simulation*	−0.170	−0.112	−0.040	−0.022	−0.290	−0.208
	KELPIE	–	−0.230	−0.185	−0.090	−0.057	−0.310	−0.248
	KELPIE++	*bisimulation*	−0.230	−0.185	−0.090	−0.057	−0.310	−0.248
	KELPIE++	*simulation*	**−0.310**	**−0.246**	**−0.190**	**−0.121**	**−0.400**	**−0.329**
TRANSE	W-IMAGINE	–	−0.560	−0.394	−0.290	−0.196	−0.480	−0.393
	W-IMAGINE	*bisimulation*	−0.560	−0.394	−0.290	−0.196	−0.480	−0.393
	W-IMAGINE	*simulation*	−0.500	−0.357	−0.470	−0.361	−0.400	−0.312
	KELPIE	–	**−0.710**	**−0.587**	**−0.760**	**−0.666**	−0.610	−0.501
	KELPIE++	*bisimulation*	**−0.710**	**−0.587**	**−0.760**	**−0.666**	−0.610	−0.501
	KELPIE++	*simulation*	−0.680	**−0.581**	−0.710	−0.606	**−0.680**	**−0.580**

IMAGINE, takes on average 79 seconds to explain a prediction on the environment that we employed.

Moving to a qualitative analysis, we now illustrate typical examples of explanation output, for both non top-ranked triples and those that are ranked first. We report the explanations produced by IMAGINE and I-KELPIE for a non top-ranked test triple. Specifically, we employ the triple ⟨*John Deacon, member of, Queen (band)*⟩ found in the experiments with TRANSE on YAGO4-20.

IMAGINE explains the prediction by returning the *additional triples* identified by the dashed edges in Fig. 2a. Such an explanation underscores that the LP method mostly alters the rank of the triple about band membership, if fed with information on the participation in albums and the formation of the band. In contrast, I-KELPIE returns the triples identified by edges with strike-through labels in Fig. 2b. Hence, the LP method change its outcomes mostly based on the predicates: *occupation, given name,* and *alumni of.* However, IMAGINE seems able to hypothesize connections with the object of the prediction for entities lacking such information.

Conversely, we report the explanations by W-IMAGINE and KELPIE for a given test set triple ranked first by the LP method. In particular, we consider the prediction ⟨*Renato Scarpa, has occupation, actor*⟩. KELPIE explains it by returning the triple ⟨*The Talented Mr. Ripley (film), actor, Renato Scarpa*⟩. This suggests that the LP method predicted *actor* as the occupation for *Renato Scarpa* because, during training, it had seen a triple stating his occupation as *actor* in a film. Instead, W-IMAGINE generates the *additional triples* identified by dashed edges in Fig. 3. This shows that the LP method switches its prediction to another occupation when provided with triples stating different occupations in different works. Thus, W-IMAGINE seems to be able to complement the insight given by KELPIE as it provides a further example of the dependence of occupation on triples stating such an occupation.

6   Conclusions

We introduced IMAGINE, a novel complementary approach to provide CEs for predictions made on KGs. It is based on the generation of plausible and meaningful additional triples. It generates ACEs consisting of sets of *additional triples* having the greatest impact on the predictions. We performed an experimental evaluation on three datasets and three representative LP models for demonstrating that the generated ACEs are valuable, i.e., able to improve the LP performance. We also qualitatively assessed the differences between the alternative CEs in each setting.

A natural extension of this work would be a unified approach that generates explanations that include both existing and additional triples, thus integrating SCEs and ACEs. In addition, we aim to enhance the Triple Builder component used to generate *additional triples*. For example, we may use a probabilistic generative model, such as the one proposed in [24], which would allow sampling triples from the encoded distribution. In addition, we plan to investigate the impact of distinguishing between *additional triples* that cannot be proven neither true nor false and those that can be proven false (negative statements) according to the KG schema. Finally, we may involve users in the qualitative analysis.

Acknowledgments. This work was partially supported by project *FAIR - Future AI Research* (PE00000013), spoke 6 - Symbiotic AI (https://future-ai-research.it/) under the PNRR MUR program funded by the European Union - NextGenerationEU, by PRIN project *HypeKG - Hybrid Prediction and Explanation with Knowledge Graphs* (Prot. 2022Y34XNM, CUP H53D23003700006) under the PNRR MUR program funded by the European Union - NextGenerationEU, and by project *DIXTI - DIgging into eXplainable soluTIons for link prediction on Knowledge Graphs using large language models* under ISCRA program funded by CINECA.

Disclosure of Interests. The authors declare to have no competing interests.

References

1. Amador-Domínguez, E., Serrano, E., Manrique, D.: GEnI: a framework for the generation of explanations and insights of knowledge graph embedding predictions. Neurocomputing **521**, 199–212 (2023). https://doi.org/10.1016/j.neucom.2022.12.010
2. Auer, S., et al. (eds.) The Semantic Web, vol. 4825, pp. 722–735. Springer, Heidelberg (2007). https://doi.org/10.1007/978-3-540-76298-0_52
3. Baltatzis, V., Costabello, L.: KGEx: Explaining Knowledge Graph Embeddings via Subgraph Sampling and Knowledge Distillation (2023)
4. Barile, R., d'Amato, C., Fanizzi, N.: Explanation of link predictions on knowledge graphs via levelwise filtering and graph summarization. In: Meroño Peñuela, A., et al. (eds.) Proceedings of the 26th European Semantic Web Conference (ESWC 2024), vol. 14664, pp. 180–198. Springer, Cham (2024). https://doi.org/10.1007/978-3-031-60626-7_10
5. Betz, P., Meilicke, C., Stuckenschmidt, H.: Adversarial Explanations for Knowledge Graph Embeddings. In: IJCAI, vol. 2022, pp. 2820–2826 (2022)
6. Bhowmik, R., de Melo, G.: Explainable link prediction for emerging entities in knowledge graphs. In: Pan, J.Z., et al. (eds.) ISWC 2020. LNCS, vol. 12506, pp. 39–55. Springer, Cham (2020). https://doi.org/10.1007/978-3-030-62419-4_3
7. Bollacker, K., Cook, R., Tufts, P.: Freebase: a shared database of structured general human knowledge. In: AAAI, vol. 7, pp. 1962–1963 (2007)
8. Bordes, A., Usunier, N., Garcia-Duran, A., Weston, J., Yakhnenko, O.: Translating embeddings for modeling multi-relational data. Adv. Neural Inf. Process. Syst. **26** (2013)
9. Byrne, R.M.: Counterfactuals in explainable artificial intelligence (XAI): evidence from human reasoning. In: IJCAI, pp. 6276–6282 (2019)
10. Čebirić, Š., et al.: Summarizing semantic graphs: a survey. VLDB J. **28**, 295–327 (2019)
11. d'Amato, C., Masella, P., Fanizzi, N.: An approach based on semantic similarity to explaining link predictions on knowledge graphs. In: IEEE/WIC/ACM International Conference on Web Intelligence, pp. 170–177. ACM, ESSENDON VIC Australia (2021). https://doi.org/10/gtmzqz
12. Dettmers, T., Minervini, P., Stenetorp, P., Riedel, S.: Convolutional 2d knowledge graph embeddings. In: Proceedings of the AAAI Conference on Artificial Intelligence, vol. 32 (2018). https://doi.org/10/gqjp9q
13. Ding, B., Wang, Q., Wang, B., Guo, L.: Improving knowledge graph embedding using simple constraints. In: Proceedings of the 56th Annual Meeting of the Association for Computational Linguistics (Volume 1: Long Papers), pp. 110–121 (2018)
14. Dong, X.L.: Building a broad knowledge graph for products. In: 2019 IEEE 35th International Conference on Data Engineering (ICDE), pp. 25–25. IEEE (2019). https://doi.org/10.1109/ICDE.2019.00010
15. Glimm, B., Horrocks, I., Motik, B., Stoilos, G., Wang, Z.: HermiT: an OWL 2 reasoner. J. Autom. Reason. **53**(3), 245–269 (2014). https://doi.org/10.1007/s10817-014-9305-1
16. Guidotti, R.: Counterfactual explanations and how to find them: Literature review and benchmarking. Data Min. Knowl. Disc. (2022). https://doi.org/10.1007/s10618-022-00831-6

17. Guidotti, R., Monreale, A., Ruggieri, S., Turini, F., Giannotti, F., Pedreschi, D.: A survey of methods for explaining black box models. ACM Comput. Surv **51**(5), 1–42 (2019)
18. Hogan, A., et al.: Knowledge graphs. Synth. Lect. Data Semant. Knowl. **12**(2), 1–257 (2021). https://doi.org/10.2200/S01125ED1V01Y202109DSK022
19. Ismaeil, Y., Stepanova, D., Tran, T.K., Blockeel, H.: FeaBI: a feature selection-based framework for interpreting KG embeddings. In: Payne, T.R., et al. (eds.) The Semantic Web – ISWC 2023, vol. 14265, pp. 599–617. Springer, Cham (2023). https://doi.org/10.1007/978-3-031-47240-4_32
20. Koh, P.W., Liang, P.: Understanding black-box predictions via influence functions. In: International Conference on Machine Learning, pp. 1885–1894. PMLR (2017)
21. Krishnan, N.A., Rivero, C.R.: A model-agnostic method to interpret link prediction evaluation of knowledge graph embeddings. In: Proceedings of the 32nd ACM International Conference on Information and Knowledge Management, pp. 1107–1116. ACM, Birmingham (2023). https://doi.org/10.1145/3583780.3614763
22. Kumar-Singh, N., Polleti, G., Paliwal, S., Hodos-Nkhereanye, R.: LinkLogic: A New Method and Benchmark for Explainable Knowledge Graph Predictions (2024)
23. Law, J.: Robust statistics-the approach based on influence functions (1986)
24. Loconte, L., Di Mauro, N., Peharz, R., Vergari, A.: How to turn your knowledge graph embeddings into generative models. Adv. Neural Inf. Process. Syst. **36** (2024)
25. Lundberg, S.M., Lee, S.I.: A unified approach to interpreting model predictions. Adv. Neural Inf. Process. Syst. **30** (2017)
26. Ma, T., et al.: KGExplainer: Towards Exploring Connected Subgraph Explanations for Knowledge Graph Completion (2024)
27. Nováček, V., Mohamed, S.K.: Predicting polypharmacy side-effects using knowledge graph embeddings. AMIA Summits Transl. Sci. Proc. **2020**, 449 (2020)
28. Pellissier Tanon, T., Weikum, G., Suchanek, F.: YAGO 4: a reason-able knowledge base. In: Harth, A., et al. (eds.) ESWC 2020. LNCS, vol. 12123, pp. 583–596. Springer, Cham (2020). https://doi.org/10.1007/978-3-030-49461-2_34
29. Pezeshkpour, P., Irvine, C.A., Tian, Y., Singh, S.: Investigating robustness and interpretability of link prediction via adversarial modifications. In: Proceedings of NAACL-HLT, pp. 3336–3347 (2019)
30. Rossi, A., Barbosa, D., Firmani, D., Matinata, A., Merialdo, P.: Knowledge graph embedding for link prediction: a comparative analysis. ACM Trans. Knowl. Discov. Data **15**(2), 1–49 (2021). https://doi.org/10/gjhzcz
31. Rossi, A., Firmani, D., Merialdo, P., Teofili, T.: Explaining link prediction systems based on knowledge graph embeddings. In: Proceedings of the 2022 International Conference on Management of Data, pp. 2062–2075. ACM, Philadelphia (2022). https://doi.org/10/gqjfg6
32. Shi, B., Weninger, T.: Open-world knowledge graph completion. In: Proceedings of the AAAI Conference on Artificial Intelligence, vol. 32 (2018)
33. Singhal, A.: Introducing the Knowledge Graph: Things, not strings (2012). https://blog.google/products/search/introducing-knowledge-graph-things-not/
34. Trouillon, T., Welbl, J., Riedel, S., Gaussier, É., Bouchard, G.: Complex embeddings for simple link prediction. In: International Conference on Machine Learning, pp. 2071–2080. PMLR (2016)
35. Ying, Z., Bourgeois, D., You, J., Zitnik, M., Leskovec, J.: Gnnexplainer: generating explanations for graph neural networks. Adv. Neural Inf. Process. Syst. **32** (2019)
36. Zhang, H., et al.: Data poisoning attack against knowledge graph embedding. In: Proceedings of the 28th International Joint Conference on Artificial Intelligence, pp. 4853–4859 (2019)

37. Zhang, W., Deng, S., Wang, H., Chen, Q., Zhang, W., Chen, H.: XTransE: explainable knowledge graph embedding for link prediction with lifestyles in e-commerce. In: Wang, X., Lisi, F.A., Xiao, G., Botoeva, E. (eds.) JIST 2019. LNCS, vol. 1157, pp. 78–87. Springer, Singapore (2020). https://doi.org/10.1007/978-981-15-3412-6_8

38. Zhang, W., Paudel, B., Zhang, W., Bernstein, A., Chen, H.: Interaction embeddings for prediction and explanation in knowledge graphs. In: Proceedings of the Twelfth ACM International Conference on Web Search and Data Mining, pp. 96–104. ACM, Melbourne (2019). https://doi.org/10/ggzfzp

39. Zhao, D., et al.: KE-X: towards subgraph explanations of knowledge graph embedding based on knowledge information gain. Knowl.-Based Syst. **278**, 110772 (2023)

PeGazUs: A Knowledge Graph Based Approach to Build Urban Perpetual Gazetteers

Charly Bernard[1], Solenn Tual[1(✉)], Nathalie Abadie[1],
Bertrand Duménieu[2], Joseph Chazalon[3], and Julien Perret[1]

[1] LASTIG, Université Gustave Eiffel, IGN-ENSG, 73 Avenue de Paris, 94165
Saint-Mandé Cedex, France
`{charly.bernard,solenn.tual,nathalie-f.abadie,julien.perret}@ign.fr`
[2] Centre de Recherches Historiques, École des Hautes Études en Sciences Sociales,
Paris, France
`bertrand.dumenieu@ehess.fr`
[3] LRE, EPITA, Le Kremlin-Bicêtre, France
`joseph.chazalon@lrde.epita.fr`

Abstract. Gazetteers, as compilation of named places, are central resources on the Web of data, as they provide a common ground to link and integrate many textual or structured resources on the Web. Gazetteers usually categorise and associate places names with geospatial coordinates. In more recent times, historical gazetteers, which aim to represent places from the past, have received increasing attention. The creation of these gazetteers poses specific challenges, including the definition of the identity of evolving places, the representation of their evolution through time (how they change, when the changes happen), and the population of the gazetteer based on scarce and heterogeneous historical sources. We propose an approach to create an urban historical gazetteer on the evolution of two major urban large-scale types of places, namely addresses and land plots. Our proposal is inspired by approaches for creating knowledge graphs and takes advantage of the knowledge representation and reasoning possibilities offered by Semantic Web standards to address the aforementioned challenges. The approach was applied to the *Butte aux Cailles* district of Paris, for which a variety of contemporary and historical sources were used. The resulting knowledge graph can be used for a variety of purposes, including historical geocoding of old documents, identifying the use of a plot of land at a given date, and recording the events that led to its current state.

Keywords: Historical urban gazetteer · Knowledge graph · Addresses and land plots evolution

1 Introduction

Spatial indexing of digitised archival records is a key issue to help scholars easily retrieve documents about a place of interest. State-of-the-art approaches gener-

M. Alam et al. (Eds.): EKAW 2024, LNAI 15370, pp. 364–381, 2025.
https://doi.org/10.1007/978-3-031-77792-9_22

ally compare places mentioned in documents with those in gazetteers, in order to disambiguate homonyms and get absolute or relative location information. Gazetteers are place names repositories which serve two purposes: linking place names to locations, and describing the places they list [17]. Thus, they usually gather for each place at least one name, one type and one location represented most of the time by coordinates. But old documents are very likely to mention places that have changed or even disappeared through time. In recent years, many projects have been carried out to create historical gazetteers, but rarely at urban level.

Ducatteeuw [15] defines an urban historical gazetteer as an information resource representing places on the street level and their evolution through time. This kind of resource is meant to spatially index historical sources like censuses, tax registers, directories and so on. In this article, we aim at creating a gazetteer which represents not only streets, but the addresses and land plots located on the streets, as they are key spatial entities for the fine-grained geolocation of very large historical corpora. Addresses and plots are *fiat* geographical objects, designated by phrases or identifiers based on a spatial hierarchy. They are social constructs, generally created either through peoples' practices or by an administrative authority, whose use may persist over time, even after they have been officially cancelled by the authorities. Representing such kind of spatial entities in a historical gazetteer therefore poses specific challenges.

First, representing old geographical features, for which there is no ground truth anymore, implies relying on historical sources to extract useful data. Depending on their type and their valid time, these will provide very different descriptions of the geographical features. It will therefore be necessary to link the different representations of the same geographic entity across the available sources to leverage their complementary contributions. Another difficulty is that some sources do not directly describe the state of geographical entities, but the events that happen to them. For example, a record of municipal administration decisions could provide information on street name changes. It is therefore important to take advantage of this change information to infer new facts about the state of geographical features; and conversely, to use the data available about the successive states of geographical features to infer the events between them. These events can have consequences for the properties of geographical entities or for their identity itself. It is therefore essential to have a model that can represent different successive property values for geographical entities, depending on the period of their existence and the events that have affected them.

In this article, we first present related works on the challenges posed by historical gazetteers. Then we present an ontology for representing an urban historical gazetteer containing descriptions of old addresses and plots and their evolution. We also present a generic strategy to automatically populate this ontology from scarce historical sources, that represent either the states of geographical entities or the events that affect them, and infer missing information. We evaluate the ontology and our populating strategy by constructing a geohistorical knowledge graph on the addresses and plots of land in the 19th century Paris district of *La Butte aux Cailles*.

2   Challenges for a Urban Historical Gazetteer

From early historical Geographic Information Systems ([12] or [16]) and geospatial standards for gazetteers ([7,8]), to recent works based on semantic Web standards, the question of the most suitable data model to represent historical named places dataset has been widely addressed in the literature ([17,27,28]). GIS-based solutions require precise geometries to represent the shape and location of places and are often not suitable for old places for which such information is rarely available. As pointed out by Berman [9], knowledge graphs are well suited to represent fragmented, incomplete data scattered across multiples sources.

2.1   Ontologies of Named Places for Historical Gazetteers

The World Historical Gazetteer platform and the Pelagios project have jointly developed the Linked Places JSON-LD format for historical gazetteers [18]. In its underlying ontology, a place attestation can be described by several names, types, relations, locations and temporal information. This temporal property can be represented by timespans represented according to the OWL-Time ontology [5], periods identified by URIs pointing to time gazetteers like PeriodO [3] or even labels and duration values. Besides, it can be associated either to the place attestation itself or to its name, type, relation and location properties, so as to provide versions for them. [15] builds upon this format to propose an ontology for urban gazetteers, representing streets, at a high level of detail. It is based on the classes and properties used in the Linked Places format, but they are replaced by their equivalent classes and properties from the upper level ontology designed for cultural heritage data, namely the CIDOC Conceptual Reference Model.

The two previous models have the advantage of representing successive versions of places and their properties. However, they do not include the events that cause places to change. Finally, they are intended for named places, not geographical entities designated by complex statements that include references to other named places, like old addresses or land plots.

2.2   Representing Data Changes and Real World Events

One of the first attempts to represent evolving geographical entities and the changes they undergo has been proposed by [20]. This work uses geospatial ontologies representing Finnish municipalities and their part-of relationships for successive time spans. Changes are also represented to link the successive states of the municipalities. Five types of changes are considered: establishment, merge, split, name change and part-of relationship change.

[10] proposes an ontology called TSN to represent Territorial Statistical Nomenclatures with their territorial units and their successive versions. The change bridge approach proposed by [20] to represent the changes that territorial units undergo over time is reused and extended in the TSN-Change ontology. For each change, it explicitly represents the upstream and downstream territorial units involved, the type of change (using the same types as [20]), the subchanges potentially induced by the current change and the real-world event responsible for this change at data level (designated by the *isCausedBy* predicate).

The Hierarchical Historical Territory ontology (HHT) proposed by [13] differs from the TSN/TSN-Change ontologies by focusing on representing multiple hierarchies between territories while TSN is designed to represent a single nomenclature. The chosen approach is also based on versions of territorial units, but with a temporal partitioning based on the territorial changes rather than on systematic snapshots, which are better suited to statistical data than to historical data. The change bridge principle is also adopted, but two types of changes are considered: those affecting a single territorial unit and those affecting several.

2.3 Populating an Ontology of Historical Places and Their Changes

The approach proposed by [20] includes a methodology to construct the ontology time series from metadata tables describing changes and locations (current and historical). All the data is therefore prepared in advance, possibly manually, to match the ontology's expected content. [10] proposes to populate the TSN and TSN-Change ontologies using an algorithm that takes temporal snapshots of geographical data as input, populates the TSN ontology from this data and compares the geometries of administrative units to create links between successive versions of these units. Changes are then inferred automatically by interpreting the different configurations of links between versions of territorial units. [13] propose a rule-based algorithm to automatically link each territorial unit version, already represented according to the HHT ontology, to its chronological successor and detect and classify changes between them.

2.4 Attestations and Historical Sources

As stated in [17], gazetteers do not represent places, but attestations of places: Each resource representing a place should therefore be modelled as an aggregate of sourced assertions about that place. For historical gazetteers, where ground truth is no longer available, this is of particular importance.

Historical documents are the only available sources to report traces of the past. As secondary sources of information, their content is the result of interpretations and observations whose quality and reliability are variable and often difficult to assess. It is therefore necessary to use a variety of complementary sources to populate a historical gazetteer. Finally, as the information available is incomplete, it is often necessary to infer the missing data from the knowledge and facts available. It is thus essential to document the provenance of inferred facts, to enable users to distinguish them from those based on historical sources.

Ontologies such as Prov-O [4] or the Factoid Prosopography Ontology [1] have been proposed to describe the provenance of data and can be used in conjunction with ontologies to describe historical sources such as RiC-O [2].

3   Data Sources on Land Plots and Addresses

Land plots and addresses are typical cases of geospatial entities described in multiple, fragmentary and heterogeneous sources of information that have different temporal validities, and different ways of describing geographic entities.

3.1   Contemporary Geographic Data

The municipality of Paris publishes two main geospatial datasets on the city's thoroughfares and addresses. The first one named *Dénominations des emprises des voies actuelles*[1] describes all thoroughfares with their names, geometries, dates of creation along other secondary metadata. The second one, the *Dénominations caduques des voies*[2], describes ancient thoroughfares with their date of deletion, and is structured in the same way, but no geometry is provided.

The *Base Adresse Nationale* (BAN)[3] contains all the postal addresses registered in France, each address being structured as a list of housenumber, street, city, and zip code, with a geographical position.

Lastly, volunteered geographic information is also used. OpenStreetMap is a geographic database that tends to represent the current state of geographic entities, whereas Wikidata provides the history of these entities.

3.2   Historical Large Scale City Plans

Multiple large-scale topographic maps of Paris have been produced since the 18th. They can be leveraged as valuable sources of structured geohistorical data at the cost of extensive operations of vectorization. Several digital humanities projects have carried out such works and released open datasets[4]. As an example, the *Atlas National de la ville de Paris* finely depict the streets of Paris at the scale of 1:1720 and was published between 1791 and 1799. Another example is the *Atlas municipal des vingt arrondissements de la ville de Paris*, which represents the city of Paris at the end of the 19th century. These sources have similarities with OpenStreetMap or BAN since their main goal is to describe geographical entities at a given point in time without taking into account their evolution.

---

[1] https://opendata.paris.fr/explore/dataset/denominations-emprises-voies-actuelles.
[2] https://opendata.paris.fr/explore/dataset/denominations-des-voies-caduques.
[3] https://adresse.data.gouv.fr/.
[4] E.g. Projets Time Machine (https://ptm.huma-num.fr/), ALPAGE (https://alpage.huma-num.fr) or SODUCO (https://soduco.geohistoricaldata.org).

3.3 Street Dictionaries

Unlike maps, Paris street dictionaries are not a snapshot of the city, but instead provide a historical descriptions of every streets. Such sources contain indirect spatial references: the district to which the street belongs, addresses giving the beginning and the end of the lane. The *Dictionnaire administratif et historique des rues de Paris et de ses monuments* by Félix and Louis Lazare published in 1844 is a typical example of this type of document. Another major street dictionary is the *Dictionnaire historique des rues de Paris* by Jacques Hilairet is published in 1960. Unlike the previous dictionary, it includes streets in the outer districts of Paris, which became part of the capital in 1860.

3.4 Cadastral Maps and Registers

The Napoleonic land registry is the first land registry of the entire French territory. It was created between 1808 and 1850, depending on the departement and commune. Its goal was to make the system of land taxation more efficient and to make tax collection more equitable. The Napoleonic land registry consists of two types of documents: maps and registers. Index maps represent plot division on a very large scale. Each plot is delimited and associated with a number. The initial registers are the legend of these maps at the time of their production. Maps and initial registers were not updated after their creation. The mutation registers contain all the plot updates (taxpayer, land use, tax value) over time. Plots are grouped by taxpayers in folios (numbered page or part of page describing the properties of a taxpayer) and sometimes by thoroughfares. Each table line describing a plot is a version of this plot at a given time.

4   An Ontology for Historical Urban Gazetteers

The first contribution of this work lies in the PeGazUs[5] ontology. Like a perpetual calendar that represents the day of the week on any given date, it is intended to represent the address or plot number of a location on any given date. Like those proposed by [10, 20], and [13], it is based on the *Change Bridge* concept. But unlike them, it does not impose a hierarchy between them, and above all it allows versions of geographical entities to be represented whose property values can evolve over time. This is particularly useful when the identity of geographical entities cannot always be identified *a priori* in historical sources.

4.1 Ontology Documentation and Competency Questions

To build the ontology, we followed the method called *Simplified Agile Methodology for Ontology Development*, also known as SAMOD [25]. This consists in separating a complex modeling problem into sub-problems called modelets, which are

---

[5] The PErpetual GAZetteer of approach-address UtteranceS ontology, its documentation, the data, the scripts used to build the data are available on this repository: https://github.com/umrlastig/pegazus-ontology.

easier to process. A modelet begins with a natural-language argument describing the sub-problem to be addressed, along with a glossary defining the main terms involved. Each modelet comes with a set of informal competence questions, also in natural language, which represent the questions to be answered by the knowledge base. We defined these questions, by interviewing historians, archivists, and archaeologists about their needs regarding old addresses and plots. Finally, each question is associated with a set of example answers, which serve to validate the modelet once implemented.

The SAMOD method enables us to operate in rapid cycles, with regular testing of the ontology under construction. Different modelets were identified: address, temporal evolution, sources, land registry documents use and taxpayers.

4.2   Modelets Structure

An address is an indirect spatial reference described by a structured statement that unambiguously designates a place [14]. The way addresses are structured is through an ordered sequence of spatial relationships between geographical entities (also called landmarks) [11]. Several models for addresses exist, such as *locn* [24] or ISO 19133 [6], but they focus on postal addresses and are not suited for less structured utterances typically found in historical sources, like *In the center of the capital city, between the Palais-Royal and the Tuileries, close to the main theaters.*

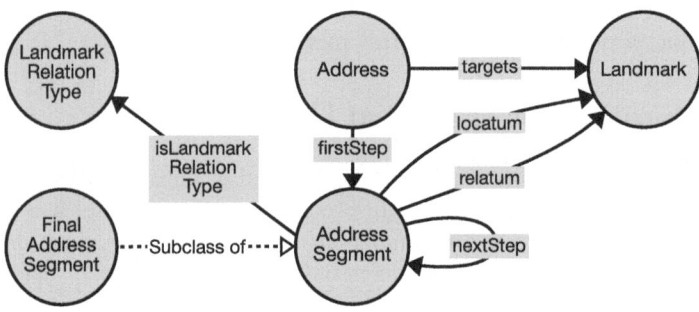

Fig. 1. Part of the ontology to describe addresses.

Addresses We proposed a common model for historical and postal addresses in [11] (see Fig. 1). An instance of `Address` corresponds to a structured statement. As it designates a place, it targets a `Landmark` which is a geographical entity (administrative unit, thoroughfare, building number, building...). An instance of `AddressSegment` is a spatial relationship between multiple instances of `Landmark` and its nature is given by `LandmarkRelationType`. To define the roles of landmarks for this relationship, `locatum` and `relatum` predicates are used [29]. In the spatial relation "Rue Gérard **is in** Paris", Rue Gérard is the locatum, Paris is the relatum, and "is in" is the landmark relation type. These segments form an ordered sequence which is described here by `firstStep` and `nextStep` predicates.

To validate this modelet, we selected the following competency questions: (1) What addresses are listed along a given street? (2) What are the coordinates of the target of a given address? (3) Which addresses are located in a given area?

An additional set of classes and properties are defined to describe plots depicted in the land registry. To validate this extension of the modelet, we defined the following questions: (4) Which are the plots located in a given commune or section of a commune? (5) What are the values of attribute X (nature, taxpayer or address) associated with a given plot ?

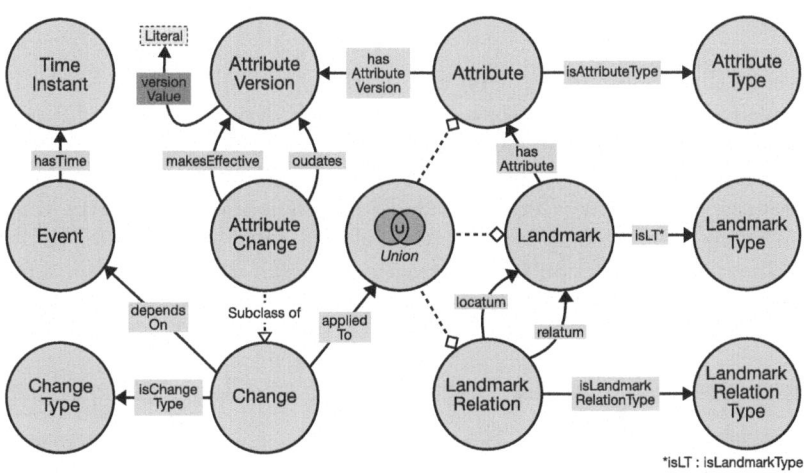

Fig. 2. Part of the ontology to describe geographical entities and their evolution.

Temporal Evolution The landmarks mentioned in the addresses may change through time, which means that the addresses also change. To capture both changes, landmarks have to be modelled so that we can apply changes to them (see Fig. 2). For landmark evolution, [10] and [13] use an approach based on the representation of states (also called versions). Each landmark has a set of its temporally ordered versions, each of which has a valid period. This implies that the attributes of the territorial units must have constant values for each version. But when landmarks evolve, changes do not always apply to all their attributes. For example, the extension of a street only affects its geometry but has no effect on its name. This is why we opt for a modelling approach in which each attribute is represented independently of the identity of the landmark to which it is related. Each attribute has a type (name, geometry, length, plot nature...)

and may have versions, like in the Linked Places format. `LandmarkRelation`, a super-class of `AddressSegment`, allows to describe relations between landmarks such as spatial one. Figure 3 shows an example of this modelling for the street called *Rue Gérard*.

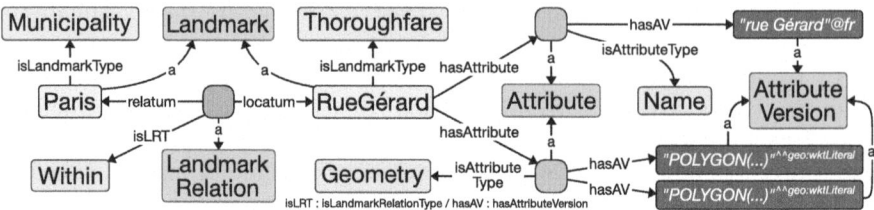

Fig. 3. *The Rue Gérard* modelled according the ontology.

Events that cause changes on the territory are also represented. These may involve one or more changes, each of which describes the evolution of a resource (`Landmark`, `LandmarkRelation`, `Attribute`). For example, Fig. 4 models an event that happened on August 30, 1978, whose textual description is "the eastern part of Rue Gérard located in Paris is now named Rue du Père Guérin". This event is composed of several changes: new geometry for the street called *Rue Gérard*, creation of the street called *Rue du Père Guérin*, and the appearance of a version for the name and geometry attributes of this street.

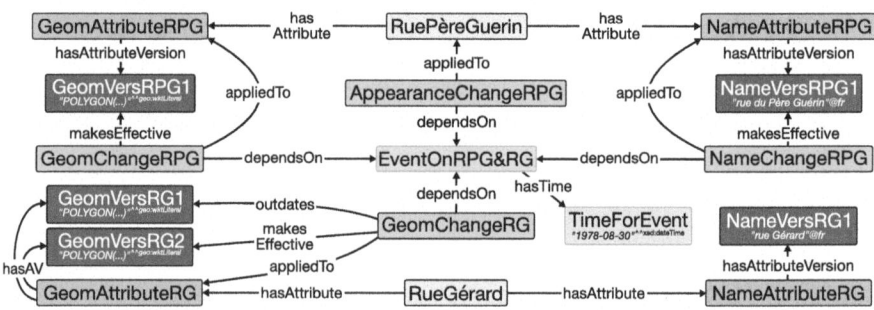

Fig. 4. Modelling of the event "the eastern part of Rue Gérard located in Paris is now named Rue du Père Guérin".

Four competency questions have been pointed out for this modelet: (1) what landmarks exist at a given time? (2) in what time interval(s) is an address of a given name valid? (3) what is the history of a landmark? In other words, what events are associated with it? (4) What states and events are missing from the history of an address? In the case of plots, we added the following questions: (5) What is the nature (or taxpayer or address) of a given plot at a given date? (6) What are the successive natures (or taxpayers or addresses) of a given plot?

Sources. Each version of an attribute of a landmark can originate from one or more sources. Those sources might contain contradictory and/or false information. Thus, the link between data and primary sources must be preserved to be able to detect and explain these inconsistencies. Each source has to be described according to its nature. This description might also include its author, its creation date, its valid period (e.g. the period during which a register is updated). In the case of archival records made up of several parts, the articulation of these parts must also be described. Furthermore, if relevant, the tools and processes used to transform primary sources into structured data should be documented. The main competency questions for this modelet are: (1) From which source does information X come from? (2) What is the description of source Y?

Land Registry Documents. This modelet has been developed to deal with land registry documents specificities. The Napeoleonic land registry is a sort of old fashion spatio-temporal database, printed and filled in manually. This modelet defines the capture rules that enable to follow one plot from one page to another within a register and from one register to another. It also lists the special values that appear in table cell of the registers (described in the SpecialCellValuesList skos concept list). These values are used to infer events (like construction, destruction) and changes, to follow the lifeline of each plot and to keep information like typography metadata (e.g. strikethrough) that provide a better understanding of how these document were filled out. The competency questions for this modelet are: (1) Which folios mention plot (or taxpayer) X? (2) Which resource have been created using a crossed-out table line?

Taxpayers. Taxpayers are associated with the land they own or exploit. They can be natural or legal persons and are designated with a name, one or more given names, an activity and an address. The identity of the taxpayer is crucial information to link the plots mentioned in the different registers. It is also a relevant domain to study the land registry with a socio-economic point of view and to link the land registry with other significant archival resources like census or civil registration. The associated competency questions are as follow: (1) Who are the taxpayers of a given area? (2) What are the plots associated with a given taxpayer? (3) Who are the taxpayers living in a given commune? (4) Who are the taxpayers of a given commune with a given activity?

4.3   Alignments with Existing Vocabularies

According to good practices for the development of ontologies, existing vocabularies are reused. To describe the documents of the Napeolenic land registry, we integrate concepts of the *Records in Contexts Ontology* [2]. This ontology has been developed to describe archival records, including their structure, relations between their sub-parts, distinction between concepts of records and their instanciation. It also enables the description of derived data from primary sources. Combined with the Prov-Ontology [4] as described by [19], we can precisely detail the treatments that are used to create these new instanciations. Valid periods of attributes versions and temporal relationships are represented with the OWL-Time Ontology [5].

5   The *Butte aux Cailles* neighbourhood geohistorical KG

The Butte aux Cailles neighbourhood is located in the southeast of Paris. Formerly part of the neighbouring commune of Gentilly, it was incorporated into Paris in 1860. This event led to a major transformation of the local urban fabric and is one of the reasons why this study area was chosen.

The second contribution of this work lies in the algorithm proposed to populate the ontology. It is based on the factoid approach [23]. For each source, its associated data is structured according to the ontology to form a so-called factoid graph. A factoid graph contains versions of landmarks describing them according to the source, possibly with various valid time intervals. Since factoids graphs are the base of construction of the knowledge graph, SHACL rules check if there are inconsistencies within its explicit triples [21]. Once all factoid graphs are built, we build a unique graph of facts whose goal is to rebuild each geographic entity identity from the factoid versions, and integrate all the attribute values from factoids graphs. We develop a six-steps process to reach this objective. This algorithm is iterative: some steps can be executed many times, while new knowledge is discovered at each step and is added to the final graph of facts.

5.1   Identity-Based Landmark Versions Rooting

The first step of the process aims to root similar landmarks from all the factoid graphs to a specific resource based on an identity criterion. Similar landmarks are linked to a root landmark using the `hasRoot` object property. Root landmarks can be landmarks from a given factoid graph (for plots) or a new empty resource initialised from one of these graphs (for addresses).

Districts and thoroughfares are linked based on name equality. In addition to having the same value, house number are equivalent if they are the locatum of a `Belongs` landmark relation whose relatum is the same thoroughfare (or district). Plots are linked if they have the same number and are in the same section of a commune. In the Napoleonic cadaster, when plots are split or merged, the new plots keep the same number as the previous objects, so that plot numbers are only a pseudo-identification criterion. In this situation a disambiguation step has to be performed later in the process (see Sect. 5.4).

5.2   Ordering Landmark Versions and Their Attributes Versions

Landmark versions linked to the same root are ordered using their valid time. For each version noted ?lv, pairs of landmarks are formed with all versions that match this requirement: $start(?lv_i) \geq start(?lv)$. Then, gaps between their valid times are calculated (i.e. difference between the end of the interval of $?lv_i$

and the start of the interval of ?lv). If the gap is negative (not null), ?lv hasOverlappingVersion ?lv$_i$ relation is inferred. In case of positive gap (or null), ?lv hasNextVersion ?lv$_i$ is inferred if this gap is equal to the smallest gap (positive or null) related to ?lv.

This step can be repeated with attribute versions. Indeed, the algorithm deals with heterogeneous sources that might not represent the same attributes (e.g. some have a geometry attribute and others do not). Thus, using global ordering of landmark versions could create disruptions in the sequence of attribute versions.

5.3   Inferring Changes and Events Related to Landmarks

This step aims at inferring LandmarkAppearance and LandmarkDisappearance changes depending on real-world Events. In the case of addresses, changes and events associated with the root landmarks are inferred using statements of the related factoids. For plots, this step is first done to infer these changes and events on the factoids. This first iteration of the method is mandatory to detect changes and events that impact landmark identity, and consequently to disambiguate versions that have the same root, the same plot number but might not be the same real-world object. A real-world plot is considered as a piece of land with a stable geometry between two events of type Split or Merge. Changes of type LandmarkAppearance or LandmarkDisappearance are detected using registers capture rules and the associated Event are created. As an example, two or more Folios in a cell of the *Next folio* column of the mutation register tables are interpreted as a Split event.

5.4   Inferring Landmarks Identity

If the identity of landmarks is already known at the beginning of the process, as for the addresses, this step aims only at adding hasTrace relations between landmark versions and their root landmark in order to retrieve the information that composed them. If the identity is not known, as for the plots, this step aims to create Landmark resources that aggregate landmark versions that are likely to describe the same real-world object. Additional ordering relations between landmark versions are created based on the order of their mentions in the documents. The position of the line in the document used to create the landmark version is taken into account. ?lv1 hasNextVersionInSRCOrder ?lv2 and ?lv1 hasOverlappingVersionInSRCOrder ?lv2 are created, as well as their inverse properties.

According to these properties, ordered series of landmark versions without a version associated with a LandmarkAppearance or a LandmarkDisappearance change in the middle of their lifeline are merged to create a new Landmark resource corresponding to a real-world geographic entity. This new resource is linked to the versions that make it up with the hasTrace relationship.

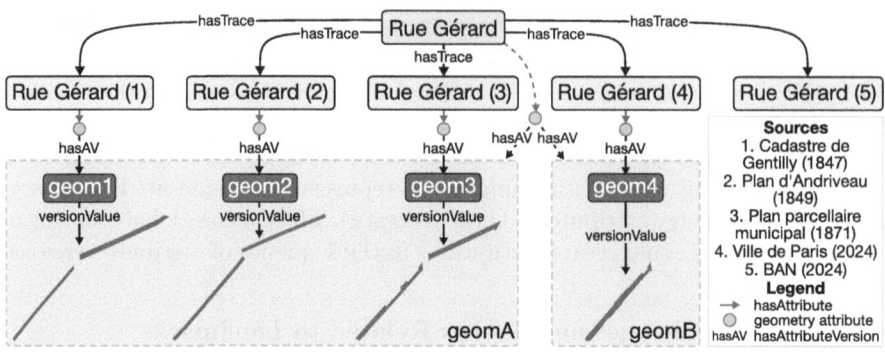

Fig. 5. Aggregation of geometry attribute versions from several sources for the street *Rue Gérard*. The dotted elements are those that have been inferred.

5.5 Inferring Attribute Versions

The previous steps of the method lead to the linking of factoids resources describing the same real-world landmark (called fact landmark) and their temporal ordering. This fifth step aims to build attribute versions of facts landmarks using the attribute versions from the associated factoids. To do so, their values need to be compared and they have to be ordered temporally. Then, successive similar versions are aggregated.

Versions of the same attribute are ordered temporally using the strategy described in Sect. 5.2. Their value are compared using different criteria according to the type of attribute. This results in creating `sameVersionValueAs` and `differentVersionValueFrom` relations between attribute versions. For instance, the comparison of two geometry attribute versions of a landmark has two be adapted to two situations. If geometries are lines or polygons, areas of the intersection and of the union between their bounding box can be computed. If the ratio between these areas is greater than a given value, they are considered as similar. In the case of points, they are considered as similar if the distance between them is below a threshold value. Taking the example of `?geom1` described in Fig. 5: the following relations are infered `?geom1 sameVersionValueAs ?geom2, ?geom3; differentVersionValueFrom ?geom4`.

Finally, similar attribute versions can be merged. This is done according to two criteria: value similarity and valid time order. In other words, attribute versions are aggregated if they form a continuous sequence of versions with the same value and that follow each other in time. By taking the example for Fig. 5, `?geom1`, `?geom2` and `?geom3` are similar and the follow each other so they are aggregated in a root attribute called `?geomA`. On the other hand, `?geom4` is only similar to itself so it forms an aggregation of one version noted `?geomB`.

5.6   Inferring Changes and Events Related to the Attributes

Finally, changes and events that affect the attribute versions are inferred. They are ordered based on attribute valid time values. When successive versions with different values are detected, then changes are created. Each change depends on an event whose time can be derived from the validity of the versions.

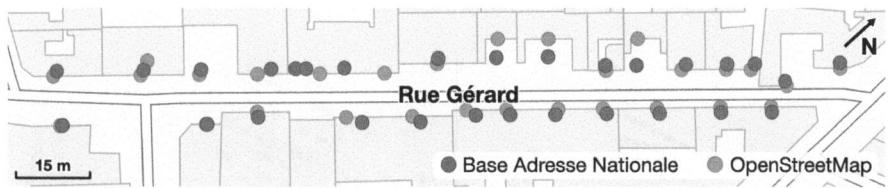

Fig. 6. Map of current addresses whose target is along the street *Rue Gérard*.

6   Evaluation

6.1   Ontology Evaluation

We checked the consistency and compliance with good design practices of the proposed ontologies using the HermiT reasoner integrated into Protégé[6] [22] as well as OOPS! tool[7] [26].

Besides, the evaluation of the ontology consists in checking if we are able to answer competency questions of each modelet. Figure 6 represents the answers of multiple questions: coordinates of addresses along a defined street during at a given instant (here 2024) with the source of each house number.

6.2   Knowledge Graph Evaluation

First of all, to evaluate the knowledge graph, we tested the internal consistency of the data using SHACL rules or SPARQL queries to see if the iterative construction had generated any inconsistencies. These inconsistencies can be the description of the appearance of a landmark that occurs after its disappearance. The only detected inconsistencies deal with attribute versions having different values but whose valid time intervals overlap. In this case, the method presented in Sect. 5.3 induces that an event occurs in $[a; b]$ where $a > b$.

The consistency of the graph with the truth on the ground is particularly difficult to assess because the truth is no longer accessible. Dictionaries cited in Sect. 3.3 provide a basis to estimate the quality of the graph. Indeed, we can check our graph (restricted to thoroughfares) built from different sources

---

[6] http://protege.stanford.edu/.
[7] https://oops.linkeddata.es/.

describing the state of the territory is coherent with the truth. By taking the example of Sect. 5 with the geometry of the street *Rue Gérard*, we deduced a change appeared between 1888 and 2024 while the truth is this change appeared on August 30, 1978. It means what we deduced is coherent with the truth, the lack of precision is only due to the lack of sources describing the neighbourhood during the XXth century. Eventually, comparisons say the graph does not contain many contradictions with the ground truth.

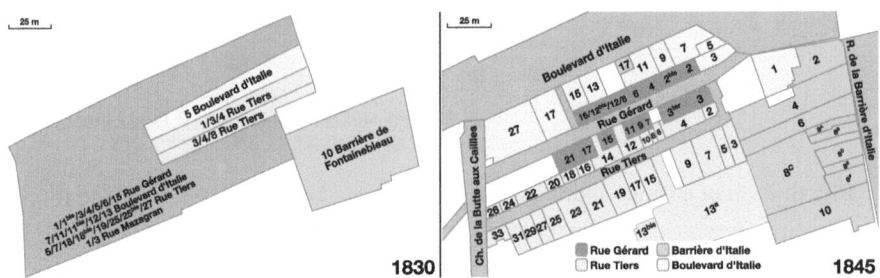

Fig. 7. Snapshots of the *Butte aux Cailles* district in 1830 and 1845 built from the graph representing the relation between plots and addresses.

In order to qualitatively assess the temporal and spatial coherence of the graph, two temporal snapshots representing the district of *Butte aux Cailles* in 1830 and 1845 were created (Fig. 7). These maps describe the addresses, roads and the associated plots. In order to produce these maps, we searched for all the landmarks that were valid in 1830 and 1845 in the sources. In the Napoleonic land registers, each plot is associated with a location (a place, a street or a full address). The locations associated with a given plot change over time for various reasons, such as road construction or operator error. In the first snapshot (1830) there are no geometries for thoroughfares, as historical sources only provide geometries from the 1840 s. Some plots are associated with more than one address. In fact, the plot geometries were extracted from the 1810 cadastral index maps that were not updated until 1845 (split and merged plots have no geometric representation). In the 1845 snapshot, the data shown on the map represent a coherent urban structure of the district. The house numbers are placed along the streets and follow the rules of Parisian house numbering. Although there were no metrics, we can assume that the map represents plots with consistent addresses. Furthermore, the approximate street locations provided by the plot addresses are consistent with the 1845 snapshot.

7   Conclusion and Future Work

In this article, we presented an approach to construct a knowledge graph to create a multi-scale historical gazetteer from multiple and heterogeneous sources. The first contribution of this work is the PeGazUs ontology, which was proposed to model addresses, land plots, other geographical entities and their evolution, as well as sources. The second contribution is the algorithm defined to populate this ontology with data from different sources describing the Butte aux Cailles district from the end of the 18th century to the present day. It enables to automatically link, temporally order and merge data fragments extracted from different sources and representing different attributes of the same real-world geographical entity. It also infers change and events from the available state data and vice-versa.

Although there was no ground truth with which to compare our graph, we were able to assess the quality of the graph qualitatively. The integration work presented in this article is restricted to a small district of Paris. The subsequent aim is to extend the spatial coverage of the graph and publish it. There are other interesting data to integrate, particularly those from directories containing a large number of addresses. In addition, some of these sources, such as very complete old address data (e.g. notaries' minute books), could be leveraged to assess the exhaustivity of the graph. Both contributions are available on the repository https://github.com/umrlastig/pegazus-ontology.

Acknowledgements. This work is supported by the French Ministry of the Armed Forces - Defence Innovation Agency (AID).

References

1. The factoid prosopography ontology. https://www.kcl.ac.uk/factoid-prosopography/ontology. Accessed 12 July 2024
2. International council on archives records in contexts ontology. https://www.ica.org/standards/RiC/ontology. Accessed 12 July 2024
3. Periodo, a gazetteer of periods for linking and visualizing data. https://perio.do/. Accessed 24 Jun 2024
4. Pro-o, the prov ontology. https://www.w3.org/TR/prov-o/. Accessed 12 July 2024
5. Time ontology in owl. https://www.w3.org/TR/owl-time/. Accessed 24 Jun 2024
6. International standards organisation. ISO 19113: geographic information - location-based services. tracking and navigation (2005)
7. Gazetteer service - application profile of the web feature service implementation spacification (2006)
8. International standards organisation. ISO 19112: geographic information - spatial referencing by geographic identifiers (2019)
9. Berman, M.L., Mostern, R., Southall, H.: Placing Names: Enriching and Integrating Gazetteers. Indiana University Press (2016). https://doi.org/10.2307/j.ctt2005zq7
10. Bernard, C., Villanova-Oliver, M., Gensel, J., Dao, H.: Modeling changes in territorial partitions over time: ontologies TSN and TSN-change. In: Proceedings of the 33rd Annual ACM Symposium on Applied Computing, pp. 866–875 (2018). https://doi.org/10.1145/3167132.3167227
11. Bernard, C., Abadie, N., Perret, J., Duménieu, B.: Création d'un référentiel géo-historique d'adresses à partir de sources multiples. In: GAST - Gestion et l'Analyse de données Spatiales et Temporelles. Dijon, France (2024). https://hal.science/hal-04490732

12. Bol, P.K.: The china historical geographic information system (CHGIS). choices faced, lessons learned. In: Conference on Historical Maps and GIS, vol. 23 (2007)
13. Charles, W., Aussenac-Gilles, N., Hernandez, N.: HHT: an approach for representing temporally-evolving historical territories. In: European Semantic Web Conference, pp. 419–435. Springer Nature Switzerland (2023). https://doi.org/10.1007/978-3-031-33455-9_25
14. Coetzee, S., Cooper, A.K., Ditsela, J.: Towards good principles for the design of a national addressing scheme. In: 25th International Cartographic Conference (ICC 2011). French Committee of Cartography, Paris, France (2011). http://hdl.handle.net/10204/5101
15. Ducatteeuw, V.: Developing an urban gazetteer: a semantic web database for humanities data. In: Proceedings of the 5th ACM SIGSPATIAL International Workshop on Geospatial Humanities pp. 36–39. Beijing, China (2021). https://doi.org/10.1145/3486187.3490204,
16. Gregory, I.N., Bennett, C., Gilham, V.L., Southall, H.R.: The great Britain historical GIS project: from maps to changing human geography. Cartogr. J. **39**(1), 37–49 (2002). https://doi.org/10.1179/caj.2002.39.1.37
17. Grossner, K., Janowicz, K., Keßler, C.: Place, period, and setting for linked data gazetteers. In: Berman, M.L., Mostern, R., Southall, H. (eds.) Placing names: Enriching and integrating gazetteers, pp. 80–96. Indiana University Press (2016). https://doi.org/10.2307/j.ctt2005zq7
18. Grossner, K., Mostern, R.: Linked places in world historical gazetteer. In: Proceedings of the 5th ACM SIGSPATIAL International Workshop on Geospatial Humanities, pp. 40–43. Beijing, China (2021). https://doi.org/10.1145/3486187.3490203
19. Hersent, M., Abadie, N., Duménieu, B., Perret, J.: Modèles et outils pour la publication de métadonnées d'archives géographiques et de leurs données dérivées. In: Humanistica 2023, Association francophone des humanités numériques. Geneva, Switzerland (2023). https://hal.science/hal-04110787
20. Kauppinen, T., Väätäinen, J., Hyvönen, E.: Creating and using geospatial ontology time series in a semantic cultural heritage portal. In: Bechhofer, S., Hauswirth, M., Hoffmann, J., Koubarakis, M. (eds.) ESWC 2008. LNCS, vol. 5021, pp. 110–123. Springer, Heidelberg (2008). https://doi.org/10.1007/978-3-540-68234-9_11
21. Knublauch, H., Kontokostas, D.: Shapes constraint language (SHACL) (2017). https://www.w3.org/TR/shacl/
22. Musen, M.A.: The protégé project: a look back and a look forward. AI Matters **1**(4), 4–12 (2015)
23. Pasin, M., Bradley, J.: Factoid-based prosopography and computer ontologies: towards an integrated approach. Digit. Scholarship Hum. **30**(1), 86–97 (2013). https://doi.org/10.1093/llc/fqt037
24. Perego, A., Lutz, M.: Isa programme location core vocabulary (2015). https://www.w3.org/ns/legacy_locn
25. Peroni, S.: A simplified agile methodology for ontology development. In: Dragoni, M., Poveda-Villalón, M., Jimenez-Ruiz, E. (eds.) OWLED/ORE -2016. LNCS, vol. 10161, pp. 55–69. Springer, Cham (2017). https://doi.org/10.1007/978-3-319-54627-8_5
26. Poveda-Villalón, M., Gómez-Pérez, A., Suárez-Figueroa, M.C.: Oops ! (ontology pitfall scanner !): an on-line tool for ontology evaluation. Int. J. Semantic Web Inf. Syst. (IJSWIS) **10**(2), 7–34 (2014). https://doi.org/10.4018/ijswis.2014040102

27. Schneider, P., Jones, J., Hiltmann, T., Kauppinen, T.: Challenge-derived design practices for a semantic gazetteer for medieval and early modern places. Seman. Web **12**(3), 493–515 (2021). https://doi.org/10.3233/SW-200394

28. Southall, H., Mostern, R., Berman, M.L.: On historical gazetteers. Int. J. Hum. Arts Comput. **5**(2), 127–145 (2011). https://doi.org/10.3366/ijhac.2011.0028

29. Tenbrink, T., Kuhn, W.: A model of spatial reference frames in language. In: Egenhofer, M., Giudice, N., Moratz, R., Worboys, M. (eds.) Spatial Information Theory, pp. 371–390. Springer, Berlin, Heidelberg (2011). https://doi.org/10.1007/978-3-642-23196-4_20

Ontology-Constrained Generation
of Domain-Specific Clinical Summaries

Gaya Mehenni[1,2(✉)] and Amal Zouaq[1,2]

[1] LAMA-WeST, Montreal, Canada
[2] Polytechnique Montreal, Montreal, Canada
{gaya.mehenni,amal.zouaq}@polymtl.ca

Abstract. Large Language Models (LLMs) offer promising solutions for text summarization. However, some domains require specific information to be available in the summaries. Generating these domain-adapted summaries is still an open challenge. Similarly, hallucinations in generated content is a major drawback of current approaches, preventing their deployment. This study proposes a novel approach that leverages ontologies to create domain-adapted summaries both structured and unstructured. We employ an ontology-guided constrained decoding process to reduce hallucinations while improving relevance. When applied to the medical domain, our method shows potential in summarizing Electronic Health Records (EHRs) across different specialties, allowing doctors to focus on the most relevant information to their domain. Evaluation on the MIMIC-III dataset demonstrates improvements in generating domain-adapted summaries of clinical notes and hallucination reduction.

1 Introduction

Large Language Models (LLMs) have shown major improvements in their extraction and summarization capabilities. In the medical field, these models offer the potential to automate the summarization process of complex medical data, such as Electronic Health Records (EHRs) and clinical notes [22]. These documents, which contain an overwhelming amount of information, are a significant contributor to clinician burnout [27]. Thus, the generation of more focused and domain-specific summaries would help alleviate this task. Multiple challenges arise when LLMs are applied to the medical domain. Not only is the information condensed in a domain-specific terminology, but clinical notes are unstructured and do not follow specific formats. Additionally, these models are prone to hallucinations which can have serious consequences in healthcare settings. These issues become even more complex when the generated content must be adapted to different medical contexts. For example, the critical information required for cancer treatments differs significantly from that needed for diagnosis imaging analysis. Thus, ideally, different summaries should be generated for different areas of focus. To address these challenges, medical ontologies can be utilized to extract and prioritize information relevant to certain domains, specialties or fields. These ontologies provide a structured representation of medical knowledge, allowing for the identification of key concepts and relationships within a particular field aka *domain*. This information can

M. Alam et al. (Eds.): EKAW 2024, LNAI 15370, pp. 382–398, 2025.
https://doi.org/10.1007/978-3-031-77792-9_23

be used to extract relevant information from clinical notes and leverage it to produce domain-adapted summaries and to reduce hallucinations. In this context, a subsequent challenge is to constrain the generation of LLMs to specific domain concepts and properties and avoid the generation of non factual non-grounded information. Following the above, this paper focuses on three main research questions:

– How can we adapt LLM-generated clinical summaries to different domains?
– Can we leverage ontologies to constrain the generation of LLMs?
– How can we constrain the generation process of LLMs to reduce hallucinations?

Our contributions can be summarized into the following aspects:

– We show how ontologies can be harnessed by LLMs for a domain-aware and constrained generation;
– We design a new ontology-guided decoding process that utilizes the knowledge embedded in an ontology to reduce hallucinations. We constrain the LLM's output to align with the relationships and concepts defined in the ontology. Thus, we reduce the likelihood of generating information that contradicts established domain knowledge, leading to more relevant answers and less hallucinated content;
– We create a new summarization process that can generate domain-tailored summaries of clinical notes.

To the best of our knowledge, our work is the first approach that employs ontologies in conjunction with LLMs to constrain generation. Figure 1 shows how our approach can generate ontology-structured summaries of clinical notes and domain-adapted summaries [1].

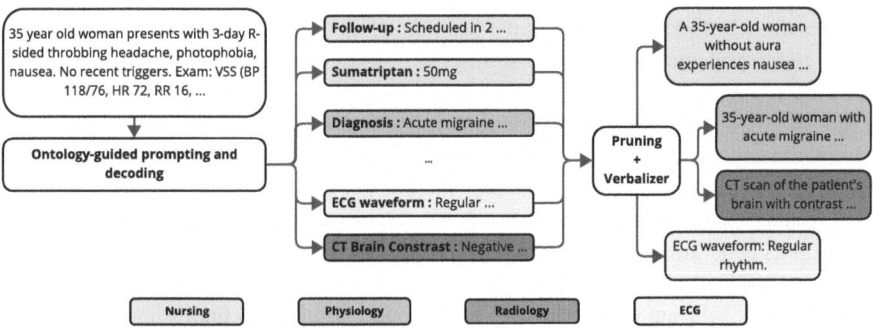

Fig. 1. Overview of our general architecture to generate domain-tailored summaries.

2   Related Work

Summarization. Summarization is the process of generating a smaller text from a larger input text. The main objective of that process is to grasp the information of the input text in the most condensed way possible. Existing work on the task of summarization

---

[1] Code is available at https://github.com/Lama-West/Ontology-based-decoding_EKAW2024.

can be divided into two main groups: *extractive* and *abstractive* summarization. Extractive summarization involves retrieving sections where the information is mostly present [18,19]. Abstractive summarization relies on neural architectures to generate a reformulated version of the original text. While the latter usually works better due to its context-dependent summarization properties, recent work have tried utilizing a hybrid approach [30,38]. These works usually obtain better performance [23] since they guide abstractive models towards relevant sections of the input while keeping the advantages of extractive methods. Our work, which is also based on this hybrid approach, adds an ontological component to the abstractive and extractive modules to improve the final summarization capabilities.

Clinical Text Summarization. The development of clinical note summarization is especially slow due to the absence of reference summaries [22]. Prior work on clinical summarization has mainly focused on summarizing radiology reports [7,29] and generating sections of the discharge summary of an admission [2,22]. In contrast to current methods that simply generate an output text, we aim to generate a more structured report tailored to the needs of a given specialist. This is particularly useful as clinical narratives often exhibit significant variation in how the information is structured and presented [24]. This would allow the clinician to quickly query on certain attributes of the patient while having access to the full report if needed. In this study, we rely on a medical ontology to structure the summaries.

Hallucination. One problem that arises from current summarization techniques is that LLMs are still prone to hallucinations [12], slowing down their deployment in the medical domain. Multiple techniques have been developed to mitigate this problem. They can be regrouped into design-time solutions, training-time solutions, generation-time solutions and external tools. Design time solutions include modifying the base architecture of LLMs to reduce hallucinations [10]. Training-time solutions include reinforcement learning [36], latent space understanding [17] and loss function modification [5]. As for generation-time solutions, they include pre-generation prompting techniques [21,33] and post-generation evaluation [8,9]. The latter measures the confidence of a model on its generation. Finally, external tools solutions incorporate search engines and vector databases into the loop [16]. In this work, our contribution is to propose ontologies to identify relevant beams during generation to improve factuality.

Constrained Generation. Constrained generation has seen significant advancements in recent years, addressing the need for generated content to follow specific constraints. These constraints can, for example, be utilized to impose a certain structure in the generation process. While some [34] have utilized finite state machines to generate a structured output, others [39] improved the efficiency of the process by fully exploiting the constraints to reduce the number of inference passes. [26] have also applied constraints for a structured output through grammars, making their method even more modular. [11] extended this process with a framework allowing grammars to be input-dependent.

Lexically constrained generation methods often rely on logits re-weighting or beam pruning to guide the model towards certain words or concepts [37]. Recent work have improved these methods through the use of heuristic estimates [20] and knowledge graphs [6]. To our knowledge, very few, if any, state-of-the-art approaches have leveraged ontologies for constrained generation. Our work is a step in this direction.

3 Methodology

Our research explores the potential of using ontologies to guide a language model towards relevant information using prompting and constrained generation. By constraining the generation using ontological structures, we aim to improve the summarization capabilities of language models and to reduce their hallucinations. To do this, we propose to utilize the ontology in conjunction with the beam search algorithm to assess the relevance and factual accuracy of potential beam candidates in relation to the input. By implementing this ontology-guided beam search, we expect to enhance the overall coherence and reliability of the generated text, ensuring that it aligns more closely with the knowledge represented in the ontology. To reduce hallucinations, we propose to also evaluate the beam paths based on the clinical note to favor those that resemble it the most. As a proof of concept, we develop a new method which divides the summarization task into multiple simple inference passes guided by an ontology-based prompting approach. Figure 2 shows how, given multiple clinical notes about the same patient, our method outputs a text summary and a structured summary whose structure is defined by ontology concepts. This structured representation of the clinical notes can be leveraged afterwards to adapt the final summary to various domains (medical fields such as cardiology, oncology, etc.). Finally, our approach can be applied to any model since it only requires token probabilities.

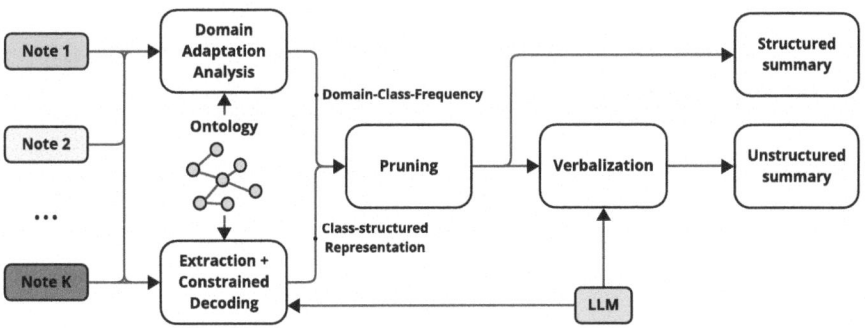

Fig. 2. Overall architecture of our method: Multiple notes about the same patient are passed to the framework and structured and unstructured summaries are generated

3.1 Domain Adaptation Analysis

We define a domain to be a set of ontology classes of interest related to a specific medical field. To adapt the generation to multiple domains, an initial analysis is performed using texts from different domains to identify their key concepts. Given texts that are linked to a certain domain D, we identify important concepts, or classes, in each text using an ontology-based annotator and create a set S based on a minimum occurrence threshold. We presume that the annotator can detect different formulations of the same concept (abbreviations, plurals, etc.). Then, using the ontology, we retrieve all ancestors of each class and add the ancestors to S. Subsequently, the frequency of each class in S is computed and stored in a class-to-frequency dictionary. We define this dictionary as the Domain-Class-Frequency dictionary (DCF). Its goal is to store the most relevant classes in a domain to guide the generations towards these classes. Then, each DCF is normalized according to the average DCF, computed by averaging the classes frequencies of all domains. This normalizing step allows the DCF to only contain relevant classes to the domain and reduce the weight of general medical concepts (which are more frequent). The DCF stores the important domain concepts to specialize the summary at the next stage.

3.2   Information Extraction Using Ontology-Based Prompting

Our next step is to extract medical concepts, properties and their values from clinical notes about a patient using a large language model (LLM), an ontology-based prompting process and a constrained decoding strategy. The overall process of the extraction phase is shown in Fig. 3. The main goal of this step is to generate a structured version of clinical notes allowing doctors to easily query information based on concepts. This structured version will also be used in conjunction with the DCF dictionary (see Sect. 3.1) in latter steps to specialize the summary to a given domain.

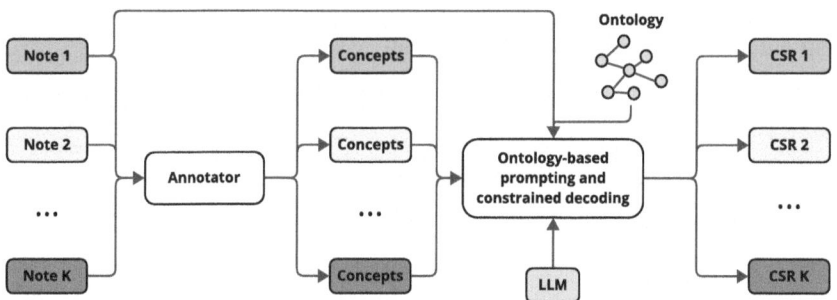

Fig. 3. Extraction Phase

To improve the information extraction capabilities of the LLM, we adapted [32]'s summarization technique to the medical domain by incorporating an ontology-guided prompting process. Given multiple clinical notes of a patient during an admission and a medical ontology, we start by annotating each clinical note to retrieve all the medical concepts mentioned in each note. This process can easily be parallelized since, at this step, each note is considered independently. We then prompt the model to summarize each note around the concepts in multiple inference passes. Following a RAG-like [16] approach, we also augment the prompt with all relevant restriction properties that can be inferred using the ontology class. This augmentation provides the model with more context about what the ontology class is referring to. For instance, if the model does not know what an electrocardiogram is, augmenting the prompt with characteristics of this concept can improve its extraction capabilities since a certain definition is given. The following prompt is used in that case:

Here is a clinical note about a patient: [clinical note]. In a short sentence, summarize everything related to the "[concept]" concept mentioned in the clinical note. "[concept]" is characterized by [properties]. If nothing is mentioned, answer with "N/A".

For example, the prompt applied to the "electrocardiogram" concept would be:

Here is a clinical note about a patient: [clinical note]. In a short sentence, summarize everything related to the "electrocardiogram" concept mentioned in the clinical note. "electrocardiogram" is characterized by Evaluation - action AND Heart Structure AND Electrocardiographic monitor and recorder, device. If nothing is mentioned, answer with "N/A".

Then, for each clinical note, a class-structured representation (CSR) is created. The CSR is a mapping of all ontology classes (concepts) detected in the note to short summaries - extracted from the notes - of the information related to each class. These summaries, which we refer to as "extracted values", are generated by a model using the mentioned prompt template.

3.3 Constrained Decoding

The decoding strategy during the extraction step is designed to guide the model's answer towards responses that are more relevant to the prompt. This strategy serves two primary purposes: leveraging the knowledge embedded within the ontology to steer the model towards a more relevant answer to the prompt (*relevance*), and reducing the occurrence of hallucinations (*groundedness*) during the generation process to ensure that the model's responses align with both the prompt and the provided notes. The overall constrained decoding process is shown in Fig. 4. This process is used in conjunction with the ontology-based prompting (see Sect. 3.2) to obtain a final class-structured representation (CSR) for each clinical note.

We propose an algorithm based on diverse beam search [31], wherein grouped beam search is employed to diversify the results. In a nutshell, our algorithm favors beams that textually resemble the input and that contain concepts that are related in the ontology through hierarchical relations(subclasses and superclasses) and restriction properties. Since extracting information from generated beam candidates is not an operation that can be done trivially token by token, our algorithm computes the beam scores after a certain number of tokens, defined as the *generation window*. After this threshold, the newly generated tokens are analyzed using the same annotator used to tag the clinical notes. Once the beam is tagged, a score is calculated, as detailed below, to favor beam paths that are aligned with the internal structure of the ontology and with the note content.

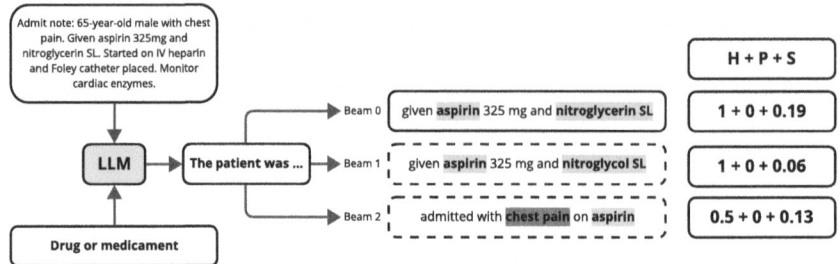

Fig. 4. Constrained decoding process: Each beam rectangle represents the current generation window associated with the beam. The concepts in green are concepts that are associated to a children class of the base class (Drug or medicament) in the ontology. Green concepts improve the hierarchy score which augments the probability that the beam will be chosen as a final output. The similarity score is computed using the ROUGE-2 score between the generation window and the clinical notes.

To calculate the score of a beam, we compute three sub-scores: the hierarchy score H, the property score P and the similarity score S. For all scores, we define b to be the base class corresponding to the ontology class used in the prompt shown in Sect. 3.2. This corresponds to the ontology class replacing the "[concept]" tag. We also define T to be the newly generated tokens in the beam, C the set of classes retrieved from T and $A(c)$ the set of ancestors of a class c in C.

Hierarchy Score. The hierarchy score H computes a score based on the number of descendants of the base class that are present in the generated beam:

$$H = H_{bf} \frac{1}{|C|} \sum_{c \in C} \mathbb{1}\{b \in A(c)\}$$

where H_{bf} is the hierarchy boost factor, a hyperparameter controlling how relevant we want the hierarchy score to have an impact on the final beam score. The primary goal of the hierarchy score is to guide beams towards the ontology's hierarchy to make sure that the generation is relevant to what is expected. For instance, when asked to summarize the patient's diseases, we would expect the model's answer to contain ontology classes that inherit the disease class.

Property Score. The property score P evaluates how relevant a beam is to the base class, based on restriction properties associated to the base class. While this can be generalized to any class property, we only consider restriction properties as they are the most frequent in the ontology used in our case. This score allows the decoding process to incorporate knowledge about concepts that can be inferred from the ontology. We thus want to favor beams that mention classes present in the restriction property values. For example, when asked about the "Fever" concept, we want the model's answer to ideally specify, if mentioned in the notes, that the patient has a body temperature above the normal range. We would then favor beams that mention the "Body Temperature" class. Following this intuition, the property score is given by:

$$P = \frac{P_{bf} \sum_{c \in C} \mathbb{1}\{c \in P(b)\}}{|C||P(b)|} + \text{R2}(T, P'(b))$$

where P_{bf}, similar to H_{bf}, is the property boost factor and R2 is the ROUGE-2 score. $P(c)$ is the set of classes related to c through restriction properties and $P'(c)$ is a natural language formulation of $P(c)$. In practice, we only take into account *And* and *Or* restrictions. Given an object property restriction of the form { *property1* : *value1, ...* }, to compute $P'(c)$, we simply concatenate all values in the case of an *And* restriction. For example, if $P(\text{Fever}) = AND(Interprets: Body Temperature, Has Interpretation: Above Reference Range)$, then $P'(\text{Fever}) = Body Temperature Above Reference Range$. In the case of an *Or* restriction, we add *or* between every value. Computing the ROUGE-2 score between the natural language formulation of the restrictions and the newly generated tokens allows us to favor the beam if the annotator did not tag concepts correctly.

Similarity Score. Finally, the similarity score S aims to measure how similar a beam is to the clinical notes. Since the model's answer about a single class should be short and resemble the clinical notes in an extractive manner, we hypothesize that the ROUGE-2 score is a good measure of the similarity of the answer to the notes. Its formula is simply given by:

$$S = S_{bf} \, \text{R2}(b, N)$$

where S_{bf} is the similarity boost factor, N is the clinical note and b the current beam.

Overall Score. The final beam score BS is given by :

$$BS = \text{LogSoftmax}(H + P + S)$$

3.4 Pruning and Verbalization

Pruning. Once a class-structured representation of each clinical note is obtained using the extraction process, we prune each representation to adapt the summary to a given domain (or specialty) by keeping only the domain-relevant ontological classes. This pruning step aims to adapt the summaries to a given domain by focusing only on the extracted information that is relevant to the domain. This is done using the DCF dictionary computed initially (see Sect. 3.1). In practice, for computational reasons, we retrieve the top-k most frequent classes in the DCF dictionary. However, upper-level classes tend to be more frequent. To account for this problem, we also retrieve all classes that are within α nodes from a frequent class in the ontology using hierarchical relationships from superclasses to subclasses. In this case, k is a hyperparameter controlling the length of generated summaries in terms of the number of concepts covered and α is a hyperparameter controlling the preciseness of the generated summary.

Verbalization. We also use a final forward pass using the same LLM that transforms the structured output into an unstructured format. This process ensures that the output aligns with the requirements of the task and allows us to compare the efficiency of the method in terms of summarization.

4   Experiments

4.1   Dataset and Ontology

The Medical Information Mart for Intensive Care (MIMIC-III) database [14] was used to retrieve clinical notes. This database regroups over 45,000 de-identified patient admissions to critical care in the Beth Israel Deaconess Medical Center between 2001 and 2012. Each admission contains multiple clinical notes associated to the same patient and is linked to a Brief Hospital Course (BHC) section of the patient's discharge summary. Overall, the dataset contains over 1.4 million single clinical notes. For computational reasons, we only use a test set called Ω of 400 randomly chosen admissions, regrouping over 3000 clinical notes, to perform our evaluations.

Clinical notes are associated to several categories ranging from electrocardiogram (ECG) reports to nursing notes and discharge summaries. Each note category serves a specific purpose and is associated to its own set of medical terms. For instance, ECG reports offer detailed insights about the patient's cardiovascular functions and structures, and nursing notes provide a continuous narrative of the patient's day-to-day care. These categories (ECG, Nursing, ...) are used to define our domains. For the domain adaptation of summaries, we employed the SNOMED-CT ontology [25], given its comprehensive representation of diverse medical fields. To analyze the occurrence of the SNOMED concepts in the MIMIC dataset, we computed the most frequent concepts in each domain (category) of MIMIC's notes. As shown in Fig. 5, the domains differ by a lot on their class representation. Note that to perform this analysis, we pruned some branches of the SNOMED-CT ontology as they do not correspond to medical concepts (Linkage concepts, Qualifier values, ...).

4.2   Models

To generate our tailored summaries, we used the Phi-3 model, a 3.5 billion parameter model matching GPT3.5's performance [1] as well as Zephyr-7b-beta [28]. Both models were used for the extraction phase and as verbalizers. As for the annotator linking text sequences and SNOMED-CT classes, we utilized the MedCAT annotator [15]. This annotator was used during the decoding process and during evaluation to tag concepts (see Sect. 5.3).

Generation Window. We evaluate empirically that a generation window of 5 to 15 tokens is a good trade-off between extraction capabilities and performance. We use a generation window of 10 tokens.

Hyperparameters. For the constrained decoding process, we use $H_{bf} = 3, P_{bf} = 10, S_{bf} = 10$, a beam size of 10 and a group size of 2 for diverse beam search. As for the pruning phase, we use $k = 30$ and $\alpha = 2$.

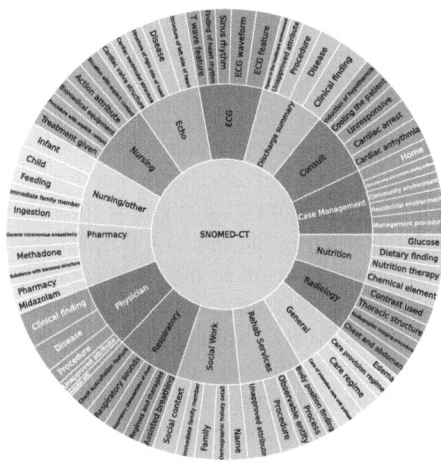

Fig. 5. Most frequent classes in SNOMED-CT based on different domains of clinical notes. This plot was computed by automatically annotating 1000 clinical notes from each category in MIMIC-III and associating each concept to a SNOMED-CT class. We then performed a domain adaptation analysis as shown in Sect. 3.1. Only the 5 most frequent concepts were kept in this figure.

5 Evaluation

5.1 Domain Adaptation

Evaluator. Since no ground truth exists for this task, we needed to train an evaluator to compute a domain adaptation score. For this purpose, we fine-tuned a BERT classifier trained on the medical domain [4] to predict the domain of a clinical note. We denote this model as the evaluator model. To do so, we utilized the *CATEGORY* column of MIMIC-III which indicates the medical field of a clinical note. In practice, since not all domains are equally present, we focus on the *Nursing, ECG, Radiology* and *Physician* domains as they are the most frequent. To train the model, we use 400k individual clinical notes with their respective categories. For validation, 40k clinical notes were randomly selected. These training and validation sets are disjoints from the test set used to evaluate the summaries in other experiments. After training, the model achieved a 99% accuracy on the validation set. However, we found that the model tended to rely on each domain's note format instead of relying entirely on the underlying concepts to predict the domain. Thus, we further fine-tuned the model on a custom-made dataset generated by passing 4k clinical notes not seen during the initial training through a BART paraphraser.

Domain Score. We compute the domain score D as the average logit score of the expected domain:

$$D = \frac{1}{N} \sum_{i=0}^{N} \text{EVALUATOR}(d(x_i))[d_i] \quad (1)$$

where $d(x_i)$ is the domain adapted summary of the admission x_i, d_i is the expected domain and N is the number of samples. This metric is used to evaluate how tailored the generated summary is based on the domain.

Baselines. To assess the effectiveness of domain adaptation, we evaluated three distinct approaches: (1) greedy search (2) diverse beam search and (3) our method. For the greedy and diverse beam search methods, we augment the prompt with a prefix specifying the desired domain, instructing the model to tailor the summary accordingly[2]. For our method, we simply pass the result of the pruning phase to the verbalizer without specifying the desired domain. We then pass each of the summaries (greedy, beam search, our method) through our BERT Evaluator. By comparing these three methods, we aim to demonstrate the relative efficiency of our proposed technique in producing domain-tailored summaries without the need for explicit prompt engineering. We evaluate these methods on the Ω test set. Since each admission can be used to generate 4 domain-adapted summaries (*Nursing, ECG, Radiology* and *Physician*), the final test set, in this experiment, contains 1600 pairs (admission, domain-adapted summary).

Table 1. Domain score of generated summaries using greedy search, diverse beam search and our method

Method	Phi-3	Zephyr
Greedy Search	0.41	0.56
Diverse Beam Search	0.43	0.59
Ours (Extraction+Pruning)	**0.86**	**0.78**

Table 1 shows the domain score of greedy and diverse beam search summaries versus our approach. The results highlight an improvement in domain adaptation when using our method. In the case of Phi-3, it achieves more than twice the domain score of greedy search.

5.2 Hallucination Reduction

Evaluator. We also focus on evaluating the impact of our proposed constrained decoding process on reducing hallucinations. More precisely, we aim to evaluate how the constrained decoding process improves the extraction capabilities of the model. We evaluate two aspects: the *groundedness* and the *relevance* of the answers. We formulate this problem as an entailment task where we leverage Natural Language Inference (NLI) models to determine the entailment between our results and the clinical notes. We ran this experiment on 1000 randomly extracted values of the Ω test set and used a DEBERTA model [13] fine-tuned on the MNLI dataset [35] as the NLI model.

---

[2] See the GitHub repository for a prompt example.

Groundedness. For an admission, we apply the extraction phase (see Sect. 3.2) using all methods, resulting in multiple CSRs (one per clinical note). In a CSR, each extracted value prepended by its concept ([concept]: [extracted value]) is considered as a hypothesis of the clinical note (the premise). We then compute the groundedness score by taking the entailment score as determined by an NLI model. The final groundedness score is calculated by averaging over all CSRs.

Relevance Score. We also compute a relevance score where keys from a CSR (concepts) are considered the hypotheses and the extracted values the premises. Since the keys are used to generate the prompt question, computing this score allows us to evaluate how relevant the answer was based on the concepts.

Baselines. Similar to the domain score, we employ a comparative analysis of three distinct generation methods during the extraction phase to quantify the effectiveness of our approach: (1) greedy search (2) diverse beam search, and (3) our novel constrained generation.

Table 2. Groundedness (Gnd) and Relevance (Rel) of extracted sequences using greedy search, diverse beam search and our ontology-constrained generation

Method	Phi-3			Zephyr		
	Gnd	Rel	Avg	Gnd	Rel	Avg
Greedy Search	0.83	0.57	0.70	0.83	0.82	0.83
Diverse Beam Search	0.85	0.58	0.72	0.87	0.83	0.85
Constrained Generation	**0.90**	**0.64**	**0.78**	**0.90**	**0.88**	**0.89**

Table 2 presents the groundedness and relevance scores for different generation methods. The results mirror those of the domain scores but more moderately. Our method improves the groundedness score of both models by over 7% compared to greedy search and by 5 to 3% compared to beam search. As for the relevance score, we boost it by 7% compared to greedy search in the case of the Phi-3 model. These improvements indicates that generations based on our decoding process contain slightly less hallucinated content.

5.3 Summarization

To compare our approach to state-of-the-art techniques, we evaluate its performance on generating the BHC section of the discharge summary of a MIMIC-III's patient admission, a well known task in the literature of clinical summarization [2,22]. We will define this task as the BHC task. We evaluate it on the Ω test set. As our method only requires a set of ontology classes to define a domain, we extended our approach to consider the BHC section as a domain. Just as presented in Sect. 3.1, we performed a domain adaptation analysis on the BHC section of 1000 admissions to build our DCF dictionary.

Baselines. We compare our method against a standard single-pass generation using a predefined prompt and other state-of-the-art techniques that underwent fine-tuning such as SPEER [3,22]'s dual Transformer. Following the literature, we compute the ROUGE score as it is the most used for this task.

Metrics. Similarly to [2], we also compute an hallucination score which measures how much the generated summary tends to mention concepts that are not mentioned in the original notes. Given the set N of concepts in the clinical notes and the set S of concepts in the generated summary, this score is defined as :

$$ \text{HS} = \frac{|S - N|}{|S|} \tag{2} $$

However, this score is not entirely perfect since it does not take into account the uncertainty of the annotator used to tag the concepts. We thus compute the adjusted hallucination score (AHS) which incorporates the concepts from the reference summary into the metric. Given the set R of concepts in the reference summary, this metrics is defined as :

$$ \text{AHS} = \frac{|S - (N \cup R)|}{|S|} \tag{3} $$

The AHS will give room for concepts that can be inferred from the notes.

Table 3. Results for the BHC summarization task

Fine-tuned	Method	R1↑	R2↑	RLSum↑	HS (%) ↓	AHS (%) ↓
✓	Zephyr [2]	25.00	6.90	–	–	–
	SPEER [2]	**25.90**	7.10	–	–	–
	Dual Transformer [22]	–	**11.50**	**34.90**	–	–
✗	Baseline (Phi-3)	26.59	4.75	13.85	45.70	38.68
	Ours (Extraction)	25.94	4.70	13.30	38.47	33.54
	Ours (Extraction+Pruning)	**28.41**	**5.47**	**13.93**	**37.95**	**33.08**
✗	Baseline (Zephyr)	20.32	4.01	11.56	45.43	41.60
	Ours (Extraction)	20.01	3.10	11.24	45.00	41.18
	Ours (Extraction+Pruning)	**23.44**	**4.20**	**12.78**	**43.76**	**40.39**

The results shown in Table 3 show that our prompt-based method does not outperform fine-tuned models overall. Both the extraction and pruning step are needed to achieve the best performance. Likewise, our technique showed improvements in the hallucination and ajusted hallucination scores for the Phi-3 model. This effect is not as important with Zephyr. Possible explanations for these results include that the prompt structure was specifically tailored to maximize Phi-3's and that Zephyr has a higher number of parameters.

6 Discussion and Limitations

6.1 Domain Adaptation

The results in Table 1 suggest that our approach is really effective on its intended application. It is more efficient at producing more domain-adapted summaries than simply prompting the model to do so. These findings underscore the significance of the pruning step as it is the main step performing the domain adaptation. This also shows that the set of relevant concepts for a domain can be inferred from the data, as the pruning step is completely based on the initial domain adaptation analysis. These findings also implicitly confirm that our ontology-based decoding process improves model performance. Indeed, the effectiveness of the pruning step heavily relies on the values in the CSR which are determined using our custom decoding process. Furthermore, our method significantly enhances the summarization process interpretability by separating the extraction and adaptation phases, allowing for a more modular, interpretable process. This transparency allows for easier verification and validation of the generated content. Once the extraction phase is performed, the pruning phase can be used to adapt to any domain without the need to repeat the extraction.

6.2 Reduction of Hallucinations

As shown in Table 3 by the hallucination score (HS) and adjusted hallucination score (AHS), our method handles hallucinations better in the generated summaries compared to the baselines, especially with Phi-3. While the extraction phase seems to have a large impact on the performance of Phi-3, in the case of Zephyr, it is the pruning phase that creates a bigger jump in performance. The improvements of our method shown in Table 2 also suggest that summaries generated using our custom ontology-guided decoding process are more likely to be factually consistent. This improvement in factuality and relevance is especially pronounced in the case of the Phi-3 model. Grounding each beam with the input and restraining the decoding process using an ontology contributes to hallucination reduction especially for a smaller model.

6.3 Summarization Performance

As Table 3 suggests, our technique demonstrates slight improvements in the summarization performance for clinical notes while providing a structured version of the same summary. This format is easier to query for clinicians. While this is not the best task to evaluate our method as it requires formulations specific to BHC summaries, it allows us to compare our results to state-of-the-art methods. Additionally, we also show that our definition of domains can be generalized since it is applicable to other tasks such as the BHC task. Finally, is not surprising that our results did not outperform fine-tuned models since we used a prompting-based approach on models that were not specifically pre-trained to handle medical data.

6.4 Limitations

Since our approach relies on summarizing the notes around multiple concepts, the primary challenge is the computational overhead associated with the inference passes required by this method. While, they can easily be parallelized across all clinical notes and ontology classes, the overhead associated to beam search on all these passes poses challenges for scalability and deployment. Plus, our method is highly sensible to hyperparameters (prompt format, k, α, ...) which makes it hard to optimize. Additionally, it heavily relies on a good annotator. While such an annotator exists for SNOMED-CT, it might not be available for other ontologies. Finally, a significant limitation of our work is the absence of human assessment and gold standards with domain-related summaries. Due to time constraints in accessing medical experts, our evaluation only relies on automated evaluation methods which may not be always reliable. For instance, the NLI model employed to detect hallucinations in our generated content may not be optimal, as it lacks specific training on medical data. Consequently, the scores about hallucinated content may not be always trustworthy. Our future work will provide a qualitative evaluation of our results.

7 Conclusion

In this paper, we introduced a novel approach capable of generating domain-adapted summaries. We leveraged an ontology-guided decoding process to improve the factuality and relevance of the summarization process. We showed that guiding the model towards ontological concepts creates better domain-adapted summaries. Furthermore, our method improves performance in summarization, domain-adaptation, and groundedness on text input while providing a structured version of the summary. This structured version can easily be queried to extract relevant information in any use cases.

Applied to the medical domain, our approach highlights the potential for generating tailored summaries across various medical fields. While computational overhead remains a challenge, this work represents a step towards reducing clinician burnout and improving patient care through more efficient information synthesis.

Acknowledgements. We gratefully acknowledge the financial support provided by the FRQS (Fonds de Recherche du Québec - Santé) Chair in Artificial Intelligence and Digital Health. This research also benefited from the computational resources and GPU infrastructure provided by the Digital Research Alliance of Canada (DRAC) and Calcul Québec. We extend our sincere thanks to Prof. John Kildea for his valuable input during initial discussions on the challenge of adapting summaries for specialists.

References

1. Abdin, M., Jacobs, S.A., et al.: Phi-3 technical report: a highly capable language model locally on your phone. arXiv preprint arXiv:2404.14219 [cs] (2024)
2. Adams, G., Zucker, J., Elhadad, N.: SPEER: sentence-level planning of long clinical summaries via embedded entity retrieval. arXiv preprint arXiv:2401.02369 (2024)

3. Adams, G., Zucker, J., Elhadad, N.: A meta-evaluation of faithfulness metrics for long-form hospital-course summarization. arXiv preprint arXiv:2303.03948 [cs] (2023)
4. Alsentzer, E., et al.: Publicly available clinical BERT embeddings. arXiv preprint arXiv:1904.03323 [cs] (2019)
5. Chern, I.C., Wang, Z., Das, S., Sharma, B., Liu, P., Neubig, G.: Improving factuality of abstractive summarization via contrastive reward learning. arXiv preprint arXiv:2307.04507 [cs] (2023)
6. Choi, S., Fang, T., Wang, Z., Song, Y.: KCTS: knowledge-constrained tree search decoding with token-level hallucination detection. arXiv preprint arXiv:2310.09044 [cs] (2023)
7. Chuang, Y.N., Tang, R., Jiang, X., Hu, X.: SPeC: a soft prompt-based calibration on performance variability of large language model in clinical notes summarization. arXiv preprint arXiv:2303.13035 [cs] (2023)
8. Cohen, R., Hamri, M., Geva, M., Globerson, A.: LM vs LM: detecting factual errors via cross examination. arXiv preprint arXiv:2305.13281 [cs] (2023)
9. Dhuliawala, S., et al.: Chain-of-verification reduces hallucination in large language models. arXiv preprint arXiv:2309.11495 (2023)
10. Gao, Y., et al.: Leveraging a medical knowledge graph into large language models for diagnosis prediction. arXiv:2308.14321 [cs] (2023)
11. Geng, S., Josifoski, M., Peyrard, M., West, R.: Grammar-constrained decoding for structured NLP tasks without finetuning. arXiv:2305.13971 [cs] (2023)
12. Guerreiro, N.M., et al.: Hallucinations in large multilingual translation models. arXiv:2303.16104 [cs] (2023)
13. He, P., Liu, X., Gao, J., Chen, W.: Deberta: decoding-enhanced bert with disentangled attention. In: International Conference on Learning Representations (2021). https://openreview.net/forum?id=XPZIaotutsD
14. Johnson, A.E., et al.: MIMIC-III, a freely accessible critical care database. Scientific Data **3**(1), 160035 (2016). https://doi.org/10.1038/sdata.2016.35
15. Kraljevic, Z., et al.: Multi-domain clinical natural language processing with MedCAT: the medical concept annotation toolkit. Artif. Intell. Med. **117**, 102083 (2021). https://doi.org/10.1016/j.artmed.2021.102083
16. Lewis, P., et al.: Retrieval-augmented generation for knowledge-intensive NLP tasks. arXiv preprint arXiv:2005.11401 (2020)
17. Li, K., Patel, O., Viégas, F., Pfister, H., Wattenberg, M.: Inference-time intervention: eliciting truthful answers from a language model. arXiv preprint arXiv:2306.03341 [cs] (2023)
18. Liu, M., Zhang, D., Tan, W., Zhang, H.: DeakinNLP at ProbSum 2023: clinical progress note summarization with rules and language ModelsClinical progress note summarization with rules and languague models. In: The 22nd Workshop on Biomedical Natural Language Processing and BioNLP Shared Tasks, pp. 491–496. Association for Computational Linguistics, Toronto (2023).https://doi.org/10.18653/v1/2023.bionlp-1.47
19. Logan IV, R.L., Liu, N.F., Peters, M.E., Gardner, M., Singh, S.: Barack's Wife Hillary: using knowledge-graphs for fact-aware language modeling. arXiv preprint arXiv:1906.07241 [cs] (2019)
20. Lu, X., West, P., Zellers, R., Bras, R.L., Bhagavatula, C., Choi, Y.: NeuroLogic decoding: (un)supervised neural text generation with predicate logic constraints. arXiv preprint arXiv:2010.12884 [cs] (2021)
21. Press, O., Zhang, M., Min, S., Schmidt, L., Smith, N.A., Lewis, M.: Measuring and narrowing the compositionality gap in language models. arXiv preprint arXiv:2210.03350 [cs] (2023)
22. Searle, T., Ibrahim, Z., Teo, J., Dobson, R.J.: Discharge summary hospital course summarisation of in patient Electronic Health Record text with clinical concept guided deep pre-trained Transformer models. J. Biomed. Inform. **141**, 104358 (2023)

23. Shi, J., Gao, X., Kinsman, W.C., Ha, C., Gao, G.G., Chen, Y.: DI++: a deep learning system for patient condition identification in clinical notes. Artif. Intell. Med. **123**, 102224 (2022)

24. Sorita, A., et al.: The ideal hospital discharge summary: a survey of U.S. physicians. J. Patient Safety **17**(7) (2021). https://journals.lww.com/journalpatientsafety/fulltext/2021/10000/the_ideal_hospital_discharge_summary__a_survey_of.16.aspx

25. Stearns, M.Q., Price, C., Spackman, K.A., Wang, A.Y.: SNOMED clinical terms: overview of the development process and project status. In: Proceedings of the AMIA Symposium, pp. 662–666 (2001)

26. Stengel-Eskin, E., Rawlins, K., Van Durme, B.: Zero and few-shot semantic parsing with ambiguous inputs. arXiv preprint arXiv:2306.00824 [cs] (2024)

27. Tajirian, T., et al.: The influence of electronic health record use on physician burnout: cross-sectional survey. J. Med. Internet Res. **22**(7), e19274 (2020)

28. Tunstall, L., et al.: Zephyr: direct distillation of LM alignment. arXiv:2310.16944 [cs] (2023)

29. Van Veen, D., et al.: Clinical text summarization: adapting large language models can outperform human experts. arXiv preprint arXiv:2309.07430 [cs] (2023)

30. Van Veen, D., et al.: Adapted large language models can outperform medical experts in clinical text summarization. arXiv preprint arXiv:2309.07430 [cs] (2024)

31. Vijayakumar, A.K., et al.: Diverse beam search: decoding diverse solutions from neural sequence models. arXiv preprint arXiv:1610.02424 [cs] (2018)

32. Wang, Y., Zhang, Z., Wang, R.: Element-aware summarization with large language models: expert-aligned evaluation and chain-of-thought method. arXiv preprint arXiv:2305.13412 [cs] (2023)

33. Wei, J., et al.: Chain-of-thought prompting elicits reasoning in large language models. arXiv preprint arXiv:2201.11903 (2022)

34. Willard, B.T., Louf, R.: Efficient guided generation for large language models. arXiv preprint arXiv:2307.09702 [cs] (2023)

35. Williams, A., Nangia, N., Bowman, S.: A broad-coverage challenge corpus for sentence understanding through inference. In: Proceedings of the 2018 Conference of the North American Chapter of the Association for Computational Linguistics: Human Language Technologies, Volume 1 (Long Papers), pp. 1112–1122. Association for Computational Linguistics (2018). http://aclweb.org/anthology/N18-1101

36. Wu, Z., et al.: Fine-grained human feedback gives better rewards for language model training. arXiv preprint arXiv:2306.01693 [cs] (2023)

37. Yang, K., Klein, D.: FUDGE: controlled text generation with future discriminators. In: Proceedings of the 2021 Conference of the North American Chapter of the Association for Computational Linguistics: Human Language Technologies, pp. 3511–3535 (2021). https://doi.org/10.18653/v1/2021.naacl-main.276

38. van Zandvoort, D., Wiersema, L., Huibers, T., van Dulmen, S., Brinkkemper, S.: Enhancing summarization performance through transformer-based prompt engineering in automated medical reporting. arXiv preprint arXiv:2311.13274 [cs] (2024)

39. Zheng, L., et al.: SGLang: efficient execution of structured language model programs. arXiv preprint arXiv:2312.07104 [cs] (2024)

Contextualizing Entity Representations for Zero-Shot Relation Extraction with Masked Language Models

Riley Capshaw[✉] and Eva Blomqvist

Linköping University, 581 83 Linköping, Sweden
{riley.capshaw,eva.blomqvist}@liu.se

Abstract. Knowledge graphs (KGs) and their related ontologies constitute a key component in modern knowledge-based systems. However, hand-crafting these is not scalable, particularly due to the rate at which knowledge changes in many real-world applications. Partially automating the process of extracting and even modelling knowledge has therefore been a subject of research for many years. Nevertheless, accurate and reliable KG construction from natural language documents still remains a difficult task with many challenges, even in light of the impressive recent advances in language modelling. This paper focuses on one of those challenges, namely the extraction of accurate entity representations from text documents in order to facilitate relation extraction (RE). We present a novel method for generating document-contextualized input representations for entities using a masked language model (MLM) without the need for any sort of fine-tuning. These representations are then used as inputs to the same MLM that generated them, alleviating the need to include entire documents when prompting. Our results show that these representations 1) improve the ability of the MLMs BERT and RoBERTa to identify statements that represent correct relations between two entities; and 2) allow BERT to perform on par with the fine-tuned MLMs BioBERT and PubMedBERT.

Keywords: Knowledge Graphs · Masked Language Models · Machine Reading · Entity Embedding · Document-level Relation Extraction

1 Introduction

A Knowledge Graph (KG) is a labeled, directed graph where the nodes represent real world entities of interest and edges represent the existence of a typed (named) relation between those entities. Hence, a KG can also be viewed as a set of subject-predicate-object triples representing the edges of the graph. Such KGs, and their related ontologies describing the types of entities and relations in the domain, constitute a key component in modern knowledge-based systems.

However, hand-crafting large-scale KGs is not feasible, particularly due to the rate at which knowledge changes in many real-world applications, requiring

M. Alam et al. (Eds.): EKAW 2024, LNAI 15370, pp. 399–415, 2025.
https://doi.org/10.1007/978-3-031-77792-9_24

continuous KG maintenance and updates. Automating the process of extracting, and even modelling, knowledge has therefore been a subject of research for many years. The recent advances in natural language processing and understanding have also benefited the area of KG construction from natural language documents. Nevertheless, accurate and reliable KG construction from text still remains a difficult task with many challenges, even in light of the impressive recent advances of large language models (LLMs). These LLMs offer suffer from hallucinations since the are generally indifferent to the factuality of their output [10]. This can result in the extraction of triples which are inaccurate or irrelevant with respect to the knowledge in the input texts.

One important sub-task of KG construction from text is *relation extraction* (RE). The KG construction process starts by extracting the relevant entities from a document. The RE step then identifies which entity pairs have a relationship and what that relationship is. Many high-quality RE systems already exist, but their performance relies on framing the task as a classification problem. As a consequence, these classifiers only work for a pre-defined set of relations (i.e., a schema) and their training requires accurately annotated data with many examples for each relation type. In real-world scenarios where the relations of interest can change frequently, it is infeasible to annotate hundreds of new examples to retrain that classifier. This may also be computationally or financially costly depending on the number of relations and the training data size, especially if proprietary (and power-hungry) LLM-based systems are used.

To begin addressing some of these challenges, we therefore narrow our focus to studying a zero-shot RE setting where there is no need for training data or fine-tuning. We argue that such a setting also supports other aspects of KGs beyond just their construction, e.g., KG evolution, by simplifying the process of remodeling that data. We additionally focus on smaller language models, in particular masked language models (MLMs) like BERT [5] and RoBERTa [15], due to their relatively low computational requirements and energy consumption.

In this paper, we therefore present a novel method for generating document-contextualized input representations for entities using MLMs, and demonstrate how they improve results on a zero-shot RE task. These representations are used when prompting the same MLM, alleviating the need to include the full context of the document when prompting, further decreasing computation costs. The intuition behind our approach is two-fold. First, when a MLM has seen mentions of an entity in its training data, it will have captured patterns about how that entity occurred. Second is that when a MLM has not seen an entity in its training data (or rarely did so), its tokenizer will often tokenize that entity into a longer sequence of sub-word tokens. Such sequences often represent patterns not present in the MLM's training data. For both cases, we adjust the embeddings used to represent the entities based on the patterns *within a particular document*, rather than exclusively on the background knowledge of the MLM. We show experimentally that this contextualization method improves the MLM's ability to rank statements in a zero-shot setting for a relation extraction task. We also show how our method allows a general-purpose MLM to perform on par with

MLMs that have been fine-tuned on domain-specific biomedical data, to support our claim of reducing the need for fine-tuning.

The remainder of this paper is organized as follows. First, we provide background information and position our work with regard to related literature in Sect. 2. Then we discuss our method in detail in Sect. 3. In Sect. 4 we describe our experimental setup and analyze the results in Sect. 5. A thorough discussion of these results and their limitations is presented in Sect. 6, with concluding remarks in Sect. 7.

2   Background and Related Work

The largest body of related work focuses on fact extraction by treating MLMs as knowledge bases in their own right [9,20]. Some approaches use hand-crafted cloze-style (fill-in-the-blank) prompts to explore which token an MLM predicts is missing [20], while others automatically find the best prompt to maximize the score for a downstream task like knowledge base completion [1]. We follow the former methodology and use manually-written prompts. For both prompting methodologies, the MLM performing the infilling can either be used out-of-the-box in a zero-shot fashion [8], or be fine-tuned to enhance results for a particular data set [6]. Given that MLMs capture various types of biases from their training data [11,13,19], we avoided any fine-tuning and use only openly-available models.

For each prompt, we do not ask a MLM to fill in the two blanks. Instead, we fill in the blanks with entities and use the MLM to rank the resulting statements by their pseudo-log-likelihood (PLL) scores. We focus primarily on statements regarding a single relationship between two entities. These statements can be mapped to a standard triple from a KG, e.g. "Amsterdam is the capital of the Netherlands." refers to the triple ⟨Amsterdam, capitalOf, Netherlands⟩. Given a list of candidate statements, the goal is to rank those that correspond to facts supported by a document higher (closer to zero) than the unsupported statements. Shin [22] and Wang and Cho [24] show how to calculate PLL using MLMs as follows. Let M be a MLM and $\mathbf{V}^M$ be its vocabulary of size W that maps an input token t to an index k: $\mathbf{V}_t^M = k$. Let the sequence $\mathbf{S} = (t_1, t_2, \ldots, t_N)$ be a tokenized sentence of length N and $\mathbf{S}_{\backslash i}$ be $\mathbf{S}$ with t_i replaced by a mask token. The output of $M(\mathbf{S})$ is a matrix $\mathbf{M}^{\mathbf{S}}$ of size $(1 + N) \times W$, where row j corresponds to the output logits for $t_j \in \mathbf{S}$. Row 0 corresponds to a special [CLS] token prepended to all sequences. The pseudo-likelihood of t is then the likelihood of replacing the masked token in $\mathbf{S}_{\backslash i}$ with t if the replacement were randomly sampled from $\mathbf{V}^M$ weighted by $\mathbf{M}_i^{\mathbf{S}_{\backslash i}}$ normalized with softmax:

$$P_M(t \mid \mathbf{S}_{\backslash i}) = \text{softmax}\left(\mathbf{M}_i^{\mathbf{S}_{\backslash i}}\right)_k . \tag{1}$$

PLL is then calculated by taking the average of the logarithm of the values for every $t \in \mathbf{S}$:

$$\text{PLL}_M(\mathbf{S}) = \frac{1}{N} \sum_{i=1}^{N} \log\left(P_M(t \mid \mathbf{S}_{\backslash i})\right) . \tag{2}$$

Salazar et al. [21] demonstrate a strong correlation between PLL and the concept of linguistic acceptability [4]. They use PLL to improve existing methods for scoring translations, supporting their argument that this correlation enables MLMs to be useful across a wide variety of tasks without fine tuning. Our prior work showed several additional factors that affect PLL scores, such as the background knowledge of a given MLM and behaviors of the MLM's tokenizer [3].

When statements are ranked without any grounding in a document, this PLL scoring scheme becomes a probing task that analyzes the background knowledge captured by the MLM, similar to treating the MLM as a knowledge base. A straightforward way to improve results using this scheme would then be to fine-tune the MLM on the documents [6]. However, the process of fine-tuning can be costly in terms of time and power, and risks stability issues such as catastrophic forgetting [18]. Perhaps most importantly to our study, fine-tuning mixes the information held within the new documents, affecting all scores for all future statements. This makes it unclear how to determine exactly which document supports a given statement (provenance), as well as how to capture and model contradictory information, for instance where a document incorrectly claims that The Hague is the capital of the Netherlands. Even though this is factually incorrect, we still want to model that a particular document claims it.

3   Method

To avoid fine-tuning entirely, but still reap its benefits, we propose a method for generating and using *contextualized input embeddings* for entity mentions within documents. Given a document $\mathbf{S}$ of length N, we first use the MLM M to generate contextualized output representations $\mathbf{t}_i^{\text{out}}$ for every token $t_i \in \mathbf{S}$. Since MLMs are generally not trained to use their own output as input, these contextualized representations are not yet enough. As in Sect. 2 for generating PLL scores, we use the MLM's language-modeling head to produce $\mathbf{M^S}$, the pseudo-probabilities over the vocabulary $\mathbf{V}^M$. For a given token t_i, we can use the vector $\mathbf{M}_i^{\mathbf{S}}$ as a *weighting* of all the input embeddings to calculate the contextualized input embedding for t_i by taking the weighted average over all input embeddings:

$$\mathbf{t}_i' = f(\mathbf{M}_i^{\mathbf{S}}) \cdot \mathbf{E}^M \tag{3}$$

Here, $\mathbf{E}^M$ is M's pre-trained embedding matrix, f is some nonlinear function (usually softmax), and $\mathbf{t}_i'$ is the input embedding for token t_i from $\mathbf{S}$ after contextualization. This is a generalization of the usual embedding operation, which can be calculated as the dot product of a one-hot vector and $\mathbf{E}^M$.

Initially, $\mathbf{M}_i^{\mathbf{S}}$ is not guaranteed to sum to one and may contain negative values. This can affect the magnitude of the resulting vector, which may impact the MLM's ability to use it effectively. We account for this effect by applying some function f to $\mathbf{M}_i^{\mathbf{S}}$ as seen in Eq. 3. The choices of f which we tried were the identity function (no change), ReLU, softmax, and two custom variants of softmax: rectified softmax and top-k softmax. Viewing $\mathbf{M}_i^{\mathbf{S}}$ as a weighting over $\mathbf{E}^M$, ReLU ensures that only positive weights are used. Softmax enforces that all weights are positive and nonzero, and that their sum is one, giving the weights the appearance of a probability distribution. *Rectified softmax* calculates softmax only for the positive elements of a vector and sets all other elements to zero, while still ensuring that the sum of the vector is one:

$$\text{rectsoft} \left(\mathbf{x} \right)_i = \frac{\text{rectexp} \left(x_i \right)}{\sum_{j=1}^{K} \text{rectexp} \left(x_j \right)} \tag{4}$$

$$\text{rectexp} \left(x_i \right) = \begin{cases} 0 & \text{if } x_i \leq 0 \\ e^{x_i} & \text{otherwise.} \end{cases} \tag{5}$$

Here, $\mathbf{x} = (x_1, x_2, \ldots, x_K)$ is a real-valued vector of length K. *Top-k softmax* works similarly to rectified softmax, but only calculates softmax for the k largest elements of $\mathbf{x}$ and sets the remaining elements to zero, again ensuring that the sum is one:

$$\text{topksoft} \left(\mathbf{x}, k \right)_i = \frac{\text{topkexp} \left(x_i, k \right)}{\sum_{j=1}^{K} \text{topkexp} \left(x_j, k \right)} \tag{6}$$

$$\text{topkexp} \left(x_i, k \right) = \begin{cases} e^{x_i} & \text{if sorted-index} \left(i \right) \leq k \\ 0 & \text{otherwise.} \end{cases} \tag{7}$$

Here, sorted-index(i) returns the index of x_i in $\mathbf{x}'$, where $\mathbf{x}'$ is $\mathbf{x}$ with its values sorted in descending order.

3.1 Recontextualization

One extension to this method for contextualizing input embeddings is to repeat it, essentially re-reading the document but with the contextualized input embeddings in place of the original embeddings. We refer to the outputs of this second pass as *recontextualized input embeddings*. In principle this operation could be repeated to some manner of convergence, but we only explore the effects of one recontextualization step.

Here, there are at least two possible methods for performing this second pass. One is to replace all tokens in $\mathbf{S}$ with their contextualized forms:

$$\mathbf{S}' = \left\{ \mathbf{M}_i^{\mathbf{S}} \cdot \mathbf{E}^M \mid t_i \in \mathbf{S} \right\} \tag{8}$$

For our experiments, we instead opt to replace only the tokens in $\mathbf{S}$ which are part of an entity mention:

$$\mathbf{S}'_i = \begin{cases} \mathbf{M}_i^{\mathbf{S}} \cdot \mathbf{E}^M & \text{if } t_i \text{ is part of a mention,} \\ \mathbf{t}_i & \text{otherwise.} \end{cases} \tag{9}$$

Then the recontextualized input embeddings are calculated as one might expect:

$$\mathbf{t}_i'' = \mathbf{M}^{\mathbf{S}'} \cdot \mathbf{E}^M \tag{10}$$

3.2   Partial PLL for Soft Prompts

Soft prompting is the broad term for the technique of using input vectors not present in a language model's embedding matrix [17]. The use of contextualized input embeddings as inputs to a MLM is a form of soft prompting. When applying soft prompting, PLL cannot be calculated, as it relies on there being a one-to-one mapping from all inputs to tokens. Approximations for PLL in these situations have been proposed [3], but we chose a simpler approach to limit our study to the effects of the embeddings. For contextualization and recontextualization experiments, we calculate a *partial PLL* (pPLL) score for each statement by averaging the pseudo-likelihoods only for the tokens that come from the prompt. For the example sentence "Amsterdam is the capital of the Netherlands.", pPLL would only be based on the tokens in "is the capital of .". Entity tokenization is known to impact PLL scores [12], so pPLL has the added benefit of avoiding this effect by skipping the tokens of the entities altogether.

4   Experimental Setup

We evaluate the use of the contextualized input embeddings in a zero-shot relation extraction setting using two relation extraction data sets by extracting candidate statements from those data sets, scoring those statements according to their PLL or pPLL scores, ranking the statements according to those scores, and examining how often correct statements are among the highest-ranked statements. All data and code are available in our GitHub repository[1], including further analyses that could not be included in the paper due to space restrictions.

4.1   Data

We derive our first set of experiments from the development portion of the Document-Level Relation Extraction Data set (DocRED) [25]. Table 1 shows an example document from this data set. DocRED is based on Wikidata [23] and covers 96 unique relation types. We only use the ten most frequently occurring relations to reduce the overall computation costs, as we felt that these were sufficient to highlight the effects of using contextualized input representations.

DocRED is intended for measuring the accuracy of relation-extraction systems in a challenging setting where support for some relations spans multiple sentences in a document. In line with our prior works [2,3], rather than solving the problem as DocRED presents it, we instead use DocRED to extract many short candidate statements, some of which are true according to the documents, while most are not. We then use an MLM rank these statements with different settings and analyze the changes in the rankings between those settings to draw conclusions about what affects such a ranking scheme in a zero-shot setting. For this experiment, our focus is on the effects of the contextualization technique from Sect. 3 and its configuration.

---

[1] https://github.com/LiUSemWeb/entity-recontextualization.

Table 1. Document 963 from the DocRED dev set and its annotated facts.

Text	Facts
Cornelis or Cornelius Ketel (18 March 1548-8 August 1616) was a Dutch Mannerist painter, active in Elizabethan London from 1573 to 1581, and in Amsterdam from 1581 to the early 17th century, now known essentially as a portrait-painter, though he was also a poet and orator, and from 1595 began to sculpt as well. According to Ketel's biography, written by his contemporary Karel van Mander, he seems to have wanted to concentrate on the most prestigious of the hierarchy of genres, history painting, which included mythological subjects, but after he left France he is known almost entirely as a portrait-painter. Neither England nor Holland had much demand for large history paintings during his lifetime, ...	⟨Cornelius Ketel, P27, Dutch⟩ ⟨Cornelius Ketel, P27, Holland⟩ ⟨Cornelius Ketel, P569, 18 March 1548⟩ ⟨Cornelius Ketel, P570, 8 August 1616⟩ ⟨Cornelius Ketel, P937, Amsterdam⟩

Additionally, we extend our work into domain-specific relation extraction by using the training portion of the Biomedical Relation Extraction Data set (BioRED) [16]. BioRED contains abstracts of articles published on PubMed[2] that have been annotated with the same kinds of information as DocRED, covering seven unique relation types. We only examined four of these relations, as the others did not occur often enough in the data to produce reliable results. The main challenge with this data is its domain. Biomedical entities like diseases, chemicals, and phenotypes tend to tokenize poorly; i.e., into long sequences of sub-word tokens. For this reason, MLMs that are only trained on general texts, like BERT, will have trouble accurately ranking statements that use those entities. With this experiment, we show how contextualization alleviates this issue, allowing the general-purpose BERT model to perform on-par with MLMs specifically pre-trained or fine-tuned for the biomedical domain. Given the large number of available MLMs to choose from, we limited our selection to BioBERT [14] and PubMedBERT [7], the same two models used by Luo et al. [16] in their original evaluation of BioRED.

Some documents needed to be excluded from the analysis according to very strict criteria. First, documents that did not fit into the context window of all the models we tested after tokenization were excluded, in order to ensure a fair comparison. Second, in order to focus our analysis on the more challenging documents, we excluded documents if we could only extract fewer than ten candidate statements, or fewer than five unsupported statements. Finally, we only report results for relations which appeared in at least ten documents after this filtering. This is why we only analyze four relations from BioRED; the remaining four only occurred in between three and seven documents, which was too few to draw conclusions from. The scores for these excluded relations are reported in the supplementary material available in our GitHub repository. All

---

[2] https://pubmed.ncbi.nlm.nih.gov/.

Table 2. Prompts for the relations in DocRED and BioRED.

Data set	Relation	Prompt
DocRED	P17: country	?x is in the country of ?y.
	P27: country of citizenship	?x is a citizen of ?y.
	P131: located in A.T.E.	?x is located in ?y.
	P150: contains A.T.E.	?x contains ?y within its borders.
	P161: cast member	The cast of ?x includes ?y.
	P175: performer	?x was performed by ?y.
	P527: has part	One part of ?x is ?y.
	P569: date of birth	?x was born on ?y.
	P570: date of death	?x died on ?y.
	P577: publication date	?x was published on ?y.
BioRED	Association	There is an association between ?x and ?y.
	Bind	?x binds ?y.
	Negative Correlation	?x is negatively correlated with ?y.
	Positive Correlation	?x is positively correlated with ?y.

results additionally report the number of documents which were analyzed, to show a measure of support for the results.

4.2 Statement Extraction

To extract the candidate statements, we first wrote a short fill-in-the-blanks prompt for each relation type in the two data sets. These are shown in Table 2. Each prompt contains two blanks that are to be filled in using the entity mentions in the documents. In the spirit of the zero-shot setting, the prompts were not tuned to improve results in any way except where otherwise stated. The prompts for DocRED are taken directly from our prior work [2], while the prompts for BioRED were written specifically for this study based on information provided in the annotation guidelines for the data set. The exact prompts for all relations, including those we did not analyze, are present in our GitHub repository. While a prompt for any given relation can be written in many ways, each yielding different PLL scores, we made the assumption that keeping the prompts short and concise was sufficient to capture the patterns we wanted to examine. Our results then highlight the effects both for good and for poor prompt formulations.

We extract a set of candidate statements for each document by taking the prompts for each relation type present in it and populating them with every plausible pair of entity mentions. Here, plausibility is based on the domain and range of the relation as mined from the data. For a given document, this usually yields many more unsupported statements than supported statements. For the document 963 from Table 1, we would only extract candidate statements for the relations country of citizenship (P27), date of birth (P569), and date of death (P570). For country of citizenship, there are two correct relations to be extracted. Given that the domain of P27 is PER (people) and its range is LOC (locations), there are five possible mentions that could serve as the subject of the triple and seven total possible objects. In particular, Cornelius Ketel's name appears four times throughout the document in various forms. As a result, a total of thirty-five candidate statements are extracted, of which twenty-seven are unsupported and eight are supported. We then rank those statements and analyze changes in the ranking to understand to what extent various models and configurations affect it.

4.3   Ranking and Evaluation Metrics

We view this experimental setup as mimicking a situation where a user is attempting to model the information found within a document, not just extract it. When a user inputs a prompt plus some simple information, they are effectively defining a new relation in their model. Given that the model is assumed to have seen neither the prompt nor the documents before, there is no way to know the ground truth. If a user does not easily find any statements that look correct in those presented by the system, then they should be able to safely assume that the document does not have any information modeled by that relation. We therefore evaluate each configuration using top-k metrics with values of k from 1 to 5, and additionally use mean reciprocal rank (MRR) as a summarization metric. These metrics represent how many incorrect statements a user on average needs to sift through to find a correct one. Note that Tables 3 and 4 report these values as percentages, and so range from 0 (worst) to 100 (best).

We do not focus on more concrete evaluation metrics like F_1-score or predictive accuracy in our analysis for two main reasons. The first is that in order to calculate such metrics, the system would need an additional step which determines a threshold at which it labeled each candidate statement as being either true or false based on PLL scores. One way to do so, as in our prior work, would be to use the ground truth to determine what this threshold should be for each relation [2]. However, in the document-modeling scenario envisioned before, there may be no ground-truth to draw from, meaning that it becomes unclear how to determine such a threshold for unseen relations. While most data sets understandably assume a fixed set of relations, we wish to support any relation that can be expressed in text. The second reason is that the PLL scores for the candidate statements are heavily reliant upon the formulation of the prompts. We wished to avoid scenarios where it became important to tune these prompts simply to increase the metrics.

5   Results and Analysis

This section present the results of our experiments along with our analysis of those results. We first discuss the general effects of contextualization, then discuss its effects when handling domain-specific texts. It should be noted that all ranking results were observed to be deterministic during experimentation and that there is no inherent randomness in the experimental setup. That is, multiple runs of the same configuration always yielded the same ordering across statements.

5.1   Contextualization

Both Tables 3 and 4 show how contextualization affects the ranking results. The Pass column describes how many times contextualization is applied, with 0 meaning that only background knowledge is used, while 2 means that recontextualization from Sect. 3.1 is applied. Each table additionally reports the number of documents that contained each relation type, as well as the average ratio of supported to unsupported statements. For instance, a ratio of 0.1 means that, on average, there was one true statement for every ten false statements, so the top-1 score for a random-order baseline should be about $\frac{1}{11} = 0.09$, or 9%.

Table 3. MRR and Top-k scores for the ten most common relations in the dev portion of the DocRED data set. Scores for passes 1 and 2 use the top-50 nonlinearity method. For P131 and P150, "the administrative territorial entity" is abbreviated to "ATE" Rows where Passes is 0 represents ranking via the background knowledge from the MLM's training data.

| Relation info | Pass | BERT large | | | | | RoBERTa large | | | | |
		MRR	$k=1$	$k=3$	$k=5$	$k=10$	MRR	$k=1$	$k=3$	$k=5$	$k=10$
P17: country	0	53.1	**40.3**	58.1	68.5	81.7	38.0	22.9	44.2	**55.9**	72.4
546 documents	1	52.7	38.1	61.0	**70.1**	81.7	35.3	22.9	39.7	45.9	62.5
Ratio: 0.05	2	**53.3**	39.2	**61.7**	70.0	**82.4**	40.3	**26.7**	44.8	55.9	**72.8**
P27: country of cit.	0	35.6	18.4	42.3	57.3	76.5	34.0	17.5	36.8	56.0	74.8
234 documents	1	**44.5**	**27.4**	**52.1**	63.7	**84.6**	32.9	16.7	37.6	50.9	73.5
Ratio: 0.09	2	43.8	26.9	50.4	**64.5**	84.2	**44.4**	**26.9**	**53.0**	**66.7**	**83.8**
P131: loc. in ATE	0	42.0	27.7	47.8	56.7	73.1	22.4	9.7	22.6	34.4	50.8
372 documents	1	52.1	36.3	61.8	72.0	**84.1**	13.0	3.0	10.5	19.1	35.5
Ratio: 0.03	2	**52.2**	**36.6**	**62.4**	**72.8**	**84.1**	22.8	10.2	23.9	34.9	51.3
P150: contains ATE	0	**20.8**	**9.3**	**22.5**	**28.9**	**44.3**	18.9	8.6	18.2	27.9	43.6
280 documents	1	16.3	4.6	15.7	24.3	42.9	21.5	9.3	21.1	31.1	53.2
Ratio: 0.02	2	16.0	4.3	16.1	23.9	41.1	**24.8**	**11.4**	**26.1**	**38.2**	**58.2**
P161: cast member	0	44.9	**26.0**	**58.0**	66.0	84.0	36.1	20.0	44.0	56.0	72.0
50 documents	1	38.0	20.0	48.0	62.0	76.0	**40.8**	**24.0**	**48.0**	**60.0**	**80.0**
Ratio: 0.11	2	**46.2**	**26.0**	**58.0**	**76.0**	**86.0**	31.3	16.0	32.0	54.0	70.0
P175: performer	0	39.3	24.2	46.5	53.5	68.7	31.6	19.2	34.3	43.4	59.6
99 documents	1	46.8	**32.3**	51.5	**65.7**	76.8	**36.9**	**25.3**	**40.4**	**47.5**	**64.6**
Ratio: 0.05	2	**47.2**	**32.3**	**55.6**	64.6	**79.8**	22.7	10.1	25.3	33.3	49.5
P527: has part	0	13.9	**6.7**	8.9	18.9	**32.2**	19.0	12.1	20.9	25.3	29.7
90 documents	1	6.8	1.1	5.6	11.1	15.6	14.6	7.7	13.2	20.9	29.7
Ratio: 0.01	2	6.2	1.1	4.4	7.8	13.3	5.4	1.1	4.4	7.7	12.1
P569: date of birth	0	43.2	25.1	53.8	63.8	84.9	39.7	**23.1**	43.7	56.3	80.4
199 documents	1	**54.6**	**36.7**	**65.3**	**77.4**	89.4	**43.0**	**23.1**	51.3	69.3	**88.9**
Ratio: 0.05	2	49.7	30.2	60.8	76.4	**90.5**	36.4	19.1	42.2	56.3	77.9
P570: date of death	0	36.9	22.4	40.6	53.8	76.2	32.0	**16.8**	33.6	43.4	69.2
143 documents	1	**51.2**	**34.3**	**58.0**	**74.1**	**90.2**	32.2	11.9	37.8	58.0	**81.1**
Ratio: 0.05	2	50.1	**34.3**	57.3	72.0	87.4	**33.4**	**16.8**	**36.4**	**55.2**	73.4
P577: pub. date	0	28.3	11.7	30.7	44.5	70.1	30.0	13.9	30.7	46.0	73.0
137 documents	1	**34.8**	**19.7**	**39.4**	**50.4**	**73.0**	**34.3**	16.1	**42.3**	**58.4**	**73.7**
Ratio: 0.09	2	29.8	14.6	32.1	46.0	72.3	28.1	12.4	30.7	45.3	69.3

Table 4. MRR and Top-k scores for relations present in at least ten documents from the train portion of BioRED. BERT and BioBERT use the 'large-cased' variant, while PubMedBERT uses 'large-uncased'. Scores for passes 1 and 2 use the top-50 nonlinearity. Each relation reports how many documents it occurred in and the average ratio of supported to unsupported candidate statements. Rows where Passes is 0 represents ranking via the background knowledge of a given MLM, where the two models fine-tuned on biomedical data tend to score similarly to or higher than BERT.

Relation info	Model	Passes	MRR	k=1	k=2	k=3	k=5	k=10
		0	24.6	12.2	15.9	29.7	35.8	54.5
	BERT	1	42.8	28.9	41.1	47.2	58.5	**72.4**
		2	**43.5**	**29.7**	**41.9**	**50.0**	**60.2**	72.0
Association		0	33.4	20.7	27.6	38.6	48.4	60.2
246 documents	PubMedBERT	1	32.9	19.1	26.0	37.4	49.6	65.0
Ratio: 0.11		2	32.9	19.1	26.0	37.8	49.6	65.0
		0	37.8	26.8	30.5	39.4	52.0	65.4
	BioBERT	1	33.1	17.5	30.9	39.4	50.8	64.2
		2	33.2	18.7	29.7	38.6	50.0	64.6
		0	24.4	13.0	17.4	26.1	30.4	60.9
	BERT	1	15.2	0.0	8.7	17.4	26.1	60.9
		2	15.4	0.0	8.7	17.4	26.1	65.2
Bind		0	40.0	30.4	30.4	39.1	52.2	69.6
23 documents	PubMedBERT	1	**42.2**	**26.1**	**43.5**	47.8	**60.9**	**78.3**
Ratio: 0.02		2	**42.2**	**26.1**	**43.5**	47.8	**60.9**	**78.3**
		0	30.0	21.7	21.7	30.4	39.1	56.5
	BioBERT	1	19.3	4.3	13.0	21.7	39.1	56.5
		2	24.5	13.0	21.7	26.1	34.8	43.5
		0	22.4	10.6	18.4	23.4	36.9	45.4
	BERT	1	29.4	**17.0**	**25.5**	30.5	40.4	57.4
		2	**29.5**	**17.0**	24.1	30.5	**41.1**	**59.6**
Negative Corr.		0	20.3	9.9	15.6	23.4	29.1	41.8
141 documents	PubMedBERT	1	26.3	15.6	19.9	23.4	36.2	53.9
Ratio: 0.06		2	26.4	15.6	19.9	24.1	36.2	53.9
		0	21.5	9.2	18.4	22.0	32.6	47.5
	BioBERT	1	22.8	9.2	17.7	25.5	34.8	53.2
		2	26.9	13.5	23.4	30.5	39.7	55.3
		0	23.0	11.5	17.2	24.2	32.2	52.0
	BERT	1	26.8	14.1	22.0	27.3	39.2	56.4
		2	27.1	14.1	22.9	27.3	38.8	58.6
Positive Corr.		0	13.4	4.8	7.5	10.1	15.9	33.9
227 documents	PubMedBERT	1	22.5	10.1	17.6	23.3	31.7	50.7
Ratio: 0.07		2	22.4	10.1	17.2	23.3	31.3	51.1
		0	21.7	11.0	13.7	22.0	31.7	48.5
	BioBERT	1	27.8	13.7	22.9	31.7	**41.9**	**61.2**
		2	**29.4**	**16.3**	**27.8**	**33.5**	39.6	56.4

Contextualization results for DocRED (Table 3) show that it generally improves the overall scores with either one or two passes. For BERT, half of the relations improved by at least 8 points in MRR scores, with P570 improving by over 13 points. The relations P150, P161, and P527 showed a decrease in scores when applying contextualization, though a second pass ultimately improved the scores for P161. The prompts we used for P150 and P527 were shown to be sub-optimal in earlier work [2], which may be why the scores were drastically impacted here. Additionally, these two relations had the lowest ratios, which meant that the ranking task was generally more difficult.

For RoBERTa, a larger, more robustly-trained MLM, we saw an overall decrease in scores compared to BERT. This may be attributable to its tokenizer, which will tokenize a word differently depending on whether it is preceded by a space. In practical terms, many of our prompts expect the first entity to not be preceded by a space, and the second entity to be preceded by one. There was not a clear way to take this into consideration when performing contextualization. As such, rather than comparing RoBERTa with BERT, we again focus on the different between the number of passes of contextualization. For RoBERTa, we generally saw the same trends as with BERT, though the increases in scores were more moderate. One aspect to note is that for five of the relations, recontextualization performed the best across all metrics, while that was only true for three of the relations when using BERT. This implies that RoBERTa may be better able to handle more complex transformations of the input tokens.

The contextualization results for BioRED (Table 4) additionally show a general improvement for all models, though a second pass only really benefited BERT. The stark exception to this result is the `Bind` relation, which again had a low ratio and a likely poor prompt, resulting in all but PubMedBERT scoring lower after contextualization.

5.2   Domain Adaptation

As a form of zero-shot domain adaptation, contextualization allows BERT to outperform the domain-specific models, but only for some cases. This is particularly evident for the `Association` relation, which was the most common relation present in the data set. Using only background knowledge (Passes=0), BERT scored at least nine points lower than the other models in MRR. After contextualization, it outperformed BioBERT and PubMedBERT by roughly five and nine points, respectively, with a further improvement after the second pass. Interestingly, BERT did have the best MRR score for both `Positive Correlation` and `Negative Correlation` when only using background knowledge. For `Negative Correlation`, BERT saw an improvement of 7 points in MRR when using contextualization, PubMedBERT improved by 6 points, and BioBERT only improved by about 1 point.

Some of this effect may be due to the fact that, while the entities are domain-specific, the text of the prompts and thus the interpretation of the relations they represent is mostly general knowledge. For instance, "?x is positively correlated with ?y." is not specific to the biomedical domain. Since the contextualization phases use pPLL to score the statements, they in fact only look at the tokens from the prompt, meaning that they only examine the effect on the scores for tokens from the general-knowledge portion of each statement. Since BERT is a general-knowledge MLM, this may explain why it tended to score slightly better in our setup.

5.3   Nonlinearities

For Eq. 3, we discussed several choices for the nonlinearity function f in Sect. 3. Due to space constraints, the full results for this experiment are only included in the supplementary material in our GitHub repository. However, in short, we found that the function that consistently scored the best for both contextualization passes was top-k softmax (Eq. 7) with $k = 50$. This function consistently scored better than $k = 10$ and either equal to or better than $k = 100$. We believe this to be related to the fact that normal softmax will still assign a small value to every entry in a vector, which leads to a large amount of noise when the vectors are long.

5.4   Sentence-Level Provenance

One benefit that our contextualization approach offers comes from the fact that it gives different representations to each entity mention in a document. As mentioned earlier, using MLMs to rank statements without any other context is a probing task examining the patterns learned during training. In that scenario, the provenance for any statement can only be narrowed down to the MLM itself. When including a document as context, the rankings are more likely to be attributable to that particular document, offering some degree of provenance. Since the contextualization process creates representations for every mention of an entity, the sentences that contain the mentions that produced the highest-ranking statement could be attributed as provenance. However, since the contextualization process mixes the MLM's background knowledge with the content of the documents, this provenance may still be unreliable.

6   Discussion and Future Work

In this section, we discuss our results with a particular focus on their limitations, how future work could address those limitations, and the larger aim that this work is building toward. Despite this work addressing limitations from our prior work by incorporating context into the ranking scheme, the ranking is still assumed to be heavily influenced by the background knowledge of the MLMs we tested. While we did not explore this effect directly, it is most likely difficult for a MLM to rank a statement highly if it contradicts patterns seen in its training data, despite the inclusion of a document supporting it. Future work should address this by incorporating some degree of background knowledge removal into the scheme, for instance by making the ranking invariant to the way in which entity mentions are written, instead focusing exclusively on the context in which they are used. We feel that this is the most important next step.

Another limitation of this work is that we only used four relations from one biomedical data set when examining domain adaptation capabilities. While the results indicate that this contextualization method does generalize, future work will need to explore whether those results are consistent for documents from other domains or data sets, and identify why some domains may present greater challenges. Additionally, our work focuses only on English-language documents and language models, primarily due to their availability. While transformer-based MLMs like BERT and RoBERTa should not be biased by most characteristics of a language (e.g. text direction or syntactic structure), care should still be taken when extrapolating these results to other languages, as the quality and availability of text for those languages will have an impact. Another possible limitation of this work is that the task as we presented it is entity-centric. We make the assumption that the difficulty in performing RE lies in the MLM's ability to understand the entities, rather than the relations. This puts a heavy burden on the quality of the named entity recognition system used to identify those entities in the first place.

The idea of recontextualization grew from a deeper study of BERT's masked language modeling pre-training objective, and as such our experiments have focused primarily on MLMs. These language models attempt to produce the exact same text as they are given, which is not true for other classes of language models with different pre-training objectives, such as autoregressive language models. Future work should examine whether the idea of recontextualization can be applied to other classes of language model to avoid fine-tuning. Also important would be to explore whether the enhanced entity representations given by recontextualization improve results for downstream tasks other than statement ranking, in particular tasks which require understanding a body of text, such as reading comprehension and question answering.

7 Conclusions

In this work, we present a method that uses the contextualization abilities of MLMs to augment PLL scores for short statements to perform zero-shot relation extraction. We first illustrated how this contextualization improves the ranking over simply using the background knowledge of a MLM by incorporating documents into the scores. We then showed how this can be extended to multiple passes, which in many cases improved the scores further, though only by a small margin. Finally, we showed that this effect enabled a general-domain model like BERT to perform on par with domain-specific models despite lacking any fine-tuning.

We believe that this work is an important step toward a robust yet flexible system for zero-shot RE built on smaller LMs which are more cost and time effective than popular proprietary LLMs. As part of a larger pipeline, such a system would support the extraction and maintenance of a KG based on natural language documents. By incorporating contextualization, fine-tuning can be avoided entirely, which further reduces the cost of updating such a KG, better supporting its evolution over time. By not using any fine-tuning, results for one document are not affected by text from any other documents, besides those already present in the MLM's training data. This, for instance, allows for knowledge extracted from a set of documents to be deleted from the KG simply by removing the relations and entities attributed to it, rather than re-training and re-running the pipeline. Finally, by avoiding fine-tuning, we ensure that the MLM will never succumb to catastrophic forgetting and will always maintain the same level of generality that it started with.

Acknowledgments. This work was funded by the Swedish National Graduate School in Computer Science (CUGS). Portions of this work were carried out using the AIOps/ Stellar facilities funded by the Excellence Center at Linköping-Lund in Information Technology (ELLIIT).

Disclosure of Interests. The authors have no competing interests to declare that are relevant to the content of this article.

References

1. Alivanistos, D., Santamaría, S.B., Cochez, M., Kalo, J.C., van Krieken, E., Thana-palasingam, T.: Prompting as probing: using language models for knowledge base construction. In: 2022 Semantic Web Challenge on Knowledge Base Construction from Pre-trained Language Models, LM-KBC 2022, pp. 11–34. CEUR-WS. org (2022)
2. Capshaw, R., Blomqvist, E.: Towards tailored knowledge base modeling using masked language models. In: Proceedings of TEXT2KG, Co-located with ESWC 2023. CEUR-WS (2023)
3. Capshaw, R., Blomqvist, E.: Understanding and estimating pseudo-log-likelihood for zero-shot fact extraction with masked language models. In: Proceedings of the 12th International Joint Conference on Knowledge Graphs (2023)
4. Chomsky, N.: Syntactic Structures. Mouton de Gruyter (1957)
5. Devlin, J., Chang, M.W., Lee, K., Toutanova, K.: BERT: pre-training of deep bidirectional transformers for language understanding. In: Proceedings of the 2019 Conference of the North American Chapter of the Association for Computational Linguistics: Human Language Technologies, Volume 1 (Long and Short Papers), pp. 4171–4186. Association for Computational Linguistics, Minneapolis (2019). https://doi.org/10.18653/v1/N19-1423
6. Fichtel, L., Kalo, J.C., Balke, W.T.: Prompt tuning or fine-tuning–investigating relational knowledge in pre-trained language models. In: 3rd Conference on Automated Knowledge Base Construction (2021)
7. Gu, Y., et al.: Domain-specific language model pretraining for biomedical natural language processing. ACM Trans. Comput. Healthc. **3**(1), 1–23 (2021)
8. Haviv, A., Berant, J., Globerson, A.: BERTese: learning to speak to BERT. In: Proceedings of the 16th Conference of the European Chapter of the Association for Computational Linguistics: Main Volume, pp. 3618–3623 (2021)

9. Heinzerling, B., Inui, K.: Language models as knowledge bases: On entity representations, storage capacity, and paraphrased queries. In: Proceedings of the 16th Conference of the European Chapter of the Association for Computational Linguistics: Main Volume, pp. 1772–1791. Association for Computational Linguistics (2021). https://doi.org/10.18653/v1/2021.eacl-main.153

10. Hicks, M.T., Humphries, J., Slater, J.: ChatGPT is bullshit. Ethics Inf. Technol. **26**(2), 38 (2024)

11. Kaneko, M., Bollegala, D.: Unmasking the mask–evaluating social biases in masked language models. Proc. AAAI Conf. Artif. Intell. **36**(11), 11954–11962 (2022)

12. Kauf, C., Ivanova, A.: A better way to do masked language model scoring. In: Proceedings of the 58th Annual Meeting of the Association for Computational Linguistics, pp. 925–935. Association for Computational Linguistics (2023). https://aclanthology.org/2023.acl-short.80.pdf

13. Kwon, B.C., Mihindukulasooriya, N.: An empirical study on pseudo-log-likelihood bias measures for masked language models using paraphrased sentences. In: Proceedings of the 2nd Workshop on Trustworthy Natural Language Processing (TrustNLP 2022), pp. 74–79. Association for Computational Linguistics, Seattle (2022). https://doi.org/10.18653/v1/2022.trustnlp-1.7

14. Lee, J., et al.: BioBERT: a pre-trained biomedical language representation model for biomedical text mining. Bioinformatics **36**(4), 1234–1240 (2020)

15. Liu, Y., et al.: RoBERTa: a robustly optimized BERT pretraining approach. arXiv preprint arXiv:1907.11692 (2019)

16. Luo, L., Lai, P.T., Wei, C.H., Arighi, C.N., Lu, Z.: BioRED: a rich biomedical relation extraction dataset. Briefings Bioinform. **23**(5), bbac282 (2022). https://doi.org/10.1093/bib/bbac282

17. Lv, B., et al.: DSP: discriminative soft prompts for zero-shot entity and relation extraction. In: Findings of the Association for Computational Linguistics: ACL 2023, pp. 5491–5505. Association for Computational Linguistics, Toronto (2023). https://doi.org/10.18653/v1/2023.findings-acl.339

18. Mosbach, M., Andriushchenko, M., Klakow, D.: On the stability of fine-tuning BERT: misconceptions, explanations, and strong baselines. arXiv preprint arXiv:2006.04884 (2020)

19. Nangia, N., Vania, C., Bhalerao, R., Bowman, S.: CrowS-Pairs: a challenge dataset for measuring social biases in masked language models. In: Proceedings of the 2020 Conference on Empirical Methods in Natural Language Processing (EMNLP), pp. 1953–1967 (2020)

20. Petroni, F., et al.: Language models as knowledge bases? In: Proceedings of the 2019 Conference on Empirical Methods in Natural Language Processing and the 9th International Joint Conference on Natural Language Processing (EMNLP-IJCNLP). pp. 2463–2473 (2019)

21. Salazar, J., Liang, D., Nguyen, T.Q., Kirchhoff, K.: Masked language model scoring. In: Proceedings of the 58th Annual Meeting of the Association for Computational Linguistics. pp. 2699–2712 (2020)

22. Shin, J., Lee, Y., Jung, K.: Effective sentence scoring method using BERT for speech recognition. In: Lee, W.S., Suzuki, T. (eds.) Proceedings of The Eleventh Asian Conference on Machine Learning. Proceedings of Machine Learning Research, vol. 101, pp. 1081–1093. PMLR (17–19 Nov 2019), https://proceedings.mlr.press/v101/shin19a.html

23. Vrandečić, D., Krötzsch, M.: Wikidata: a free collaborative knowledgebase. Commun. ACM **57**(10), 78–85 (2014)

24. Wang, A., Cho, K.: BERT has a mouth, and it must speak: BERT as a Markov Random Field language model. In: Proceedings of the Workshop on Methods for Optimizing and Evaluating Neural Language Generation. pp. 30–36 (2019)
25. Yao, Y., et al.: DocRED: a large-scale document-level relation extraction dataset. In: Proceedings of the 57th Annual Meeting of the Association for Computational Linguistics, pp. 764–777 (2019)

Validating a Functional Status Knowledge Graph in a Large-Scale Living Lab

Mauro Dragoni$^{(\boxtimes)}$ ⓘ, Gianluca Apriceno$^{(\boxtimes)}$ ⓘ, and Tania Bailoni$^{(\boxtimes)}$ ⓘ

Fondazione Bruno Kessler, Trento, Italy
{dragoni,apriceno,tbailoni}@fbk.eu

Abstract. Functional Status Information refers to a person's overall mental and physical health. Collecting and analyzing Function Status Information data is crucial for addressing the needs of a growing elderly population, as well as for providing effective care to those with chronic diseases, multiple health issues, or disabilities. Knowledge Graphs provide an effective method for organizing and representing Functional Status Information data in a structured way. Furthermore, they can also allow reasoning over this data to create personalized health support solutions that assist people in maintaining a healthy lifestyle and improving daily living. In this paper, we describe the integration of our Functional Status Knowledge Graph, namely FuS-KG, into a real-world application run within a large-scale living lab involving more than 4,000 people. We provide the road map of this experience including the challenges, the platform's architecture, the focus on the knowledge layer, the evaluation and the insights observed.

Keywords: Knowledge Graph · Digital Health · Large-scale Living Lab

1 Introduction

There is a growing interest in creating virtual assistants for health and well-being, aimed at supporting lifestyle management and disease prevention. This trend may be due to the increasing societal demand for better health management as well as illness prevention. A crucial point towards improving an individual's situation is represented by behavior change which makes it important to consider how a digital coach can collaborate with users to help them achieve health improvements. This collaboration may involve promoting adherence to medical guidelines, improving diet, encouraging more physical activity, and reducing stress or usage of toxic substances. To determine the best course of action, the digital coach may rely on the Functional Status Information (FSI) of the person being monitored.

In this context, three key elements are essential to allow a medical system to reason, make decisions, and act: (i) the (medical) knowledge that guides the digital assistant, (ii) an understanding of how people modify behavior, build motivation and face psychological, physical and social barriers, and (iii) the individuals' FSI data, alongside their personal experience (i.e., narrative), about behavior change process that must be handled in accordance with ethical standards and regulatory requirements. Furthermore, one of the factors contributing to the success of such systems is the integration of

M. Alam et al. (Eds.): EKAW 2024, LNAI 15370, pp. 416–433, 2025.
https://doi.org/10.1007/978-3-031-77792-9_25

strategies that are not only effective and efficient but also ethical, suggesting behavior change actions depending on the individual's specific context, needs and preferences. For example, delivering personalized motivational messages that align with an individual's current situation.

To represent this information accurately, a strategy is needed to handle the variety of data involved while ensuring compliance with privacy regulations. A conceptual model is also fundamental to make use of the information effectively. Knowledge Graphs (KGs) represent a suitable way of organizing and representing FSI, and linking it with user records, such as electronic health records allowing the development of AI-driven systems that support personalized coaching strategies, helping to prevent the decline of the Functional Status (FS) in the target population.

In this paper, we present the experience concerning the adoption into a real-world setting of a brand new KG, namely FuS-KG, supporting the representation of FSI and its exploitation to manage the generation of healthy recommendations for citizens.

Our experience is split into the three main steps presented in the remainder of the paper and summarized below. First, the construction of FuS-KG (Sect. 3), which represents the first original contribution of this work. Second, the integration of FuS-KG into a real-world digital health platform (Sect. 4), namely Salute+, is provided for completeness of understanding, but it must not be considered an original contribution of this paper. Third, the validation of the FuS-KG-enhanced Salute+ platform within a large-scale living lab involving more than 4,000 users within the Trentino territory (Sect. 6), which is the second original contribution of this work demonstrating the effectiveness of the proposed solution.

2   Related Work

Assessing a person's FS is a key step toward tailoring precise interventions and providing services that improve their health status and maximize their functional independence in daily activities. However, managing health often poses different challenges, related to healthcare systems being unable to provide adequate support to individuals with chronic illnesses or disabilities due to the scarcity of resources. As stated by the National Committee on Vital and Health Statistics (NVHCS) understanding FS is necessary to achieve optimal health outcomes[1]. The evaluation of the FS is characterized by a clinician-patient interaction to get insights into patients' lifestyles, together with standardized tests. Nevertheless, these assessments provide only a brief overview of an individual's health status, while effective behavior change aimed at improving health is a long-term process. As a result, there is a gap between an individual's health objectives and the "infrequent" health assessments conducted in healthcare settings, which may prevent the delivery of high-quality care. Therefore, frequent data collection is required to bridge this gap and ensure continual support.

Physicians and researchers have conducted a few studies on elderly people that led to the identification of the risk factors able to detect those most susceptible to functional decline [12, 23, 31]. Nonetheless, FSI is underutilized, limiting its potential impact [20].

One of the reasons may be that many physicians do not fully recognize the significance of this information [3–5, 22]. Indeed, even when informed about a patient's health status, only a minority alter patient management accordingly.

---

[1] https://www.ncvhs.hhs.gov/wp-content/uploads/2017/08/010617rp.pdf.

Studies based on self-reports of functional performance and early decline show successful prediction of actual performance and decline [13,29]. However, researchers note these reports capture only a fraction of the problems [31]. Consequently, there is a need for innovative methods to assess FS, including non-intrusive approaches that evaluate the patient's functional performance, in both normal and abnormal conditions, without relying solely on physician-patient interaction. For example, in [1] sensor-based systems can detect a decline in daily activities, thus facilitating tailored interventions. Given the increasing elderly population, such systems are crucial for early detection and intervention to prevent functional decline and slow the deterioration of functional ability.

FuS-KG enables the investigation of this research field aiming to facilitate the development of AI-powered systems capable of helping individuals monitor their FS and prevent function decline. The adoption of FuS-KG in real-world scenarios is strongly connected to the design of behavior change solutions since, given the detection of declining FS, the knowledge modeled within FuS-KG can be exploited to perform reasoning operations for discovering possible behavior change trajectory with the aims of improving the overall FS of a person. Concerning this point, FuS-KG fills a gap in the state-of-the-art since, as described below, to the best of our knowledge, KGs covering the whole FS landscape have not been proposed yet.

A recent collaborative effort aimed at creating observable and replicable interventions for influencing behavior and health can be found in the behavioral change techniques taxonomy outlined by [21]. This taxonomy facilitates the consolidation of knowledge regarding behavior and behavior change, promoting the sharing and reuse of useful sources of behavior knowledge. Additionally, as far as we are aware, there is currently no publicly available conceptual model that systematically conceptualizes barriers to behavior change.

A noteworthy taxonomy to mention is the human behavior taxonomy developed by the World Health Organization[2] (WHO) [19]. This taxonomy, rooted in the WHO's expertise and the International Classification of Functioning (ICF), Disability, and Health, provides detailed class definitions gathered extensively from sources like the U.S. National Cancer Institute (NCI) Thesaurus and the Oxford English Dictionary [30]. Another important effort aimed at understanding human behavior is the Semantic Mining of Activity, Social and Health Data (SMASH) [24] that focuses on predicting human behavior and providing explanations for these predictions.

The Health Behavior Change Ontology (HBCO) was created for a project aimed at establishing automated dialogues between a psychologist and a user for the purpose of behavioral counseling [2]. While it has the potential to connect theoretical concepts with practical applications, there are currently few practical implementations available. Additionally, specific strategies for effectively deploying a reusable behavior change ontology in real-world settings are lacking.

Since our conceptualization has to align with individuals undergoing behavioral change, it is fundamental to explore ontologies that specifically model users, encompassing their profiles, characteristics, and sometimes their behavior. Examples of such ontologies are the General User Model Ontology (GUMO) [15] (which is included in our ontology), the User Navigation Ontology (UNO) [18], and the Ontology of Personal Information Management (OntoPIM) [17].

---

[2] https://www.who.int/classifications/drafticfpracticalmanual.pdf.

Moreover, to cover the domain of physical activity behavior, notable ontologies include the HeLiS ontology [7] and the Ontology of Physical Exercises[3].

Finally, the use of KGs in the design of AI-based systems providing coaching has been researched and evaluated. [10] describes a knowledge-based solution implemented in a system that provides personalized healthy lifestyle recommendations to users, which was applied and evaluated in a real-world scenario. The article not only highlights the previously mentioned lack of publicly available structured knowledge but also provides evidence of the proposed solution's feasibility in a real-world setting, based on both performance evaluation and efficacy.

3   Building FuS-KG

The development of FuS-KG addressed a list of requirements aimed at satisfying the need to provide a KG able to describe the domain of FS as-is and to support the conceptualization of an interoperable digital twin for people using applications that exploit FuS-KG. This way, FuS-KG enables the possibility to perform reasoning operations on such a digital twin also for predictive purposes by simulating possible future undesired behaviors and, in turn, by generating recommendations persuading the users to avoid them.

As we discussed in Sect. 2, ontologies available in domains connected with wellness and healthy lifestyle have been designed with different aims. For example, ontologies concerning the physical activity domain are created with a focus on classifying data without connecting each activity with potential problems or benefits associated with the whole FS of a user. Similarly, the same happened with ontologies concerning foods. FuS-KG has been built with a focus on the connection between different dimensions related to people's health representing their complete FS. The development of our ontology has been driven by the following requirements:

Requirement 1 (REQ1). The KG must conceptualize the food domain at a fine-grained level. This means that the whole knowledge chain from defining each nutrient to modeling complex recipes must be supported.

Requirement 2 (REQ2). The KG must provide a comprehensive list of activities that a person can perform. For each activity, it must be provided the knowledge required to understand the effort necessary to complete the activity to enable the inference of how much of each food defined under *REQ1* is necessary to fulfill the activity.

Requirement 3 (REQ3). The KG must include the model of the barriers that may affect a person and how such barriers obstacle the fulfillment of specific activities, the consumption of specific foods, or, in general, the following of specific guidelines.

Requirement 4 (REQ4). The KG must support the modeling of multi-modal knowledge since a MMKG may be exploited to better support users' education tasks and enable knowledge injection tasks into large foundational models. Hence, the KG must include a multi-modal knowledge representation, e.g., images of recipes and videos of how to execute activities.

---

[3] https://bioportal.bioontology.org/ontologies/OPE/?p=summary.

Requirement 5 (REQ5). Knowledge modeled under the requirements *REQ1*, *REQ2*, *REQ3*, and *REQ4* must be associated with knowledge and data gathered from users' input. Hence, the KG must include the appropriate set of concepts to enable the definition of a user model and to support the linking between such a user model and the domain knowledge defined through the requirements mentioned above.

Requirement 6 (REQ6). A KG usable for creating a behavior change solution requires a set of guidelines driving the behavior change intervention. A reasoner can exploit such guidelines to detect situations where a user is not adhering to them. To enable this feature, the KG must define the conceptual knowledge that: (i) enables the modeling of such guidelines; (ii) defines how they can be associated with the domain knowledge covered by the KG; and, (iii) allows their linking with a user profile.

Requirement 7 (REQ7). The requirements discussed above refer to the notion of *time* in different ways. For example, the merge of *REQ2* and *REQ4* concerning the modeling of an activity and how to perform it, requires the modeling of the different steps and their temporal order. Similarly, both requirements *REQ5* and *REQ6* need the notion of time associated with the knowledge gathered by the user and when users' data should be checked, respectively. Hence, the KG must include temporal knowledge to support the requirements above and enable temporal reasoning over the users' collected knowledge.

Requirement 8 (REQ8). Working with such different domains may lead to building a very large KG. Hence, the KG must be designed with a modular structure to ease its management and maintenance.

The process for building FuS-KG followed a combination of the Modular Ontology Modeling (MOMo) [26] and the METHONTOLOGY [11] methodologies. The rationale behind applying them in tandem is that, in the first phase, we worked on the modularization aspect given the expected size of FuS-KG and the purpose of easing the possible reuse of only parts of the knowledge modeled. Then, we moved to the conceptualization of each module. Concerning the decision to adopt the METHON-TOLOGY methodology, this was driven by the need to have a lifecycle organized into clearly defined steps. Additionally, since the development of FuS-KG involved direct collaboration with experts in situ, a methodology that provided a clear outline of tasks to be performed was essential. While other methodologies, such as DILIGENT [25] and NeOn [27], were considered prior to the construction of FuS-KG, their focus on decentralized engineering did not align with the specific requirements of our scenario.

The overall process involved four knowledge engineers and three domain experts from the Trentino Healthcare Department. More specifically, three of the knowledge engineers, alongside two domain experts, contributed to the ontology modeling phases (hereafter referred to as the modeling team). Meanwhile, the remaining knowledge engineer and domain expert were responsible for evaluating the ontology (hereafter referred to as the evaluators).

Due to limited space, we do not provide a detailed description of each phase, but we report only the most relevant activities of the construction process.

3.1 Modules Definition

As mentioned above, the first step focused on the application of the MOMo method-ology to define FuS-KG modules. This step aims to address *REQ8* and was necessary due to the huge amount of knowledge generated from the selected unstructured sources. Hence, to ease the maintenance of FuS-KG, the split into a set of modules was a manda-tory step. Figure 1 provides an overview of the modules composing the current version of FuS-KG.

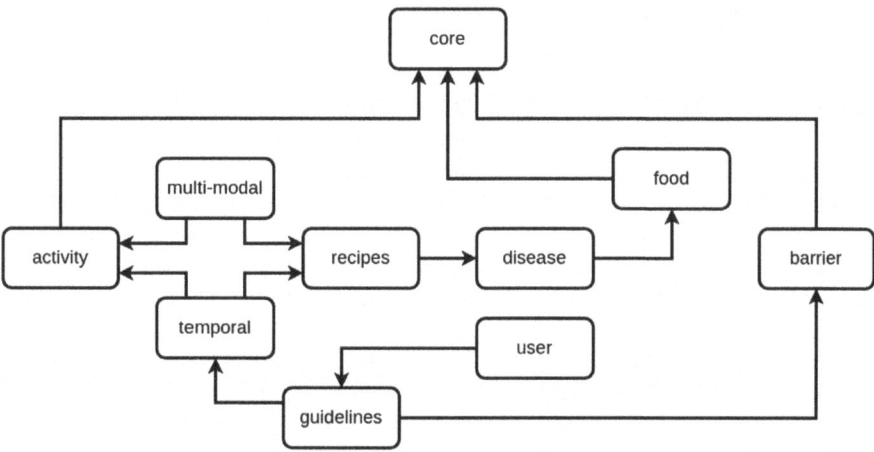

Fig. 1. Summary of the modules composing the current version of FuS-KG. Each block repre-sents a module of FuS-KG, while each arrow corresponds to the *owl:imports* property.

The knowledge contained in each module is the following:

– *core*: this module includes the upper level of FuS-KG, i.e., the set of abstract con-cepts defining the main type of knowledge covered by FuS-KG.
– *food*: this modules imports the *core* module and it contains all the knowledge about the *BasicFood*, *ComposedFood*, and *Nutrient* categories. As *BasicFood*, we mean those foods for which fine-grained nutritional information is provided within the sources we adopted to build FuS-KG (e.g., Bread). Then, as *ComposedFood*, we mean those foods that are aggregations of instances of *BasicFood* but for which fine-grained nutritional information is available as well (e.s., Tomato Sauce). While *Nutrient* represents the specific nutritional information associated with a *BasicFood*.
– *recipes*: this module imports the *diseases* and (indirectly) the *food* module since a *Recipe* is defined as a group of *BasicFood* each with the associated quantity. This module includes a set of sub-modules containing (i) the list of the recipes collected from the different sources (we created one sub-module for each source); and, (ii) the alignments between the different sources (i.e., we preserve the fact that a recipe may be defined within more than one of the sources we used).

- *diseases*: this module imports the *food* module and it models the association between each *BasicFood* and nutritional-wise *Disease*. Each association is modeled through a *DiseaseRiskLevel* entity associating to each pair *BasicFood-Disease* a risk level.
- *activity*: this module imports the *core* module and it includes the taxonomy of all activities covered by FuS-KG together with the knowledge related to the effort required to fulfill each activity. The effort is represented through the Metabolic Equivalent of Task (MET) coefficient and the calories required to perform 1 min of each activity for each kilogram of body weight.
- *barrier*: this module imports the *core* module and it contains the type of barriers defined within the SIS manual and how these barriers may affect the fulfilment of specific activities.
- *multi-modal*: this module imports the *recipes* and *activity* modules. The module contains, when available within the adopted sources, the multi-modal knowledge associated with specific *Recipe* and *Activity*. In particular, the current version of FuS-KG contains knowledge about image and video modalities.
- *temporal*: this module imports the *recipes* and *activity* modules. Then, this module is split into two sub-modules. The first one defines the temporal intervals that may be of interest to model the guidelines (e.g., *Day*, *Meal*), while the second one contains the knowledge related to steps to prepare a specific *Recipe* or perform a specific *Activity*.
- *guidelines*: this module imports the *barrier* and *temporal* modules. This module aims to model behavioral guidelines that may be associated with users to support reasoning tasks about possible recommendations related to the user data that can be collected [9].
- *user*: this module imports the *guidelines* and *recipes* modules. This way, the *user* module may access the full knowledge of FuS-KG to enable the storage of any type of data that can be collected in several ways (e.g., sensors or mobile applications).

3.2   Modules Conceptualization

FuS-KG aims to provide a conceptualization with a *high* granularity level. For example, for each recipe modeled within FuS-KG, we provided its composition to the micro-nutrient level. Thanks to this granularity level, we support the integration of FuS-KG into various solutions going from simple mobile applications to expert systems.

The acquisition of the knowledge necessary for building FuS-KG has been done in two steps: (i) the discussion with domain experts for deciding how to model the core entities of FuS-KG (i.e., abstract classes and properties); and, (ii) the acquisition, analysis, and processing of unstructured resources containing information to include in FuS-KG.

The first step consisted of defining the set of entities addressing the list of requirements described above. Here, the modeling team started from the conceptual model built to create the HeLiS ontology [8] since it provides the basic elements to start satisfying requirements *REQ1*, *REQ2*, *REQ5*, and *REQ6*. Then, such a conceptual model has been extended with further entities defined by the modeling team to cover the remaining requirements, i.e., *REQ3*, *REQ4*, and *REQ7*. This way, the final description of the core FuS-KG conceptual model defines the barrier domain and it has been equipped with the capabilities of accepting knowledge related to both multi-modal resources and temporal information about entities.

The second step consisted of identifying the sources for building and populating FuS-KG. In particular, such sources were related to the following domains: food, activity, and barriers. The HeLiS ontology has been used as a starting point to build the T-Box of FuS-KG. Concerning the food domain, we imported into the FuS-KG schema, the model of the recipes already defined within the HeLiS ontology, i.e., (i) the archives of the Italian Minister of Agriculture[4] and the Italian Epidemiological department[5]; and, (ii) the Turconi's atlas [28]. Then, we selected the following four (1 structured and 3 unstructured) further sources concerning the food and nutrition domain: (i) the USDA database[6], enriching the lists of basic foods and nutrients provided by the HeLiS ontology; (ii) the Recipe1M[7] dataset, which provides both images and step-wise description of recipes; (iii) the Tasty[8] dataset, which provides descriptions and videos of recipe preparations; and, (iv) the RecipeDb[9] dataset, which provides a comprehensive set of recipes still missing in FuS-KG. During the acquisition of such information, we also worked on the alignment between the INRAN and USDA models to have a common representation of basic foods and nutrients. This way, it was possible to reconcile the different sources exploited to build FuS-KG.

Concerning the activity domain, we started from the Compendium of Physical Activities[10] to create the taxonomy of physical activities and model all information concerning the associated effort. Finally, concerning the barrier domain, we relied on the Supported Intensity Scale (SIS) manual[11] that provided all the knowledge necessary to model barriers used to measure the functional status of a person. Here, we integrated object properties defining which barriers may affect the capability of fulfilling specific activities.

As a final step, we focused on the refinement of the conceptual model adopted within FuS-KG and the definition of the ontology design patterns (ODP) [16] to adopt. From the ODP catalog[12], we adopted several patterns, in particular: the alignment pattern *Class Equivalence*, the logical patterns *Tree* and *N-Ary Relation*, and the content patterns *Classification*, *Action*, *Time Interval*, and *Parameter*.

Finally, the main metrics related to the content of FuS-KG are summarized to give an overview of its size: 588 Concepts, 128 Object Properties, 49 Data Properties, 58 Annotation Properties, 1879205 Individuals, and 8668859 Axioms.

---

[4] http://nut.entecra.it/.

[5] http://www.bda-ieo.it/.

[6] https://fdc.nal.usda.gov/.

[7] http://pic2recipe.csail.mit.edu/.

[8] https://cvml.comp.nus.edu.sg/tasty/.

[9] https://cosylab.iiitd.edu.in/recipedb/.

[10] https://pacompendium.com/.

[11] https://www.aaidd.org/sis.

[12] http://ontologydesignpatterns.org/wiki/Community:ListPatterns.

4   Integration Into the Salute+ Platform

FuS-KG has been integrated into the Salute+ platform to enable its usage. In this section, we provide a brief description of how FuS-KG has been integrated and the role it plays. This part is reported for completeness of the in-use experience described in this paper, but it is not an original contribution of this work. The reader may refer to [9] for a more in-depth description of the general architecture of the platform and the message generation pipeline.

The Salute+ platform is composed of four layers, as shown in Fig. 2. The *Input Layer*, is responsible for storing events that trigger the platform activities and accounts for the system's ability to sense the context of interaction. These events are categorized into two types: (i) data input, where information is passed from the *Input Layer* to the *Knowledge Layer*, and (ii) context communication, where contextual data is sent from the *Input Layer* to the *Communication Layer* for further use (e.g., communication purposes). The *Knowledge Layer* includes the FuS-KG resource described in Sect. 3. The *Communication Layer* exploits the output of the *Knowledge Layer* (i.e., reasoning operations) to determine the language strategies to include in the natural language-generated messages. This layer also focuses on selecting, ordering and phrasing the arguments to include in the message. While further details are provided below, a detailed description of this layer is beyond the scope of this paper. Finally, the *Output Layer* completes the process by delivering the generated message to users. It encompasses various devices capable of receiving the data produced by the *Communication Layer* and provides physical feedback to users.

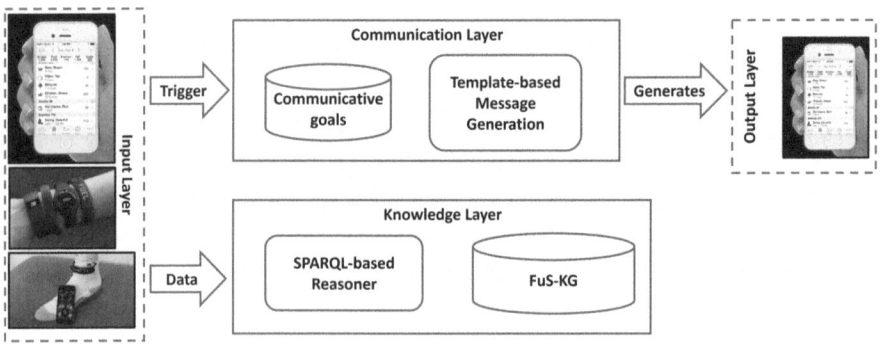

Fig. 2. Overview of the Salute+ architecture.

FuS-KG is integrated into a SPARQL-based reasoner to monitor a user's FS. The primary objective of this reasoner is to detect undesired situations in users' behaviors. When inconsistencies occur between the user's reported data and the corresponding guidelines, the reasoner generates an individual of type *UndesiredEvent* that is then exploited by the *CommunicationLayer* to generate feedback to users.

The activity of the reasoner can be triggered in two ways. First, when new or updated data packages are provided by users or acquired from IoT devices. Second, at the end of a specific timespan, such as daily or weekly. In this latter scenario, the reasoner checks and analyzes a collection of timestamped data related to each user's behavior within the specified timespan. The architecture of the integrated reasoner is based on RDFpro [6]. This choice was made for two reasons. First, RDFpro's architecture supports the integration of custom methods into its reasoning processes, enabling (i) mathematical calculation on user data and (ii) allowing the use of real-time information from external sources without storing it in the knowledge repository. Second, RDFpro has shown superior performance compared to other state-of-the-art reasoners in real-time contexts [6]. In this work, RDFpro has been adapted and extended to better align with the proposed solution, incorporating new methods that support the real-time reasoning of sensor data, thereby improving the efficiency of user data processing.

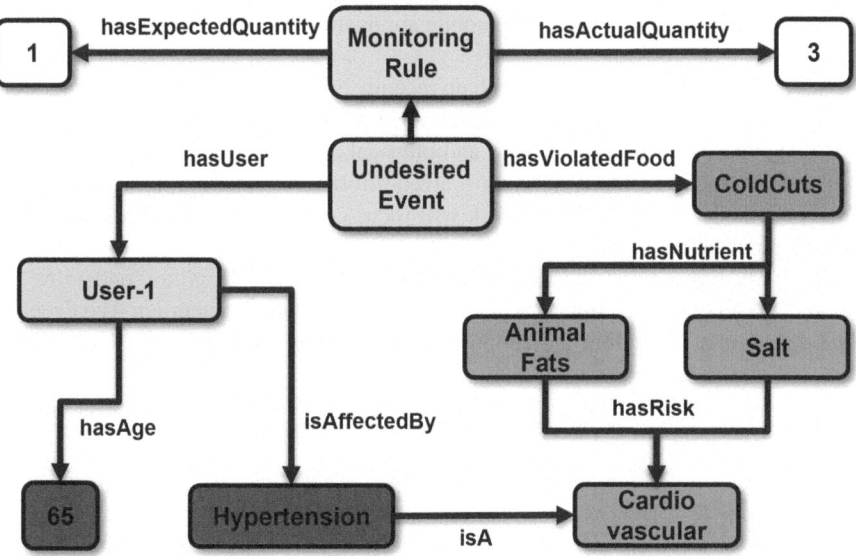

Fig. 3. Example of an *UndesiredEvent* individual generated by the reasoner.

Figure 3 shows an example of the knowledge that is generated when an undesired event is detected. A new individual of type *UndesiredEvent* is created and it is linked to the guideline that has been violated, the data led to the undesired event, and data coming from the user profile.

The generated knowledge is given as input to the *Communication Layer* that applies a template-based strategy to generate the recommendation sent to the user. The example shown in Fig. 3 is translated into the following message: "This week you consumed too much (3 portions of a maximum 1) cold cuts. Cold cuts contain animal fats and salt that can cause cardiovascular diseases. People over 60 years old are particularly at risk especially if they suffer from hypertension. Next time try with some fresh fish".

The large-scale living lab described in Sect. 6 aimed to demonstrate the effectiveness of the messages generated using FuS-KG, in comparison to pre-defined canned texts and messages generated using the HeLiS ontology (i.e., the ontology we presented in [8], containing only knowledge about food and physical activities).

5   Large-Scale Living Lab: The Salute+ Project

As introduced in Sect. 1, this work is part of the Salute+[13] project. The project consists of a set of innovative interventions/initiatives aimed at providing the general population, as well as specific target groups, with different functions with the objective of creating a healthy decision-making environment. By offering concrete tools and opportunities, the project seeks to mitigate risk factors for chronic diseases, such as poor nutrition and sedentary behavior. This project is part of a set of citizen health promotion initiatives run by *Trentino Salute 4.0*[14]. Within this framework, we ran a territorial living lab that enrolled 4,274 citizens using the mobile application integrated within the Salute+ platform for seven weeks.

User enrollment was voluntary, with no incentives provided, as all participants were already motivated to take part in the study. Each participant was equipped with smart bands that synchronized data on steps and physical activity information with our system. Additionally, to validate the synchronized information, users were asked to manually report their activities. Future work will focus on reducing the effort required for acquiring physical activity data.

This study represented the first large-scale living lab in Trentino concerning the adoption of knowledge-equipped AI tools supporting health prevention. The *Trentino Salute 4.0* aims at extending the adoption of the Salute+ solution to at least 30,000

Table 1. Distribution of demographic information of the users involved in the evaluation campaign.

Dimension	Property	Value
Gender	Male	57%
	Female	43%
Age	25–35	12%
	36–45	49%
	46–55	23%
	56–65	16%
Education	High-school or lower	42%
	University Degree	58%
Type of Occupation	Sedentary	39%
	Active Indoor	34%
	Active Outdoor	27%

---

[13] https://trentinosalutedigitale.com/blog/portfolio/trentinosalute/.
[14] https://trentinosalutedigitale.com/.

citizens of the Province of Trento by the end of 2024, with plans to introduce this platform at the national level in Italy during the years 2025–2026.

For completeness, Table 1 shows the main demographic information concerning the citizens involved in the evaluation proposed in this work. We want to highlight that in this living lab, all users presented a healthy status, as we specifically chose not to include individuals with chronic or other diseases for this initial pilot study.

6   Evaluation

In this Section, we report the evaluation activities we conducted on the Salute+ solution during the 49-day timespan of the large-scale living lab, as described in Sect. 5. The results observed from the collected data are presented in Sect. 6.1, while Sect. 6.2 discusses the key lessons learned from this experience.

6.1   Results

This user study consisted of providing a group of participants with a mobile application that we developed, based on the services offered by the Salute+ solution. In particular, we measure the effectiveness of the recommendations generated using FuS-KG and compare them with the results we reported in [9], where a non-randomized experiments setup [14] was adopted. In that setup, a Control Group received predefined, canned text messages, while an Intervention Group received messages generated using the same methodology adopted in Salute+, but relying on the HeLiS ontology instead of FuS-KG.

The sets of guidelines implemented in this living lab were the same as those used in the baselines, ensuring a fair comparison of the results:

- MEAL-Rules (related to single meals) check the correct quantity of a specific food category to be consumed in a single meal. Users were asked to fill in 4 meals every day: breakfast, lunch, snack, and dinner.
- DAY-Rules (related to a single day) check the maximum (or minimum) quantity (or portion) of a specific food category that can (or should) be consumed daily.
- WEEK-Rules (related to a single week) check the maximum (or minimum) quantity (or portion) of a specific food category that can (or should) be consumed weekly.

Figures 4, 5, and 6 present the evolution of the average number of undesired events per user detected concerning the MEAL-Rules, DAY-Rules, and WEEK-Rules sets, respectively. The green lines represent the trends of the Control Group, the blue lines represent the trends of the Intervention Group received recommendations generated by using the HeLiS ontology, and, finally, the red lines represent the trends observed on the users who adopted the Salute+ solution.

As mentioned, MEAL-Rules are verified every time a user enters a meal within the system; DAY-Rules are verified at the end of the day; while WEEK-Rules are verified at the end of each week. The increasing gap between the green lines and the others demonstrates the positive impact of the knowledge-based generated recommendations sent to users. We can observe how the average number of detected undesired events

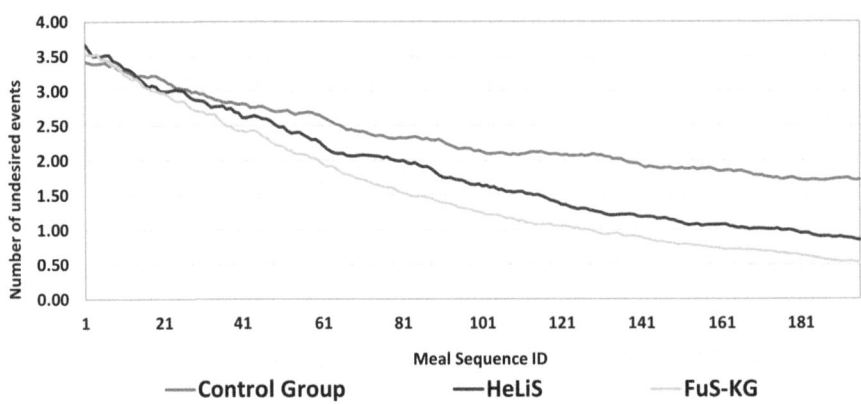

Fig. 4. Trend of the number of undesired events observed on the MEAL-Rules.

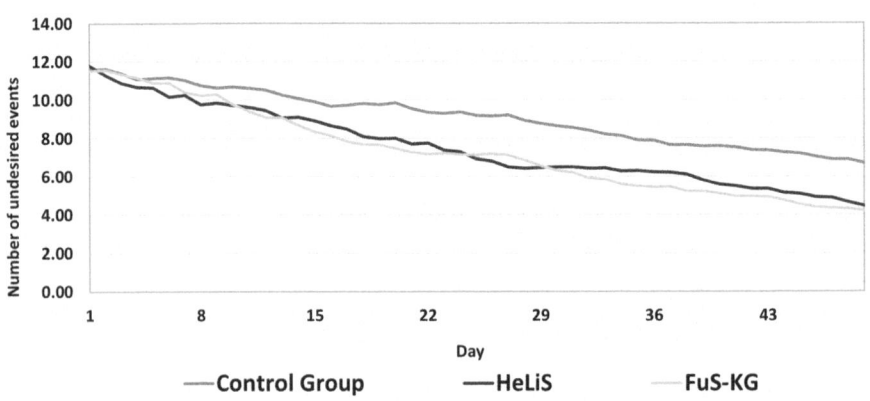

Fig. 5. Trend of the number of undesired events observed on the DAY-Rules.

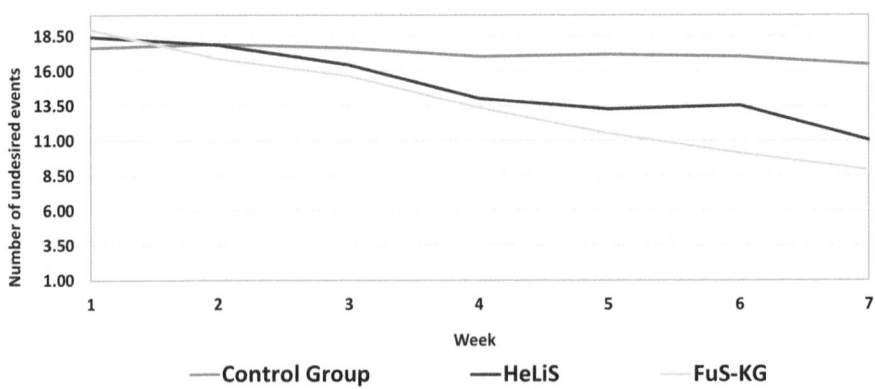

Fig. 6. Trend of the number of undesired events observed on the WEEK-Rules.

for the MEAL-Rules is below 1.0 after the first 7 weeks of the project. We can also notice how the recommendations generated by using FuS-KG led to a better drop of the detected undesired events compared to the ones generated by using HeLiS. A positive result has been obtained also for the DAY-Rules and the WEEK-Rules. However, we can observe how for the DAY-Rules the blue and red lines remained close for the entire timespan and, even if the FuS-KG group obtained a better drop, the improvement is not significant. This is a point of attention that triggered a more in-depth analysis of the data by combining the results observed for the MEAL-Rules with the ones observed for the DAY-Rules since it is expected that an improvement in the quality of single meals, an improvement should be observed also for the entire day. Instead, the fact that MEAL-Rules are more focused on food quantity and DAY-Rules are more focused on food categories highlighted how users found more easy-to-follow guidelines about the amount of food to consume during a single meal instead of distributing food categories appropriately during the entire day. This result will drive the analysis of how to refine the recommendations associated with DAY-Rules.

For completeness, we report in Table 2 the drop values of the observed undesired events at the beginning and at the end of the 7-week timespan of the living lab.

Table 2. Drop of violations at the end of the observation period. The first column lists the type of rules observed, while the second, third, and fourth columns indicate the drops observed within the FuS-KG, HeLiS, and Control groups, respectively. The most significant drop in violations occurs with the more frequent rules.

	FuS-KG Group	HeLiS Group	Control Group
QB-Rules	**85.06%**	76.63%	50.00%
DAY-Rules	**63.48%**	62.18%	41.98%
WEEK-Rules	**53.02%**	40.12%	6.68%

6.2   Lessons Learned

The integration of FuS-KG within the Salute+ platform and the large-scale living lab we ran, provided interesting insights, summarized in the four key lessons learned reported below, which can drive future enhancement of the overall solution.

Reasoning Effectiveness. The effectiveness of the reasoner was one of the most sensitive aspects of our living lab given the unpredictability of the number of contemporary reasoning tasks that could be launched. However, the work performed on the optimization of rules design and rules evaluation schedule allowed us to maintain the time for each reasoning task below 6 s making the interaction with the users acceptable for a real-time context. The rule analysis and optimization process was performed in two phases. In the first phase, we designed a set of complex rules to cover all monitoring activities. While this approach allowed us to cover multiple constraints with a single rule, it also resulted in high computational time for rule evaluation. Consequently, in the second phase, we chose to decompose these complex rules into simpler ones and scheduled their evaluations based on their timing property. This strategy led to an improvement in the overall reasoning performance and allowed us to have easier control over the overall reasoning process, including the exactness of the Violation instances and debugging operations. In the scenario addressed by the current deployment of Salute+, reasoning operations are executed on sets of triple that describe only specific user events.

Perception of Personalization from the User's Perspective The second key lesson relates to user perception of the personalization capabilities in our proposed solutions, as revealed through feedback from participants in our large-scale living lab. To explore this, a focus group was conducted at the end of the study, involving 30 users who were selected based on their different levels of adherence to the provided guidelines. This selection allowed us to gather insights from users with different levels of compliance. The goal was to get qualitative feedback on their perception of the system's personalized recommendations, specifically identifying when the system performed well and where improvements were needed. While the overall feedback was generally positive, the discussion revealed that some users found the combination of some rules difficult to follow and the Salute+ platform has been perceived as not very effective in explaining the rationale behind certain rules adoption. Users who encountered this issue agreed that the Salute+ system should be able to detect scenarios in which some of the rules cannot be followed by users and to automatically update the user profile accordingly. This suggestion will be further discussed with the domain experts to better understand if a priority mechanism on rules can be integrated to discriminate scenarios in which the Salute+ system should discard some *UndesiredEvent* individuals.

Abandon Rate. Further considerations can be made regarding the system's abandonment rate, as an overly intrusive notification system could lead to higher user drop-off. In our study, 87% of users engaged with the Salute+ application during the entire monitoring period of seven weeks, and no complaints about notifications were reported during the focus group discussions. A common suggestion to enhance user engagement was the more effective exploitation of geographical data acquired through smartphone sensors. Participants considered this information valuable for encouraging behavior change in real-life situations, such as avoiding vending machines during walks. Additional examples include sending notifications about nearby healthy food stores, restaurants with menu options that align with users' health goals, and sports events that match their interests. These suggestions will be taken into account in the development of the next version of the personalization component of Salute+, aiming to enhance users' perception that the system provides real-time and context-aware support.

Long-Term Adoption. At the end of the large-scale living lab, a discussion on the long-term adoption of the Salute+ was necessary, focusing on FuS-KG due to the effort required to keep the KG updated and the vision of making it a reference point for the research community. We discussed this aspect through the analysis of three main perspectives: (i) the *availability and reusability* of FuS-KG; (ii) the *sustainability* plan; and, (iii) the *maintenance* plan.

Concerning availability and reusability, FuS-KG is licensed under the Creative Commons Attribution-NonCommercial-ShareAlike 4.0[15]. It is available for download from the FuS-KG website[16] and, additionally, we have created a GitHub repository[17] to ease version control and issue tracking of the resource. The choice of publishing the KG open-source is to foster its adoption within the community enabling also the collection of feedback about its conceptualization to refine it. Moreover, by increasing its adoption, sustainability will benefit as well.

Concerning sustainability, as mentioned in the previous section, the presented ontology is the result of collaborative work between several experts in the context of the framework Trentino Salute 4.0. The main goals of this framework are to "combine efforts of employers, employees, and society to improve the mental and physical health and well-being of citizens". This long-term objective is aligned with the Sustainable Development Goal 3 (i.e., "Ensure healthy lives and promote well-being for all at all ages") of the United Nations, and it aims at preventing the onset of chronic diseases related to an incorrect lifestyle through organizational interventions directed to citizens. The overall sustainability plan for the continuous update and expansion of the FuS-KG ontology is granted by this framework since it is considered a strategic asset within the AI Strategy of the Trentino Local Government.

Finally, concerning the maintenance perspective, we opted to create a collaborative environment on GitHub enabling the research community to collaborate in refining and expanding FuS-KG. This way, through the *Issues* facility all community members may open new discussions concerning specific aspects related to FuS-KG ranging from integrating new information sources to suggesting novel modeling patterns.

7 Conclusions and Future Work

In this paper, we presented the modeling pathway of FuS-KG and how it has been integrated into the Salute+ platform representing a real-world solution aiming at supporting the promotion of healthy lifestyles among citizens. We discussed the role of FuS-KG within the Salute+ platform and we presented the results observed within a large-scale living lab in the Trentino territory involving more the 4,200 users. Results demonstrated the possibility of adopting the system in real-world scenarios, and the lessons learned provide valuable insights for future developments aimed at improving the overall efficiency of the system, thus allowing the deployment of the Salute+ platform in more challenging environments.

References

1. Alexander, G.L., et al.: Sensor systems for monitoring functional status in assisted living facility residents. Res. Gerontol. Nurs. **1**(4), 238–244 (2008). https://doi.org/10.3928/19404921-20081001-01

---

[15] https://creativecommons.org/licenses/by-nc-sa/4.0/.

[16] https://w3id.org/fuskg.

[17] https://github.com/IDA-FBK/FuS-KG.

2. Bickmore, T.W., Schulman, D., Sidner, C.L.: A reusable framework for health counseling dialogue systems based on a behavioral medicine ontology. J. Biomed. Inform. **44**(2), 183–197 (2011). https://doi.org/10.1016/j.jbi.2010.12.006

3. Calkins, D.R., et al.: Failure of physicians to recognize functional disability in ambulatory patients. Ann. Intern. Med. **114**(6), 451–454 (1991)

4. Calkins, D.R., et al.: Functional disability screening of ambulatory patients. J. Gen. Intern. Med. **9**(10), 590–592 (1994)

5. Cassell, E.J.: Doctoring: The Nature of Primary Care Medicine. Oxford University Press, Oxford (2002)

6. Corcoglioniti, F., Rospocher, M., Mostarda, M., Amadori, M.: Processing billions of RDF triples on a single machine using streaming and sorting. In: ACM SAC, pp. 368–375 (2015)

7. Dragoni, M., Bailoni, T., Maimone, R., Eccher, C.: Helis: an ontology for supporting healthy lifestyles. In: International Semantic Web Conference (2). Lecture Notes in Computer Science, vol. 11137, pp. 53–69. Springer, Cham (2018)

8. Dragoni, M., Bailoni, T., Maimone, R., Eccher, C.: Helis: an ontology for supporting healthy lifestyles. In: Vrandecic, D., et al. (eds.) The Semantic Web - ISWC 2018 - 17th International Semantic Web Conference, Monterey, CA, USA, 8–12 October 2018, Proceedings, Part II. Lecture Notes in Computer Science, vol. 11137, pp. 53–69. Springer, Cham (2018). https://doi.org/10.1007/978-3-030-00668-6_4

9. Dragoni, M., Donadello, I., Eccher, C.: Explainable AI meets persuasiveness: translating reasoning results into behavioral change advice. Artif. Intell. Med. **105**, 101840 (2020). https://doi.org/10.1016/J.ARTMED.2020.101840

10. Dragoni, M., Rospocher, M., Bailoni, T., Maimone, R., Eccher, C.: Semantic technologies for healthy lifestyle monitoring. In: The Semantic Web - ISWC 2018: 17th International Semantic Web Conference, Monterey, CA, USA, 8–12 October 2018, Proceedings, Part II, pp. 307–324. Springer, Heidelberg (2018). https://doi.org/10.1007/978-3-030-00668-6_19

11. Fernández-López, M., Gómez-Pérez, A., Juristo, N.: Methontology: from ontological art towards ontological engineering. In: Proceedings of Symposium on Ontological Engineering of AAAI (1997)

12. Fried, L.P., Bandeen-Roche, K., Chaves, P., Johnson, B.A., et al.: Preclinical mobility disability predicts incident mobility disability in older women. J. Gerontol.-Biol. Sci. Med. Sci. **55**(1), M43 (2000)

13. Fried, L.P., Young, Y., Rubin, G., Bandeen-Roche, K., Group, W.I.C.R., et al.: Self-reported preclinical disability identifies older women with early declines in performance and early disease. J. Clin. Epidemiol. **54**(9), 889–901 (2001)

14. Harris, A., et al.: Preclinical mobility disability predicts incident mobility disability in older women. J. Am. Med. Inform. Assoc. JAMIA **13–1**, 16–23 (2006)

15. Heckmann, D., Schwartz, T., Brandherm, B., Schmitz, M., Wilamowitz-Moellendorff, M.: Gumo - the general user model ontology, pp. 428–432 (2005)

16. Hitzler, P., Gangemi, A., Janowicz, K., Krisnadhi, A., Presutti, V. (eds.): Ontology Engineering with Ontology Design Patterns - Foundations and Applications, Studies on the Semantic Web, vol. 25. IOS Press (2016)

17. Katifori, A., Golemati, M., Vassilakis, C., Lepouras, G., Halatsis, C.: Creating an ontology for the user profile: method and applications. In: RCIS (2007)

18. Kikiras, P., Tsetsos, V., Hadjiefthymiades, S.: Ontology-based user modeling for pedestrian navigation systems. In: ECAI 2006 Workshop on Ubiquitous User Modeling (UbiqUM), Riva del Garda (2006)

19. Larsen, K.R.T., et al.: Behavior change interventions: the potential of ontologies for advancing science and practice. J. Behav. Med. **40**, 6–22 (2016)

20. Li, I., Ms, G.: Capturing and classifying functional status information in administrative databases. Health Care Financ. Rev. **3**(24), 61–76 (2003). https://pubmed.ncbi.nlm.nih.gov/12894635/

21. Michie, S., et al.: From theory-inspired to theory-based interventions: a protocol for developing and testing a methodology for linking behaviour change techniques to theoretical mechanisms of action. Ann. Behav. Med. **52**(6), 501–512 (2017). https://doi.org/10.1007/s12160-016-9816-6

22. Nelson, E., et al.: Functional health status levels of primary care patients. JAMA **249**(24), 3331–3338 (1983)

23. Onder, G., Penninx, B.W., Ferrucci, L., Fried, L.P., Guralnik, J.M., Pahor, M.: Measures of physical performance and risk for progressive and catastrophic disability: results from the women's health and aging study. J. Gerontol. A Biol. Sci. Med. Sci. **60**(1), 74–79 (2005)

24. Phan, N., Dou, D., Wang, H., Kil, D., Piniewski, B.: Ontology-based deep learning for human behavior prediction with explanations in health social networks. Inf. Sci. **384**, 298–313 (2017). https://doi.org/10.1016/j.ins.2016.08.038

25. Pinto, H.S., Staab, S., Tempich, C.: DILIGENT: towards a fine-grained methodology for distributed, loosely-controlled and evolving engineering of ontologies. In: de Mántaras, R.L., Saitta, L. (eds.) Proceedings of the 16th European Conference on Artificial Intelligence, ECAI'2004, Including Prestigious Applicants of Intelligent Systems, PAIS 2004, Valencia, Spain, 22–27 August 2004, pp. 393–397. IOS Press (2004)

26. Shimizu, C., Hammar, K., Hitzler, P.: Modular ontology modeling. Semant. Web **14**(3), 459–489 (2023). https://doi.org/10.3233/SW-222886

27. Suárez-Figueroa, M.C.: NeOn methodology for building ontology networks: specification, scheduling and reuse. Ph.D. thesis, Technical University of Madrid (2012). http://d-nb.info/1029370028

28. Turconi, G., Roggi, C.: Atlante fotografico alimentare. Uno strumento per le indagini nutrizionali (2007)

29. Wakefield, B.J., Holman, J.E.: Functional trajectories associated with hospitalization in older adults. West. J. Nurs. Res. **29**(2), 161–177 (2007)

30. Weisz, J., Ng, M.Y., Bearman, S.: Odd couple? Reenvisioning the relation between science and practice in the dissemination-implementation era. Clin. Psychol. Sci. **2**, 58–74 (2014)

31. Wolinsky, F.D., Miller, D.K., Andresen, E.M., Malmstrom, T.K., Miller, J.P.: Further evidence for the importance of subclinical functional limitation and subclinical disability assessment in gerontology and geriatrics. J. Gerontol. B Psychol. Sci. Soc. Sci. **60**(3), S146–S151 (2005)

Human Evaluation of Procedural Knowledge Graph Extraction from Text with Large Language Models

Valentina Anita Carriero$^{(\boxtimes)}$ ⓘ, Antonia Azzini ⓘ, Ilaria Baroni ⓘ,
Mario Scrocca ⓘ, and Irene Celino ⓘ

Cefriel – Politecnico di Milano, viale Sarca 226, 20126 Milano, Italy
{valentina.carriero,antonia.azzini,ilaria.baroni,mario.scrocca,
irene.celino}@cefriel.com

Abstract. Procedural Knowledge is the know-how expressed in the form of sequences of steps needed to perform some tasks. Procedures are usually described by means of natural language texts, such as recipes or maintenance manuals, possibly spread across different documents and systems, and their interpretation and subsequent execution is often left to the reader. Representing such procedures in a Knowledge Graph (KG) can be the basis to build digital tools to support those users who need to apply or execute them.

In this paper, we leverage Large Language Model (LLM) capabilities and propose a prompt engineering approach to extract steps, actions, objects, equipment and temporal information from a textual procedure, in order to populate a Procedural KG according to a pre-defined ontology. We evaluate the KG extraction results by means of a user study, in order to qualitatively and quantitatively assess the perceived quality and usefulness of the LLM-extracted procedural knowledge. We show that LLMs can produce outputs of acceptable quality and we assess the subjective perception of AI by human evaluators.

Keywords: Knowledge Engineering · Knowledge Extraction · Ontology · Large Language Models · Procedural Knowledge · Knowledge Graphs

1 Introduction

Procedural Knowledge (PK) is knowing how to perform some tasks. Such knowledge is usually expressed in the form of sequences of steps needed to achieve an overall goal, as in the case of recipes and maintenance activities. Making PK explicit is not trivial: identifying all actions to be performed, and splitting them into separate steps including all relevant information (e.g., the needed equipment), is often subject to interpretation and commonsense. Moreover, even when documented, this is usually done in an unstructured format, by means of natural language, across heterogeneous documents and systems. This makes it hard to

© The Author(s), under exclusive license to Springer Nature Switzerland AG 2025
M. Alam et al. (Eds.): EKAW 2024, LNAI 15370, pp. 434–452, 2025.
https://doi.org/10.1007/978-3-031-77792-9_26

access procedures for those who need to execute them, leading to partial or poor compliance with processes and best practices. As argued in [40], formally representing such procedures as Knowledge Graphs (KGs) can be the basis to build digital tools that provide automatic support to the users, and foster procedural knowledge documentation, reuse, and interoperability.

However, developing high-quality KGs is time-consuming: making this task even partially automatic may be a game changer. Large Language Models (LLMs) – probabilistic models trained on huge natural language corpora – exhibit remarkable capabilities in natural language processing and generation. LLMs have recently gained significant attention to support different tasks of the whole knowledge engineering process, as emphasized in the report of the Dagstuhl Seminar 22372 [15] and the LM-KBC Challenge[1].

In this paper, we leverage LLM capabilities for extracting PK from text; differently from other extraction tasks like entity recognition or relation extraction, identifying procedural steps (the actual sentences describing actions to be performed) is a potentially more ambiguous task: as we will show in our experiments, different human annotators may produce different yet valid outcomes, thus precluding the existence of a definitive ground truth for such task.

Our contribution is therefore twofold. On the one hand, we experiment different prompting strategies to employ LLMs for extracting PK from unformatted textual descriptions, and generating a valid KG thereof according to a predefined procedural ontology. On the other hand, we evaluate the knowledge extraction outcomes by means of a user study, in order to evaluate the (perceived) quality and usefulness of the resulting KG, comparing it to human annotation and assessing the user-acceptance of the LLM-produced results on such a subjective task; we also evaluate a potential bias in human evaluation with respect to AI. All materials, code, results of this work are available on GitHub[2].

The paper is organised as follows: after introducing the state of the art in Sect. 2, we better define the problem and the data we used in Sect. 3; then we report the findings of a preliminary study in Sect. 4 which led us to the definition of our research questions in Sect. 5; Sect. 6 describes our iterative prompt engineering process, and the results of our proposed prompt-based pipeline; in Sect. 7 we describe our human evaluation design and we present and discuss our quantitative and qualitative findings in Sect. 8; Sect. 9 concludes the paper with some future work.

2 Related Work

Our work is at the intersection of the LLM application to Knowledge Engineering and its result assessment from a human point of view. Therefore, we take into account several aspects from the state of the art, as explained in the following.

---

[1] Cf. https://lm-kbc.github.io/.

[2] Cf. https://github.com/cefriel/procedural-kg-llm.

2.1 LLMs for Knowledge Engineering and Knowledge Extraction

As reported in [1], LLMs provide a powerful tool for mapping natural language to formal language, a fundamental activity in knowledge engineering. Existing efforts in unifying LLMs and KGs are summarized in [37,38]. The LM-KBC challenge explores how to build disambiguated knowledge bases from LLMs, given a subject and a relation (see, among others, [47,50]).

Several recent works cover the use of LLM prompting to address specific knowledge engineering tasks, like ontology engineering [5,8,19,49], ontology learning [2,11], named entity recognition and linking [9,21,43], knowledge graph construction including mapping generation [18,25]; some works specifically focus on benchmark, metrics and evaluation of such methods [12,31].

Specifically on (procedural) knowledge extraction, several studies (including [51–53]) investigate some extraction and reasoning tasks on procedures, such as the relation between a step and the procedure's goal, and the temporal relation between steps. Other works focus on the business process management domain: a corpus of business processes annotated by humans with activities, gateways, actors, and flow information [4] is exploited in [3] for extracting, from such processes, activities and their participants, using an LLM; an activity recommender to support business process modeling is introduced in [44]. A semi-structured dataset of repair manuals related to *Mac Laptops* is annotated with the required tools and the parts that are disassembled during the process, along with two methods for extracting them [33]. The procedures included in the dataset focus on opening the device and removing/repairing a broken component. Reusing such datasets as-is was not applicable for our experiments, since they did not fully covered our requirements, e.g., by taking into account only actions related to repairs [33], and by missing annotations about objects and tools [4] (cf. Section 3). Micrographs storing relevant entities and actions from technical support web pages are created in [22]. A prompting-based pipeline to extract a list of ordered steps from a procedure expressed as a numbered/bullet/indented list is proposed in [41]; differently, we give as input a text with no formatting and include additional descriptive sentences that the LLM needs to discard.

2.2 Evaluation of Generative AI

The evaluation of Generative AI and specifically LLMs is an open research problem, in that it is not easy to objectively assess the generated (natural language) output [26]. In the area of text annotation, different approaches exist [6,45], because LLMs are expected to act like human annotators [29]. Indeed, among other things, they can be employed for fact checking [35] or to identify claims and sub-claims in text entailment [20].

The approach to assess LLM outputs can combine human and LLM annotations [24], and several authors propose to use LLMs to perform also the evaluation task: by simulating human feedback via LLMs [10], by employing a (LLM) debate method [7], or by making LLMs adopt a human-like comparative evaluation approach [48].

While a unified approach to the assessment of natural language generation has not yet emerged, some work started to propose evaluation frameworks [13], also focusing on the definition of evaluation metrics for specific tasks like summarization [27]. We based our human evaluation approach on these works, as well as on general usability methods. It is important to note that a holistic evaluation should not only assess the quality of the LLM outcome, but also the human interaction experience with the LLM [23].

2.3 Human Bias on (Generative) AI

Finally, whenever involving humans in the evaluation of an artificial system, the subjective perception and the potential cognitive biases of users must be carefully considered. Indeed, some studies suggest that human attitudes towards AI are largely negative [34]. In the area of Generative AI, several concerns emerge in relation to the creativity potential of such models [28], especially in the context of art [32,39] and music [54].

In our work, we also investigate the potential bias of human subjects against LLMs that play the role of annotators, by applying an A/B testing approach, as done by [17]: they performed several experiments on how people judge humans and machines differently, in several scenarios (e.g., natural disasters, labor displacement, policing, privacy, algorithmic bias); they demonstrate that people tend to favour humans or machines in different scenarios, revealing the biases in human-machine interaction.

3 Problem Definition and Data

As mentioned in the introduction, our goal is to build a knowledge graph out of textual descriptions of procedures. Our vision is therefore to extract procedural knowledge (PK) from those documents, and to build a knowledge graph (KG) according to an ontology. This KG can then be used by different downstream applications to facilitate the access and use of such procedures by human operators. Examples of such applications could be (KG-empowered) search applications or intelligent assistant: we expect the users to feel the need to be helped to find a specific procedure (or a part thereof) and to be guided step-by-step in its execution, for example by being informed about the action they have to perform (e.g., *turning off* a switch), the equipment they may need to use (e.g., wearing *protective gloves*) or the time it may take to perform a specific step (e.g., approximately *15 min*).

In order to fulfill such requirements, the procedural KG extracted from text should (1) preserve the intended meaning of the original document and (2) contain enough information to guide a user in correctly executing the procedure. The extracted KG should therefore be evaluated, respectively, on the basis of its *quality* and *usefulness* (cf. Sect. 7). Based on this scenario, we define a simple ontology and we identify a general-purpose dataset to be used in our LLM-powered PK extraction and KG building experiments.

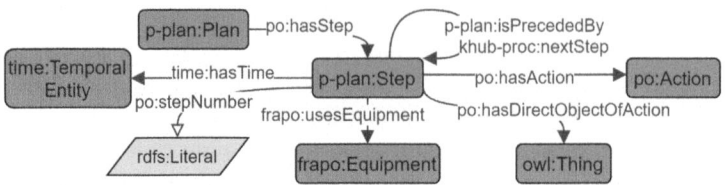

Fig. 1. Ontology used in the experiments.

Ontology. We reuse existing ontologies when applicable, namely: P-Plan [14] and K-Hub [42], that address plans and related concepts, FRAPO, one of the SPAR ontologies[3], and the Time Ontology[4] while creating a few new classes and properties when needed (po:). Specifically, as depicted in Fig. 1, a procedure is represented by the class p-plan:Plan and is linked to its p-plan:Steps (po:hasStep), which are sequentially ordered (property p-plan:precededBy and its inverse khub-proc:nextStep). Each step is then linked to its frapo:Equipment, if any, via the property frapo:usesEquipment, and with the action(s) to be performed while executing it (po:hasAction), along with the direct object of the action (po:hasDirectObjectOfAction). The information about the time needed for executing the step is represented by the class time:TemporalEntity.

Input Procedures. For our experiments, we use as a reference dataset Wiki-How, one of the largest online databases of PK, which includes how-to articles on multiple domains. We reuse the JSON dataset built by [52], crawled from the WikiHow website[5], and focus on atomic procedures, that do not include methods/parts as sub-procedures. Out of each selected procedure, we build an unformatted text[6] (also partially removing punctuation), by concatenating (i) procedure title, representing the overall goal, (ii) general procedure description, (iii) headline of each step, (iv) description of each step. Thus, we also include in the text irrelevant sentences, which are supposed to be discarded during the extraction phase. In total, we used four procedures randomly selected from WikiHow, one as working example for our prompt engineering phase, and three for our method replication and human assessment. Such procedures were chosen to be as diverse as possible in terms of content and complexity (length, topic, actions, tools, etc.): *how to clean a computer monitor* (working example), *how to fix a rubbing door*, *how to make honey glazed parsnips*, and *how to plant a bare root tree.*

---

[3] Cf. http://www.sparontologies.net/.

[4] p-plan: http://purl.org/net/p-plan#
   khub-proc: https://knowledge.c-innovationhub.com/k-hub/procedure#
   frapo: http://purl.org/cerif/frapo/
   time: http://www.w3.org/2006/time#.

[5] Cf. https://github.com/zharry29/wikihow-goal-step.

[6] The actual input texts of the procedures used for the method replication and human assessment can be found on GitHub.

4   Preliminary Study

In a preliminary formative study, we wanted to kick-start the LLM-powered procedure extraction from text, by identifying the best performing prompt chaining [46] to execute the task. We evaluated both the prompt instructions and the LLM outputs ourselves, but we also wanted to involve human evaluators outside our research team to get an external assessment. Employing the same procedural texts described in Sect. 3 and a prompting approach similar to the one described in Sect. 6, we extracted the procedural knowledge from text according to the predefined ontology; the procedure annotation task was explained (both to the human evaluators and the LLM) as a verbatim extraction from the text of the relevant sentences explaining the procedure steps and the identification of specific parts of those steps (e.g., action, tools). Then, we involved a group of people to judge the annotation of each procedure (n=24, 8 evaluators for each procedure; 79% male, 21% female; aged between 20 and 60; from our company, but outside our group, so to be able to ask follow-up questions); each participant was asked to first manually perform the entire extraction (generating a semi-structured representation, not the final RDF format) and then to evaluate the LLM results on the same procedure and to give us suggestions on how to improve the process. The results of this study were partially unexpected and were very useful to design the experiment described in this paper. Our preliminary observations were as follows.

Humans Highly Disagree in Annotating Procedural Knowledge. The number of annotated steps, tools, and actions varied quite a lot, thus the level of agreement of people in executing such a task was very low (no significant inter-rater agreement indicator). This suggests that *not all annotation/extraction tasks are equal*: in our case, the step identification is quite different from an entity recognition or relation extraction task, because it involves the extraction of entire sentences, thus our goal is somehow more "subjective" and it is not amenable to have an actual "ground truth. This also means that *the LLM can be considered as "yet another annotator"*.

Humans Tend to Adjust the Task Regardless of the Instructions and Example. Analysing their annotations, we realised that humans did not closely followed the given instructions (opposed to the LLM "obedient" execution) based on what they consider more useful, using their creativity or reasoning. In some cases, new steps that were absent from the input procedure were introduced; some actions were inferred by the participants; some users tended to rephrase or summarise the text of the step, to make it clearer based on their interpretation, even if we asked for verbatim extraction of sentences. Moreover, some participants were quite critical with respect to the prompt instructions, commenting that they were too long and not immediate. This suggests that the prompt can be less detailed and *leaves room for the more generative capabilities of the LLM*, as humans do the same.

Humans are Highly Critical When Evaluating the LLM. Also in the evaluation of LLM results, performed through a list of Likert-scale ratings on various aspects of the generated output, there was no agreement between the participants (computed as Krippendorff's alpha metrics [16]), as they directly compared their work with that of the LLM which was quite different. Some evaluators assigned very low scores, which means that the LLM is still not perceived by some as "good enough" to solve the task. This suggests that those *evaluations may have been influenced by a bias on AI* and that this hypothesis was worth being tested.

5   Research Questions

Our goal is to create a procedural knowledge graph by extracting structured knowledge from unstructured text; we would like to apply LLMs to the *procedural knowledge extraction* task and assess their results from a human point of view, taking into consideration the findings of our preliminary study.

Our research questions are formulated as follows:

RQ1. *What is the quality of the extracted procedural knowledge (graph) as perceived by human evaluators?* In this respect, we would like to understand if people judge the LLM-extracted knowledge as correctly representing the meaning of the original text and what influences their evaluation.

RQ2. *What is the usefulness of the extracted procedural knowledge (graph) as perceived by human evaluators?* In this respect, we would like to assess the "fitness for use" of the output of the LLM extraction, when people are explained its potential downstream use (i.e. are the extracted steps/actions/tools *good enough* to guide a user in executing the procedure?).

RQ3. *Do human evaluators show any systematic bias if they are told that the extraction task was executed by a LLM rather than an expert human annotator?* In this respect, we would like to check any difference in people judgment about the extracted procedural knowledge (graph).

RQ4. *Qualitatively, are there any differences in the way human annotators and the LLM extract explicit knowledge and infer implicit knowledge from the text?* In this respect, we would like to assess if the LLM behaves like human annotators in interpreting and extracting procedural knowledge.

6   Prompt Engineering Solution

The expected result for each procedure was an ordered list of sentences summarizing the steps extracted from the text, with their respective actions, direct objects, equipment items, and temporal information, expressed in RDF according to the given ontology. We tested different prompting approaches, as detailed below, and in an initial phase we manually evaluated the LLM results, finding an agreement between us with majority voting to decide whether each result was qualitatively satisfactory; the prompting approach that revealed to yield the best results was then used in the subsequent human evaluation experiment (cf. Sect. 7).

Prompting Approach. We tested different prompting approaches on the working procedure (cf. Sect. 3) to find the best solution. Initially, we tried with a single prompt with zero/one/few-shot and we found that one prompt was not enough to achieve an acceptable result and a correct RDF, while giving one example in the prompt (one-shot learning) was the best trade-off between zero-shot and few-shots; then, as demonstrated in [46], we tried a Chain-of-Thought (CoT) prompting approach, i.e. decomposing the global problem into intermediate steps (prompts) and solve each of them sequentially, because this significantly improves the ability of LLMs to perform complex reasoning tasks, including commonsense. We tested our prompts with or without the definitions of the entities to be extracted, and found that very short definitions worked better in obtaining a more accurate result than both longer definitions and no definition at all. Moreover, allowing the LLM to generate the step sentence descriptions gave better results in the extraction of steps as opposed to strictly report the verbatim sentence from the text without modifications, as in the preliminary experiment (cf. Sect. 4). After multiple refinement iterations, we came up with a CoT prompting approach decomposed in 2 steps, both applying a one-shot learning, as depicted in Fig. 3. Each step is associated with a different agent playing a different role.

P1: Generate Steps Descriptions and Annotations. In this prompt, the LLM is given the role of "expert in information extraction with a special background in procedures" and is asked to generate a semi-structured output, composed of a list of sentences describing the procedure steps, each including the performed action(s), its direct object(s), the used equipment, and any execution time information (if any). Each piece of information to be included in the semi-structured output is explained as follows:

- an *action* is defined as "the verb or phrasal verb or idiomatic expression that represents an action to be performed",
- the *direct object* of the action is defined as "the noun or pronoun being acted upon by the verb of the action",
- the *equipment* is an "item that is needed to perform the action",
- the *time* is "any temporal information relevant to perform the action".

It is further specified in the prompt that each generated step may include one or more actions, and that the direct object of each action cannot be considered an equipment itself. Finally, we asked the LLM to include also possibly implicit equipment that is not mentioned in the text, if relevant to perform the action. By asking the LLM to generate a list of steps from an input procedure, that summarizes the input and contains a specific subset of information, we managed to leverage its generative capabilities. The prompt also includes an example of procedure with its respective output (see Fig. 2). The example is crucial to show the LLM not only how to generate the semi-structured output, but also how to produce the annotations that are preliminary to the RDF KG construction.

P2: Generate the Ontology-Based Knowledge Graph of the Procedure. In order to obtain the intended output – i.e. a KG of the procedure, linked to its steps, actions, direct objects, equipment and temporal information, according to the

For example, from the following excerpt of procedure

"""

Clean a Computer Keyboard
In this article, we'll walk you through how to clean a computer keyboard effectively.
Unplug and turn off your computer to operate safely.
In case of an accident, you want to ensure nothing is harmed.
Remove the batteries, if they are removable, from a wireless keyboard, and you should be set to go.
[...]
Press the power button of the computer for 3 seconds to switch it on.

"""

You generate this output:

"""

title: clean a computer keyboard
1. unplug and turn off computer
   - **action**: unplug ; **direct object of action**: computer
   - **action**: turn off ; **direct object of action**: computer
2. remove batteries from wireless keyboard
   - **action**: remove ; **direct object of action**: batteries ; **equipment**: wireless keyboard
[...]
7. press power button for 3 seconds to switch on computer
   - **action**: press ; **direct object of action**: power button of computer ; **time**: 3 seconds

"""

Fig. 2. Example procedure and output provided in prompt *P1*.

given ontology – we assign to the LLM the new role of "expert in knowledge graph construction, with a special background in ontologies on procedural knowledge", and we ask it convert the semi-structured output of the first prompt into RDF formatted in Turtle syntax (similarly to [12]). Rather than providing the entire ontology to be used, we showed the language model an example translation from its initial output to RDF, so that the LLM could find in the example all classes and properties to be used, and how they needed to be mapped to the annotation.

While designing *P2*, we started by providing the whole ontology, asking the LLM to translate its initial output to Turtle according to such ontology. However, relevant triples were missing: many inverse relations included in the ontology were never used (e.g., two consecutive steps were linked only with `nextStep`, but not its inverse `precededBy`), and the procedure was linked only to the first step, but not all other subsequent steps. Moreover, additional and very specific instructions would have been needed in the prompt for defining all prefixes and how to design local names, to prevent ambiguous individuals. Instead, by showing an example on how to map the textual list of steps and annotations to Turtle, we avoid possible errors in complying with ontological constraints, with a greater guarantee of the final KG to be complete. Moreover, this method may be particularly useful when we work with multiple, and possibly large, ontologies, that would exceed the context window of the LLM.

Results. For our experiments, we use the GPT 4o model, and rely on the LangChain framework[7]. We also tested with GPT 3.5 Turbo model, however, we obtained worse results than GPT 4o, with respect to both *P1* and *P2*.

---

[7] Cf. https://www.langchain.com/.

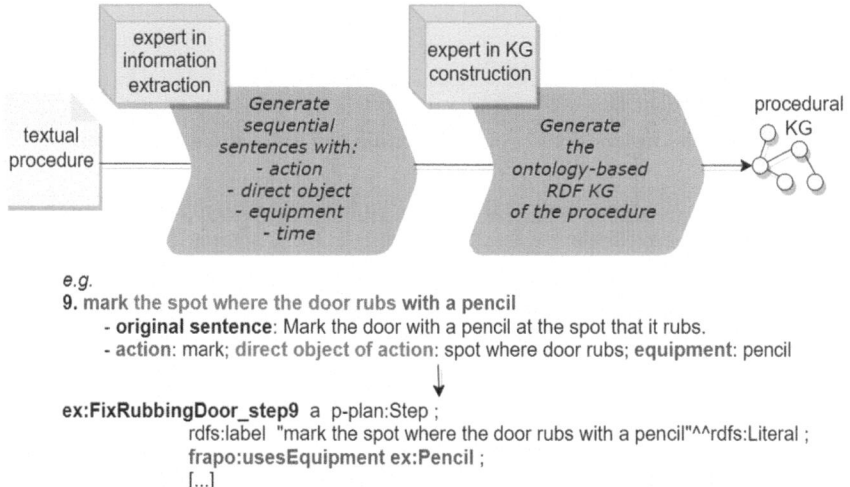

e.g.
9. mark the spot where the door rubs with a pencil
  - **original sentence**: Mark the door with a pencil at the spot that it rubs.
  - **action**: mark; **direct object of action**: spot where door rubs; **equipment**: pencil

ex:FixRubbingDoor_step9 a p-plan:Step ;
        rdfs:label "mark the spot where the door rubs with a pencil"^^rdfs:Literal ;
        frapo:usesEquipment ex:Pencil ;
        [...]

Fig. 3. Illustration of the Chain of Prompt for Procedural KG extraction, along with an example of extracted step.

Specifically, GPT 3.5 Turbo tended to collapse some objects and equipment items instead of keeping them separate as in the example (e.g. *bucket filled with water* in place of *bucket* and *water*), and it did not always comply with the provided template, as opposed to GPT 4o. The results of both prompts are available on GitHub. All produced KGs are valid RDF in Turtle syntax. Properties' domains and ranges are correctly used, and individuals are typed flawlessly. The LLM perfectly mimics our conventions for local names. The KGs generated with *P2* are complete, meaning that they include all entities that have been extracted in *P1*, thus addressing all requirements we defined. Therefore, we use the results of *P1* in our user experiment.

7   Experiment Design

In order to test our hypotheses, we extracted PK from 3 procedure texts applying the prompting approach described in Sect. 6. Then, we setup a crowdsourcing campaign to collect feedback from human evaluators as follows.

Input to the Human Evaluator. Each participant was given the annotation instructions (corresponding to the prompt *P1*, including an example of the expected input-output), the procedure text and the extracted semi-structured PK as generated by the LLM.

Expected Output. Each participant was presented with a form (implemented with Microsoft Forms) and asked to perform two tasks (presented in two following pages in the form): (1) a short manual annotation exercise, to get familiar with the task and to provide feedback on the task itself, and (2) the assessment

Table 1. Evaluation items used to assess the manual annotation task.

I found this task very easy.
I am confident that my output is correct.
I found the instructions provided for this annotation task very clear.

Table 2. Evaluation items used to assess the LLM output. Items marked with * are expressed on a reversed scale.

Dimension	Item
Quality (Q1)	The annotator was accurate in rephrasing and extracting the procedure from the text
Quality (Q2)	The steps identified and rephrased by the annotator are relevant to the task to be executed
Quality (Q3)	The steps identified and rephrased by the annotator provide sufficient details to execute the
Comparative Quality (CQ1)	I would have identified a different set of equipments from those suggested by the annotator *
Comparative Quality (CQ2)	I would have identified a different set of actions from those suggested by the annotator *
Comparative Quality (CQ3)	I would have identified a different set of direct objects of action from those suggested by the annotator *
Comparative Quality (CQ4)	I think that the annotator missed some important parts of the procedure in their processing *
Usefulness (U1)	Following only the instructions identified by the annotator would make it easier to execute the procedure (with respect to the original text)
Usefulness (U2)	Following only the instructions identified by the annotator would make it quicker to execute the procedure (with respect to the original text)
Usefulness (U3)	I believe that an automatic system providing information for executing a procedure could use the extraction from this annotator as an input
Usefulness (U4)	The instructions identified by the annotator don't include the proper information to support procedure execution *
Usefulness (U5)	I think I would correctly execute the original procedure based only on the instructions identified by the annotator

of the LLM annotation, according to the evaluation criteria explained in the following.

Instructions. For the manual annotation task, participants were given exactly the same instructions included in the *P1* prompt used with the LLM; it is worth noting that, as explained in Sect. 6, the second prompt *P2* always yielded a correct RDF representation of the semi-structured output of the first prompt, therefore we gave the participants the LLM results in response to *P1*, to avoid the need for people to understand and validate Turtle. For the LLM assessment, participants were asked to express their evaluation with a set of Likert-scale ratings according to our evaluation dimensions.

Evaluation Dimensions. All quantitative evaluation items were expressed as statements to be assessed on a 1 to 5 Likert scale of agreement (from strongly disagree to strongly agree); we also asked for free-text feedback to also collect some more qualitative feedback. Regarding the manual annotation task performed by the participants, we asked feedback about the task difficulty and the instructions clarity, as per the items in Table 1. Regarding the LLM output assessment, we collected ratings about three main dimensions: *perceived quality*, intended as the perceived value of the knowledge extraction, *perceived comparative quality*, intended as the value compared to what the evaluator would have done on the same task, and *perceived usefulness*, intended as the characteristics of the output to be used as intended; the respective items are listed in Table 2. Additionally, we also asked some profiling question, to collect user characteristics, to check their possible influence on the answers.

Experimental Setting. For each of the procedures used in the experiment, we set up two crowdsourcing campaigns on the Prolific platform[8] [36], in a A/B testing with a between-subject design: in one campaign, we told the participants that the annotations they were evaluating were performed by a LLM, while the other group was said that the annotations were produced by an experienced human annotator. Each campaign involved 30 respondents, accounting for a total of n = 180 participants in 6 campaigns. The inclusion criteria were: age between 20 and 60, residence in Europe, fluency in English and the use of a desktop for the study execution, while, to implement an A/B testing, the exclusion criterion was the participation to the other five experimental campaigns. No inclusion or exclusion criteria were imposed on other personal characteristics (gender, nationality, etc.).

8   Human Assessment and Discussion

In this section, we summarise and discuss the results and main findings of our quantitative and qualitative evaluation with the crowdsourcing participants.

Human Evaluators. As explained before, we involved 180 participants from the Prolific crowdsourcing platform, between June and July 2024. Excluding participants who revoked the consent to register demographic data, the others were aged between 20 and 60, 37% male, 63% female; their employment status was 14% student, 47% employed and 39% unemployed. There was no significant difference between the 6 groups that were involved in the 6 campaigns, in terms of demographic and profiling characteristics.

Perceived Quality. The overall distribution of the ratings given by the human evaluators on the quality-related dimensions are displayed in the first 7 boxplots in Fig. 4 and the first 7 rows in Table 3. On the 1–5 Likert scale, the human judgement is generally positive (median value is always 4). We tested with ANOVA the possible variability of the ratings with respect to the 3 different procedures, but we did not find any strongly significant difference, which means that the results are not dependent on the different input. With respect to *perceived quality*, i.e. assessment of the extracted knowledge accuracy and relevance, we notice that the evaluators gave generally high scores and those scores are quite positively

---

[8] Cf. https://www.prolific.com/.

Table 3. Distribution of evaluation scores by the human participants (n = 180) on the items (cf. Table 2).

Evaluation Item	Average	Median	Std dev.
Q1	3,94	4	0,88
Q2	4,16	4	0,79
Q3	3,82	4	0,97
CQ1	3,83	4	1,15
CQ2	3,71	4	1,18
CQ3	4,02	4	1,09
CQ4	3,55	4	1,29
U1	2,90	3	1,09
U2	3,04	3	1,17
U3	3,35	3	1,06
U4	3,39	3	1,18
U5	3,31	3	1,13

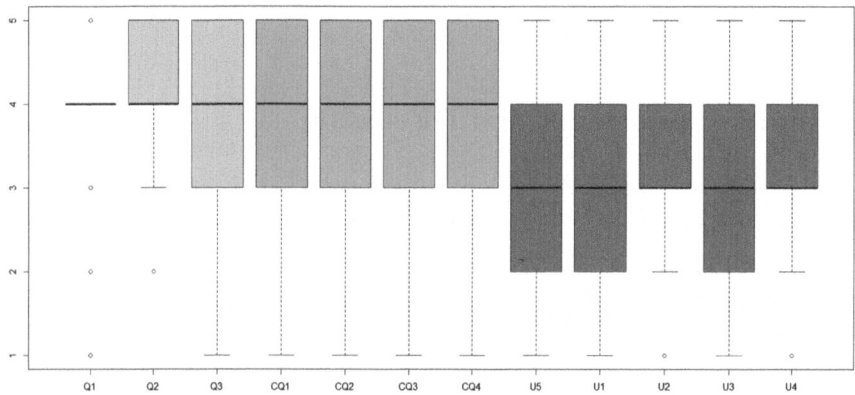

Fig. 4. Distribution of human ratings on the evaluation items (cf. Table 2).

correlated (r between 0.56 and 0.64). With respect to *perceived comparative quality*, i.e. assessment of the extracted knowledge quality with respect to what the evaluator would have done on the same task, the participants were still positive, but slightly less generous in their ratings and those scores are a bit less positively correlated (r between 0.26 and 0.59); this means that the participants still think that the quality of the extracted knowledge is quite high, but some of them suppose that they could have done a better job, confirming the findings of our preliminary study (cf. Sect. 4).

We can therefore conclude that, in relation to our research question **RQ1**, on average, human evaluators judge the extracted knowledge as correctly representing the meaning of the original text.

Perceived Usefulness. The overall distribution of the ratings given by the human evaluators on the usefulness items are displayed in the last 5 boxplots in Fig. 4 and the last 5 rows in Table 3. On the 1–5 Likert scale, the human judgement ratings are generally lower that the quality-related items (median value is always 3) and the scores are weakly or moderately correlated (r between 0.15 and 0.62). Those results indicate that the evaluators were more doubtful when they had to judge the usefulness; this may be due to an unclear explanation or an imprecise understanding of the intended use of the procedural knowledge extracted from the text. Indeed, when asked to assess their manual annotation and the clarity of the instructions they gave central-range scores (average values between 2.84 and 3.19 on a 1–5 scale, cf. items in Table 1): this may have negatively impacted their interpretation of the "fitness for use" of the extracted knowledge.

Therefore, in relations to our research question **RQ2**, we cannot conclude that the procedural knowledge extracted from text is perceived as highly useful by our evaluators.

Potential Bias About AI. As explained in Sect. 7, we performed an A/B testing to check for any difference in people judgment, if they are told that the extraction task was executed by a LLM rather than an expert human annotator; in other words, we tested for any indication of a systematic bias about humans vs. machines. For each of the evaluation items, we applied the Kruskal-Wallis test[9] [30] on the two groups (LLM vs. human annotator) to check for any difference. We run the test on the entire dataset (n=90 for each group), and the data displays only a few statistically significant differences between the two groups, in relation to items Q2, Q3, CQ1, CQ4 and U4 (p-values of the Kruskal-Wallis test between 0.007 and 0.04); in all those cases, the evaluators gave lower ratings to the LLM with respect to the expert human annotator. This result, even if not emerging on all items, seems to confirm the considerations out of our preliminary study.

In relations to our research question **RQ3**, therefore we cannot conclude that there is any systematic bias about LLMs, but a slight tendency of human evaluators to be more forgiving towards people than towards machines.

Qualitative Comparison Between Human and LLM Knowledge Extraction. From a qualitative comparison between the annotations produced by human annotators and the knowledge extracted by the LLM, it emerges that both humans and LLMs benefit from the possibility to rephrase/simplify the original text of the procedure in order to generate instructions that are more concise and *useful* from their point of view, even if this leads to a greater variability of the output. However, even when differently rephrased, the vast majority of

---

[9] We preferred a non-parametric test (instead of the well-known parametric *t-test*), because we could not assume a normal distribution of the population.

annotators detected the same steps as the LLM. Humans tended to rephrase also some verbs (e.g. *turn on* in place of *preheat, prune* instead of *nip off*) or some objects (e.g. *packaging* instead of *container*). Furthermore, we noticed that both humans and the LLM took advantage of the option to include in their annotations implicit equipment items (e.g. inferring *screwdriver* to check if screws are tight) – increasing variability again. However, the LLM is more compliant to the given instructions, since it only extracted implicit equipment, while some human annotators introduced also implicit time indications when they consider it relevant to execute the procedure.

In relations to our research question **RQ4**, we can conclude that the extraction of both explicit and implicit knowledge from procedural texts shows multiple similarities between the LLM and the human annotators, with the only difference that humans are less likely to strictly follow the given instructions.

9 Conclusions

In this paper, we provided our results in the use of LLMs to extract procedural knowledge from textual descriptions, in order to build procedural knowledge graphs, adopting a suitable prompt chaining approach. Moreover, we performed an extensive human evaluation study to assess the LLM output in terms of its perceived quality and usefulness. We showed that the evaluators rated quite positively the quality of the extracted PK, while they were more doubtful about its usefulness; moreover, we highlighted that in some cases, people tend to be more critical towards an AI system rather than another human annotator: even if we did not find evidence of a systematic bias, this phenomenon is worth exploring in future LLM evaluation studies. Finally, we qualitatively studied the ability of both humans and LLMs to extract explicit and implicit knowledge from text, coming to the conclusion that, in extraction tasks where a definitive ground truth does not exist, an LLM displays abilities similar to those of people.

All in all, we believe that LLMs are a promising technology to address this complex task, still the human intervention – through a human-in-the-loop approach – is very likely required to verify that the generated output is "good enough" to be used in real settings, especially when compliance is critical, like for example in the support to execute industrial procedures.

Our future work will be oriented towards a broader and more systematic evaluation of the proposed prompting approach, extending the assessment to industrial procedures in multiple formats (e.g. PDF or spreadsheets documents), addressing the challenge of extracting more complex cases of procedures (e.g. sub-procedures, optional or alternative steps, identification of additional entities), extending the approach to include fine-tuning or retrieval-augmented generation with the support of background knowledge, and comparing and benchmarking the outcomes of different LLMs.

Acknowledgements. This work is partially supported by the PERKS project, co-funded by the European Commission (Grant id 101070186). The authors would like to thank all the participants to the human assessment reported in this paper.

References

1. Allen, B.P., Stork, L., Groth, P.: Knowledge engineering using large language models. TGDK (2023). https://doi.org/10.4230/TGDK.1.1.3
2. Babaei Giglou, H., DSouza, J., Auer, S.: LLMs4OL: large language models for ontology learning. In: Payne, T.R., et al. (eds.) ISWC 2023. LNCS, vol. 14265, pp. 408–427. Springer, Cham (2023). https://doi.org/10.1007/978-3-031-47240-4_22
3. Bellan, P., Dragoni, M., Ghidini, C.: Extracting business process entities and relations from text using pre-trained language models and in-context learning. In: Almeida, J.P.A., Karastoyanova, D., Guizzardi, G., Montali, M., Maggi, F.M., Fonseca, C.M. (eds.) EDOC 2022. Lecture Notes in Computer Science, vol. 13585, pp. 182–199. Springer, Cham (2022). https://doi.org/10.1007/978-3-031-17604-3_11
4. Bellan, P., Dragoni, M., Ghidini, C., van der Aa, H., Ponzetto, S.P.: Process extraction from text: benchmarking the state of the art and paving the way for future challenges. arXiv preprint arXiv:2110.03754 (2021)
5. Bischof, S., Filtz, E., Parreira, J.X., Steyskal, S.: LLM-based guided generation of ontology term definitions. In: Proceedings of EKAW 2024 - Industry Track. Springer, Cham (2024)
6. Brown, T., et al.: Language models are few-shot learners. In: Advances in Neural Information Processing Systems, vol. 33, pp. 1877–1901 (2020)
7. Chan, C.M., et al.: ChatEval: towards better LLM-based evaluators through multi-agent debate. arXiv preprint arXiv:2308.07201 (2023)
8. Ciroku, F., de Berardinis, J., Kim, J., Meroño-Peñuela, A., Presutti, V., Simperl, E.: RevOnt: reverse engineering of competency questions from knowledge graphs via language models. J. Web Semant. 100822 (2024)
9. Ding, N., et al.: Few-nerd: a few-shot named entity recognition dataset. arXiv preprint arXiv:2105.07464 (2021)
10. Dubois, Y., et al.: AlpacaFarm: a simulation framework for methods that learn from human feedback. In: Advances in Neural Information Processing Systems, vol. 36 (2024)
11. Fathallah, N., Das, A., De Giorgis, S., Poltronieri, A., Haase, P., Kovriguina, L.: Neon-GPT: a large language model-powered pipeline for ontology learning. In: Proceedings of ESWC 2024 - Special Track on Large Language Models for Knowledge Engineering (2024)
12. Frey, J., Meyer, L.P., Arndt, N., Brei, F., Bulert, K.: Benchmarking the abilities of large language models for RDF knowledge graph creation and comprehension: how well do LLMs speak turtle?. arXiv:2309.17122 (2023)
13. Fu, J., Ng, S.K., Jiang, Z., Liu, P.: GPTscore: evaluate as you desire. arXiv preprint arXiv:2302.04166 (2023)
14. Garijo, D., Gil, Y.: Augmenting PROV with plans in P-PLAN: scientific processes as linked data. In: Kauppinen, T., Pouchard, L.C., Keßler, C. (eds.) Proceedings of LISC2012 Workshop - Co-located with ISWC 2012. CEUR Workshop Proceedings, vol. 951. CEUR-WS.org (2012)
15. Groth, P., Simperl, E., van Erp, M., Vrandečič, D.: Knowledge graphs and their role in the knowledge engineering of the 21st century (Dagstuhl Seminar 22372). DROPS-IDN/v2/document/10.4230/DagRep.12.9.60 (2023). https://doi.org/10.4230/DagRep.12.9.60, publisher: Schloss-Dagstuhl - Leibniz Zentrum für Informatik
16. Hayes, A.F., Krippendorff, K.: Answering the call for a standard reliability measure for coding data. Commun. Methods Meas. $\mathbf{1}(1)$, 77–89 (2007)

17. Hidalgo, C.A., Orghian, D., Canals, J.A., De Almeida, F., Martin, N.: How Humans Judge Machines. MIT Press, Cambridge (2021)
18. Hofer, M., Frey, J., Rahm, E.: Towards self-configuring knowledge graph construction pipelines using LLMs - a case study with RML. In: Proceedings of KGCW 2024 co-located with ESWC2024. CEUR Workshop Proceedings, vol. 3718. CEUR-WS.org (2024)
19. Hoseini, S., Burgdorf, A., Paulus, A., Meisen, T., Quix, C., Pomp, A.: Towards LLM-augmented creation of semantic models for dataspaces. In: Proceedings of SDS 2024 Workshop Co-located with ESWC 2024 (2024)
20. Kamoi, R., Goyal, T., Rodriguez, J.D., Durrett, G.: Wice: real-world entailment for claims in Wikipedia. arXiv preprint arXiv:2303.01432 (2023)
21. Kumar, A., Pandey, A., Gadia, R., Mishra, M.: Building knowledge graph using pre-trained language model for learning entity-aware relationships. In: Proceedings of 2020 IEEE International Conference on Computing, Power and Communication Technologies (GUCON), pp. 310–315 (2020). https://doi.org/10.1109/GUCON48875.2020.9231227
22. Kumar, A., Gupta, N., Dana, S.: Constructing micro knowledge graphs from technical support documents. In: Verborgh, R., et al. (eds.) ESWC 2021. LNCS, vol. 12739, pp. 249–253. Springer, Cham (2021). https://doi.org/10.1007/978-3-030-80418-3_37
23. Lee, M., et al.: Evaluating human-language model interaction. arXiv preprint arXiv:2212.09746 (2022)
24. Li, M., et al.: Coannotating: uncertainty-guided work allocation between human and large language models for data annotation. arXiv preprint arXiv:2310.15638 (2023)
25. Li, X., et al.: Knowledge-centric prompt composition for knowledge base construction from pre-trained language models. In: KBC-LM/LM-KBC@ ISWC (2023)
26. Liu, Y., Iter, D., Xu, Y., Wang, S., Xu, R., Zhu, C.: G-Eval: NLG evaluation using GPT-4 with better human alignment. arXiv preprint arXiv:2303.16634 (2023)
27. Liu, Y., et al.: Revisiting the gold standard: grounding summarization evaluation with robust human evaluation. arXiv preprint arXiv:2212.07981 (2022)
28. Magni, F., Park, J., Chao, M.M.: Humans as creativity gatekeepers: are we biased against AI creativity? J. Bus. Psychol. 1–14 (2023)
29. Marreddy, M., Oota, S.R., Gupta, M., Flek, L.: Large language models are human-like annotators (2024). https://sites.google.com/view/lllms-as-human-like-annotators/. Tutorial at KR 2024
30. McKight, P.E., Najab, J.: Kruskal-Wallis test. The Corsini Encyclopedia of Psychology, p. 1 (2010)
31. Mihindukulasooriya, N., Tiwari, S., Enguix, C.F., Lata, K.: Text2KGBench: a benchmark for ontology-driven knowledge graph generation from text. In: Payne, T.R., et al. (eds.) ISWC 2023. LNCS, vol. 14266, pp. 247–265. Springer, Cham (2023). https://doi.org/10.1007/978-3-031-47243-5_14
32. Millet, K., Buehler, F., Du, G., Kokkoris, M.D.: Defending humankind: anthropocentric bias in the appreciation of AI art. Comput. Hum. Behav. **143**, 107707 (2023)
33. Nabizadeh, N., Kolossa, D., Heckmann, M.: Myfixit: an annotated dataset, annotation tool, and baseline methods for information extraction from repair manuals. In: Proceedings of LREC 2020, pp. 2120–2128. European Language Resources Association (2020)

34. Neudert, L.M., Knuutila, A., Howard, P.N.: Global Attitudes Towards AI, Machine Learning & Automated Decision Making. Oxford Commission on AI and Good Governance, Oxford (2020)
35. Ni, J., Shi, M., Stammbach, D., Sachan, M., Ash, E., Leippold, M.: Afacta: Assisting the annotation of factual claim detection with reliable llm annotators. arXiv preprint arXiv:2402.11073 (2024)
36. Palan, S., Schitter, C.: Prolific.ac a subject pool for online experiments. J. Behav. Exp. Finance **17**, 22–27 (2018)
37. Pan, J.Z., et al.: Large language models and knowledge graphs: opportunities and challenges. DROPS-IDN/v2/document/10.4230/TGDK.1.1.2 (2023). https://doi.org/10.4230/TGDK.1.1.2
38. Pan, S., Luo, L., Wang, Y., Chen, C., Wang, J., Wu, X.: Unifying large language models and knowledge graphs: a roadmap. IEEE Trans. Knowl. Data Eng. 1–20 (2024)
39. Ragot, M., Martin, N., Cojean, S.: AI-generated vs. human artworks. a perception bias towards artificial intelligence? In: Extended Abstracts of the 2020 CHI Conference on Human Factors in Computing Systems, pp. 1–10 (2020)
40. Rula, A., Calegari, G.R., Azzini, A., Bucci, D., Baroni, I., Celino, I.: Eliciting and curating procedural knowledge in industry: challenges and opportunities. In: Proceedings of (Qurator 2022). CEUR Workshop Proceedings, vol. 3234. CEUR-WS.org (2022)
41. Rula, A., D'Souza, J.: Procedural text mining with large language models. In: Proceedings of the 12th Knowledge Capture Conference 2023. K-CAP '23, pp. 9–16. Association for Computing Machinery, New York, NY, USA (2023). https://doi.org/10.1145/3587259.3627572
42. Rula, A., Re Calegari, G., Azzini, A., Baroni, I., Celino, I.: K-hub: a modular ontology to support document retrieval and knowledge extraction in industry 5.0. In: Pesquita, C., et al. (eds.) ESWC 2023. LNCS, vol. 13870, pp. 454–470. Springer, Cham (2023). https://doi.org/10.1007/978-3-031-33455-9_27
43. Shi, S., Xu, Z., Hu, B., Zhang, M.: Generative multimodal entity linking. In: Proceedings of LREC-COLING 2024, pp. 7654–7665. ELRA and ICCL (2024)
44. Sola, D., van der Aa, H., Meilicke, C., Stuckenschmidt, H.: Activity recommendation for business process modeling with pre-trained language models. In: Pesquita, C., et al. (eds.) ESWC 2023. LNCS, vol. 13870, pp. 316–334. Springer, Cham (2023). https://doi.org/10.1007/978-3-031-33455-9_19
45. Su, H., et al.: Selective annotation makes language models better few-shot learners. arXiv preprint arXiv:2209.01975 (2022)
46. Wei, J., et al.: Chain-of-thought prompting elicits reasoning in large language models. In: Advances in Neural Information Processing Systems, vol. 35, pp. 24824–24837 (2022)
47. Yang, D., Wang, X., Celebi, R.: Expanding the vocabulary of BERT for knowledge base construction (2023). https://doi.org/10.48550/arXiv.2310.08291
48. Yuan, P., et al.: BatchEval: towards human-like text evaluation. arXiv preprint arXiv:2401.00437 (2023)
49. Zhang, B., et al.: OntoChat: a framework for conversational ontology engineering using language models. In: Proceedings of ESWC 2024 - Special Track on Large Language Models for Knowledge Engineering (2024)
50. Zhang, B., Reklos, I., Jain, N., Peñuela, A.M., Simperl, E.: Using large language models for knowledge engineering (LLMKE): a case study on Wikidata (2023). https://doi.org/10.48550/arXiv.2309.08491

51. Zhang, L.: Reasoning about procedures with natural language processing: a tutorial (2022). https://doi.org/10.48550/arXiv.2205.07455, publication Title: arXiv e-prints ADS Bibcode: 2022arXiv220507455Z

52. Zhang, L., Lyu, Q., Callison-Burch, C.: Reasoning about goals, steps, and temporal ordering with WikiHow. In: Proceedings of the 2020 Conference on Empirical Methods in Natural Language Processing (2020). https://doi.org/10.18653/v1/2020.emnlp-main.374

53. Zhou, Y., Shah, J., Schockaert, S.: Learning household task knowledge from WikiHow descriptions. In: Espinosa-Anke, L., Declerck, T., Gromann, D., Camacho-Collados, J., Pilehvar, M.T. (eds.) Proceedings of SemDeep-5 2019 Workshop, pp. 50–56. Association for Computational Linguistics, Macau, China (2019). https://aclanthology.org/W19-5808

54. Zlatkov, D., Ens, J., Pasquier, P.: Searching for human bias against AI-composed music. In: Johnson, C., Rodríguez-Fernández, N., Rebelo, S.M. (eds.) EvoMUSART 2023. LNCS, vol. 13988, pp. 308–323. Springer, Cham (2023). https://doi.org/10.1007/978-3-031-29956-8_20

Modelling and Mining Knowledge About Computational Complexity

Anton Gnatenko[1]([✉]) [iD], Oliver Kutz[1] [iD], and Nicolas Troquard[2] [iD]

[1] Free University of Bozen-Bolzano, Bolzano, Italy
{agnatenko,oliver.kutz}@unibz.it
[2] Gran Sasso Science Institute, L'Aquila, Italy
nicolas.troquard@gssi.it

Abstract. We present an ontology of computational complexity that allows for a representation of research findings on the subject and supports query answering and reasoning tasks to help students and researchers in finding known facts and deriving new ones. The facts about decision problems and complexity classes are organised as a knowledge graph. The relationships between them are axiomatised using the FOWL framework that allows one to combine OWL 2 and first-order logic to balance between reasoning efficiency and expressive power.

While the axioms were created through ontological analysis based on received 'textbook knowledge', the facts were extracted from the textual corpus of the 'Complexity Zoo' website, a human-curated 'encyclopedia' of complexity classes, in a human-supervised process employing large language models. We discuss, on the one hand, the modelling choices in relation to the previous work on knowledge representation in mathematics and, on the other hand, the peculiarities of using language models for the mining of complex symbolic facts. Finally, we illustrate some of the features of the hybrid reasoning system by providing a usage example.

Keywords: Domain Modelling · Knowledge Mining · Representation of Mathematical Knowledge · LLM · Computational Complexity

1 Introduction

Computer science, as a saying goes, is no more about computers than astronomy is about telescopes.[1] Theoretical computer science deals with two questions: (i) what can be computed and what can not, and (ii) if something can be computed, how hard it is to do so? The former question originated from Hilbert and Ackermann's 1928 decision problem (Entscheidungsproblem): is there a general algorithm that, given a first-order sentence, tells whether it is a theorem of first-order logic. The Entscheidungsproblem was systematically addressed by

---

[1] This is usually attributed to E. Dijkstra, although the story is more involved (quoteinvestigator.com/2021/04/02/computer-science).

M. Alam et al. (Eds.): EKAW 2024, LNAI 15370, pp. 453–470, 2025.
https://doi.org/10.1007/978-3-031-77792-9_27

Church [5] and Turing [35], who provided a formal meaning to the term 'algorithm': Church by developing his λ-calculus and Turing by defining a class of machines that now bear his name. The eventual finding was that, for the Entscheidungsproblem, no algorithm existed.

More generally, a *decision problem* is a function that, given an input, returns 'yes' or 'no'. Following the results of Church and Turing, the decision problems were classified into two types: *decidable*, that is, those for which there is a general algorithm to determine the answer, and *undecidable* (like the theoremhood of first-order logic formulae) for which no such algorithm exists. The rise in the use of computers led to further refinement of this classification. Indeed, some of the decidable problems appeared to necessitate more computational resources than others, rendering them 'harder to solve' in practice. Hartmanis and Stearns [10] proposed to measure the amount of time used by an algorithm as a function of the size of the input. These ideas marked the origin of the *theory of computational complexity*, or *complexity theory*, for short, that studies decision problems based on which and 'how much' of certain computational resources are required to solve them. Decision problems were divided into complexity classes, that is, collections of problems that feature the same inherent complexity. The results in the field range from establishing general properties of whole families of complexity classes to discovering concrete facts about relationships between two classes or between a class and a problem. Since 1965, our understanding of computational complexity has grown tremendously into a refined and detailed web of knowledge. Diverse models of computation (deterministic, nondeterministic, and probabilistic machines, alternating and parallel computations, Boolean and quantum circuits, interactive protocols, etc.) as well as different measures of complexity (time and space, circuit size and depth, the number of random bits used in a computation, the number of messages exchanged between communication agents, and many more) were introduced and studied in-depth. These results have found numerous applications in the theoretical foundations of many areas of computer science: design and analysis of algorithms, databases, machine learning, cryptography, etc.

However, this totality of complexity theoretic results needs a unified representation accessible by all working computer scientists. Textbooks usually cover only the basics (plus an eclectic choice of additional results) and are difficult to keep up to date. New results are scattered across various research papers that a rare expert may claim to have read all. Searching among these diverse sources is not convenient. Something else was needed, and it was created in the form of the Complexity Zoo, in the following called 'Zoo' for brevity, a web-encyclopaedia of complexity classes. In a welcome statement[2], its creator Scott Aaronson writes:

> "Some theorists seem able to hold in their minds, in one instant, the results of every FOCS, STOC, and Complexity[3] paper ever published. [...] I am not one of those theorists. The sprawling web of known relations among complexity classes—containments, oracle separations, random oracle separations, lowness results, the occasional inequality—is not fixed in my

---

[2] complexityzoo.net/Complexity_Zoo_Introduction.
[3] The leading conferences in the field.

memory like the English language. And so it's largely for my own benefit that I recorded a chunk of what's known in one unwieldy HTML file."

The Zoo is indeed a valuable reference for many computer scientists, including the authors of this paper. However, it is plain text and thus has all the associated drawbacks. First, the only way to query the Zoo data is textual search: even though the text contains information on relations between classes, one can not use relational queries to extract them. Second, it is prone to errors and typos, as the consistency of the information presented there cannot be checked automatically. Third, the Zoo is only a reference for humans, and the theorems described there cannot be used by an automated reasoner.

We are making a step towards creating a *semantic* Complexity Zoo, an ontology populated with decision problems and complexity classes with lemmas and theorems about them formalised by the use of axioms. In fact, some complexity phenomena, such as infinite hierarchies of classes, *require* ontological axioms to be represented, since such results cannot be fully represented just in a dataset. A modest target is a knowledge base capable of answering simple queries and making visualisations, helping students in their studies and researchers in their work. An ambitious goal is creating a system that can help discovering, by automated or computer-aided reasoning, previously unknown facts of the complexity world that follow from the existing results, but that have been overlooked so far.

In this article, we present the first version of Ontoplex[4], our *onto*logy of computational com*plex*ity, which allows one to employ off-the-shelf reasoners for query answering, consistency checking, and inference tasks. It is based on OWL 2 [9] and first-order logic, unified under the framework of FOWL [8], which allows alternating between OWL 2 reasoners and more powerful, but more resource-demanding, first-order solvers. The challenges that arise when building an ontology of this kind are two-fold. First, we face the standard modelling dilemma of expressing more content versus keeping the reasoning tasks tractable. The choice is even more difficult in the case of a domain whose description already comes in the form of axioms, sometimes inexpressible in popular fragments of first-order logic. Second, given the nature of the subject, the task of populating the ontology with objects and facts requires a considerable level of mathematical knowledge.

The first challenge is addressed in Sect. 2. We briefly describe the domain, formulate the competency questions for our ontology, review the literature on formal representation of mathematical knowledge, and describe the modelling approach that we adopt. We conclude the section with a discussion of the schema and axioms of Ontoplex. In Sect. 3, we deal with the second challenge, namely how to populate the ontology with facts. We test and discuss the usage of large language models for fact mining from the original Complexity Zoo texts. Finally, in Sect. 4, we provide examples and use cases for the Ontoplex approach and conclude in Sect. 5 with a summary and discussion.

---

2   Ontological Modelling of Computational Complexity

2.1   The Domain of Complexity Theory

For a detailed presentation of the domain, the reader is invited to consult one of the standard textbooks [1,17,29]. Here, we outline the main definitions and highlight phenomena that pose a challenge for ontological modelling.

We will rely on the concept of Turing machines, which are the most common model of computation, for the definition of the term 'algorithm'. The primary definitions are those of decision problems and complexity classes. Let f be a mapping $\{0,1\}^* \rightarrow \{0,1\}^*$ of finite bit-strings to finite bit-strings (which are traditionally called 'words'). A Turing machine M computes a function f if on any input word w it outputs $f(w)$. The *time complexity* of M is a function $T_M \colon \mathbb{N} \rightarrow \mathbb{N}$, such that on an input word of length n the machine performs at most $T_M(n)$ steps of computation. *Space complexity* is a similar function that bounds the number of tape cells (i.e. the memory space) sufficient to process any input of length n. Apart from (standard) deterministic Turing machines, these notions can be defined for machines equipped with additional powers, such as nondeterministic machines (that can make correct 'guesses' in the process of computation), or probabilistic machines (that employ randomised computations), etc. The fundamental observation of Hartmanis and Stearns [10] is that bit-string functions themselves have a kind of inherent time complexity, so that no machine that computes such a function can do it faster. Formally, the time complexity of a function f is defined as the minimal time complexity of an algorithm that computes it. An analogous definition is given for the so-called space complexity, which corresponds to the minimal required amount of memory.

When it comes to studying inherent complexity, theorists usually consider a simpler class of computational problems—decision problems. A *decision problem* is a bit-string function whose output consists of just one bit, i.e. a mapping of the form $a \colon \{0,1\}^* \rightarrow \{0,1\}$. To compute a means to 'decide' whether the input word yields '1' as the answer. Thus, a can be seen as a characteristic function of a set of words, and, by slight abuse of notation, we write $w \in a$ whenever $a(w) = 1$, and $w \notin a$, otherwise. Here is an example of a decision problem called SAT: given a Boolean formula, decide whether it is satisfiable.

Complexity classes are sets of decision problems. For example, P is the class of problems decidable by a deterministic Turing machine in polynomial time. Another class, NP is defined in the same way for non-deterministic Turing machines. It is known that SAT belongs to NP, but whether it is in P constitutes a cornerstone question of complexity theory. Indeed, if SAT $\in$ P, then P $=$ NP [6]. This example shows the challenging nature of ontological modelling in this domain. While the former relationship, i.e. SAT $\in$ NP, can be represented fairly easily in a knowledge graph, the latter relationship is conditional, so conditional reasoning is needed to work with it.

Moreover, this conditional relationship is, in fact, a manifestation of a deeper phenomenon, crucial in complexity theory: the existence of NP-hard problems. This is how it is defined: A problem a is *reducible* to a problem b if there is a function r such that for every word w we have $w \in a \iff r(w) \in b$ (Karp reducibility [15]). The function r is called a *reduction*. Thus, if we have an algorithm for b and another one that computes r, we can combine them to solve a. The complexity of this compound solution, however, will depend on the respective complexities of the two components. Therefore, reducibility in complexity theory is a ternary relation: it features a, b, and the complexity of computing the reduction between them. Standard 'textbook' reductions are usually computable by deterministic Turing machines in polynomial time or logarithmic space [1], but many other types of reductions appear in specific fields, such as first-order reductions in database theory [19]. Furthermore, a problem a is *hard* for a complexity class A with respect to a type of reduction, if every problem in A is reducible to a by a reduction of that type. The relation 'a is hard for A with respect to the class of reductions R', is therefore also a ternary one. It is thus beyond the power of OWL and similar 'binary' languages to fully represent those theorems of complexity theory that talk about Karp reducibility and hardness.

Another relation of this kind is Turing reducibility. A problem a is Turing-reducible to a problem b if a can be solved by an algorithm that is allowed to make calls to an algorithm that solves b and get answers within one step of computation. We say that we solve a with an *oracle* b. A complexity class A^b is defined as the class of problems solvable within the original restrictions for class A but with access to the oracle b. This gives rise to even more rich and refined varieties of complexity classes.

Yet another complex phenomenon that we would like to mention are infinite constructions, such as hierarchies of complexity classes. For example, the polynomial-time hierarchy, where the class PH is defined as a union of two infinite chains of complexity classes, denoted Σ_n^p and Π_n^p, $n \geqslant 0$, both of which start with P, and which have various properties, ranging from 'local', such as $\Sigma_n^p \subseteq \Pi_{n+1}^p$, for all n, to 'global', e.g. if $\Sigma_n^p = \Sigma_{n+1}^p$ then $\Sigma_n^p = \text{PH}$. This latter statement, known as the 'collapse theorem' for PH, requires talking simultaneously about an infinite sequence of classes, individual classes in that sequence, and the union of all classes in that sequence, the PH itself. This PH is designed to contain exactly those decision problems contained in at least one of the classes in the sequence. An exact ontological representation of this theorem goes beyond the power of first-order logic, although there are some approximations to it that we could achieve, already in OWL 2. We highlight that stating the existence of the polynomial hierarchy, as opposed to 'standard' classes like P, requires the use of axioms, since it cannot be represented in a finite knowledge graph.[5]

The big picture of complexity theory, comprising, in our opinion, the core knowledge, is thus as follows. There are objects of different types: decision problems, complexity classes, and reductions. These objects form a refined relational structure that features 'simple' binary set-theoretic relations such as 'to be contained in', as in $\text{SAT} \in \text{NP}$, or 'to be a subset of', as in $\text{P} \subseteq \text{NP}$, along with

---

[5] Strictly speaking, P is also a hierarchy, but it is usually considered in complexity theory 'as a whole', which is not the case for PH, where classes like Σ_2^p have their own importance.

relations of higher arity, infinite chains, and systems of chains. In the following section, we summarise our modelling task in the form of competency questions.

2.2 Competency Questions

We now formulate competency questions that we consider essential for the ontology of complexity theory. Potential users of the ontology can be classified into three groups. *Students* could use the ontology to visualise and conceptualise the entities and the relations that they are learning. *Researchers* could query for facts, both stored and inferred, related to their research question. *Complexity theorists* could benefit from computer-aided inference to discover corollaries of the known results that are hard to detect 'by hand' and employ it as an assistant to find and formulate new conjectures. To accommodate these three groups, we impose two sets of requirements on our ontology, one regarding the fact that it should represent the core knowledge of complexity theory, and the other stating that it should support key reasoning tasks. These requirements are summarised in Table 1.

Table 1. Competency questions for a complexity theory ontology

#	Does it represent?	Examples
1	Set-theoretic relations	$SAT \in NP$, $P \subseteq NP$
2	Reducibility and hardness relations	3COLOR is reducible to SAT, SAT is complete for NP
3	Relations depending on a condition	if SAT is in P, then $P = NP$
4	Infinite hierarchies of classes	PH, the polynomial-time hierarchy
5	Theorems about hierarchies	the collapse theorem for PH
#	Does it support?	Examples
6	Consistency checking	Is the fact $SAT \in L$ consistent with current knowledge?
7	Ontology-mediated queries	Does $P^{SAT} \subseteq PSPACE$ follow from the current knowledge? What are the named problems that are known to be hard for NP?
8	Automated inference of new facts	Is $NP = PH$?

2.3 Related Work and Modelling Approaches

To our knowledge, complexity-theoretic phenomena have received little attention from the perspective of ontological modelling. In 2005, Sack [26] created an ontology of NP-complete problems to support bibliographic search on the topic. More recently, Bodily and Ventura [2] developed a knowledge base of reductions between NP-complete problems.

However, substantial work has been done in the more general context of mathematical KR&R. Lange [18] provides a comprehensive review of the developments in the field from the 2000s till early 2010s, featuring RDF-, XML-, and

OWL-based ontologies. The first wave of development featured digital libraries of mathematical works, such as MathNet [40], and more sophisticated systems, such as HELM [38], which allowed integration with reasoning tools. MONET [28] used Semantic Web technologies to query digital libraries of mathematics.

In general, mathematical knowledge is difficult to model within the Semantic Web paradigm for numerous reasons, the most obvious being that it is hard to describe the details of fields like algebra or geometry in terms of knowledge graphs. Further research focused on either cataloguing of mathematical notions in standard mark-up languages [4], or on pure reasoning with proof assistants like Coq [36]. Another direction was to stick with Semantic Web tools, but limit the scope to ontologies that describe entities like theorems, lemmas, proofs and research papers, rather than mathematical objects themselves [41].

Recently, the quest of describing mathematics in OWL was restarted by Elizarov et al. [7]. Their OntoMath^{PRO} defines both general concepts (Problem, Method, Statement, Formula) and actual mathematical objects (Set, Operator, Mapping, etc.). However, due to the limited expressiveness of RDF in the context of general mathematics, they have to retreat to textual annotations to give specific definitions, which limits the applicability of reasoners.

A more concrete expressiveness issue is that of relations of higher arity that are ubiquitous in mathematics, while OWL only supports binary relations directly. Lange [18] proposes to combine an RDF layer and an XML layer into one ontology, with the latter expressing what is beyond the power of the former, but admits that using information from both layers would require a separate implementation. Elizarov et al. [7] use reification for relations of higher arity. However, with reification it is hard to express properties such as

$$isHard(a, A, R) \wedge isReducible(a, b, R) \implies isHard(b, A, R) \tag{1}$$

in the OWL setting. We argue, however, that an OWL-based ontology is suitable for the objectives that we pursue, specifically also for the field of mathematics that we are focusing on, namely complexity theory, and, in general, theoretical computer science. Complexity classes are discrete objects describable by a finite number of parameters: the model of computation employed, the type of complexity measure, the bound on that measure, etc. For example, PSPACE is the class of problems decidable by a deterministic Turing machine using a polynomial amount of (memory), while NEXPTIME contains those decidable by nondeterministic machines in exponential time. Such objects can be readily described in a knowledge graph by attaching appropriate labels, in contrast to, say, object of calculus that are functions on $\mathbb{R}$.[6] Theorems of complexity theory can be formulated as structural restrictions on this graph, for which OWL and other Semantic Web formalisms were invented.

The problem of relations of higher arity remains in complexity theory. For example, formula (1) has to be expressed as: if a is hard for A with respect to reductions of type R, and a is reduced to b via a reduction of that type, then b is also hard for A. To this end, we employ FOWL. Similarly to the solution

---

[6] Note that also Aaronson refers to the set of results in the field as 'sprawling web'.

proposed by Lange [18], FOWL allows us to write axioms on two levels: that of OWL 2 and that of first-order logic. The reasoners, both for OWL and for first-order logic, are abundant, well-optimised, and ready to be used.

2.4   Axioms

We discuss the axiomatisation of the domain described in Sect. 2.1. The axioms below are written in description logic $\mathcal{SROIQ}$ [11,25], the basis of OWL 2 DL. When we need to say more than is possible in $\mathcal{SROIQ}$, we retreat to first-order logic. We provide several examples of axiomatising the phenomena mentioned in Sect. 2.1 to give the reader an idea of how it is done. The axioms can be found in full, in OWL Functional Syntax, in the Ontoplex Github repository.

Certain phenomena of complexity theory, such as the existence of infinite hierarchies, require one to treat complexity classes as individual objects rather than OWL classes. To this end, we need to distinguish between different types of objects by labelling them with disjoint concept names, e.g.

$$Prob \sqcap Class \sqsubseteq \bot \tag{2}$$

For the same reason, we simulate the set-theoretic relations with binary relations ('roles' in the description logic jargon), such as *in* and *subset*, that are axiomatised according to the desired behaviour. For example, the relation 'being a subset of' is transitive, and the relation 'being contained in' is distributed over the subset relation:

$$subset \circ subset \sqsubseteq subset \qquad\qquad in \circ subset \sqsubseteq in \tag{3}$$

Some 'sanity' constraints are also useful, e.g. that if $a \in A$, then a must be a decision problem, and A a complexity class:

$$\exists in.\top \sqsubseteq Prob \qquad\qquad \exists in^-.\top \sqsubseteq Class \tag{4}$$

Already at this point, we need expressiveness beyond the power of $\mathcal{SROIQ}$ to define the relation of equality for two classes. Partially, we can axiomatise it in $\mathcal{SROIQ}$ saying that $A = B$ implies both $A \subseteq B$ and $B \subseteq A$:

$$equals \sqsubseteq subset \qquad\qquad equals \sqsubseteq subset^- \tag{5}$$

However, the converse statement, that $A \subseteq B$ and $B \subseteq A$ *together* imply the equality, is forbidden in $\mathcal{SROIQ}$. We thus add a first-order axiom:

$$\forall x,y \,.\, equals(x,y) \iff subset(x,y) \wedge subset(y,x) \tag{6}$$

We can add more interesting set-theoretic relations, using the negation that is available on the first-order level, e.g. the notion of a *strict subset*:

$$\forall x,y \,.\, strictSubset(x,y) \implies \neg equals(x,y) \tag{7}$$
$$\forall x,y \,.\, strictSubset(x,y) \implies \exists z \,.\, in(z,y) \wedge notIn(z,x) \tag{8}$$

However, some connections between *strictSubset* and *subset* is still expressible in $\mathcal{SROIQ}$, such as that a strict subset is still a subset:

$$strictSubset \sqsubseteq subset \qquad (9)$$

The latter is useful when we work with data instances involving strict subsets but want to perform reasoning on the $\mathcal{SROIQ}$ level, using as much information as possible.

We use the same trick of combining $\mathcal{SROIQ}$ and first-order axioms to give a coherent treatment to non-binary relations of reducibility and hardness. First, we define the relation *isReducible(a, b, R)*, where R is a type of reductions in first-order. Then, in $\mathcal{SROIQ}$, we define subcases of this relation for particular types of reductions that are popular in the literature, such as *isReducible$_{Log}$* and *isReducible$_{Poly}$*, reductions computable in logarithmic space or in polynomial time. In $\mathcal{SROIQ}$, we axiomatise particular properties of these relations (e.g. both *isReducible$_{Log}$* and *isReducible$_{Poly}$* are transitive, but this is not true for all projections of *isReducible(a, b, R)* on the third component):

$$isReducible_F \circ isReducible_F \sqsubseteq isReducible_F, \text{ for } F \in \{Log, Poly\} \qquad (10)$$
$$isReducible_{Log} \sqsubseteq isReducible_{Poly} \qquad (11)$$

These roles can be connected back to their first-order original as follows:

$$\forall x, y \ . \ isReducible_{Log}(x, y) \iff isReducible(x, y, Log) \qquad (12)$$
$$\forall x, y \ . \ isReducible_{Poly}(x, y) \iff isReducible(x, y, Poly) \qquad (13)$$

Constructions such as the infinite polynomial-time hierarchy can be described in $\mathcal{SROIQ}$ via intricate role axioms. We first create a chain of hierarchy levels, where the ith level is represented by an object ℓ_i, and the objects ℓ_k and ℓ_{k+1} are connected by the specific role *next*, which is functional in both directions and has the property of creating an infinite chain:

$$\geqslant 2 \ next.\top \sqsubseteq \bot \qquad \geqslant 2 \ next^-.\top \sqsubseteq \bot \qquad \exists next^-.\top \sqsubseteq \exists next.\top \qquad (14)$$

We start the chain at the level ℓ_0 by assertions $\exists next(\ell_0)$ and $\neg \exists next^-(\ell_0)$. To access any object in the chain, say ℓ_3, it suffices to list all the previous:

$$next(\ell_0, \ell_1) \qquad next(\ell_1, \ell_2) \qquad next(\ell_2, \ell_3) \qquad (15)$$

The respective classes Σ_k^p and Π_k^p are attached to ℓ_k by dedicated roles *sigma* and *pi*. For example, we will have P attached to ℓ_0 by both of these roles. Then the property $\Sigma_k^p \subseteq \Pi_{k+1}^p$ is expressed as 'if you go along *sigma*, then along *next*, and then step back via *pi*, you get a subset relation':

$$sigma \circ next \circ pi \sqsubseteq subset \qquad (16)$$

Many other things can be expressed about this infinite chain. For example, every Σ_k^p and Π_k^p is a subset of PSPACE, the class of problems that are decidable in polynomial space. To express it, we add a special connection between ℓ_0 and PSPACE: $connection(\ell_0, \text{PSPACE})$ and propagate this connection along the chain, so that if ℓ_k has it with PSPACE, then ℓ_{k+1} also has it:

$$next^- \circ connection \sqsubseteq connection \qquad (17)$$

Finally, say that this connection means the subset relation for the classes on that level:

$$sigma \circ connection \sqsubseteq subset \qquad (18)$$
$$pi \circ connection \sqsubseteq subset \qquad (19)$$

Note that all this is expressed in $\mathcal{SROIQ}$, without the use of first-order tools. We believe that many properties of hierarchies can be represented in this way. However, some things remain unreachable, even within first-order logic. Namely, defining an object PH that is, as a complexity class, the union of all classes in the polynomial hierarchy. In FOWL, only a partial solution is possible. Similarly to the case of PSPACE, we can state that $\Sigma_k^p, \Pi_k^p \subseteq \text{PH}$ for all k. However, it remains to say that PH is contained in the union of Σ_k^p. That is, we want to express that for any a, if $in(a, \text{PH})$ is true, then there exists such k that Σ_k^p contains a. Expressing this would require a second-order layer on top of FOWL, and we leave it for future development.

3   Fact Mining

Having formulated the axioms, we face the task of populating the ontology with *individuals* and *facts*. By individuals we mean concrete decision problems and complexity classes: $\text{SAT}, \text{P}, \text{NP}, \text{PSPACE}$, etc. By facts we intend relational data atoms, or ABox assertions in description logic terminology. Examples are $subset(\text{P}, \text{NP})$, $in(\text{SAT}, \text{NP})$, $isHard_{Log}(\text{SAT}, \text{P})$, etc.

The number of facts known in complexity theory is large enough to make this task laborious: Complexity Zoo has roughly 3MB of plain text, most of which are sentences describing one or more relation instance at once, as in:

$$\text{"Important subclasses of P include L, NL, NC, and SC"}, \qquad (20)$$

which yields $subset(\text{L}, \text{P})$, $subset(\text{NL}, \text{P})$, $subset(\text{NC}, \text{P})$, $subset(\text{SC}, \text{P})$.

We attempt to solve this task with the help of natural language processing tools based on neural networks. Unlike standard textbooks [1, 17, 29] or research papers, which contain explanatory remarks, general suggestions, examples, and complex mathematical formulae, Complexity Zoo seems a good fit for automated text processing. It is plain text and mostly consists of sentences of simple structure that speak directly about complexity theory facts. We thus focus on extracting individuals and facts from the Complexity Zoo texts.

3.1   Related Work

Neural network-based approaches to ontology building or relation extraction from text can be classified into two broad domains: using models trained specifically for relation extraction, or employing large language models (LLMs) [44]. There are various models specialised in extraction of entities and relations. Some have a general scope, such as REBEL [13] and its successor mREBEL [14], trained on Wikipedia texts, and some are tuned to concrete domains, such as medicine [30]. These models were usually trained to extract relations from a specific schema (e.g. REBEL has that of WikiData), and adapting them to new domains or subdomains requires fine-tuning, i.e. additional training on a dataset from the target domain. LLMs, given their size, are expensive to fine-tune but benefit from general 'text understanding' and can adapt their behaviour after seeing a few examples and being provided with additional advice from a carefully designed prompt. Mateiu and Groza [20] exposed GPT-3 [3] to a set of 150 examples of prompt-to-OWL translations and achieved promising performance of automatically building ontologies from simple sentences like "Anna is a girl", "Anna and Lena are sisters", etc. In a more recent work, Saeedizade and Blomqvist [27] conducted a comprehensive study of performance of several LLMs (various versions of Llama [33] and Llama2 [34], other open-source models, as well as proprietary models Gemini [32], GPT-3 and 4 [23]), using an elaborate prompt structure. Furthermore, they analyse several interactive strategies that allow a human expert to guide the model in performing the task step be step. Their key findings are the following. First, when tasked with OWL code generation from the given text, GPT-3 and 4 outperform all other tested models according to the evaluation criteria they provide. Second, step-by-step interactive approaches generally yield better results than zero-shot prompting when the output to the initial prompt is taken as the final result. Finally, the model outputs still contain minor errors, so a supervision and correction by a human expert are needed to achieve good results.

3.2   The Case of the Complexity Zoo

Despite the 'simplicity' of the Zoo's text as described above, it is still complicated enough to pose challenges for automated processing. The subject is an abstract and rigorous mathematical theory, so a slight alternation in presentation, say due to a hallucination of an LLM [12], can ruin the meaning and render the ontology inconsistent. On the other hand, these are still informal natural language texts, which sometimes leave precisification as a task for the reader. To see an example, go back to the sentence (20). It says that L is a *subclass* of P, while we use the relation *subset*. There are numerous other ways to state this relation ('subsumed by', 'contained in', 'is in', etc.) which have to be dealt with. Moreover, it is stated that NC is a subclass of P. A knowledgeable reader would understand that this is about the *uniform* NC (with a version of uniformity appropriate for the situation), since the nonuniform version of NC is not contained in P. Mathematicians frequently sweep this kind of things under the carpet, claiming

Table 2. The structure of the master prompt.

Context:	This is a text about objects of OWL classes "Decision Problem" and "Complexity Class".
Task:	Extract: (i) Named individuals of classes "Decision Problem" and "Complexity Class". (ii) Object properties of them.
Requests:	Provide output in OWL 2 Functional Syntax. Use temperature = 0.1
Warnings:	If a complexity class A is contained in a complexity class B, use object property :isSubClassOf. If a decision problem D is contained in a complexity class A, use object property :isContainedIn.
Examples:	Input: \<Example input\>, Output: \<Expected Output\>
Text:	The ICD

that they are 'clear from the context'. Apart from that, the Zoo contains purely informal and even humorous remarks, as in "NP: The class of dashed hopes and idle dreams", which are irrelevant for the task and may easily put the model off-track. These considerations imply that obtaining a large enough dataset for fine-tuning a model like REBEL would in our case be no easier than just processing the whole Zoo by hand. We thus have to retreat to the use of LLMs.

The text of Complexity Zoo is divided into chunks of 50–150 sentences each, describing a single complexity class and its known relationships with other classes and problems. We call such chunks *individual class descriptions*, or ICDs. Our approach relies on building a small OWL 2 ontology out of each ICD. These ontologies can be later modularly combined together, provided that they share the same schema or at least that their schemas are easily translatable to one another. Adopting one of the techniques described by Saeedizade and Blomqvist [27], we work with models GPT4-o [37] and Gemini 1.5 Pro [31]. The processing of each ICD consists of a *task formulation* and a *correction phase*. The task formulation is given by the *master prompt* which introduces the domain (complexity theory, featuring named individuals of types 'decision problem' and 'complexity class'), poses the task (provides the text and asks to extract facts out of it), sets up the output format (e.g. OWL functional syntax) and the desired temperature parameter (lower values correspond to more regular answers, higher to more 'creative'), and warns of common pitfalls, such as using the same relation *isContainedIn* for SAT $\in$ NP and P $\subseteq$ NP. Thus, the master prompt for each ICD consists of six parts, given in Table 2.

Given such a prompt, the models were usually able to extract most of the mentioned complexity classes and decision problems, correctly declare them as named individuals of the appropriate OWL class, extract and declare object properties such as *isSubClassOf* or *isContainedIn*, correctly restrict their domain and range, and finally list the instances of these relations mentioned in the text. Table 3 provides an example of the resulting list.

However, the ontologies produced in response to the master prompt, apart from very simple cases, contained a number of errors and needed careful human assessment. Those errors can be roughly classified into *syntactic*, such as new names for the same relations or creating a reversed relation instead of a direct one, and *semantic*, featuring problems with understanding the meaning of the text. Examples are provided in Table 4.

Table 3. (A part of) the ICD for class L and object properties extracted by GPT4-o.

L: Logarithmic Space. Is contained in NC1 and in contained in generalisations including SL, parityL, and Mod_kL. Remarkably, L = SL.	`ObjectPropertyAssertion(:isSubClassOf :L :NC1)` `ObjectPropertyAssertion(:isSubClassOf :L :SL)` `ObjectPropertyAssertion(:isSubClassOf :L :ParityL)` `ObjectPropertyAssertion(:isSubClassOf :L :ModkL)` `ObjectPropertyAssertion(:isEqualTo :L :SL)`

Table 4. Output errors examples

Type	Description	Examples
Syntactic	OWL syntax error	Wrong output syntax, a forgotten prefix, etc.
Syntactic	Wrong relation name	Using *subset* and *isSubClassOf* for the same relation when processing different ICDs.
Syntactic	Wrong relation direction	Using Contains(NP, SAT) instead of *isContainedIn*(SAT, NP).
Semantic	Treating conditionals as affirmative	Extracting *equals*(NP, P) from the sentence 'If P = NP, then...'
Semantic	General misunderstanding	Extracting *subset*(AM, NP) from 'Generalisations of NP include AM and MA'.

While syntactic errors were generally easy to detect, fix and avoid in the future by appending the 'warning' section of the master prompt, the semantic errors tend to be more persistent with both models continuing to infer *subset*(AM, NP) even when specifically told in the prompt that generalisations usually contain the base class, and provided explicit examples. They are also easy to oversee and hard to detect when doing a consistency check of the ontology, as opposed, for example, to misusing *in* in place of *subset*, which is caught by a domain and range restrictions check.

LLMs still significantly reduce the workload on the task of obtaining facts by mining *fairly* correct facts with a *practically* correct schema. However, careful supervision is necessary so that the resulting OWL code conforms to the general schema and to the semantics of the domain. By now, we have curated 7 single-class ontologies for the most popular complexity classes: L, NL, P, NP, PSPACE, EXPTIME, NEXPTIME and an ontology extracted from the general description of the domain provided on the page called 'Petting Zoo'[7]. These files can be found on the Github Ontoplex repository. It remains a subject of future work to develop a less error-prone methodology of fact mining for complexity theory.

4   Working with Ontoplex

Ontoplex is available as an alpha-version on Github[8]. It is a FOWL ontology, i.e. an OWL 2 DL ontology with additional first-order axioms presented as OWL

---

[7] https://complexityzoo.net/Petting_Zoo.
[8] https://github.com/gnatenko/ontoplex/.

annotations (see [8] for details). It can be used either as a standard OWL 2 knowledge base, for more efficient reasoning, or as a first-order knowledge base, for a more detailed picture of the domain representation.

Through standard tools such as Protégé [22] and HermiT [39], Ontoplex supports description logic queries to its data. A simple query may be to show all known problems that are hard for a certain class. A less trivial query would involve some reasoning, e.g. one may want to see all complexity classes known to be between NL and PSPACE. To answer such a query, the reasoner will have to take into account the transitivity of the relation *subset*, given in the axioms (Fig. 1).

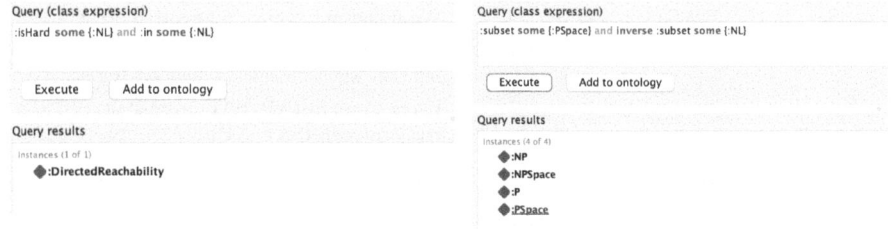

Fig. 1. DL Queries to Ontoplex with Protégé

Yet less trivial tasks can be performed with a first-order reasoner, such as Vampire [24]. Scott Aaronson says: "...I spent a week trying to put AM outside QMA relative to an oracle, only to learn that this followed trivially from two known results: that of Vereshchagin [42] that AM is outside PP relative to an oracle, and that of Kitaev and Watrous (unpublished, but mentioned in [43]) that QMA is in PP".[9] AM, PP, and QMA are, as the reader has probably guessed, complexity classes. Aaronson was trying to show that there existed an oracle b such that $AM^b \not\subseteq QMA^b$, unaware that it was already known that $AM^c \not\subseteq PP^c$, for some c, and $QMA \subseteq PP$. The target statement follows by setting b equal to c, given that for any classes A, B and any problem a, if $(A \subseteq B) \Rightarrow (A^a \subseteq B^a)$, i.e. that the subset relation is *relativisable* [1]. We show how Ontoplex models these results and saves a week of work. Assume that the known facts are already presented in Ontoplex. Currently, it is done using disjoint relations *subset* and *notSubset*, a special individual referring to the oracle c, and relations $hasBase(B, A)$ and $hasPower(B, c)$ to say $B = A^c$. Certain properties such as the relativisability of *subset* require first-order formalisation. We translate this FOWL knowledge base into a first-order one using the tool of Flügel et al. [8], and query it for the existence of an oracle with the desired property, using Vampire. Figure 2 illustrates the reasoning.

---

[9] https://complexityzoo.net/Complexity_Zoo_Introduction.

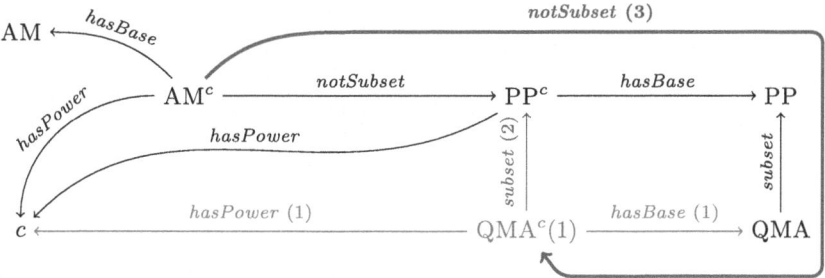

Fig. 2. Reasoning steps: the known facts are in **black**; on the (1)st step the existence of QMA^c is inferred; on the (2)nd step the *subset* relation is proved by relativisation; and on the (3)rd step the **notSubset** relation is produced by the composition of *notSubset* and *subset⁻*. It follows that there exists an oracle, c, such that $AM^c \not\subseteq QMA^c$.

5   Conclusions

The current version of Ontoplex is a proof-of-concept framework that features: (i) a good part of the knowledge about complexity theory can be organised in the ontology (using FOWL for competency questions 1–2 and pure OWL 2 for 3–5); (ii) the ontology can be populated by computer-aided fact extraction from natural language texts; and (iii) it is capable of supporting useful reasoning tasks (competency questions 6–8).

The directions for future work are numerous. On the ontological side, there are many more phenomena to represent, ranging from various computation models (probabilistic, quantum, parallel) and sub-fields (descriptive complexity, formal languages) to detailed properties of currently represented classes and problems (more hierarchies, more results concerning oracle computations, etc.) Some of them would require adding a second-order level of axioms, and adopting a framework able to orchestrate all these levels (e.g. in DOL [21], see [16] for a related effort). On the fact mining side, there is a need to continue the experiments and gradually represent the whole Complexity Zoo in Ontoplex. Finally, on the practical side, we would like to develop a unified and user-friendly application for querying, reasoning, and visualisation, which should provide a front-end for the various tools mentioned in Sect. 4.

References

1. Arora, S., Barak, B.: Computational Complexity: A Modern Approach. Cambridge University Press, Cambridge (2009)
2. Bodily, P.M., Ventura, D.: Open computational creativity problems in computational theory. In: Hedblom, M.M., Kantosalo, A.A., Confalonieri, R., Kutz, O., Veale, T. (eds.) Proceedings of the 13th International Conference on Computational Creativity, Bozen-Bolzano, Italy, 27 June–1 July 2022, pp. 112–120. Association for Computational Creativity (ACC) (2022). http://computationalcreativity.net/iccc22/papers/ICCC-2022_paper_61.pdf

3. Brown, T.B., et al.: Language models are few-shot learners. In: Proceedings of the 34th International Conference on Neural Information Processing Systems. NIPS '20. Curran Associates Inc., Red Hook, NY, USA (2020)

4. Carlisle, D., et al.: Mathematical Markup Language (MathML) Version 2.0. World Wide Web (2000)

5. Church, A.: A Note on the Entscheidungsproblem. J. Symb. Log. **1**(1), 40–41 (1936). https://doi.org/10.2307/2269326

6. Cook, S.A.: The complexity of theorem-proving procedures. In: Proceedings of the Third Annual ACM Symposium on Theory of Computing. STOC '71, pp. 151–158. Association for Computing Machinery, New York, NY, USA (1971). https://doi.org/10.1145/800157.805047

7. Elizarov, A.M., Kirillovich, A.V., Lipachev, E.K., Nevzorova, O.A.: OntoMath-PRO: an ontology of mathematical knowledge. Dokl. Math. **106**(3), 429–435 (2022). https://doi.org/10.1134/S1064562422700016

8. Flügel, S., Glauer, M., Neuhaus, F., Hastings, J.: When one logic is not enough: integrating first-order annotations in OWL ontologies. Semant. Web J. 1–16 (2024). https://doi.org/10.3233/SW-243440

9. Grau, B.C., Horrocks, I., Motik, B., Parsia, B., Patel-Schneider, P., Sattler, U.: OWL 2: the next step for OWL. J. Web Semant. **6**(4), 309–322 (2008). https://doi.org/10.1016/j.websem.2008.05.001

10. Hartmanis, J., Stearns, R.E.: On the computational complexity of algorithms. Trans. Am. Math. Soc. (1965)

11. Horrocks, I., Kutz, O., Sattler, U.: The even more irresistible SROIQ. In: Proceedings of the 10th International Conference on Principles of Knowledge Representation and Reasoning (KR2006), pp. 57–67. AAAI Press (2006)

12. Huang, L., et al.: A survey on hallucination in large language models: principles, taxonomy, challenges, and open questions (2023). https://arxiv.org/abs/2311.05232

13. Huguet Cabot, P.L., Navigli, R.: REBEL: relation extraction by end-to-end language generation. In: Findings of the Association for Computational Linguistics: EMNLP 2021, pp. 2370–2381. Association for Computational Linguistics, Punta Cana, Dominican Republic (2021). https://aclanthology.org/2021.findings-emnlp.204

14. Huguet Cabot, P.L., Tedeschi, S., Ngonga Ngomo, A.C., Navigli, R.: Red^{fm}: a filtered and multilingual relation extraction dataset. In: Proceedings of the 61st Annual Meeting of the Association for Computational Linguistics: ACL 2023. Association for Computational Linguistics, Toronto, Canada (2023). https://arxiv.org/abs/2306.09802

15. Karp, R.M.: Reducibility among combinatorial problems, pp. 85–103. Springer, Boston (1972). https://doi.org/10.1007/978-1-4684-2001-2_9

16. Keet, C.M., Kutz, O.: Orchestrating a network of mereo(topo)logical theories. In: Proceedings of the 9th Knowledge Capture Conference. K-CAP '17. Association for Computing Machinery, New York, NY, USA (2017). https://doi.org/10.1145/3148011.3148013

17. Kozen, D.C.: Theory of Computation. Texts in Computer Science, Springer, Heidelberg (2006). https://doi.org/10.1007/1-84628-477-5

18. Lange, C.: Ontologies and languages for representing mathematical knowledge on the semantic web. Semant. Web **4**(2), 119–158 (2013)

19. Libkin, L.: Elements of Finite Model Theory. Springer, Heidelberg (2004). https://doi.org/10.1007/978-3-662-07003-1

20. Mateiu, P., Groza, A.: Ontology engineering with large language models. In: 2023 25th International Symposium on Symbolic and Numeric Algorithms for Scientific Computing (SYNASC), pp. 226–229. IEEE Computer Society, Los Alamitos, CA, USA (2023). https://doi.org/10.1109/SYNASC61333.2023.00038

21. Mossakowski, T., Kutz, O., Codescu, M., Lange, C.: The distributed ontology, modeling and specification language. In: Vescovo, C.D., Hahmann, T., Pearce, D., Walther, D. (eds.) Proceedings of the 7th International Workshop on Modular Ontologies Co-located with the 12th International Conference on Logic Programming and Non-monotonic Reasoning (LPNMR 2013), Corunna, Spain, 15 September 2013. CEUR Workshop Proceedings, vol. 1081. CEUR-WS.org (2013). http://ceur-ws.org/Vol-1081/womo2013_invited_paper_1.pdf

22. Musen, M.A.: The Protégé project: a look back and a look forward. AI Matters **1**(4), 4–12 (2015). https://doi.org/10.1145/2757001.2757003

23. OpenAI: GPT-4 Technical report (2024)

24. Riazanov, A., Voronkov, A.: The design and implementation of VAMPIRE. AI Commun. **15**(2,3), 91–110 (2002)

25. Rudolph, S.: Foundations of Description Logics, pp. 76–136. Springer, Heidelberg (2011). https://doi.org/10.1007/978-3-642-23032-5_2

26. Sack, H.: Npbibsearch - an ontology augmented bibliographic search. In: Bouquet, P., Tummarello, G. (eds.) SWAP 2005 - Semantic Web Applications and Perspectives, Proceedings of the 2nd Italian Semantic Web Workshop, University of Trento, Trento, Italy, 14–16 December 2005. CEUR Workshop Proceedings, vol. 166. CEUR-WS.org (2005). https://ceur-ws.org/Vol-166/28.pdf

27. Saeedizade, M.J., Blomqvist, E.: Navigating ontology development with large language models. In: Meroño Peñuela, A., et al. (eds.) ESWC 2024. LNCS, vol. 14664, pp. 143–161. Springer, Cham (2024). https://doi.org/10.1007/978-3-031-60626-7_8

28. Schena, I., Asperti, A.: Towards a semantic web for formal mathematics. Ph.D. thesis (2002)

29. Sipser, M.: Introduction to the Theory of Computation, 3rd edn. Course Technology, Boston (2013)

30. Sousa, D., Lamurias, A., Couto, F.M.: Using Neural Networks for Relation Extraction from Biomedical Literature, pp. 289–305. Springer, New York (2021). https://doi.org/10.1007/978-1-0716-0826-5_14

31. Team, G.: Gemini 1.5: Unlocking multimodal understanding across millions of tokens of context (2024). https://arxiv.org/abs/2403.05530

32. Team, G.: Gemini: A family of highly capable multimodal models (2024). https://arxiv.org/abs/2312.11805

33. Touvron, H., et al.: Llama: open and efficient foundation language models (2023). https://arxiv.org/abs/2302.13971

34. Touvron, H., et al.: Llama 2: open foundation and fine-tuned chat models (2023). https://arxiv.org/abs/2307.09288

35. Turing, A.M.: On computable numbers, with an application to the Entscheidungsproblem. Proc. Lond. Math. Soc. **s2-42**(1), 230–265 (1937). https://doi.org/10.1112/PLMS/S2-42.1.230

36. CoQ proof assistant. https://coq.inria.fr

37. GPT4-o web page. https://openai.com/index/hello-gpt-4o/

38. HELM. http://helm.cs.unibo.it

39. Hermit reasoner. http://www.hermit-reasoner.com

40. Math-net. http://www.math-net.org

41. OMDoc Project. http://omdoc.org
42. Vereschchagin, N.: On the power of pp. In: [1992] Proceedings of the Seventh Annual Structure in Complexity Theory Conference, pp. 138–143 (1992). https://doi.org/10.1109/SCT.1992.215389
43. Watrous, J.: Succinct quantum proofs for properties of finite groups. In: Proceedings 41st Annual Symposium on Foundations of Computer Science, pp. 537–546 (2000). https://doi.org/10.1109/SFCS.2000.892141
44. Zhao, et al.: A survey of large language models (2023). https://arxiv.org/abs/2303.18223

Generating a Question Answering Dataset About Geographic Changes in a Knowledge Graph

Michalis Mitsios[1] , Dharmen Punjani[2] , Sara Abdollahi[3(✉)] ,
Simon Gottschalk[3] , Eleni Tsalapati[4] , Elena Demidova[5,6] ,
and Manolis Koubarakis[1,7(✉)]

[1] AI Team, Department of Informatics and Telecommunications, National and
Kapodistrian University of Athens, Athens, Greece
koubarak@di.uoa.gr
[2] Hubert Curien Laboratory, Université Jean Monnet, Saint Etienne, France
[3] L3S Research Center, Leibniz Universität Hannover, Hanover, Germany
gottschalk@L3S.de
[4] Athens Technology Center, Chalandri, Greece
[5] Data Science and Intelligent Systems Group (DSIS), University of Bonn, Bonn,
Germany
[6] Lamarr Institute for Machine Learning and Artificial Intelligence, Bonn, Germany
[7] Archimedes/Athena RC, Marousi, Greece

Abstract. Most studies on semantic question answering (QA) are predominantly focused on encyclopedic knowledge graphs like DBpedia and Wikidata. These studies cover, if at all, the spatial and temporal characteristics of geospatial entities in isolation, not addressing them simultaneously. In this paper, we introduce a pipeline for creating question answering datasets for evaluating the reasoning capabilities of QA models in the context of geographic changes over time. This pipeline generates questions, GeoSPARQL queries, and corresponding answers by leveraging subgraph and query template extraction techniques.

We exemplify this pipeline with the creation of the *GeoChangesQA* dataset with questions over a knowledge graph of US counties and states and their changes from 1629 to 2000. By evaluating *GeoChangesQA* using a Transformer-based model, we demonstrate that historical geospatial questions pose a substantial challenge for semantic question answering.

1 Introduction

Geospatial knowledge is crucial in many aspects of our lives, from urban planning and environmental monitoring to transportation and disaster management. Hence, the need to organise, store, and retrieve geospatial data constantly grows, often addressed by utilising geospatial knowledge graphs (KGs) [11,25] which contain not only *thematic* knowledge (e.g., the population of Los Angeles is 3.9 million) but also *geospatial* knowledge (e.g., the geometry of Los Angeles as a polygon). However, geospatial knowledge is often not static [3,8]: events such as

M. Alam et al. (Eds.): EKAW 2024, LNAI 15370, pp. 471–489, 2025.
https://doi.org/10.1007/978-3-031-77792-9_28

geopolitical conflicts, referendums, or legislations can affect the geospatial extent of a geographical area. The existence of such geographic changes makes querying spatiotemporal knowledge a challenging task, where both the geospatial and the temporal dimensions need to be considered when building semantic question answering (QA) systems. To develop and evaluate such systems, there is a need for the creation of spatiotemporal QA datasets.

An example of a historical geospatial change is illustrated in Fig. 1, which shows the US American county Los Angeles before and after April 2nd, 1866, where the county Kern was created and took large parts of Los Angeles. Consequently, answering questions regarding Los Angeles' area is time-sensitive and requires temporal inference over the geospatial knowledge. An example question to evaluate the availability of QA systems to deal with such spatiotemporal requirements is "What was the area of Los Angeles before its contraction on April 2nd, 1866?" which can only be answered with spatiotemporal reasoning.

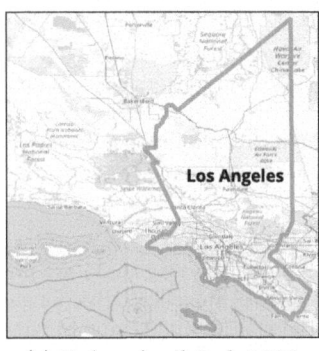 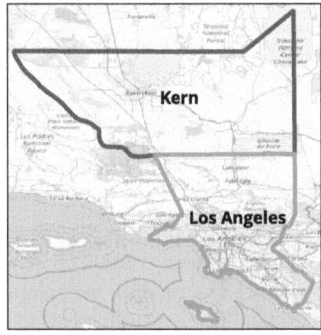

(a) Before April 2nd, 1866.   (b) After April 2nd, 1866.

Fig. 1. Area of the county Los Angeles (in red) before and after the event "Los Angeles lost to creation of Kern" on April 2nd, 1866. The answer to the question "What was the area of Los Angeles?" is time-sensitive. © OpenStreetMap contributors. (Color figure online)

Well-known examples of geospatial knowledge graphs include YAGO2 [18], YAGO2geo [21], WorldKG [13] and KnowWhereGraph [19]. Geospatial KGs typically use geospatial RDF stores like Strabon [23] for storage, and query languages such as GeoSPARQL [29] and stSPARQL [23] for retrieving knowledge present in the KG. Notable QA engines for geospatial KGs are GeoQA [31], GeoQA2 [30] and the system of Hamzei et al. [17]. To evaluate QA engines such as these, the GeoQuestions1089 dataset was proposed [22] and was used to compare the effectiveness and efficiency of GeoQA2 and the engine of [17]. However, these QA datasets do not contain spatiotemporal questions about changing geographies and the events causing them.

To fill the gap of spatiotemporal QA datasets, we propose a semi-automated spatiotemporal QA dataset generation pipeline based on four major steps: (i) subgraph generation to find meaningful structures in the target ontology, (ii) question template generation by converting these subgraphs into natural language questions with placeholders, (iii) query generation where the subgraphs are instantiated with elements in the target knowledge graph and (iv) generation of natural language questions by automatically filling the questions templates according to the generated queries. The whole pipeline is guided by our goal of creating spatiotemporal questions, specifically by adding spatiotemporal relations and filter expression and by considering both geospatial nodes and date expressions in the target knowledge graph.

As a demonstration of this pipeline, we introduce *GeoChangesQA*, a dataset of 8, 900 questions over a new historical geospatial knowledge graph – the Historical County Boundaries KG (*HCB-KG*) that contains all 50 US states and their counties along with their geometries, from 1629 to 2000. *GeoChangesQA* comes with questions, their GeoSPARQL queries, and answers. To the best of our knowledge, this is the first QA dataset that targets geographic changes.

Overall, the main contributions of this work are summarised as follows:

- We develop a question generation pipeline capable of generating complex, randomised SPARQL and GeoSPARQL queries alongside natural language questions and their answers.
- We introduce the Historical County Boundaries ontology (*HCB-O*) and knowledge graph (*HCB-KG*)[1] to represent US counties and states that have undergone continuous changes in their geometries over time.
- We introduce *GeoChangesQA*[2], a QA dataset with 8, 900 ⟨*question, query, answer*⟩ tuples.
- We evaluate and discuss the performance of the T5-Base language model [32] on generating SPARQL queries for the questions in *GeoChangesQA*.

This is the first approach for generating QA datasets with spatiotemporal questions that specifically target at KGs with geographies changing over time. While QA datasets have focused on the single aspects in isolation (time and space), changing geographies heavily complicate QA. Therefore, we envision the use of our pipeline and provided resources to evaluate the capability of QA systems to deal with spatiotemporal questions.

2   Related Work

We review spatiotemporal ontologies, knowledge graphs, and QA datasets.

2.1   Ontologies for Geographic Changes

The TSN and TSN-Change ontologies [4] model territorial partitions over time. The TSN Ontology is a geographic ontology that allows for the description of territorial partitions, where a territory is defined as a "bounded geographical space delimited by and under the control of a human group". The TSN-Change ontology allows for the description of territorial changes through an extensive

---

[1] https://hcbkg.l3s.uni-hannover.de/.
[2] https://zenodo.org/records/10989422.

typology and the formal representation of boundary changes from one territory version to its next. Bernard et al. [6] utilised the TSN and TSN-Change ontologies to develop the GeoChangeKG based on administrative French data, and developed a web application for visualisation of this data. The Hierarchical Historical Territory (HHT) [8] ontology allows the representation of historical territories, by taking into account multiple overlaying hierarchies.

In this paper, we reuse the ontologies TSN and TSN-Change for the creation of *HCB-KG*, our example KG of geographic changes.

2.2   Question Answering Datasets and Their Creation

There are many datasets for QA over KGs, that differ in their underlying KG, the way they are created, the size and the question phenomena they involve. The datasets WebQuestions [2] and SimpleQuestions [5], both targeting Freebase[3], were the first considerably large datasets that appeared in the literature. Other significant benchmarks that contain complex questions are QALD-9 [37], LC-QuAD [36] and LC-QuAD 2.0 [14] which target DBpedia. LC-QuAD was created in a backward manner: the queries were generated semi-automatically by extracting subgraphs containing tuples within a 2-hop distance. The generation of the questions was done automatically, using templates, and, then, refined manually. KQA Pro [7] provides nearly 120,000 questions based on Wikidata. The natural language questions were generated using templates and paraphrases through crowdsourcing. ParaQA [20] contains 5,000 template-based question-answer pairs with unique paraphrased responses for each question using a back-translation mechanism. Mintaka [34] is composed of 20,000 question-answer pairs, written by crowd workers.

2.3   Spatiotemporal QA Datasets

POIReviewQA[4] contains questions about points of interest. Other datasets about geospatial knowledge include Geo-Query [35], GeoAnQu [38], GeoQuestions201 [31] and GeoQuestions1089 [22]. GeoQuestions201, created for the evaluation of the QA engine GeoQA [31], focuses on a linked geospatial dataset comprising DBpedia, GADM[5], and OpenStreetMap subsets limited to the United Kingdom and the Republic of Ireland. As for temporal events, Event-QA [9] contains 1,000 semantic queries and the corresponding verbalisation for EventKG, an event-centric KG [16]. CRONQuestions [33] is a large QA dataset with natural language questions requiring temporal reasoning. STQAD [10] provides 10,000 questions (paraphrased through ChatGPT) with a spatial and temporal intent based on quintuples extracted from YAGO2, but does not consider geographic changes and complex geometries.

---

[3] SIMPLEQUESTIONS were later reformulated to target also Wikidata [12].

[4] http://stko.geog.ucsb.edu/poireviewqa/.

[5] https://gadm.org/.

From the above, we conclude that there is no QA dataset or dataset creation method for questions that relate geospatial and temporal information together and consider changing geographies over time.

3   *HCB-KG* and *HCB-O*: The Historical County Boundaries Knowledge Graph and its Ontology

As a basis of our spatiotemporal QA dataset generation pipeline, we require a knowledge graph backed by an ontology specifically targeted at spatiotemporal dynamics. Therefore, we first introduce *HCB-KG* and its ontology *HCB-O*, as well as the underlying data source and the *HCB-KG* creation process.

3.1   The Atlas Historical County Boundaries Dataset (*AHCB*)

We exemplify our QA dataset creation process using $AHCB^6$ as our data source of historical geospatial knowledge[7]. It contains the *polygons* of all 50 US states and their counties per period of time, the description of the *change* that occurred at the start date (e.g., "The United States created Mississippi Territory; it covered the south-central portions of Alabama and Mississippi.") along with a *reference* to the source of data for the described change (e.g., "(U.S. Stat., vol. 1, ch. 28 [1798], secs. 1, 3, 5/pp. 549-550)". Further attributes of the counties described in the dataset include: the *state name* of the county's current or most recent affiliation, and the *version number* that tracks sequential and chronological changes in the county name or configuration.

3.2   *HCB-O*: The Historical County Boundaries Ontology

We create the Historical County Boundaries Ontology (*HCB-O*, prefix hbc-o), illustrated in Fig. 2, as a lightweight ontology (OWL2QL[9] profile of OWL2) that enables the semantic representation of *AHCB*, i.e., of geospatial data of US American states and counties that have changed over time, recording also the reasons for each occurring change. For this purpose, we reuse the TSN[10], TSN-Change[11], SEM[12] and Time[13] ontologies. To ensure that the external ontologies are not violated, *HCB-O* is developed in such a way that it is a conservative extension [15] of these ontologies.

---

[6] https://digital.newberry.org/ahcb/downloads/states.html.

[7] US states serve as an example domain thanks to availability of such data. Spatiotemporal questions also appear in other contexts such as country border changes[8], traffic (floating car data), ship movements (AIS Data) or animal tracking data.

[9] https://www.w3.org/TR/owl2-profiles/#OWL_2_QL.

[10] Territorial Statistical Nomenclature Ontology (TSN), http://lig-tdcge.imag.fr/tsn/index.html, prefix tsn: http://purl.org/net/tsn#.

[11] Territorial Statistical Nomenclature Ontology Change (TSN-Change), https://lig-tdcge.imag.fr/tsnchange/index.html.

[12] The Simple Event Model (SEM), https://semanticweb.cs.vu.nl/2009/11/sem/.

[13] Time Ontology, https://www.w3.org/TR/owl-time/.

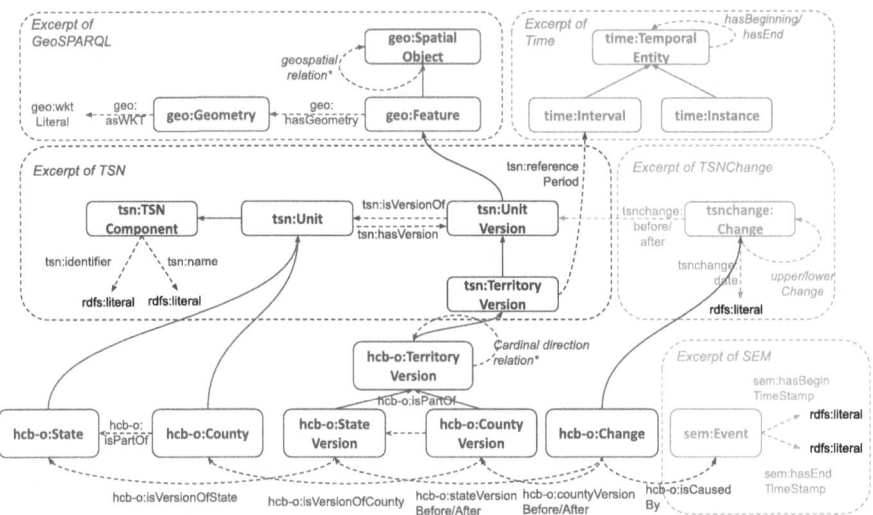

Fig. 2. The Historical County Boundaries ontology (at the bottom) along with excerpts from the SEM, TSN, TSN-Change, GeoSPARQL and Time ontologies. With solid lines we present the subclass-of relations and with dashed lines the object/data properties.

In *HCB-O*, the class hcb-o:TerritoryVersion, subclass of tsn:Territory-Version, is used to represent cardinal direction relations expressed using stSPARQL (e.g., st:northOf, st:southOf), besides the GeoSPARQL geospatial relations (e.g., geo:borders, geo:crosses) inherited by the class geo:SpatialObject.[14] As *AHCB* includes geospatial data about states and counties, tsn:TerritoryVersion is further refined with the subclasses hcb-o:CountyVersion and hcb-o:StateVersion. A county (state) version describes a unique representation of a county (state), stable for a period of time (inherited from the tsn:TerritoryVersion class). The class hcb-o:County (hcb-o:State) describes an abstract representation of a county (state) and inherits the properties tsn:identifier and tsn:name from the class tsn:SpatialComponent.

We extend the TSN-Change ontology by introducing the hcb-o:Change class (subclass of tsnchange:Change), which is related to the sem:Event class through the property hcb-o:isCausedBy. In this way, we can formally represent the events causing the change of state of the counties or states. Additionally, we can further classify the types of change that have occurred from one county (state) version to another by reusing the classes tsnchange:Extraction, tsnchange:Appearance, etc., of the TSN-Change ontology.

---

[14] These stSPARQL cardinal direction relations using stSPARQL cannot be modelled through tsn:TerritoryVersion directly.

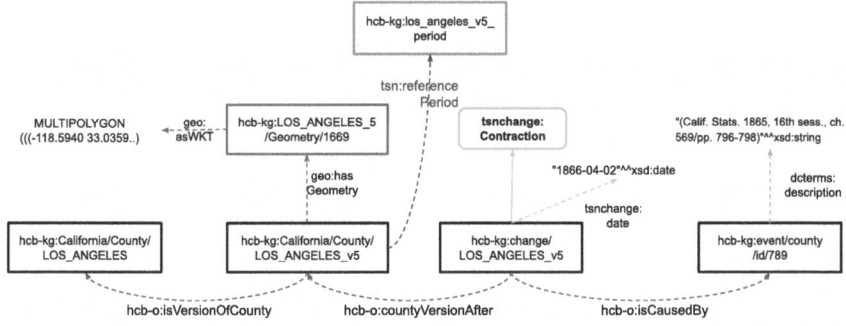

Fig. 3. Sample of *HCB-KG* that illustrates the contraction of Los Angeles on April 2nd, 1866 according to the California Statutes in 1865.

3.3 *HCB-KG*: The Historical County Boundaries Knowledge Graph

HCB-KG results from mapping *AHCB* to *HCB-O* in a straightforward process. To model events, we make use of the descriptions (modelled via `dcterms:de-scription`) and the references of changes provided in *AHCB* (modelled via `rdfs:label`). For example, a change referenced as "Orleans Terr. Acts 1811, 2d sess., p. 210" does actually refer to an event, namely, the Act passed at the second session of the Third Legislature of the Territory of Orleans, which began and was held in the city of New-Orleans on the 23rd of January 1811.[15]

In Fig. 3, we illustrate a sample of *HCB-KG*, which is related to our motivating example (Fig. 1), i.e., the contraction of Los Angeles, on April 2nd, 1866. The contraction was result of the California Statutes in 1865.

For the geospatial data, we utilise the GeoTriples tool [24] to transform them into RDF triples. Additionally, to increase the performance of the query answering process, we pre-computed and materialised the GeoSPARQL geospatial relations: `geo:sfWithin`, `geo:sfCrosses`, `geo:sfIntersects`, `geo:sfTouches`, `geo:sfOverlaps`, `geo:sfCovers` and `geo:sfEquals`, which require expensive spatial computations. As we have shown in previous work [22], this approach can greatly boost the performance of computationally demanding queries with insignificant cost to system memory. The calculation of the implied relations is facilitated by utilising a distributed implementation of the algorithm GIA.nt [27], implemented in the system DS-JedAI [28][16].

Having created a preliminary version of *HCB-KG* following the process described above, we then utilise the rich geospatial knowledge for classifying the changes according to the subclasses of `tsnchange:Change` (e.g., `tsnchange:`

---

[15] Most changes refer to legal acts not represented in existing KGs such as Wikidata or EventKG. We leave the linking of events which are also apparent in other KGs for future work.

[16] GIA.nt is a holistic geospatial interlinking algorithm that uses the DE-9IM topological model to discover all the topological relations between the geometries of two geospatial datasets.

Table 1. Statistics of change types in *HCB-KG*.

Change	Count	Description
Contraction	4,011	The size of a county has decreased.
Extraction	3,668	The county is split and at least one new county is created.
Expansion	1,893	The size of a county has increased.
Appearance	393	A new county has been created.
Merge	241	A county that has been merged with some other county.
Fusion	141	A county that has been fused with some other county.
Integration	100	A county disappears and is integrated in another one.
NameChange	21	The name of a county changed.

Table 2. Statistics of selected items in *HCB-KG*.

	Count
Triples	917,281
Changes	17,925
County Versions	17,714
Events	8,200
State Versions	220

`Appearance` and `tsnchange:Extraction`). Using geospatial relations such as `geo:sfContains` and geospatial operations such as `strdf:area`, we figure out how different counties and states have changed over time. For each change class, we manually create a SPARQL query (for example, to select all county versions that have decreased in area in comparison to their previous version) and extend *HCB-KG* with the identified change types. Table 1 provides an overview of change types detected like this and how frequently they occur in *HCB-KG*.

Table 2 shows selected statistics of *HCB-KG*. On the *HCB-KG* website[17], we provide links to the definition of *HCB-O*[18], the *HCB-KG* data dump on Zenodo[19] and our code repository[20].

4   Question Generation Pipeline

In this section, we present our semi-automatic pipeline for creating QA datasets over spatiotemporal knowledge, exemplified via *GeoChangesQA*, our dataset over historical geospatial knowledge in *HCB-KG*. In this example, *HCB-KG* is the target KG and *HCB-O* the target ontology.

---

[17] https://hcbkg.l3s.uni-hannover.de/.
[18] https://github.com/AI-team-UoA/GeoChanges-Dataset/blob/main/Geo_Changes/ resources/ontology/hcb_ontology.rdf.
[19] https://zenodo.org/records/11508199.
[20] https://github.com/AI-team-UoA/GeoChanges-Dataset.

4.1   Overview

A QA dataset contains tuples of natural language questions, their corresponding SPARQL queries, and their answers. Our pipeline describing the dataset
generation is presented in Fig. 4 exemplified by the creation of *GeoChangesQA*.

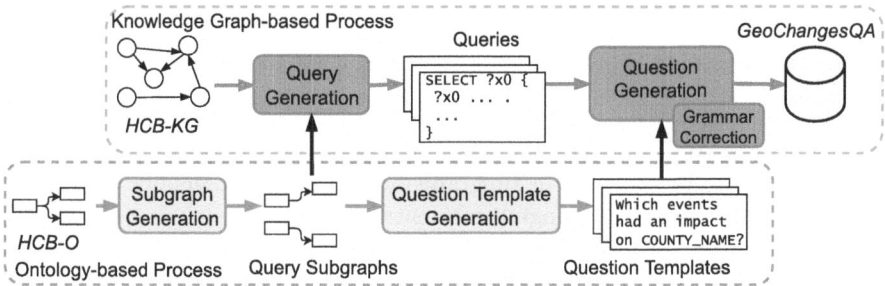

Fig. 4. Question generation pipeline exemplified by *GeoChangesQA* with its two inputs
HCB-KG and *HCB-O*.

The depicted pipeline consists of two major steps: In the *ontology-based process*, we take the target ontology (i.e., in our example: *HCB-O*) to create query
subgraphs and question templates. Then, in the *knowledge graph-based process*,
we utilise the target KG (*HCB-KG*) to instantiate these subgraphs and question templates to derive SPARQL queries and natural language questions. The
application of this process on multiple subgraphs and for multiple instantiations
leads to the creation of the QA dataset (*GeoChangesQA*).

   This process aims to generate a dataset of questions over spatiotemporal
knowledge which is not biased through manual selection of questions but reflects
a large diversity of questions on the target KG. Instead of relying on a set of seed
entities and subsequent graph traversal (as, for example, in [9,14]), we follow
an ontology-guided approach, which gives us more control over the generated
queries and their convenience and the possibility of generating natural language
questions semi-automatically. This pipeline specifically address the challenges of
geographic changes by the inclusion of geospatial entities (subgraph generation),
time and geo filters (query generation) and its exemplary application on *HCB-O*.

4.2   Subgraph Generation

In the first step, we generate *subgraphs* from the target ontology, where each
subgraph represents a set of potential SPARQL queries. Here, a subgraph is a
connected graph where each node is assigned to a class in the target ontology,
and the edges are unlabelled[21].

---

[21] We only consider an edge between two nodes if there exists at least one property
that can be used on the classes of its adjacent nodes.

Typically, the more complex a query (e.g., nested, aggregated or with many query values), the more unnatural the respective natural language question, and the more difficult to provide a valid question template. Therefore, the subgraph generation is performed under the following constraints, which ensure the generation of subgraphs that potentially generate meaningful questions:

- Each subgraph must contain at least two nodes and at most nine.
- Each subgraph contains at most two geospatial entities (geo:Geometry).
- One class can only be assigned to two nodes at maximum.

Under the constraints, potentially extended with further domain-specific constraints, and by following the connections between classes in the target ontology, we automatically create a set of subgraphs which is then manually filtered for subgraphs which reflect meaningful information needs.

For the creation of *GeoChangesQA*, in addition to the listed constraints, we require that each subgraph contains either a node assigned to the class hcb-o: County or hcb-o:State, since our goal is to generate questions about counties or states and their geospatial properties over time. In addition, if a subgraph contains the hcb-o:CountyVersion (hcb-o:StateVersion) class, then it must also contain the hcb-o:County class (hcb-o:State) since county versions (state versions) are not an intuitive concept when not considering the respective county (state) as well.[22] Figure 5 shows a subgraph example, consisting of 4 nodes and 3 edges, that correlates a county to one of its versions, the geometry of this version and a change involved.

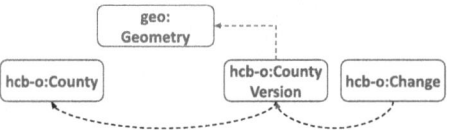

Fig. 5. An example subgraph of *HCB-O*.

Following the described procedure, we extract 17 subgraphs from *HCB-O*. 11 of these subgraphs involve geospatial entities that are counties only (as in Fig. 5), and 7 involve both counties and states. From the subgraphs only with counties, we automatically generate 11 similar subgraphs but this time only for states (e.g., a similar subgraph to Fig. 5 contains the nodes hcb-o:State, hcb-o:StateVersion, geo:Geometry, and hcb-o:Change). In this way, the total number of generated subgraphs is 28.

---

[22] For brevity, we focus on subgraphs about counties in the remainder of this section; subgraphs, queries and questions about states are created analogously.

4.3 Question Template Generation

Given a subgraph, a *question template* is a natural language question, where several terms are placeholders. In addition, each question template comes with (i) a target node (?x0), i.e., the expected result type[23] (for example, hcb-o:County when asking for a specific county), and (ii) seed nodes (?s0, ?s1, ...). These seed nodes are replaced with actual values when instantiating the subgraph (query generation step). The text placeholders in the question template are replaced to generate a proper natural language question (question generation step).

Two possible question templates for the subgraph in Fig. 5 are:

- *"What was COUNTY_NAME's geometry after its CHANGE_TYPE.noun?"*.
  - Target node: geo:Geometry (?x0)
  - Seed nodes: hcb-o:County (?s0), a subclass of tsnchange:Change (?s1)
  - Placeholders: COUNTY_NAME, CHANGE_TYPE
- *"How did the area of COUNTY_NAME CHANGE_TYPE DATE_NAME?"*.
  - Target node: the area of geo:Geometry (?x0)
  - Seed nodes: hbc:County (?s0), a subclass of tsnchange:Change (?s1) and its datatype property tsnchange:date (?s2)
  - Placeholders: COUNTY_NAME, CHANGE_TYPE, DATE_NAME

We manually create a set of question templates per subgraph based on the potential roles each node can have (i.e., a geometry node can take the role of the target node, so we created a question template for this scenario). For grammatical correctness and linguistic variety, we add information regarding the grammatical case (e.g., CHANGE_TYPE.**noun**) or conjugation (e.g., COUNTY_RELATION.**past**), if required.

While creating *GeoChangesQA*, the question template generation step results in 200 unique templates along with several paraphrases.

4.4 Query Generation

Given a subgraph, we aim to create corresponding SPARQL queries through random instantiation, where instantiation refers to the process of selecting an entity in the target KG for each seed node in the subgraph and a property from the target ontology for each edge.

The following steps describe the process of building such a query:

1. Target node determination: We randomly select a target node which is the expected result of the query.
2. Property selection: For each edge in the subgraph, we randomly select a property which can be used on the adjacent node classes.
3. Extension with filters: Following random decisions, we can add SPARQL filter conditions to the query:

---

[23] To generate meaningful queries, we only consider nodes of classes such as hcb-o: County as potential target nodes, but not nodes that are unlabelled, e.g., hcb-o: CountyVersion.

Listing 1.1. SPARQL query generated for instantiation of the subgraph in Fig. 5.

```
SELECT DISTINCT ?s0 ?s1 ?s2 ?s3 {
?x0 a geo:Geometry .
?x1 a hcb-o:CountyVersion .
?s0 a hcb-o:County .

?s0 tsn:hasName ?s3 .
?s0 hcb-o:hasCountyVersion ?x1 .

?x1 geo:hasGeometry ?x0 .
?x2 a ?s2 ,
    hcb-o:outputCountyVersion ?x1 ,
    tsnchange:date ?s1 .

FILTER (
    ?s1 < "1872-04-11"^^xsd:date) .
}
ORDER BY RAND()
LIMIT 1
```

Listing 1.2. Example SPARQL query generated for the subgraph in Fig. 5.

```
SELECT DISTINCT
         (strdf:area(?x4) as ?area) {
?x0 a geo:Geometry .
?x1 a hcb-o:CountyVersion .
?x5 a hcb-o:County .

?x5 tsn:hasName
       "Los Angeles"^^xsd:string .
?x5 hcb-o:hasCountyVersion ?x1 .

?x1 geo:hasGeometry ?x0 .
?x0 geo:asWKT ?x4 .

?x2 a tsnchange:Contraction ;
    tsnchange:date ?x3 ;
    hcb-o:outputCountyVersion
                               ?x1 .

FILTER (
    ?x3 < "1872-04-11"^^xsd:date)
}
```

- Time filter: To ensure that the query considers the temporal dimension, we can add a temporal filter based on a time interval within the subgraph (e.g., to select county versions created after 1866-04-02).
- Geo filter: If the subgraph contains two geometry nodes, we can add a geo filter to require geometrical constraints between them (e.g., to select only counties north of Los Angeles). The supported geospatial relations are west of, east of, south of and north of represented by the functions `strdf:right`, `strdf:left` `strdf:below`, and `strdf:above`.
4. Node instantiation: The seed nodes of the subgraph are replaced with instances by querying the target KG.

For instance, given the example subgraph of Fig. 5, after target node selection (e.g., the node `geo:Geometry`), property selection (e.g., `hcb-o:hasCounty-Version`, `tsn:hasName`, `geo:hasGeometry`) and extension with a temporal filter (e.g., changes before 1872-04-11), we generate the SPARQL query shown in Listing 1.1 to identify potential values for the seed nodes (`?s0`,`?s1` and `?s2`).

One result of this query on *HCB-KG* is `?s0` = `hcb-kg:County/cas_los angeles`, `?s1` = 1866-04-02, `?s2` = `hcb-kg:Contraction` and `?s3` = "Los Angeles". By replacing the seed nodes accordingly, we generate the SPARQL query shown in Listing 1.2.

4.5 Question Generation

The question generation step takes as input the generated SPARQL queries and the question templates to generate the corresponding natural language questions. First, for each SPARQL query, a question template is selected based on the subgraph, the target, and the seed nodes. This question template is then used to generate the corresponding natural language question by replacing its placeholders with the values in the query. An example of this process is shown in Fig. 6. To increase the diversity of our generated questions and avoid bias towards terms used in *HCB-O*, we introduce a set of synonyms for change types and materialised relations (i.e., "intersect" as a synonym of "cross").

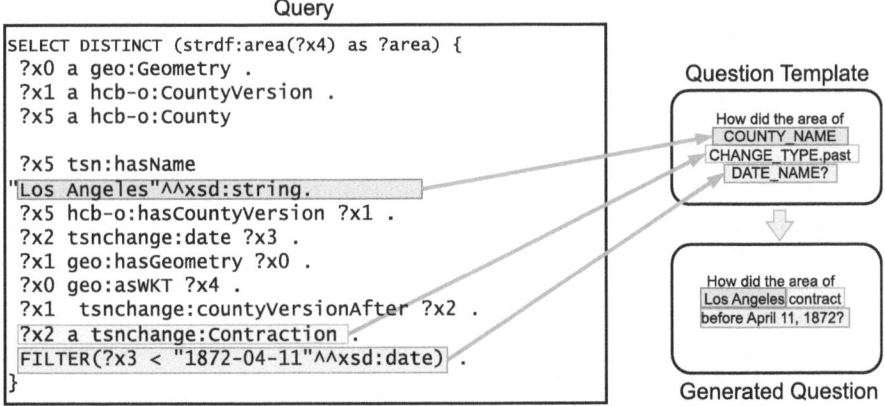

Fig. 6. Question generation where the placeholders in the question template are replaced with the values in the query.

With random entity selection, ontology-guided subgraph generation, and paraphrases of question templates, we aim to avoid biases as much as possible: in *GeoChangesQA*, most of the counties (96.72%) appear not more than in five queries/questions, while the dataset covers approx. 31% of the counties appearing in *AHCB*.

As a final step, to avoid any grammatical errors such as article mismatches (e.g., *"a expansion"* instead of *"an expansion"*), punctuation corrections, capitalisation issues, we utilise a grammar correction model. Each generated natural language question passes through a T5 grammar correction model[24] [26] and, if necessary, is replaced with a corrected version.

4.6 Scalability, Generalisability and Limitations

The application of our pipeline on the *HCB-KG* results in our *GeoChangesQA* dataset: 8,900 tuples of questions, queries and results over historical geospatial knowledge in *HCB-KG*. For generalisability to other datasets, most steps of the question generation pipeline can be directly applied on other datasets, except for the creation of domain-specific constraints, the manual subgraph selection

---

[24] https://huggingface.co/vennify/t5-base-grammar-correction.

process and question template creation. These steps require human intervention but are mandatory to ensure quality of the generated knowledge graph and questions. The complexity of queries can be changed through the constraints listed in Sect. 4.2.

Currently, our pipeline does not include aggregation operators and only supports `SELECT` queries. *GeoChangesQA* does not yet cover all types of change types in the TSN-Change ontology such as `tsnchange:Merge` and `tsnchange: Fusion`. These components noticeably complicate the questions and queries and thus require careful integration into the dataset creation pipeline.

5 Evaluation

In this section, we evaluate whether our QA generation pipeline is capable of generating a QA dataset for the training and evaluation of spatio-temporal QA models. Specifically, we evaluate how a Transformer-based model is capable of generating SPARQL queries when being trained and tested on *GeoChangesQA*.

To perform this evaluation, we follow the approach of Banerjee et al. [1] who investigated the performance of various language models on query generation. They fine-tune each language model on the SPARQL query generation task where the input consists of a natural language question along with the URIs of entities and properties mentioned in the question (i.e., they assume that entities and properties have been correctly recognised in earlier steps of the QA pipeline), and the prediction target is the corresponding SPARQL query. Fine-tuning the T5-Base model [32], the best performing model in this study, achieves F1 scores of more than 0.9 on the QA datasets LC-QuAD 1.0 [36] and LC-QuAD 2.0 [14].

By following this evaluation setting, we aim to examine how well *Geo-ChangesQA* is suited as a dataset for training and testing query generation over spatio-temporal knowledge. We utilise the T5-Base model[25], and we use the same set of hyperparameters for the model training as in [1]. Our dataset is split into 90% training and 10% validation sets. The training of T5-Base converges after around 10.000 iterations.

On average, T5-Base achieves an accuracy of 0.72, i.e., for 72% questions in the test data, the model's output exactly matches the target SPARQL query.[26] Fig. 7[27] shows the accuracy scores when grouping the questions by the subgraph they were derived from, further split into questions about counties and states. For some subgraphs, specifically those with fewer nodes (i.e., less complex), accuracy reaches 100%. In contrast, for some other subgraphs, we observe lower accuracy scores of less than 70%, including the two subgraphs containing two geospatial entity nodes. There is one subgraph for which no correct answer is found, with an example question being "What event had led Hampshire to expand with its

---

[25] https://huggingface.co/docs/transformers/model_doc/t5#transformers. T5ForConditionalGeneration.

[26] We expect lower accuracy when not inputting the entities and properties.

[27] Due to conversions between dates and date intervals, the reported numbers of nodes in the subgraphs partially exceed 9 (as requested in the constraints in Sect. 4.2).

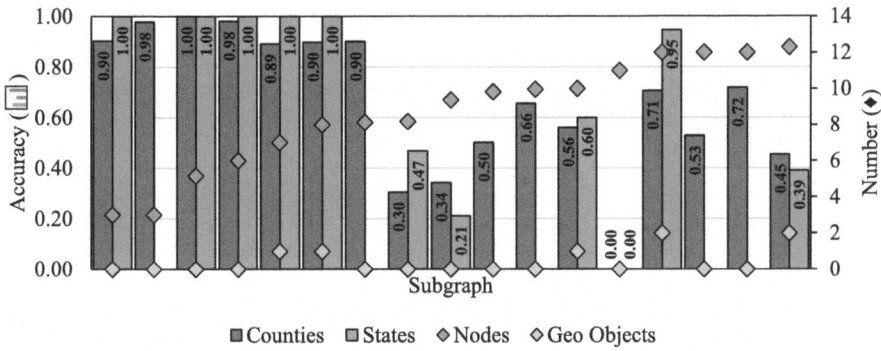

Fig. 7. Accuracy of the query generation over questions grouped by the subgraphs they were derived from, ordered by their size (number of nodes). Blue bars denote subgraphs with counties, orange bars with states. If no orange bar is shown for a subgraph, there are no distinguished county and state subgraphs. (Color figure online)

neighbor counties?", i.e., a highly complex query involving 11 nodes, an event, a change and a spatial relation. Another observation is that the accuracy for subgraphs about states is slightly larger than about counties, potentially since the pre-trained language model has a greater knowledge of states.

From this evaluation, we make the following conclusions:

– With an average accuracy of 0.72, the same evaluation methodology on *Geo-ChangesQA* reaches lower results than on LC-QuAD 1.0 and LC-QuAD 2.0 (both 0.91). This confirms the complexity of our question answering dataset about historical geospatial knowledge.
– For 26 out of 28 subgraphs from which we derive questions, the T5-Base model achieves an accuracy of more than zero, demonstrating the feasibility of using *GeoChangesQA* for training and testing spatiotemporal QA systems.
– A particular challenge lies in the correct identification of relations between two geometric objects (`geo:Geometry`) in the query.

6 Summary

We introduced a pipeline for generating a Question Answering dataset for spatiotemporal knowledge exemplified by *GeoChangesQA*, a QA dataset about geographic changes of US American states and counties between 1629 and 2000. We demonstrated that (i) our pipeline is capable of generating a QA dataset for training and testing the capability of QA models to deal with spatiotemporal questions and geographic changes and (ii) that historical geospatial questions pose a substantial challenge for the next generation of semantic QA systems.

Acknowledgments. This work was partially supported from the Horizon Europe project STELAR (GA No 101070122), the Hellenic Foundation for Research and Innovation project GeoQA (HFRI-FM17-2351), the Horizon Europe Marie Sklodowska-Curie Action QuAre (GA No 101032307), the project MIS 5154714 of the National

Recovery and Resilience Plan Greece 2.0 funded by the European Union under the NextGenerationEU Program, ATTENTION! (01MJ22012D and 01MJ22012C), a project of the Federal Ministry for Economic Affairs and Climate Action (BMWK), Germany, and the project MoToRes (19F2271A and 19F2271C) of the Federal Ministry for Digital and Transport (BMDV), Germany.

Disclosure of Interests. The authors have no competing interests to declare that are relevant to the content of this article.

References

1. Banerjee, D., Nair, P.A., Kaur, J.N., Usbeck, R., Biemann, C.: Modern baselines for SPARQL semantic parsing. In: SIGIR '22: The 45th International ACM SIGIR Conference on Research and Development in Information Retrieval, Madrid, Spain, 11–15 July 2022, pp. 2260–2265. ACM (2022). https://doi.org/10.1145/3477495.3531841
2. Berant, J., Chou, A., Frostig, R., Liang, P.: Semantic parsing on freebase from question-answer pairs. In: Proceedings of the 2013 Conference on Empirical Methods in Natural Language Processing, EMNLP 2013, 18–21 October 2013, Grand Hyatt Seattle, Seattle, Washington, USA, A Meeting of SIGDAT, a Special Interest Group of the ACL, pp. 1533–1544. ACL (2013). https://aclanthology.org/D13-1160/
3. Bereta, K., Smeros, P., Koubarakis, M.: Representation and querying of valid time of triples in linked geospatial data. In: Cimiano, P., Corcho, O., Presutti, V., Hollink, L., Rudolph, S. (eds.) ESWC 2013. LNCS, vol. 7882, pp. 259–274. Springer, Heidelberg (2013). https://doi.org/10.1007/978-3-642-38288-8_18
4. Bernard, C., Villanova-Oliver, M., Gensel, J., Dao, H.: Modeling changes in territorial partitions over time: ontologies TSN and TSN-Change. In: Proceedings of the 33rd Annual ACM Symposium on Applied Computing, pp. 866–875 (2018)
5. Bordes, A., Usunier, N., Chopra, S., Weston, J.: Large-scale simple question answering with memory networks. CoRR abs/1506.02075 (2015). http://arxiv.org/abs/1506.02075
6. Camille Bernard, Matthieu Viry, M.V., Gensel, J.: GeoChangeViz: visualizing knowledge graphs about changes in geographical divisions. In: Proceedings of the ISWC 2023 Posters and Demonstrations and Industry Tracks Co-located with 22nd International Semantic Web Conference (ISWC 2023), Athens, Greece, 6–10 November 2023. CEUR Workshop Proceedings (2023)
7. Cao, S., et al.: KQA pro: a dataset with explicit compositional programs for complex question answering over knowledge base. In: Proceedings of the 60th Annual Meeting of the Association for Computational Linguistics (Volume 1: Long Papers), ACL 2022, Dublin, Ireland, 22–27 May 2022, pp. 6101–6119. Association for Computational Linguistics (2022). https://doi.org/10.18653/V1/2022.ACL-LONG.422
8. Charles, W., Aussenac-Gilles, N., Hernandez, N.: HHT: an approach for representing temporally-evolving historical territories. In: Pesquita, C., et al. (eds.) ESWC 2023. LNCS, vol. 13870, pp. 419–435. Springer, Cham (2023). https://doi.org/10.1007/978-3-031-33455-9_25
9. Costa, T.S., Gottschalk, S., Demidova, E.: Event-QA: a dataset for event-centric question answering over knowledge graphs. In: CIKM '20: The 29th ACM International Conference on Information and Knowledge Management, Virtual Event, Ireland, 19–23 October 2020, pp. 3157–3164. ACM (2020). https://doi.org/10.1145/3340531.3412760

10. Dai, X., Li, H., Qi, G.: Question answering over spatio-temporal knowledge graph. arXiv preprint arXiv:2402.11542 (2024)
11. Demidova, E., Dsouza, A., Gottschalk, S., Tempelmeier, N., Yu, R.: Creating knowledge graphs for geographic data on the web. SIGWEB Newsl. **2022**(Winter), 4:1–4:8 (2022). https://doi.org/10.1145/3522598.3522602
12. Diefenbach, D., Tanon, T.P., Singh, K.D., Maret, P.: Question answering benchmarks for Wikidata. In: Proceedings of the ISWC 2017 Posters and Demonstrations and Industry Tracks Co-located with 16th International Semantic Web Conference (ISWC 2017), Vienna, Austria, 23–25 October 2017. CEUR Workshop Proceedings, vol. 1963. CEUR-WS.org (2017). https://ceur-ws.org/Vol-1963/paper555.pdf
13. Dsouza, A., Tempelmeier, N., Yu, R., Gottschalk, S., Demidova, E.: WorldKG: a world-scale geographic knowledge graph. In: CIKM '21: The 30th ACM International Conference on Information and Knowledge Management, Virtual Event, Queensland, Australia, 1–5 November 2021, pp. 4475–4484. ACM (2021). https://doi.org/10.1145/3459637.3482023
14. Dubey, M., Banerjee, D., Abdelkawi, A., Lehmann, J.: LC-QuAD 2.0: a large dataset for complex question answering over Wikidata and DBpedia. In: Ghidini, C., et al. (eds.) ISWC 2019, Part II. LNCS, vol. 11779, pp. 69–78. Springer, Cham (2019). https://doi.org/10.1007/978-3-030-30796-7_5
15. Ghilardi, S., Lutz, C., Wolter, F.: Did I damage my ontology? A case for conservative extensions in description logics. In: Proceedings, Tenth International Conference on Principles of Knowledge Representation and Reasoning, Lake District of the United Kingdom, 2–5 June 2006, pp. 187–197. AAAI Press (2006). http://www.aaai.org/Library/KR/2006/kr06-021.php
16. Gottschalk, S., Demidova, E.: EventKG: a multilingual event-centric temporal knowledge graph. In: Gangemi, A., et al. (eds.) ESWC 2018. LNCS, vol. 10843, pp. 272–287. Springer, Cham (2018). https://doi.org/10.1007/978-3-319-93417-4_18
17. Hamzei, E., Tomko, M., Winter, S.: Translating place-related questions to geosparql queries. In: WWW '22: The ACM Web Conference 2022, Virtual Event, Lyon, France, 25–29 April 2022, pp. 902–911. ACM (2022). https://doi.org/10.1145/3485447.3511933
18. Hoffart, J., Suchanek, F.M., Berberich, K., Weikum, G.: YAGO2: a spatially and temporally enhanced knowledge base from Wikipedia: extended abstract. In: IJCAI 2013, Proceedings of the 23rd International Joint Conference on Artificial Intelligence, Beijing, China, 3–9 August 2013. , pp. 3161–3165. IJCAI/AAAI (2013). http://www.aaai.org/ocs/index.php/IJCAI/IJCAI13/paper/view/6864
19. Janowicz, K., et al.: Know, know where, knowwheregraph: a Densely connected, cross-domain knowledge graph and geo-enrichment service stack for applications in environmental intelligence. AI Mag. **43**(1), 30–39 (2022). https://doi.org/10.1609/AIMAG.V43I1.19120
20. Kacupaj, E., Banerjee, B., Singh, K., Lehmann, J.: ParaQA: a question answering dataset with paraphrase responses for single-turn conversation. In: Verborgh, R., et al. (eds.) ESWC 2021. LNCS, vol. 12731, pp. 598–613. Springer, Cham (2021). https://doi.org/10.1007/978-3-030-77385-4_36
21. Karalis, N., Mandilaras, G., Koubarakis, M.: Extending the YAGO2 knowledge graph with precise geospatial knowledge. In: Ghidini, C., et al. (eds.) ISWC 2019, Part II. LNCS, vol. 11779, pp. 181–197. Springer, Cham (2019). https://doi.org/10.1007/978-3-030-30796-7_12
22. Kefalidis, S., et al.: Benchmarking geospatial question answering engines using the dataset geoquestions1089. In: Payne, T.R., et al. (eds.) ISWC 2023, Part II.

LNCS, vol. 14266, pp. 266–284. Springer, Cham (2023). https://doi.org/10.1007/978-3-031-47243-5_15

23. Kyzirakos, K., Karpathiotakis, M., Koubarakis, M.: Strabon: a semantic geospatial DBMS. In: Cudré-Mauroux, P., et al. (eds.) ISWC 2012, Part I. LNCS, vol. 7649, pp. 295–311. Springer, Heidelberg (2012). https://doi.org/10.1007/978-3-642-35176-1_19

24. Kyzirakos, K., et al.: Geotriples: transforming geospatial data into RDF graphs using R2RML and RML mappings. J. Web Semant. **52–53**, 16–32 (2018). https://doi.org/10.1016/J.WEBSEM.2018.08.003

25. Koubarakis, M. (ed.): Geospatial Data Science: A Hands-on Approach for Building Geospatial Applications Using Linked Data Technologies. ACM Books (2023)

26. Napoles, C., Sakaguchi, K., Tetreault, J.R.: JFLEG: a fluency corpus and benchmark for grammatical error correction. In: Proceedings of the 15th Conference of the European Chapter of the Association for Computational Linguistics, EACL 2017, Valencia, Spain, 3–7 April 2017, Volume 2: Short Papers. pp. 229–234. Association for Computational Linguistics (2017). https://doi.org/10.18653/V1/E17-2037

27. Papadakis, G., Mandilaras, G.M., Mamoulis, N., Koubarakis, M.: Progressive, holistic geospatial interlinking. In: WWW '21: The Web Conference 2021, Virtual Event / Ljubljana, Slovenia, 19–23 April 2021, pp. 833–844. ACM / IW3C2 (2021). https://doi.org/10.1145/3442381.3449850

28. Papamichalopoulos, M., Papadakis, G., Mandilaras, G., Siampou, M.D., Mamoulis, N., Koubarakis, M.: Three-dimensional geospatial interlinking with JedAI-spatial. CoRR abs/2205.01905 (2022). https://doi.org/10.48550/ARXIV.2205.01905

29. Perry, M., Herring, J.: GeoSPARQL - a geographic query language for RDF data (2012). http://www.opengis.net/doc/IS/geosparql/1.0, open Geospatial Consortium, OGC 11-052r4

30. Punjani, D., Kefalidis, S.A., Plas, K., Tsalapati, E., Koubarakis, M., Maret, P.: The question answering system GeoQA2. In: The 2nd International Workshop on Geospatial Knowledge Graphs and GeoAI: Methods, Models, and Resources at the GIScience Conference (2023)

31. Punjani, D., et al.: Template-based question answering over linked geospatial data. In: Proceedings of the 12th Workshop on Geographic Information Retrieval, GIR@SIGSPATIAL 2018, pp. 7:1–7:10. ACM (2018). https://doi.org/10.1145/3281354.3281362

32. Raffel, C., et al.: Exploring the limits of transfer learning with a unified text-to-text transformer. J. Mach. Learn. Res. **21**, 140:1–140:67 (2020). http://jmlr.org/papers/v21/20-074.html

33. Saxena, A., Chakrabarti, S., Talukdar, P.P.: Question answering over temporal knowledge graphs. In: Proceedings of the 59th Annual Meeting of the Association for Computational Linguistics and the 11th International Joint Conference on Natural Language Processing, ACL/IJCNLP 2021, (Volume 1: Long Papers), Virtual Event, 1–6 August 2021, pp. 6663–6676. Association for Computational Linguistics (2021). https://doi.org/10.18653/V1/2021.ACL-LONG.520

34. Sen, P., Aji, A.F., Saffari, A.: Mintaka: a complex, natural, and multilingual dataset for end-to-end question answering. In: Proceedings of the 29th International Conference on Computational Linguistics, COLING 2022, Gyeongju, Republic of Korea, 12–17 October 2022, pp. 1604–1619. International Committee on Computational Linguistics (2022). https://aclanthology.org/2022.coling-1.138

35. Tang, L.R., Mooney, R.J.: Using multiple clause constructors in inductive logic programming for semantic parsing. In: De Raedt, L., Flach, P. (eds.) ECML 2001. LNCS (LNAI), vol. 2167, pp. 466–477. Springer, Heidelberg (2001). https://doi.org/10.1007/3-540-44795-4_40

36. Trivedi, P., Maheshwari, G., Dubey, M., Lehmann, J.: LC-QuAD: a corpus for complex question answering over knowledge graphs. In: d'Amato, C., et al. (eds.) ISWC 2017, Part II. LNCS, vol. 10588, pp. 210–218. Springer, Cham (2017). https://doi.org/10.1007/978-3-319-68204-4_22

37. Usbeck, R., Gusmita, R.H., Ngomo, A.N., Saleem, M.: 9th challenge on question answering over linked data (QALD-9) (invited paper). In: Joint proceedings of the 4th Workshop on Semantic Deep Learning (SemDeep-4) and NLIWoD4: Natural Language Interfaces for the Web of Data (NLIWOD-4) and 9th Question Answering over Linked Data challenge (QALD-9) co-located with 17th International Semantic Web Conference (ISWC 2018), Monterey, California, United States of America, 8–9 October 2018. CEUR Workshop Proceedings, vol. 2241, pp. 58–64. CEUR-WS.org (2018). https://ceur-ws.org/Vol-2241/paper-06.pdf

38. Xu, H., et al.: Extracting interrogative intents and concepts from geo-analytic questions. AGILE: GIScience Series 1, 23 (2020). https://doi.org/10.5194/agile-giss-1-23-2020, https://agile-giss.copernicus.org/articles/1/23/2020/

Author Index

M. Alam et al. (Eds.): EKAW 2024, LNAI 15370, pp. 491–492, 2025.
https://doi.org/10.1007/978-3-031-77792-9

GPSR Compliance

The European Union's (EU) General Product Safety Regulation (GPSR) is a set of rules that requires consumer products to be safe and our obligations to ensure this.

If you have any concerns about our products, you can contact us on ProductSafety@springernature.com

In case Publisher is established outside the EU, the EU authorized representative is:

Springer Nature Customer Service Center GmbH
Europaplatz 3
69115 Heidelberg, Germany

The manufacturer's authorised representative in the EU is Springer
Nature Customer Service Centre GmbH, Europaplatz 3, 69115 Heidelberg,
Germany. If you have any concerns regarding our products, please
contact ProductSafety@springernature.com

Printed and bound by CPI Group (UK) Ltd, Croydon, CR0 4YY

27/04/2026

02097845-0012